283/000114204
26-6-01 *
20 045 010 14
UNIVERSITY OF NORTHUMBRIA
AT NEWCASTLE LIBRARY
WITHDRAWN

AF598417

Analysis of Cracks in Solids

International Series on Advances In Fracture Mechanics

Aims

The science of fracture has grown considerably in power and diversity in recent years and for the engineering community the advances are matched by use in most traditional industries such as aerospace as well as new industries such as in manufacturing of micro-components.

The aim of this series is to provide a platform for gathering state-of-the-art developments in the science of fracture. The series publishes several volumes every year, covering the latest developments in the application of fracture to different engineering fields. Each volume is either authored or edited comprising several chapters written by leading researchers in the field of fracture.

The scope of the series covers the entire spectrum of computational and experimental techniques for fracture.

Series Editor

M.H. Aliabadi
Dept of Engineering
Queen Mary College
University of London
London E1 4NS
UK

Editorial Board

B. Abersek
Dept of Mechanical Engineering
University of Maribor
POB 224, Smetanova 17
62000 Maribor
Slovenia

A.G. Atkins
University of Reading
Dept of Engineering
Whiteknights
P O Box 225
Reading
UK

A.F. Blom
Aeronautical Research Institute
Structures Department
PO Box 11021
S 161 11 Bromma
Sweden

C.A. Brebbia
Wessex Institute of Technology
Ashurst Lodge
Ashurst
Southampton SO40 7AA
UK

F.G. Buchholz
Fachgruppe Angewandte Mechanik
Universitat Gesanthochschule Paderborn
D-33095 Paderbor
Germany

J. Byrne
Dept of Mechanics and Manufacturing
University of Portsmouth
Anglesea Road
Portsmouth PO1 3DJ
UK

A. Carpinteri
Dept of Structural Engineering
Politecnico di Torino
Corso Duca Degli Abruzzi, 24
10129 - Torino
Italy

D.J. Cartwright
Mechanical Engineering Department
Bucknell University
Lewisburgh
PA, 17837
USA

H. Choi
Dept of Atmospheric Environmental Sciences
Kangnung National University
San-1 Chibyundong
Kangnung
Kangwondo 210-702
Korea

R. Cook
Defence Research Agency
Structural Materials Centre
Farnborough, Hants
GU14 6NE
UK

R. De Borst
Dept Civil Engineering
Delft University of Technology
P O Box 5048
NL-2600 GA Delft
The Netherlands

J. Dominguez
ETSI de I
University of Seville
Av Reina Mercedes s/n
41012 Sevilla
Spain

W. Dover
Dept of Mechanical Engineering
University College London
Torrington Place
London WC1E 7JE
UK

F. Erdogan
Lehigh University
Mechanical Engineering and Mechanics
Bethlehem
Pennsylvania 18015
USA

E.E. Gdoutos
Directo Section of Structural Mechanics
Democritus University of Thrace
GR 67100
Xanthi
Greece

D. Gross
Technische Hochschule Darmstadt
Institut fur Mechanik
64289 Darmstadt
Hochschulstrasse 1
Germany

D.A. Hills
Dept of Engineering Science
University of Oxford
Parks Road
Oxford OX1 3PJ
UK

T.H. Hyde
Department of Mechanical Engineering
University of Nottingham
University Park
Nottingham NG7 2RD
UK

D.R.H Jones
Dept of Engineering
University of Cambridge
Trumpington Street
Cambridge, CB2 1PZ
UK

A.S. Kobayashi
University of Washington
Box 352600
Seattle
WA 98195-2600
USA

Y-W. Mai
Centre for Advanced Materials &
Mechatronic Engineering J07
University of Sydney
Sydney
NSW 2006
Australia

S.I. Nishida
Faculty of Sciences and Engineering
Saga University
Honjo-Saga-shi-Machi 1
Saga 840
Japan

H. Nisitani
Dept of Mechanical Engineering
Kyushu Sangyo University
Matsukadai 3-1-2
Higashi-ku, Fukuoka 813
Japan

P. O'Donoghue
Dept of Civil Engineering
University College Dublin
Earlsfort Terrace
Dublin 2
Ireland

J. Otegui
INTEMA
Universidad Mar del PlataFacultad de Ing.
Av Juan B Justo 4302
7600 Mar del Plata
Argentina

L.P. Pook
Dept of Mechanical Engineering
University College London
Torrington Place
London WC1E 7JE
UK

I.S. Putra
Aeronautical Engineering Department
Institut Tekhnologi Bandung
J1. Ganesha 10
Bandung 40132
Indonesia

R.A. Schapery
Dept of Aerospace Engineering
University of Texas
Austin, TX 78712-1085
USA

B. Scholtes
University Gesamthochschule Kassel
Inst. fur Werkstofftechnik
Monchebrgstr.
D-34125 Kassel
Germany

A.P.S. Selvadurai
Dept of Civil Engineering
McGill University
Macdonald Engineering Building
Montreal, QC
Canada H3A 2K6

G.C. Sih
Institute of Fracture & Solid Mechanics
Lehigh University
Bethlehem
PA 18015
USA

J. Sladek
Insititue of Construction & Architecture
Slovak Academy of Sciences
842 20 Bratislava
Slovakia

T.S. Srivatsan
Dept of Mechanical Engineering
The University of Akron
Akron
Ohio 44325-3903
USA

T.S. Sudarshan
Materials Modification Inc.
2929 PI Eskridge Centre
Fairfax
Virginia 22031
USA

C-L. Tan
Dept of Mechanical & Aero. Engineering
Carleton University
1125 Colonel by Drive
Ottowa, ON
Canada K1S 5B6

P.S. Theocaris
Academy of Athens
Inst. of Theoretical and Applied Mechanics
27 Voulis Street
105 57 Athens
Greece

A. Terranova
Politecnico di Milano
Dpto de Meccanica
Pza Leonardo da Vinci 32
21033 Milano
Italy

B. Weiss
University of Vienna
Wahringerstrasse 42,
A-1090, Vienna
Austria

Analysis of Cracks in Solids

A.M. Khludnev
V.A. Kovtunenko
Russian Academy of Sciences, Russia

UNIVERSITY OF NORTHUMBRIA AT NEWCASTLE
LIBRARY
ITEM No. 2004501014
CLASS No. 620.176 KHL

A.M. Khludnev
V.A. Kovtunenko
Lavrentyev Institute of Hydrodynamics
Russian Academy of Sciences
Russia

Published by

WIT Press
Ashurst Lodge, Ashurst, Southampton, SO40 7AA, UK
Tel: 44 (0) 238 029 3223; Fax: 44 (0) 238 029 2853
E-Mail: witpress@witpress.com
http://www.witpress.com

For USA, Canada and Mexico

Computational Mechanics Inc
25 Bridge Street, Billerica, MA 01821, USA
Tel: 978 667 5841; Fax: 978 667 7582
E-Mail: cmina@ix.netcom.com
US site: http://www.compmech.com

British Library Cataloguing-in-Publication Data

A Catalogue record for this book is available from the British Library

ISBN: 1 85312 625 X
ISSN: 1369-7323

Library of Congress Catalog Card Number: 99-65487

No responsibility is assumed by the Publisher, the Editors and Authors for any injury and/or damage to persons or property as a matter of products liability, negligence or otherwise, or from any use or operation of any methods, products, instructions or ideas contained in the material herein.

© WIT Press 2000.

Printed in Great Britain by Print in Black, Bath.

All rights reserved. No part of this publication may be reproduced, stored in a retrieval system, or transmitted in any form or by any means, electronic, mechanical, photocopying, recording, or otherwise, without the prior written permission of the Publisher.

CONTENTS

PREFACE

The need for progress in modelling and analysis of crack problems in solids has resulted in renewed attempts at using modern approaches to boundary value problems. By taking a different viewpoint on the traditional treatment of many problems, such as crack theory, the range of problems that can be resolved through mathematical tools is enlarged.

It is well known that the classical approach to crack problems is characterized by the equilibrium of loading on a crack surface, a particular case being traction free cracks. This means that displacements found as solutions to these boundary value problems do not provide a nonpenetration condition between crack faces. There is evidence to show that interpenetration of crack faces may occur in these cases. An essential feature of the book is that a restriction of Signorini type contact condition is considered along the crack faces, which does not allow the opposite crack faces to penetrate each other. The restriction can be written as an inequality for the displacement vector. As a result a complete set of boundary conditions at crack faces is written as a system of equations and inequalities. The presence of inequality type boundary conditions implies the boundary problems to be nonlinear, which requires the investigation of corresponding boundary value problems.

In the book, two- and three-dimensional bodies, plates and shells with cracks are considered. Properties of solutions are established, namely: existence of solutions, regularity up to the crack faces, convergence of solutions as parameters of a system are varying and so on. Also analysed are different constitutive laws such as elastic, thermoelastic and elastoplastic. The book gives a fresh outlook on crack problems, displays new methods of studying the problems and proposes new models for cracks in elastic and nonelastic bodies satisfying physically suitable nonpenetration conditions between crack faces.

The new approach to crack theory that is discussed is intriguing in that it fails to lead to violation of physical properties. Boundary conditions analysed are given in the form of inequalities, and they are properly nonpenetration conditions of crack faces. The above implies that similar problems may be considered from the contact mechanics standpoint.

The authors express their deep gratitude to everyone who has helped, in numerous ways, to produce this monograph. We are grateful to Professor J. Sokolowski from Nancy for valuable contributions to the book. Our cooperation with Professor M. Brokate from Kiel on plasticity problems was mutually beneficial. We are grateful to Professor K.-H.Hoffmann from Munich for many stimulating discussions on plate and shell problems, and to Professor W.Wendland from Stuttgart for his great interest and for supplying the crack problem with nonpenetration conditions. Also, we would like to express deep thanks to our colleagues from Lavrentyev Institute of Hydrodynamics in Novosibirsk, especially to Professors B. Annin, A. Kazhikhov, V. Monakhov and P. Plotnikov. Scientific collaboration with them has been fruitful and pleasurable for the authors for many years. Finally we wish to thank Professor M.H. Aliabadi from University of London for his proposal to publish this book in the series Advances in Fracture Mechanics and for his support.

In their work on the monograph the authors were partially supported by the Russian Fund for Basic Research under grant 97-01-00896 and the Alexander von Humboldt Foundation.

A.M. Khludnev and V.A. Kovtunenko
Russia, 2000.

Chapter 1

Introduction

In this chapter, principal relations of solid mechanics, elements of convex analysis and calculus of variations, and methods of approximation are considered.

Submitting the main topic, we deal with models of solids with cracks. These models of mechanics and geophysics describe the stationary and quasi-stationary deformation of elastic and inelastic solid bodies having cracks and cuts. The corresponding mathematical models are reduced to boundary value problems for domains with singular boundaries. We shall use, if it is possible, a variational formulation of the problems to apply methods of convex analysis. It is of importance to note the significance of restrictions stated a priori at the crack surfaces. We assume that non-penetration conditions of inequality type at the crack surfaces are fulfilled, which improves the accuracy of these models for contact problems. We also include the modelling of problems with friction between the crack surfaces.

1.1 Modelling of solids with cracks

1.1.1 Small deformations. Hooke's law

Let a solid body occupy the domain $\Omega \subset R^3$ with the smooth boundary Γ. The solid particle coincides with the point $x = (x_1, x_2, x_3) \in \Omega$. An elastic solid is described by the following functions:

the *displacements* $u(x) = (u_1(x), u_2(x), u_3(x))$;

the *strain tensor* $\varepsilon_{ij}(x)$, $\quad i, j = 1, 2, 3$;

the *stress tensor* $\sigma_{ij}(x)$, $\quad i, j = 1, 2, 3$

at the point $x \in \Omega$. In what follows, we do not indicate the dependence on x of the above values.

To formulate a model of the solid body, one needs the *constitutive law* $\sigma_{ij}(\varepsilon_{kl})$, the *geometrical equation* $\varepsilon_{ij} = \varepsilon_{ij}(u)$ and equilibrium or motion equations. Let $f = (f_1, f_2, f_3)$ be a given function describing an external force in the domain Ω. The *equilibrium equations* are as follows:

$$-\sigma_{ij,j} = f_i, \quad i = 1, 2, 3, \tag{1.1}$$

where $\sigma_{ij,j} = \partial\sigma_{ij}/\partial x_j$; the repeated indices j mean the sum over $j = 1, 2, 3$. In the dynamical case all functions depend also on the time variable t, $t \geq 0$, and we have the *motion equations*

$$\rho\frac{\partial^2}{\partial t^2}u_i(t) - \sigma_{ij,j}(t) = f_i(t), \quad i = 1, 2, 3.$$

We assume that the formula for $\varepsilon_{ij}(u)$ is provided by the *Cauchy law* of small deformations

$$\varepsilon_{ij} = \frac{1}{2}\left(u_{i,j} + u_{j,i}\right), \quad i, j = 1, 2, 3. \tag{1.2}$$

These are linear equations which give the symmetry of the strain tensor $\varepsilon_{ij} = \varepsilon_{ji}$. In the general case, the strain tensor is nonlinear,

$$\varepsilon_{ij} = \frac{1}{2}\left(u_{i,j} + u_{j,i} + u_{k,i}u_{k,j}\right), \quad i, j = 1, 2, 3.$$

The constitutive law $\sigma_{ij}(\varepsilon_{kl})$ has a principal meaning for the definition of solid models. The classical *Hooke law*

$$\sigma_{ij} = a_{ijkl}\varepsilon_{kl}, \quad i, j = 1, 2, 3, \tag{1.3}$$

defines the linear elasticity model (Rabotnov, 1979; Timoshenko, Goodier, 1951; Parton, Perlin, 1981). The tensor a_{ijkl} is assumed to be symmetrical,

$$a_{ijkl} = a_{jikl} = a_{klij},$$

which provides the symmetry $\sigma_{ij} = \sigma_{ji}$, and positive:

$$\exists c_1, c_2 > 0 : \quad c_1\xi_{ij}\xi_{ij} \leq a_{ijkl}\xi_{kl}\xi_{ij} \leq c_2\xi_{ij}\xi_{ij} \quad \forall\xi_{ij} = \xi_{ji}.$$

Generally, a_{ijkl} depends on x. The *isotropic* solid is characterized by the constant coefficients a_{ijkl} of the form

$$a_{ijkl} = \lambda\delta_{ij}\delta_{kl} + \mu\left(\delta_{ik}\delta_{jl} + \delta_{il}\delta_{jk}\right), \quad i, j, k, l = 1, 2, 3,$$

with the Lamé parameters $\lambda > 0$, $\mu > 0$; δ_{ij} is Kronecker's symbol. Then (1.3), (1.2) are reduced to

$$\sigma_{ij} = \lambda u_{k,k}\delta_{ij} + \mu\left(u_{i,j} + u_{j,i}\right), \quad i, j = 1, 2, 3,$$

and together with (1.1) give the *Lamé system*

$$-\mu u_{i,kk} - (\lambda + \mu)\left(u_{k,k}\right)_{,i} = f_i, \quad i = 1, 2, 3.$$

By introducing the other constants

$$E = \mu \frac{3\lambda + 2\mu}{\lambda + \mu} > 0; \quad \kappa = \frac{\lambda}{2(\lambda + \mu)}, \quad 0 < \kappa < \frac{1}{2},$$

we can rewrite the Lamé system as

$$-\frac{E}{1+\kappa}\left(\frac{1}{2}\Delta u_i + \frac{1}{1-2\kappa}(\operatorname{div} u)_{,i}\right) = f_i, \quad i = 1, 2, 3,$$

where $\Delta u_i = u_{i,kk}$, $\operatorname{div} u = u_{k,k}$, E is *Young's modulus* and κ is *Poisson's ratio.*

1.1.2 Other constitutive laws

We shall also formulate inelastic constitutive laws considered in the book (Rabotnov, 1979; Arutunyan et al., 1987).

Let us recall the dependence of solutions to dynamical and quasi-static problems on the time parameter t. Then Hooke's law (1.3) takes the form

$$\sigma_{ij}(t) = a_{ijkl}\varepsilon_{kl}(t), \quad i, j = 1, 2, 3,$$

or, denoting by A_{ijkl} an inverse tensor to a_{ijkl},

$$\varepsilon_{ij}(t) = A_{ijkl}\sigma_{kl}(t), \quad i, j = 1, 2, 3.$$

Instead of the Hooke law we assume the *creep* law

$$\varepsilon_{ij}(t) = A_{ijkl}\sigma_{kl}(t) + \int_0^t \bar{A}_{ijkl}(t-\tau)\sigma_{kl}(\tau)\,d\tau, \quad i, j = 1, 2, 3, \tag{1.4}$$

fulfilled for the given tensor $\bar{A}_{ijkl}$, and obtain the quasi-static creep model

$$-\sigma_{ij,j}(t) = f_i(t), \quad \varepsilon_{ij}(t) = \frac{1}{2}\left(u_{i,j}(t) + u_{j,i}(t)\right),$$

$$\varepsilon_{ij}(t) = A_{ijkl}\sigma_{kl}(t) + \int_0^t \bar{A}_{ijkl}(t-\tau)\sigma_{kl}(\tau)\,d\tau, \quad i, j = 1, 2, 3.$$

The following condition,

$$\sigma_{ij}(t) = a_{ijkl}\varepsilon_{kl}(t) + \int_0^t \bar{a}_{ijkl}(t-\tau)\varepsilon_{kl}(\tau)\,d\tau, \quad i, j = 1, 2, 3, \tag{1.5}$$

provides another creep model

$$-\sigma_{ij,j}(t) = f_i(t), \quad \varepsilon_{ij}(t) = \frac{1}{2}\left(u_{i,j}(t) + u_{j,i}(t)\right),$$

$$\sigma_{ij}(t) = a_{ijkl}\varepsilon_{kl}(t) + \int_0^t \bar{a}_{ijkl}(t-\tau)\varepsilon_{kl}(\tau)\,d\tau, \quad i,j = 1,2,3.$$

In the sequel, we will not indicate the dependence of functions on t for convenience.

The *viscoelastic* law

$$\sigma_{ij} = a_{ijkl}\varepsilon_{kl} + b_{ijkl}\frac{\partial}{\partial t}\varepsilon_{kl}, \quad i,j = 1,2,3, \tag{1.6}$$

together with (1.1), (1.2) give the quasi-static viscoelastic model

$$-\sigma_{ij,j} = f_i, \ i = 1,2,3,$$

$$\varepsilon_{ij} = \frac{1}{2}\left(u_{i,j} + u_{j,i}\right), \quad \sigma_{ij} = a_{ijkl}\varepsilon_{kl} + b_{ijkl}\frac{\partial}{\partial t}\varepsilon_{kl}, \ i,j = 1,2,3.$$

In thermodynamic systems we must consider the *temperature* function θ. Then, in view of the *Duhamel–Newmann law*, the constitutive equations have the form (Nowacki, 1962)

$$\sigma_{ij} = a_{ijkl}\varepsilon_{kl} - \beta_{ij}\theta, \quad i,j = 1,2,3. \tag{1.7}$$

Adding the *heat equation*

$$\frac{\partial}{\partial t}\theta - \Delta\theta + \beta_{ij}\frac{\partial}{\partial t}\varepsilon_{ij} = g \tag{1.8}$$

with the given thermal expansion coefficients β_{ij}, we obtain the quasi-static *thermoelastic* model

$$-\sigma_{ij,j} = f_i, \ i = 1,2,3; \quad \frac{\partial}{\partial t}\theta - \Delta\theta + \beta_{ij}\frac{\partial}{\partial t}\varepsilon_{ij} = g;$$

$$\sigma_{ij} = a_{ijkl}\varepsilon_{kl} - \beta_{ij}\theta, \quad \varepsilon_{ij} = \frac{1}{2}\left(u_{i,j} + u_{j,i}\right), \quad i,j = 1,2,3,$$

where g is a heat influx in Ω. If $\beta_{ij} = 0$, then one obtains two separate problems: the elastic equations (1.1)–(1.3) and the classical heat equation

$$\frac{\partial}{\partial t}\theta - \Delta\theta = g$$

for the temperature θ.

In *elastoplastic* models, it is assumed that there exist *plastic deformations* denoted by ξ_{ij}. The *Hencky law* implies that the following relations hold (Annin, Cherepanov, 1983; Duvaut, Lions, 1972):

$$\varepsilon_{ij} = A_{ijkl}\sigma_{kl} + \xi_{ij}, \quad i,j = 1,2,3. \tag{1.9}$$

Meantime, the stresses lie inside the given yield surface, namely

$$\Phi(\sigma_{ij}) \le 0 \tag{1.10}$$

and the plastic deformations are orthogonal to this surface:

$$\xi_{ij}\left(\bar{\sigma}_{ij}-\sigma_{ij}\right)\le 0 \quad \forall\bar{\sigma}_{ij},\ \Phi(\bar{\sigma}_{ij})\le 0. \tag{1.11}$$

The function Φ is assumed to be convex and continuous. Thus, we have the static elastoplastic model

$$-\sigma_{ij,j}=f_i,\ i=1,2,3;\quad \varepsilon_{ij}=A_{ijkl}\sigma_{kl}+\xi_{ij},\ i,j=1,2,3,$$

$$\Phi(\sigma_{ij})\le 0,\quad \xi_{ij}\left(\bar{\sigma}_{ij}-\sigma_{ij}\right)\le 0 \quad \forall\bar{\sigma}_{ij},\ \Phi(\bar{\sigma}_{ij})\le 0.$$

One can exclude ξ_{ij} from the last relations and obtain the inequality

$$\left(\varepsilon_{ij}-A_{ijkl}\sigma_{kl}\right)\left(\bar{\sigma}_{ij}-\sigma_{ij}\right)\le 0 \quad \forall\bar{\sigma}_{ij},\ \Phi(\bar{\sigma}_{ij})\le 0.$$

The following *flow model* is provided by the *Prandtl–Reuss law* (see Sadovskii, 1992, 1997):

$$\frac{\partial}{\partial t}\varepsilon_{ij}=A_{ijkl}\frac{\partial}{\partial t}\sigma_{kl}+\xi_{ij},\quad i,j=1,2,3. \tag{1.12}$$

Here ξ_{ij} denotes a *plastic deformation velocity*. Adding the relations (1.10), (1.11), we obtain the quasi-static elastoplastic model

$$-\sigma_{ij,j}=f_i,\ i=1,2,3;\quad \frac{\partial}{\partial t}\varepsilon_{ij}=A_{ijkl}\frac{\partial}{\partial t}\sigma_{kl}+\xi_{ij},\ i,j=1,2,3;$$

$$\Phi(\sigma_{ij})\le 0,\quad \xi_{ij}\left(\bar{\sigma}_{ij}-\sigma_{ij}\right)\le 0 \quad \forall\bar{\sigma}_{ij},\ \Phi(\bar{\sigma}_{ij})\le 0.$$

The considered constitutive laws for elastoplastic models generalize ones used in elasticity. The main peculiarity of elastoplastic models consists in an existence of inequality type restrictions imposed upon the stresses. Omitting the mentioned restrictions, elastoplastic models turn into elastic ones.

1.1.3 Linear plates and shells

In two-dimensional solids theory, the size of the solid in a fixed direction is assumed to be small as compared to the other ones. Therefore, all characteristics of the thin solid are referred to a so-called mid-surface, and one obtains the two-dimensional model. Let us give the construction of plate and shell models (Donnell, 1976; Vol'mir, 1972; Lukasiewicz, 1979; Mikhailov, 1980).

A three-dimensional body limited by two curvilinear surfaces is called a *shell* if a distance called a *thickness of the shell* between the aforementioned surfaces is small enough. We assume that the thickness is the constant $2h>0$. The surface equidistant from the surfaces is called a *mid-surface*. Thus, a shell can be uniquely defined introducing a mid-surface, a thickness and a boundary contour.

The two directions called *principal directions* could be found at every point of the mid-surface, which satisfy the following property. Orthogonal sections of the mid-surface along these directions define the *principal curvatures* k_1, k_2 of the surface. By that, the *curvature lines* could be defined on the mid-surface with tangents to them coinciding with the principal directions. Let α and β be parameters such that the coordinate net $\alpha = \text{const}$, $\beta = \text{const}$ on the mid-surface is orthogonal, and it coincides with the curvature lines. Then a position of every point x located at the mid-surface is defined by the parameters α, β:

$$x = x(\alpha, \beta), \quad (\alpha, \beta) \in \Omega, \quad \Omega \subset R^2. \tag{1.13}$$

The length ds of a linear infinitely small element can be found by the formula

$$ds^2 = a^2\, d\alpha^2 + b^2\, d\beta^2. \tag{1.14}$$

This expression is called the *first quadric* of the surface in the orthogonal coordinates α, β, where $a = a(\alpha, \beta)$, $b = b(\alpha, \beta)$. The directions of $x_{,\alpha}$ and $x_{,\beta}$ (where $x_{,\alpha} = \partial x/\partial\alpha$, $x_{,\beta} = \partial x/\partial\beta$) coincide with the principal ones, hence

$$ds^2 = |x_{,\alpha}|^2\, d\alpha^2 + |x_{,\beta}|^2\, d\beta^2.$$

Comparison of this equation and (1.14) gives

$$a^2 = x_{i,\alpha} x_{i,\alpha}, \quad b^2 = x_{i,\beta} x_{i,\beta}.$$

Let the axis z be orthogonal to the mid-surface, and $z = 0$ correspond to the mid-surface. By (1.13), functions depending on x are the functions of α, β, z. Let the displacements of the mid-surface points along $x_{,\alpha}$, $x_{,\beta}$, z be denoted by u, v, w, respectively, σ_{ij} be components of the stress tensor, and ε_{ij} denote the strains. By integrating across the thickness, let us introduce the following functions depending only on α, β:

the *integrated stresses* $N_{ij} = \int_{-h}^{h} \sigma_{ij}\, dz, \quad i, j = 1, 2;$

the *moments* $M_{ij} = \int_{-h}^{h} \sigma_{ij} z\, dz, \quad i, j = 1, 2;$

the *transverse forces* $Q_i = \int_{-h}^{h} \sigma_{i3}\, dz, \quad i = 1, 2.$

A reduction of three-dimensional models to two-dimensional ones is based on the assumptions concerning character of deformations. Thus, the *Kirchhoff–Love hypothesis* is widely used in mechanics, and is as follows. Every fibre is orthogonal to the mid-surface till the deformation remains straight and orthogonal after the deformation. Furthermore, the normal stresses along the fibre are assumed to be negligible. The utilization of the Kirchhoff–Love hypothesis leads to the following
geometrical equations:

$$\varepsilon_{11} = \frac{1}{a} u_{,\alpha} + \frac{1}{ab} a_{,\beta} v + k_1 w, \quad \varepsilon_{22} = \frac{1}{b} v_{,\beta} + \frac{1}{ab} b_{,\alpha} u + k_2 w,$$

$$\varepsilon_{12} = \frac{b}{a}\left(\frac{v}{b}\right)_{,\alpha} + \frac{a}{b}\left(\frac{u}{a}\right)_{,\beta};$$

$$\kappa_{11} = \frac{1}{a}\left(k_1 u - \frac{1}{a}w_{,\alpha}\right)_{,\alpha} + \frac{1}{ab}\left(k_2 v - \frac{1}{b}w_{,\beta}\right)a_{,\beta},$$

$$\kappa_{22} = \frac{1}{b}\left(k_2 v - \frac{1}{b}w_{,\beta}\right)_{,\beta} + \frac{1}{ab}\left(k_1 u - \frac{1}{a}w_{,\alpha}\right)a_{,\alpha},$$

$$\kappa_{12} = \frac{b}{2a}\left(\frac{1}{b}\left(k_2 v - \frac{1}{b}w_{,\beta}\right)\right)_{,\alpha} + \frac{a}{2b}\left(\frac{1}{a}\left(k_1 u - \frac{1}{a}w_{,\alpha}\right)\right)_{,\beta};$$

constitutive law equations:

$$N_{ij} = a_{ijkl}\varepsilon_{kl}, \quad M_{ij} = b_{ijkl}\kappa_{kl}, \quad i,j = 1,2 \tag{1.15}$$

with the symmetric and positively defined coefficients $a_{ijkl}(\alpha,\beta)$, $b_{ijkl}(\alpha,\beta)$; and *equilibrium equations*:

$$-\frac{1}{ab}\left((bN_{11})_{,\alpha} - N_{22}b_{,\alpha} + \frac{1}{a}(a^2N_{12})_{,\beta}\right)_{,\beta} - k_1Q_1 = f_1,$$

$$-\frac{1}{ab}\left((aN_{22})_{,\beta} - N_{11}a_{,\beta} + \frac{1}{b}(b^2N_{12})_{,\alpha}\right)_{,\alpha} - k_2Q_2 = f_2,$$

$$-\frac{1}{ab}\left((bQ_1)_{,\alpha} + (aQ_2)_{,\beta}\right) - k_1N_{11} - k_2N_{22} = f_3,$$

$$\frac{1}{ab}\left(\frac{1}{a}(a^2M_{12})_{,\beta} + (bM_{11})_{,\alpha} - M_{22}b_{,\alpha}\right) = Q_1,$$

$$\frac{1}{ab}\left(\frac{1}{b}(b^2M_{12})_{,\alpha} + (aM_{22})_{,\beta} - M_{11}a_{,\beta}\right) = Q_2.$$

The functions f_1, f_2, f_3 are given and represent exterior forces acting along the axes $x_{,\alpha}, x_{,\beta}, z$, respectively. Thus, the linear model of a shell is described by the functions u, v, w, N_{ij}, M_{ij}, Q_i, ε_{ij}, κ_{ij}, $i,j = 1,2$, satisfying the above equations in the domain Ω, $(\alpha,\beta) \in \Omega$.

The simpler model can be derived to describe a *shallow shell* which is characterized by the closeness of the mid-surface to the plane. In other words, it is assumed that $a = b = 1$ and the coordinate system (α,β) coincides with the Descartes system x_1, x_2. Then differentiating the fourth and the fifth equilibrium equations with respect to x_1 and x_2, respectively, and combining with the third equilibrium equation give

$$-N_{11,1} - N_{12,2} - k_1Q_1 = f_1, \quad -N_{12,1} - N_{22,2} - k_2Q_2 = f_2, \tag{1.16}$$

$$-M_{ij,ij} + k_1N_{11} + k_2N_{22} = f_3.$$

The geometrical equations also have the simpler form

$$\begin{gathered}\varepsilon_{11} = u_{,1} + k_1 w, \quad \varepsilon_{22} = v_{,2} + k_2 w, \quad \varepsilon_{12} = v_{,1} + u_{,2}; \\ \kappa_{11} = \left(k_1 u - w_{,1}\right)_{,1}, \quad \kappa_{22} = \left(k_2 v - w_{,2}\right)_{,2}, \\ \kappa_{12} = -w_{,12} + 1/2\left((k_2 v)_{,1} + (k_1 u)_{,2}\right);\end{gathered} \tag{1.17}$$

and the constitutive law equations (1.15) keep their form.

The obtained model (1.15)–(1.17) of a shallow shell can be simplified once more. The values of $k_1 Q_1$, $k_1 Q_2$ are small enough very often, and they can be omitted. By doing so, we obtain the simplified model of a shallow shell consisting of the following *equilibrium equations*:

$$-N_{ij,j} = f_i, \; i = 1, 2, \quad -M_{ij,ij} = f_3, \tag{1.18}$$

geometrical equations:

$$\varepsilon_{11} = u_{,1} + k_1 w, \quad \varepsilon_{22} = v_{,2} + k_2 w, \quad \varepsilon_{12} = v_{,1} + u_{,2}, \tag{1.19}$$

$$\kappa_{ij} = -w_{,ij}, \quad i, j = 1, 2, \tag{1.20}$$

and *constitutive equations*:

$$N_{ij} = a_{ijkl}\varepsilon_{kl}, \quad M_{ij} = b_{ijkl}\kappa_{kl}, \quad i, j = 1, 2.$$

We call a *plate* the shallow shell when $k_1 = k_2 = 0$. This implies that the plate mid-surface coincides with the plane $z = 0$, and the plate is limited by the two parallel planes $z = h$, $z = -h$ and a boundary contour. Let us redenote the horizontal and vertical displacements of the plate mid-surface by $u = u_1$, $v = u_2$, w. In this case, the plate horizontal and vertical displacements are not coupled. Indeed, it follows from (1.18), (1.19), that $U = (u_1, u_2)$ is described by the following *equilibrium equations*:

$$-N_{ij,j} = f_i, \quad i = 1, 2, \tag{1.21}$$

constitutive equations:

$$N_{ij} = a_{ijkl}\varepsilon_{kl}, \quad i, j = 1, 2, \tag{1.22}$$

and *geometrical equations*:

$$\varepsilon_{ij} = \frac{1}{2}\left(u_{i,j} + u_{j,i}\right), \quad i, j = 1, 2. \tag{1.23}$$

To find the normal displacements w we should consider the equilibrium equation

$$-M_{ij,ij} = f_3 \tag{1.24}$$

and the constitutive equations

$$M_{ij} = -b_{ijkl}w_{,kl}, \quad i, j = 1, 2. \tag{1.25}$$

Substituting (1.22), (1.23) into (1.21), one can see that the differential equations (1.21) of second order with respect to U have the same structure as those of the three-dimensional elasticity equations (1.1)–(1.3). The system (1.24)–(1.25) contains the fourth derivatives of w.

In the isotropic case the coefficients a_{ijkl} are as follows,

$$N_{11} = G(\varepsilon_{11} + \kappa\varepsilon_{22}), \quad N_{22} = G(\varepsilon_{22} + \kappa\varepsilon_{11}), \quad N_{12} = G(1-\kappa)\varepsilon_{12}, \tag{1.26}$$

with the constant $G = 2Eh/(1-\kappa^2)$. Substituting (1.23), (1.26) into (1.21), one gets the following system,

$$-G\left(\frac{1-\kappa}{2}\Delta u_i + \frac{1+\kappa}{2}(\operatorname{div} u)_{,i}\right) = f_i, \quad i = 1, 2,$$

to find the plate horizontal displacements $U = (u_1, u_2)$. For the isotropic coefficients b_{ijkl} we have the formulae

$$M_{11} = -D(w_{,11} + \kappa w_{,22}), \quad M_{22} = -D(w_{,22} + \kappa w_{,11}), \tag{1.27}$$

$$M_{12} = -D(1-\kappa)w_{,12}$$

with $D = 2Eh^3/3(1-\kappa^2)$. The substitution (1.27) into (1.24) provides the biharmonic equation

$$D\,\Delta^2 w = f_3. \tag{1.28}$$

The model discussed is called the *Kirchhoff model.* Meantime there are other approaches to describe the behaviour of a shell. For example, it can be assumed that the fibre is not orthogonal to the mid-surface and the corresponding angle between the mid-surface and the orthogonal direction may vary. In this case the models are called *Timoshenko* or *Reissner–Timoshenko models* (see Vol'mir, 1972; compare Ciarlet, Sanchez-Palencia, 1996). In particular, these approaches are used in Chapter 5.

1.1.4 Inelastic plates

Models of inelastic plates are introduced here, which are analysed in Chapters 2, 3 and 5.

By the constitutive law (1.4), from (1.21)–(1.25) we obtain the model of a *plate under the creep condition*:

$$-N_{ij,j}(t) = f_i(t), \ i = 1, 2, \quad -M_{ij,ij}(t) = f_3(t);$$

$$\varepsilon_{ij}(t) = A_{ijkl}N_{kl}(t) + \int_0^t \bar{A}_{ijkl}N_{kl}(\tau)\,d\tau, \quad \varepsilon_{ij}(t) = \frac{1}{2}\left(u_{i,j}(t) + u_{j,i}(t)\right),$$

$$-w_{,ij}(t) = B_{ijkl}M_{kl}(t) + \int_0^t \bar{B}_{ijkl}M_{kl}(\tau)\,d\tau, \quad i, j = 1, 2,$$

considered in Sections 2.2 and 2.3.

Utilizing the constitutive law (1.5), the other model of a plate under the creep condition follows:

$$-N_{ij,j}(t) = f_i(t), \quad i = 1, 2, \tag{1.29}$$

$$-M_{ij,ij}(t) = f_3(t), \tag{1.30}$$

$$N_{ij}(t) = a_{ijkl}\varepsilon_{kl}(t) + \int_0^t \bar{a}_{ijkl}\varepsilon_{kl}(\tau)\,d\tau, \quad i, j = 1, 2, \tag{1.31}$$

$$M_{ij}(t) = -b_{ijkl}w_{,kl}(t) - \int_0^t \bar{b}_{ijkl}w_{,kl}(\tau)\,d\tau, \quad i, j = 1, 2. \tag{1.32}$$

Let us consider the isotropic case. Instead of (1.26), the law (1.31) gives

$$N_{11}(t) = G\left(\varepsilon_{11}(t) + \kappa\varepsilon_{22}(t)\right) + G\int_0^t \left(\varepsilon_{11}(\tau) + \kappa\varepsilon_{22}(\tau)\right)\,d\tau,$$

$$N_{22}(t) = G\left(\varepsilon_{22}(t) + \kappa\varepsilon_{11}(t)\right) + G\int_0^t \left(\varepsilon_{22}(\tau) + \kappa\varepsilon_{11}(\tau)\right)\,d\tau,$$

$$N_{12}(t) = G(1-\kappa)\varepsilon_{12}(t) + G(1-\kappa)\int_0^t \varepsilon_{12}(\tau)\,d\tau.$$

Analogously, instead of (1.27), the equations (1.32) are as follows:

$$M_{11}(t) = -D\left(w_{,11}(t) + \kappa w_{,22}(t)\right) - D\int_0^t \left(w_{,11}(\tau) + \kappa w_{,22}(\tau)\right)\,d\tau,$$

$$M_{22}(t) = -D\left(w_{,22}(t) + \kappa w_{,11}(t)\right) - D\int_0^t \left(w_{,22}(\tau) + \kappa w_{,11}(\tau)\right)\,d\tau,$$

$$M_{12}(t) = -D(1-\kappa)w_{,12}(t) - D(1-\kappa)\int_0^t w_{,12}(\tau)\,d\tau.$$

Substitution of these equalities into (1.30) yields the equation

$$D\Delta^2 w(t) + D\int_0^t \Delta^2 w(\tau)\,d\tau = f_3(t).$$

Let us introduce the notations

$$u_i^\tau(t) = u_i(t) + \int_0^t u_i(\tau)\,d\tau, \ i = 1,2, \quad w^\tau(t) = w(t) + \int_0^t w(\tau)\,d\tau$$

which imply the model for an *isotropic plate under the creep condition* written in the simpler form

$$-N_{ij,j}(t) = f_i(t), \ i = 1,2; \quad D\Delta^2 w^\tau(t) = f_3(t);$$

$$N_{11}(t) = G(\varepsilon_{11}(t) + \kappa\varepsilon_{22}(t)), \quad N_{22}(t) = G(\varepsilon_{22}(t) + \kappa\varepsilon_{11}(t)),$$

$$N_{12}(t) = G(1-\kappa)\varepsilon_{12}(t); \quad \varepsilon_{ij}(t) = \frac{1}{2}\left(u_{i,j}^\tau(t) + u_{j,i}^\tau(t)\right), \ i,j = 1,2.$$

This model is analysed in Section 3.1.

For the viscoelastic law (1.6), instead of (1.24), (1.25) the same arguments guarantee a validity of the following equations for the vertical displacements w:

$$-M_{ij,ij} = f_3, \quad M_{ij} = -b_{ijkl}w_{,kl} - \bar{b}_{ijkl}\frac{\partial}{\partial t}w_{kl}, \ i,j = 1,2.$$

Substituting the moments into the equilibrium equation, we obtain the equation for an *isotropic viscoelastic plate*,

$$D\Delta^2 w + D\frac{\partial}{\partial t}\Delta^2 w = f_3,$$

which is considered in Section 2.1.

In the thermoelasticity we can get the presentation of (1.7), (1.8) for plates in the form (Nowacki, 1962)

$$N_{ij} = a_{ijkl}\varepsilon_{kl} - \alpha_{ij}\theta, \quad M_{ij} = -b_{ijkl}w_{,kl} - \beta_{ij}\theta, \quad i,j = 1,2, \tag{1.33}$$

$$\frac{\partial}{\partial t}\theta - \Delta\theta + \alpha_{ij}\frac{\partial}{\partial t}\varepsilon_{ij} - \beta_{ij}\frac{\partial}{\partial t}w_{,ij} = g.$$

Here we have assumed that the temperature θ does not depend on z. Let the plate be isotropic and $\alpha_{ij} = \beta_{ij} = \delta^2\delta_{ij}$, where δ is a constant. Then (1.33) gives the quasi-static model of a *thermoelastic plate*:

$$-N_{ij,j} = f_i, \ i = 1,2, \quad -M_{ij,ij} = f_3; \quad \frac{\partial}{\partial t}\theta - \Delta\theta + \delta^2\frac{\partial}{\partial t}(\operatorname{div} U - \Delta w) = g;$$

$$N_{11} = G(\varepsilon_{11} + \kappa\varepsilon_{22}) - \delta^2\theta, \quad N_{22} = G(\varepsilon_{22} + \kappa\varepsilon_{11}) - \delta^2\theta, \quad N_{12} = G(1-\kappa)\varepsilon_{12};$$

$$\varepsilon_{ij} = \frac{1}{2}\left(u_{i,j} + u_{j,i}\right), \quad i,j = 1,2;$$

$$M_{11} = -D(w_{,11} + \kappa w_{,22}) - \delta^2\theta, \quad M_{22} = -D(w_{,22} + \kappa w_{,11}) - \delta^2\theta,$$

$$M_{12} = -D(1-\kappa)w_{,12}.$$

Substituting the moments into the constitutive law equation, one deduces the equation

$$D\Delta^2 w + \delta^2 \Delta\theta = f_3.$$

Let us introduce the integrated stresses

$$\sigma_{11} = G(\varepsilon_{11} + \kappa\varepsilon_{22}), \quad \sigma_{22} = G(\varepsilon_{22} + \kappa\varepsilon_{11}), \quad \sigma_{12} = G(1-\kappa)\varepsilon_{12},$$

$$\varepsilon_{ij} = \frac{1}{2}\left(u_{i,j} + u_{j,i}\right), \quad i,j = 1,2.$$

Then we can write the equilibrium equations in the form

$$-\sigma_{ij,j} + \delta^2\theta_{,i} = f_i, \; i = 1,2, \quad D\Delta^2 w + \delta^2\Delta\theta = f_3,$$

$$\frac{\partial}{\partial t}\theta - \Delta\theta + \delta^2\frac{\partial}{\partial t}(\operatorname{div} U - \Delta w) = g.$$

This system is called a model of the *thermoelastic plate* which is analysed in Sections 3.3 and 3.4. For more precise and, therefore, more cumbersome relations for thermoelastic plates see (Nowacki, 1962).

Now we formulate the models for perfectly elastoplastic plates considered in Chapter 5. By the Hencky law (1.9), the vertical component w of the plate displacements satisfies the equations (Erkhov, 1978)

$$-w_{,ij} = B_{ijkl}M_{kl} + \xi_{ij}, \quad i,j = 1,2,$$

where ξ_{ij} are plastic parts of the curvatures $-w_{ij}$. Similar to (1.10), (1.11), we admit the following inequalities (Khludnev, 1988):

$$\Phi(M_{ij}) \le 0, \quad \xi_{ij}\left(\bar{M}_{ij} - M_{ij}\right) \le 0 \quad \forall \bar{M}_{ij}, \; \Phi(\bar{M}_{ij}) \le 0.$$

The given function Φ is assumed to be convex and continuous. By adding the equilibrium equation

$$-M_{ij,ij} = f_3,$$

we obtain the *Hencky model of elastoplastic plate.* One can exclude ξ_{ij} and obtain the equivalent inequality

$$\left(w_{,ij} + B_{ijkl}M_{kl}\right)\left(\bar{M}_{ij} - M_{ij}\right) \ge 0 \quad \forall \bar{M}_{ij}, \; \Phi(\bar{M}_{ij}) \le 0.$$

In this case we cannot directly substitute M_{ij} into the equilibrium equation as it was done for the previous elastic and inelastic models. So w, M_{ij} cannot be found in consecutive order, in general.

The *flow model of Prandtl–Reuss for elastoplastic plate* is as follows:

$$-M_{ij,ij} = f_3, \quad -\frac{\partial}{\partial t}w_{,ij} = B_{ijkl}\frac{\partial}{\partial t}M_{kl} + \xi_{ij}, \; i,j = 1,2;$$

$$\Phi(M_{ij}) \le 0; \quad \xi_{ij}\left(\bar{M}_{ij} - M_{ij}\right) \le 0 \quad \forall \bar{M}_{ij},\ \Phi(\bar{M}_{ij}) \le 0.$$

Here ξ_{ij} denotes a velocity of a plastic part of the curvature. This model is quasi-static, and it reduces to the equivalent inequality

$$\left(\frac{\partial}{\partial t} w_{,ij} + B_{ijkl}\frac{\partial}{\partial t} M_{kl}\right)\left(\bar{M}_{ij} - M_{ij}\right) \ge 0 \quad \forall \bar{M}_{ij},\ \Phi(\bar{M}_{ij}) \le 0.$$

In the sequel, we consider concrete boundary conditions for the above models to formulate boundary value problems. Also, restrictions of the inequality type imposed upon the solutions are introduced. We begin with the nonpenetration conditions in contact problems (see Kravchuk, 1997; Khludnev, Sokolowski, 1997; Duvaut, Lions, 1972).

1.1.5 Contact problems

The model describing interaction between two bodies, one of which is a deformed solid and the other is a rigid one, we call a *contact problem*. After the deformation, the rigid body (called also *punch* or *obstacle*) remains invariable, and the solid must not penetrate into the punch. Meanwhile, it is assumed that the contact area (i.e. the set where the boundary of the deformed solid coincides with the obstacle surface) is unknown a priori. This condition is physically acceptable and is called a *nonpenetration condition*. We intend to give a mathematical description of nonpenetration conditions to diversified models of solids for contact and crack problems. Indeed, as one will see, the nonpenetration of crack surfaces is similar to contact problems. In this subsection, the contact problems for two-dimensional problems characterizing constraints imposed inside a domain are considered.

Let a punch shape be described by the equation $z = \psi(x)$, and x_1, x_2, z be the Descartes coordinate system, $x = (x_1, x_2)$. We assume that the mid-surface of a plate occupies the domain Ω of the plane $z = 0$ in its nondeformable state. Then the nonpenetration condition for the plate vertical displacements w is expressed by the inequalities

$$w(x) \ge \psi(x) + h$$

for the below punch, or by

$$w(x) \le \psi(x) - h$$

for the above punch. Here $2h$ is the plate thickness. Redenoting $\psi + h$ by ψ, we write the first nonpenetration condition in the form

$$w(x) \ge \psi(x), \quad x \in \Omega. \tag{1.34}$$

We can write the equilibrium equation for the plate

$$-M_{ij,ij} - f = p, \tag{1.35}$$

where f is the external force and $p \geq 0$ denotes the pressure of the punch. Assume that there is no contact at the point x, that is $w(x) - \psi(x) > 0$. In this case $p(x) = 0$ and thus

$$p(x)\Big(\bar{w}(x) - w(x)\Big) = 0 \quad \forall \bar{w}, \ \bar{w} \geq \psi.$$

Conversely, let a contact occur at the point x. This means $w(x) = \psi(x)$ and $p(x) \geq 0$. Consequently, in this case

$$p(x)\Big(\bar{w}(x) - w(x)\Big) \geq 0 \quad \forall \bar{w}, \ \bar{w} \geq \psi.$$

The above arguments prove that the inequality

$$p(\bar{w} - w) \geq 0 \quad \forall \bar{w}, \ \bar{w} \geq \psi$$

always holds. Substituting here (1.35), one obtains

$$w - \psi \geq 0, \quad (-M_{ij,ij} - f)(\bar{w} - w) \geq 0 \quad \forall \bar{w}, \ \bar{w} \geq \psi. \tag{1.36}$$

In its own turn, the conditions (1.36) are equivalent to

$$w - \psi \geq 0, \quad -M_{ij,ij} - f \geq 0, \quad (w - \psi)(M_{ij,ij} + f) = 0. \tag{1.37}$$

The meaning of the relation (1.37) is the following. The punch pressure $p = -M_{ij,ij} - f$ is equal to zero if a contact is absent. If the punch pressure at the given point is positive, then we have a contact at the above-mentioned point.

Thus, the relations (1.36) or (1.37) describe the interaction between a plate and a punch. To derive the contact model for an elastic plate, one needs to use the constitutive law (1.25). Contact problems for inelastic plates are derived by the utilizing of corresponding inelastic constitutive laws given in Section 1.1.4.

Let us consider both the vertical w and horizontal $U = (u_1, u_2)$ displacements of a plate. As previously, let $z = \psi(x)$ be the equation of a punch shape. We take the point $(x_1, x_2, 0)$ on the mid-plane. The nonpenetration condition could be written in the form

$$w(x_1, x_2) \geq \psi(x_1 + u_1(x_1, x_2), x_2 + u_2(x_1, x_2)) + h.$$

We can fulfil the Taylor expansion of the function $\psi(x + U(x))$ at the point x. Retaining the linear terms, the restriction $w \geq \psi + U\nabla\psi + h$ is obtained. Redenoting $\psi + h$ by ψ, we finally have

$$w - U\nabla\psi \geq \psi. \tag{1.38}$$

Variational inequality characterizing an interaction between the punch and the plate can be written in the form

$$(-M_{ij,ij} - f)(\bar{w} - w) + (-N_{ij,j} - f_i)(\bar{u}_i - u_i) \geq 0 \tag{1.39}$$

$$\forall(\bar{U}, \bar{w}), \ \bar{w} - \bar{U}\nabla\psi \geq \psi.$$

To derive the last relation, we take the equilibrium equations

$$-M_{ij,ij} - f = p, \quad -N_{ij,j} - f_i = p_i, \ i = 1, 2,$$

for the given exterior forces f_1, f_2, f; p_1, p_2, p are punch's forces. Let us multiply the equations by $\bar{w} - w$, $\bar{u}_i - u_i$, respectively, with an arbitrary $(\bar{U}, \bar{w})$, $\bar{U} = (\bar{u}_1, \bar{u}_2)$, satisfying (1.38) and sum. One gets

$$(-M_{ij,ij} - f)(\bar{w} - w) + (-N_{ij,j} - f_i)(\bar{u}_i - u_i) = p(\bar{w} - w) + p_i(\bar{u}_i - u_i). \quad (1.40)$$

It is clear that the right-hand side of (1.40) is nonnegative. In fact, at the noncontact point x we have $p(x) = p_1(x) = p_2(x) = 0$ and, therefore, the assertion is true. Now let x be a contact point. In this case

$$w(x) - U(x)\nabla\psi(x) = \psi(x). \quad (1.41)$$

On the other hand, we have

$$\bar{w}(x) - \bar{U}(x)\nabla\psi(x) \geq \psi(x). \quad (1.42)$$

Subtracting (1.41) from (1.42), one finds

$$0 \leq \bar{w}(x) - w(x) - \left(\bar{U}(x) - U(x)\right)\nabla\psi(x) \quad (1.43)$$
$$= \left(\bar{U}(x) - U(x), \bar{w}(x) - w(x)\right) \cdot (-\nabla\psi(x), 1).$$

The normal to the surface $z = \psi(x)$ has the coordinates

$$(-\nabla\psi(x), 1)/\sqrt{1 + |\nabla\psi|^2}.$$

Taking into account that the vector p has the same direction, we conclude from (1.43) that

$$p(x)(\bar{w}(x) - w(x)) + p_i(x)\left(\bar{u}_i(x) - u_i(x)\right) \geq 0.$$

Thus, we have obtained that the right-hand side of (1.40) is always nonnegative, which gives (1.39). To derive a complete system of relations describing the interaction between the punch and the plate we should add to (1.38), (1.39) the constitutive law equations of Sections 1.1.3 and 1.1.4.

We would like to stress at this point that the derivation of (1.36) and (1.38)–(1.39) is connected with the simulation of contact problems and therefore contains some assumptions of a mechanical character. This remark is concerned with the sign of the function p in the problem (1.36) and with the direction of the vector (p_1, p_2, p) in the problem (1.38), (1.39). Note that the classical approach to contact problems is characterized by a given contact set (Galin, 1980; Kikuchi, Oden, 1988; Grigolyuk, Tolkachev, 1980). In contact problems considered in the book, the contact set is unknown, and we obtain the so called free boundary problems. Other free boundary problems can be found in (Hoffmann, Sprekels, 1990; Elliot, Ockendon, 1982; Antontsev et al., 1990; Kinderlehrer et al., 1979; Antontsev et al., 1992; Plotnikov, 1995).

1.1.6 Boundary conditions

Let a solid body occupy a domain $\Omega \subset R^3$ with the smooth boundary Γ. The deformation of the solid inside Ω is described by equilibrium, constitutive and geometrical equations discussed in Sections 1.1.1–1.1.5. To formulate the boundary value problem we need boundary conditions at Γ. The principal types of boundary conditions are considered in this subsection.

The following restriction imposed upon the boundary displacements $u = (u_1, u_2, u_3)$,

$$u = 0, \tag{1.44}$$

corresponds to the *solid clamped on the boundary.*

Let $n = (n_1, n_2, n_3)$ be a unit outer normal vector at Γ. The restriction imposed upon the boundary stresses by

$$\sigma_{ij} n_j = 0, \quad i = 1, 2, 3, \tag{1.45}$$

corresponds to the *stress free boundary.* It is convenient to rewrite (1.45) for the tangential σ_τ and the normal σ_n components of the boundary stress vector $\sigma_{ij} n_j$ defined by the decomposition

$$\sigma_{ij} n_j = \sigma_n n_i + \sigma_{\tau i}, \; i = 1, 2, 3, \quad \sigma_\tau = (\sigma_{\tau 1}, \sigma_{\tau 2}, \sigma_{\tau 3}),$$

where

$$\sigma_n = \sigma_{ij} n_j n_i, \quad \sigma_{\tau i} = \sigma_{ij} n_j - \sigma_n n_i, \quad i = 1, 2, 3.$$

Then (1.45) is equivalent to

$$\sigma_n = 0, \quad \sigma_\tau = 0.$$

We now assume a validity of the unilateral boundary constraints provided that the nonpenetration of the boundary points over the given obstacle takes place, namely

$$u_n \le 0. \tag{1.46}$$

Here u_n is a normal component of the boundary displacements vector u defined by the decomposition

$$u_i = u_n n_i + u_{\tau i}, \; i = 1, 2, 3; \quad u_\tau = (u_{\tau 1}, u_{\tau 2}, u_{\tau 3});$$

$$u_n = u_i n_i, \quad u_{\tau i} = u_i - u_n n_i, \quad i = 1, 2, 3.$$

The boundary inequality (1.46) is called a *Signorini condition* (Fichera, 1972).

Let $\mathcal{F} \ge 0$ be a given *friction coefficient*, and $g \ge 0$ be a known *friction force* at the boundary. The conditions of *given friction along the normal* implies

$$\begin{cases} |\sigma_n| \le \mathcal{F} g & \Longrightarrow \quad u_n = 0, \\ \sigma_n = \mathcal{F} g & \Longrightarrow \quad u_n \le 0, \\ \sigma_n = -\mathcal{F} g & \Longrightarrow \quad u_n \ge 0. \end{cases}$$

The last relations can be generalized in the form

$$|\sigma_n| \leq \mathcal{F}g, \quad \sigma_n u_n + \mathcal{F}g|u_n| = 0. \tag{1.47}$$

This case is considered in Section 3.6. The given *friction along the tangent* is more suitable from the standpoint of mechanics and is described by the conditions similar to (1.47),

$$|\sigma_\tau| \leq \mathcal{F}g, \quad \sigma_{\tau i} u_{\tau i} + \mathcal{F}g|u_\tau| = 0. \tag{1.48}$$

The friction force g is unknown in general. The *Coulomb law* assumes $g = |\sigma_n|$ and provides the more general relations as compared with (1.48) (Hlavaček et al., 1988; Duvaut, Lions, 1972; Demkowicz, Oden, 1982; Haslinger, Panagiotopoulos, 1984; Namm, 1995)

$$|\sigma_\tau| \leq \mathcal{F}|\sigma_n|, \quad \sigma_{\tau i} u_{\tau i} + \mathcal{F}|\sigma_n||u_\tau| = 0.$$

We formulate boundary conditions in the two-dimensional theory of plates and shells. Denote by $u = (U, w)$, $U = (u_1, u_2)$, horizontal and vertical displacements at the boundary Γ of the mid-surface $\Omega \subset R^2$. Then the horizontal displacements U may satisfy the Dirichlet-type conditions

$$U = 0$$

or the Neumann-type conditions

$$N_{ij} n_j = 0, \quad i = 1, 2.$$

Here N_{ij} denote the integrated stresses; $n = (n_1, n_2)$ is a unit outer normal at Γ.

The vertical displacements w are described by the fourth order differential equation according to the equilibrium and the constitutive laws. The following relations for w,

$$w = \frac{\partial w}{\partial n} = 0,$$

provide a *jam condition* of the boundary, where $\partial w/\partial n = w_{,i} n_i$. Let us define the *bending moment* m and *transverse force* t by the formulae

$$m = -M_{ij} n_j n_i, \quad t = -M_{ij,k} \tau_k \tau_j n_i - M_{ij,j} n_i,$$

where $\tau = (-n_2, n_1)$ is the tangent vector at the boundary Γ. The *hinge conditions* imply

$$w = m = 0,$$

and the *stress free boundary* is described by the equalities

$$m = t = 0.$$

1.1.7 Crack in a solid. Nonpenetration conditions

Let a solid occupy a bounded domain $\Omega \subset R^3$ with the smooth boundary Γ (see Fig.1.1). Let Ω contain a smooth unclosed surface Γ_c, probably intersecting Γ. We assume that Γ_c is an oriented surface such that there exists a mapping

$$x_i = x_i(y_1, y_2), \quad i = 1, 2, 3,$$

of $\overline{\gamma} \subset R^2$ onto Γ_c which sets up a one-to-one correspondence and has a positive Jacobian. Here $x \in \Gamma_c$; (y_1, y_2) are coordinates of the point $y \in \overline{\gamma}$, $\overline{\gamma} = \gamma \cup \partial\gamma$. We assume that γ is a bounded simply connected domain in R^2 with a smooth boundary $\partial\gamma$.

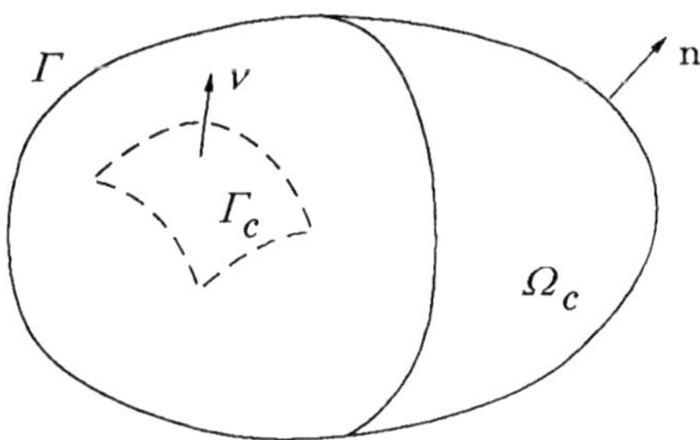

Fig.1.1. 3D-body with a crack

Let us denote by $n = (n_1, n_2, n_3)$ a unit outer normal to Γ and choose the direction $\nu = (\nu_1, \nu_2, \nu_3)$ of a unit normal vector to Γ_c. Then ν defines the positive side Γ_c^+ of the surface Γ_c with the outer normal $-\nu$ and the negative side Γ_c^- of Γ_c with the outer normal ν. Thus we get the domain $\Omega_c = \Omega \setminus \Gamma_c$ disposed between the outer boundary Γ and the inner boundary $\Gamma_c^+ \cup \Gamma_c^-$. In the sequel we call Ω_c a *solid with a crack*.

Let v be some known function defined in the domain Ω_c. If v and the boundary $\partial\Omega_c = \Gamma \cup \Gamma_c^+ \cup \Gamma_c^-$ are sufficiently smooth, then we can define values of v at the boundary (the exact smoothness conditions are studied in Section 1.4). In particular, having the values $v|_{\Gamma_c^+}$ and $v|_{\Gamma_c^-}$, we introduce the *jump* of v at Γ_c by the formula

$$[v] = v|_{\Gamma_c^+} - v|_{\Gamma_c^-}.$$

Now we are in a position to formulate suitable boundary conditions at Γ_c.

As before, let $u = (u_1, u_2, u_3)$ denote the displacement field in the domain Ω_c. If the boundary is clamped then (1.44) provides

$$u = 0$$

at Γ_c. The condition for *zero opening of the crack* is as follows:

$$[u] = 0.$$

The *stress free boundary* condition (1.45) for crack surfaces implies

$$\sigma_{ij}\nu_j = 0, \quad i = 1, 2, 3.$$

The *nonpenetration condition* of the crack faces is easily derived,

$$[u]\nu = [u_\nu] \geq 0. \tag{1.49}$$

Thus we obtain the crack model obeying the unilateral constraint (1.49) of the Signorini type.

Now we intend to derive nonpenetration conditions for plates and shells with cracks. Let a domain $\Omega \subset R^2$ with the smooth boundary Γ coincide with a mid-surface of a shallow shell. Let Γ_ψ be an unclosed curve in Ω perhaps intersecting Γ (see Fig.1.2). We assume that Γ_ψ is described by a smooth function $x_2 = \psi(x_1)$. Denoting $\Omega_\psi = \Omega \setminus \Gamma_\psi$ we obtain the description of the *shell* (or the *plate) with the crack*. This means that the crack surface is a cylindrical surface in R^3, i.e. it can be described as $x_2 = \psi(x_1)$, $-h \leq z \leq h$, where (x_1, x_2, z) is the orthogonal coordinate system, and $2h$ is the thickness of the shell. Let us choose the unit normal vector $\nu = (\nu_1, \nu_2)$ at Γ_ψ,

$$\nu = (-\psi', 1)/\sqrt{1 + |\psi'|^2},$$

which defines the positive Γ_ψ^+ and the negative Γ_ψ^- sides of the curve Γ_ψ. Then the jump $[v]$ of v at Γ_ψ is equal to $v|_{\Gamma_\psi^+} - v|_{\Gamma_\psi^-}$.

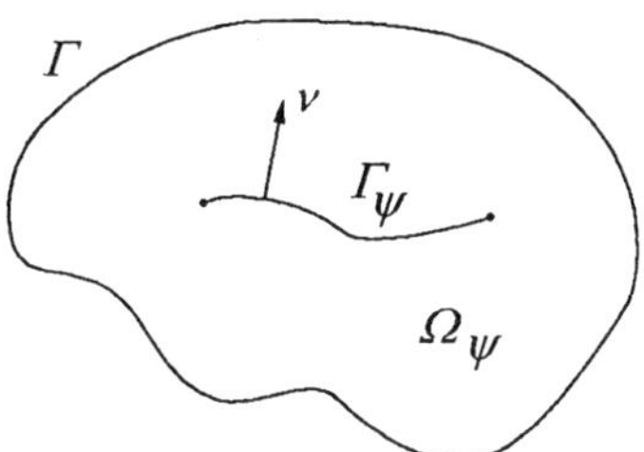

Fig.1.2. Middle surface of the plate

Studying separately the equations (1.21)– (1.23) for the shell horizontal displacements $U = (u_1, u_2)$ and (1.24), (1.25) for the shell vertical displacements w at the mid-surface, we can apply (1.49) and deduce the nonpenetration condition

$$[U_\nu] = [U]\nu = [u_i]\nu_i \geq 0 \tag{1.50}$$

for the shell horizontal displacements. On the other hand, we may consider (1.50) as a *simplified nonpenetration condition.*

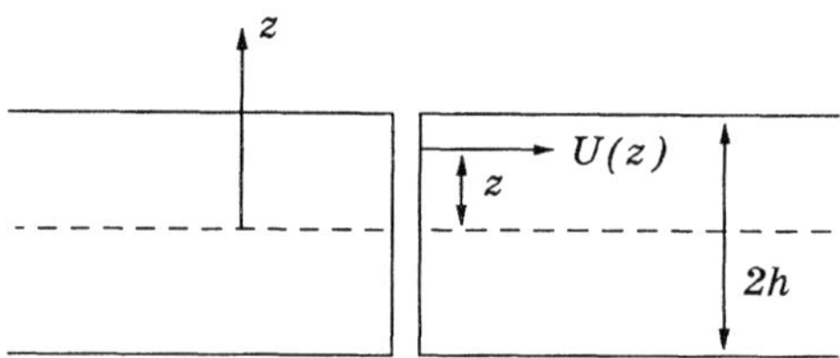

Fig.1.3. Vertical cross-section of the plate

Let us derive a condition of nonpenetrating in general case (see Fig.1.3). The Kirchhoff–Love hypothesis provides the linear dependence of the shell horizontal displacements on a distance from the mid-surface, namely

$$U(z) = U - z\nabla w, \quad w(z) = w, \quad -h \le z \le h. \tag{1.51}$$

Substituting (1.51) into (1.49), one has

$$0 \le [(U(z), w(z))] \cdot (\nu, 0) = [U - z\nabla w]\nu = [U]\nu - z\left[\frac{\partial w}{\partial \nu}\right]. \tag{1.52}$$

The arbitrariness of $-h \le z \le h$ implies that the relation (1.52) is fulfilled for $z = h$:

$$0 \le [U]\nu - h\,[\partial w/\partial \nu]$$

and for $z = -h$:

$$0 \le [U]\nu + h\,[\partial w/\partial \nu].$$

These inequalities can be written as

$$[U]\nu \ge h\left|\left[\frac{\partial w}{\partial \nu}\right]\right|. \tag{1.53}$$

Conversely, the linear dependence of (1.52) on z guarantees fulfilment of (1.52) provided that (1.53) holds.

Thus, (1.53) is a *complete nonpenetration condition* of the crack surfaces for the Kirchhoff–Love plates and shallow shells. By putting the thickness $2h$ to be zero, one reduces (1.53) to the simplified nonpenetration condition (1.50).

1.1.8 Variational formulation of the problems

Let a solid with a crack occupy the domain Ω_c in the sense shown in the previous subsection, and $f = (f_1, f_2, f_3)$ be a given external force. We define the *functional of potential energy* for the solid,

$$\Pi = \frac{1}{2}\int\limits_{\Omega_c} \sigma_{ij}\varepsilon_{ij}\, d\Omega_c - \int\limits_{\Omega_c} f_i u_i\, d\Omega_c. \tag{1.54}$$

Utilizing the geometrical equations $\varepsilon_{ij} = \varepsilon_{ij}(u)$ and the constitutive law relations $\sigma_{ij}(\varepsilon_{kl})$ discussed in Section 1.1.1, the formula (1.54) defines the functional Π depending on the displacements u, i.e. $\Pi = \Pi(u)$. Considering boundary conditions listed in Section 1.1.6 at the outer boundary Γ and the nonpenetration condition (1.49) at the crack Γ_c, we define the *admissible displacements set* K. Then the *equilibrium problem* for the elastic solid with a crack consists in the following minimization problem:

$$\Pi(u) = \inf_{\bar{u} \in K} \Pi(\bar{u}), \tag{1.55}$$

which is equivalent to the variational inequality

$$u \in K, \quad \Pi'_u(\bar{u} - u) \geq 0 \quad \forall \bar{u} \in K, \tag{1.56}$$

provided that K is convex. Here Π'_u is the derivative of the functional Π at the point u. In the sequel, we intend to show that (1.56) provides the fulfilment of the equilibrium equations (1.1),

$$-\sigma_{ij,j}(u) = f_i, \quad i = 1, 2, 3,$$

holding in the domain Ω_c, and general boundary conditions holding at the crack faces. It is of importance that the presence of the nonpenetration condition is considered. Without this constraint one obtains stress free boundary conditions at the crack faces. For displacements obeying the friction condition (1.48) at the boundary part $\Gamma_{\mathcal{F}}$, the potential energy functional takes the form

$$\Pi = \frac{1}{2} \int\limits_{\Omega_c} \sigma_{ij} \varepsilon_{ij} \, d\Omega_c - \int\limits_{\Omega_c} f_i u_i \, d\Omega_c + \int\limits_{\Gamma_{\mathcal{F}}} \mathcal{F} g |u_\tau| \, d\Gamma_{\mathcal{F}}.$$

In the two-dimensional theory of solids, the potential energy functional for the shallow shell with the mid-surface Ω_ψ is as follows:

$$\Pi = \frac{1}{2} \int\limits_{\Omega_\psi} (N_{ij} \varepsilon_{ij} + M_{ij} \kappa_{ij}) \, d\Omega_\psi - \int\limits_{\Omega_\psi} (f_i u_i + f w) \, d\Omega_\psi.$$

Substituting here the corresponding geometrical and constitutive relations of Sections 1.1.3 and 1.1.4, we obtain $\Pi = \Pi(U, w)$. The set of admissible displacements K is defined by the boundary conditions at Γ and nonpenetration conditions at the crack Γ_ψ stated in Section 1.1.7. The variational form of the equilibrium problem is the following:

$$\Pi(U, w) = \inf_{(\bar{U}, \bar{w}) \in K} \Pi(\bar{U}, \bar{w}). \tag{1.57}$$

If U and w are not coupled, one can separately analyse the following two problems:

$$\inf_{\bar{U} \in K} \Pi(\bar{U}, \bar{w}), \quad \inf_{\bar{w} \in K} \Pi(\bar{U}, \bar{w}).$$

The minimization problem (1.57) also provides the fulfilment of the equilibrium equations

$$-N_{ij,j}(U,w) = f_i,\ i = 1,2, \quad -M_{ij,ij}(U,w) = f$$

in Ω_ψ and general boundary conditions holding at the crack faces.

Let us emphasize that not every model can be presented as a minimization problem like (1.55) or (1.57). Thus, elastoplastic problems considered in Chapter 5 can be formulated as variational inequalities, but we do not consider any minimization problems in plasticity. In all cases, we have to study variational problems or variational inequalities. It is a principal topic of the following two sections. As for general variational principles in mechanics and physics we refer the reader to (Washizu, 1968; Chernous'ko, Banichuk, 1973; Ekeland, Temam, 1976; Telega, 1987; Panagiotopoulos, 1985; Morel, Solimini, 1995).

1.2 Elements of convex analysis

1.2.1 Minimization problem. Variational inequalities

Let V be a normed space, and $J : V \to R$ be an arbitrary functional. We assume that there exists a linear and continuous functional J'_u such that for each $v \in V$

$$J'_u(v) = \lim_{\lambda \to 0} \frac{J(u+\lambda v) - J(u)}{\lambda}.$$

It is said in this case that the functional J has the *derivative* J'_u at the point u. Let $V^\star$ be the space dual of V, i.e. the space of all linear continuous functionals on V. If the operator $J' : V \to V^\star$ is defined such that for each $u \in V$ the derivative J'_u can be found at the point u, then the functional J is called *differentiable.*

A set $K \subset V$ is called a *convex set* if the inclusion $\lambda u_1 + (1-\lambda)u_2 \in K$ is valid for all $u_1, u_2 \in K$, $\lambda \in (0,1)$. Let $K \subset V$ be a convex set, and J be a differentiable functional on V. We consider a *minimization problem*

$$\inf_{u \in K} J(u). \tag{1.58}$$

An element $u_0 \in K$ is called the solution of the problem (1.58) if

$$J(u_0) = \inf_{u \in K} J(u) \quad \text{(i.e.} \quad J(u) - J(u_0) \geq 0 \quad \forall u \in K).$$

We take $u_1 \in K$, $\lambda \in (0,1)$, and put the element $\lambda u_1 + (1-\lambda)u_0$ instead of u. Then the relation

$$\frac{1}{\lambda}\left(J(u_0 + \lambda(u_1 - u_0)) - J(u_0)\right) \geq 0$$

follows. Passing to the limit as $\lambda \to 0$, we find

$$u_0 \in K: \quad J'_{u_0}(u_1 - u_0) \geq 0 \quad \forall u_1 \in K. \tag{1.59}$$

The inequality like (1.59) is called a *variational inequality*. It was obtained from a minimization problem of the functional J over the set K. In the sequel we will look more attentively at a connection between a minimization problem and a variational inequality. Now we want to underline one essential point. We see that the problem (1.58) is more general in comparison with the minimization problem on the whole space V. It is well known that the necessary condition in the last problem coincides with the Euler equation. The variational inequality (1.59) generalizes the Euler equation. Moreover, for $K = V$ the Euler equation follows from (1.59). To obtain it we take $u_1 = u_0 + u$ and substitute in (1.59) with an arbitrary element $u \in V$. It gives

$$J'_{u_0}(u) = 0 \quad \forall u \in V.$$

This exactly coincides with the Euler equation.

A functional $J : V \to R$ is called a *convex functional* if

$$J(\lambda u_0 + (1 - \lambda)u_1) \leq \lambda J(u_0) + (1 - \lambda)J(u_1) \tag{1.60}$$

for all $u_0, u_1 \in V$, $\lambda \in (0,1)$. The functional J is called a *strictly convex* one if it is convex and the equality in (1.60) is nonadmissible for $u_0 \neq u_1$.

Let the functional J be convex and differentiable. We can prove the validity of the inequality

$$J(u_0) - J(u_1) \geq J'_{u_1}(u_0 - u_1) \quad \forall u_0, u_1 \in V. \tag{1.61}$$

Thus, it follows from (1.60) that

$$J(u_1 + \lambda(u_0 - u_1)) - J(u_1) \leq \lambda \left(J(u_0) - J(u_1)\right).$$

After dividing this relation by λ and passing to the limit as $\lambda \to 0$ the inequality

$$J'_{u_1}(u_0 - u_1) \leq J(u_0) - J(u_1)$$

follows. It coincides with (1.61).

We have pointed out the following fact. If $K \subset V$ is a convex set then the variational inequality

$$u_0 \in K: \quad J'_{u_0}(u_1 - u_0) \geq 0 \quad \forall u_1 \in K \tag{1.62}$$

provides a necessary condition of a minimum attainability for the functional J over the set K at the point $u_0 \in K$. It proved that the condition (1.62) is a sufficient one in the case of the convex functional J. Let us justify it. Assuming the validity of (1.62) it follows from (1.62), (1.61) that

$$J(u_1) - J(u_0) \geq J'_{u_0}(u_1 - u_0) \geq 0 \quad \forall u_1 \in K.$$

This means

$$J(u_1) - J(u_0) \geq 0 \quad \forall u_1 \in K. \tag{1.63}$$

The established statement will be formulated as a theorem.

Theorem 1.1. *The inequality* (1.62) *gives necessary and sufficient conditions of the minimum over the set K for a convex and differentiable functional J.*

The following lemma will be useful below.

Lemma 1.1. *For every convex and differentiable functional J the function*

$$\lambda^{-1}(J(u+\lambda u_0) - J(u))$$

of the variable λ is nondecreasing.

Proof. We introduce the notation

$$\phi(\lambda) = \frac{1}{\lambda}\left(J(u+\lambda u_0) - J(u)\right).$$

Our goal is to prove that $\phi'(\lambda) \geq 0$. We have

$$\phi'(\lambda) = \frac{1}{\lambda^2}\left(\lambda\frac{d}{d\lambda}J(u+\lambda u_0) - J(u+\lambda u_0) + J(u)\right). \tag{1.64}$$

It is easy to find

$$\frac{d}{d\lambda}J(u+\lambda u_0) = J'_{u+\lambda u_0}(u_0).$$

Thus, it follows from (1.64) that

$$\phi'(\lambda) = \frac{1}{\lambda^2}\left(\lambda J'_{u+\lambda u_0}(u_0) - J(u+\lambda u_0) + J(u)\right).$$

The right-hand side is nonnegative here in view of (1.61), therefore $\phi'(\lambda) \geq 0$. Lemma 1.1 is proved.

It was stated above that the inequality

$$J(u_0) - J(u_1) \geq J'_{u_1}(u_0 - u_1) \quad \forall u_0, u_1 \in V \tag{1.65}$$

is valid for convex and differentiable functionals. Let us prove the converse. We shall state that a convexity of J follows from (1.65). To verify this we take $u, u_0 \in V$ and substitute $u_1 = (1-\lambda)u + \lambda u_0$ in the inequality (1.65). This gives

$$J(u_0) - J((1-\lambda)u + \lambda u_0) \geq J'_{(1-\lambda)u+\lambda u_0}\left((1-\lambda)(u_0 - u)\right). \tag{1.66}$$

The same reasonings provide the inequality

$$J(u) - J((1-\lambda)u + \lambda u_0) \geq J'_{(1-\lambda)u+\lambda u_0}\left(\lambda(u - u_0)\right). \tag{1.67}$$

Let us multiply (1.66), (1.67) by λ, $1-\lambda$, respectively, and sum the obtained relations. The result can be written in the form

$$\lambda J(u_0) + (1-\lambda)J(u) - J((1-\lambda)u + \lambda u_0) \geq 0. \tag{1.68}$$

This means that the functional is convex.

The previous considerations can be specified. Namely, if J is a strictly convex functional then

$$J(u_0) - J(u_1) > J'_{u_1}(u_0 - u_1), \quad u_0 \neq u_1, \tag{1.69}$$

and conversely. To prove this we assume the validity of (1.69). Then the inequalities (1.66), (1.67) will be strict for $u_0 \neq u$. Consequently, the inequality (1.68) will also be strict. Conversely, let J be a strictly convex functional. Then

$$J(u + \lambda(u_0 - u)) < \lambda J(u_0) + (1-\lambda)J(u), \quad u \neq u_0.$$

Whence

$$\frac{1}{\lambda}\left(J(u + \lambda(u_0 - u)) - J(u)\right) < J(u_0) - J(u).$$

Taking into account Lemma 1.1, we conclude that the left-hand side of this inequality converges from above to $J'_u(u_0 - u)$. Thus

$$J'_u(u_0 - u) < J(u_0) - J(u).$$

The statement is proved.

1.2.2 Convex functionals

Convex functionals have a convenient description in terms of their derivatives. We briefly discuss this question.

Let V be a normed space and $V^\star$ be its dual. An operator $A: V \to V^\star$ is called a *monotonous operator* if

$$(Au - Au_1)(u - u_1) \geq 0 \quad \forall u, u_1 \in V.$$

As it was pointed out in Section 1.2.1 one can construct the operator J' : $V \to V^\star$ which assigns the derivative J'_u for each $u \in V$. The following statement is valid.

Theorem 1.2. *The functional J is convex if and only if the operator J' is monotonous.*

PROOF. By virtue of the above statements,

$$J(u) - J(u_1) \geq J'_{u_1}(u - u_1), \quad J(u_1) - J(u) \geq J'_u(u_1 - u).$$

Summing up these inequalities we find the required

$$(J'_u - J'_{u_1})(u - u_1) \geq 0 \quad \forall u, u_1 \in V. \tag{1.70}$$

Conversely. Let the inequality (1.70) be valid. We introduce the notation

$$\phi(\lambda) = J(u_1 + \lambda(u - u_1)), \quad \lambda \in (0,1)$$

for arbitrary fixed elements $u, u_1 \in V$ and prove that the value

$$\Delta \equiv \lambda J(u) + (1-\lambda)J(u_1) - J(\lambda u + (1-\lambda)u_1)$$

is nonnegative. One has

$$\Delta = \lambda(\phi(1) - \phi(\lambda)) + (1-\lambda)(\phi(0) - \phi(\lambda)) = \lambda(1-\lambda)\left(\phi'(\xi_1) - \phi'(\xi_2)\right).$$

The well-known theorem of finite differences was used here. In particular, $\xi_1 \in (\lambda, 1)$, $\xi_2 \in (0, \lambda)$, and therefore $\xi_1 > \xi_2$. At the same time

$$\phi'(\xi_i) = J'_{u_1+\xi_i(u-u_1)}(u - u_1), \quad i = 1, 2.$$

Hence, in view of (1.70),

$$\phi'(\xi_1) - \phi'(\xi_2) \geq 0.$$

This means $\Delta \geq 0$, which proves the convexity of the functional J. The proof of Theorem 1.2 is completed.

Let $K \subset V$ be a convex set, and $J : V \to R$ be a convex differentiable functional. As it was shown, the minimization problem of J over the set K is equivalent to the following variational inequality:

$$u \in K : \quad J'_u(u_1 - u) \geq 0 \quad \forall u_1 \in K. \tag{1.71}$$

It turns out that an equivalent form of the variational inequality (1.71) can be given. Namely, the following theorem is valid.

Theorem 1.3. *Let $J' : V \to V^\star$ be a continuous mapping. Then the inequality (1.71) is fulfilled if and only if*

$$u \in K : \quad J'_{u_1}(u_1 - u) \geq 0 \quad \forall u_1 \in K. \tag{1.72}$$

PROOF. By virtue of the convexity of J, the operator J' is monotonous. Thus

$$(J'_u - J'_{u_1})(u - u_1) \geq 0 \quad \forall u, u_1 \in K. \tag{1.73}$$

If the element u satisfies the inequality (1.71) then, summing (1.71) and (1.73), we obtain (1.72). Conversely, let (1.72) be fulfilled. Taking $u_1 = (1-\lambda)u + \lambda u_0$, $u_0 \in K$, we find

$$\lambda J'_{(1-\lambda)u+\lambda u_0}(u_0 - u) \geq 0.$$

Dividing this inequality by λ and passing to the limit as $\lambda \to 0$ on the basis of the continuity of J', we obtain the desired inequality

$$J'_u(u_0 - u) \geq 0 \quad \forall u_0 \in K.$$

This proves the theorem.

To conclude this subsection we consider a case of nonsmooth functionals. Let, as before, $K \subset V$ be a convex set, and $J : V \to R$ be a convex functional. We assume that J is represented as a sum of differentiable and nondifferentiable functionals. Namely, let $J = J_1 + J_2$, where J_1, J_2 are convex functionals, J_1 is differentiable and, moreover, $J_1' : V \to V^\star$ is a continuous operator. Consider the problem

$$\inf_{u \in K} J(u). \tag{1.74}$$

The following statement is valid.

Theorem 1.4. *Three conditions formulated below are equivalent:*

$$u \in K \quad \text{is the solution of } (1.74), \tag{1.75}$$

$$u \in K: \quad (J_1)'_u(u_0 - u) + J_2(u_0) - J_2(u) \geq 0 \quad \forall u_0 \in K, \tag{1.76}$$

$$u \in K: \quad (J_1)'_{u_0}(u_0 - u) + J_2(u_0) - J_2(u) \geq 0 \quad \forall u_0 \in K. \tag{1.77}$$

PROOF. First of all we prove the equivalence of (1.75) and (1.76). Let $u \in K$ be the solution of (1.74). Then

$$J(u) \leq J((1-\lambda)u + \lambda u_0) \quad \forall u_0 \in K, \quad \lambda \in (0,1).$$

By the convexity of the functional J_2, from this inequality we obtain the first relation,

$$J_1(u) + J_2(u) \leq J_1((1-\lambda)u + \lambda u_0) + (1-\lambda)J_2(u) + \lambda J_2(u),$$

and after division by λ we obtain the second one,

$$\frac{1}{\lambda}\left(J_1((1-\lambda)u + \lambda u_0) - J_1(u)\right) + J_2(u_0) - J_2(u) \geq 0.$$

Let us pass here to the limit as $\lambda \to 0$. This gives (1.76). Conversely, let u satisfy the variational inequality (1.76). Then, by the convexity of J_1, one has

$$J_1(u_0) - J_1(u) \geq (J_1)'_u(u_0 - u) \quad \forall u_0 \in K.$$

We sum this relation with (1.76) and find

$$J(u_0) - J(u) \geq 0 \quad \forall u_0 \in K.$$

This means that u is a solution of problem (1.74).

Now let us prove the equivalence of (1.76) and (1.77). We assume that (1.76) is valid. It follows from the monotonicity of J_1' that

$$\left((J_1)'_{u_0} - (J_1)'_u\right)(u_0 - u) \geq 0.$$

Summing up this inequality with (1.76), we obtain (1.77) exactly. Conversely, let (1.77) be fulfilled. We take $u_1 \in K$ and substitute $(1-\lambda)u+\lambda u_1$ as the test element which provides

$$\lambda (J_1)'_{(1-\lambda)u+\lambda u_1}(u_1 - u) + J_2((1-\lambda)u + \lambda u_1) - J_2(u) \geq 0.$$

The left-hand side of this inequality can be estimated from above by using the convexity of J_2. Then we derive the obtained inequality by λ and pass to the limit as $\lambda \to 0$. The resulting relation coincides with (1.76). Theorem 1.4 is completely proved.

1.2.3 Compactness properties

Let V be a normed space and $V^\star$ be its dual. A sequence of elements $u_n \in V$ is called *weakly converging* to an element u if for every fixed $u^\star \in V^\star$

$$u^\star(u_n) \to u^\star(u), \quad n \to \infty.$$

We can consider a dual space $V^{\star\star}$ with respect to $V^\star$. For an arbitrary fixed element $u \in V$ the functional $u^\star \to u^\star(u)$ can be defined, where $u^\star \in V^\star$. This functional is linear and continuous on $V^\star$ and therefore is an element of the space $V^{\star\star}$. For every $u \in V$ the functional $u^{\star\star} \in V^{\star\star}$ can be pointed out such that

$$u^{\star\star}(u^\star) = u^\star(u) \quad \forall u^\star \in V^\star.$$

Thus we obtain the imbedding of the space V into the second dual one $V^{\star\star}$. The imbedding operator is denoted by π. If $\pi V = V^{\star\star}$, the space V is called *reflexive* (Kantorovich, Akilov, 1984). The simplest example of reflexive spaces are $L^p(\Omega)$ for $1 < p < \infty$ since

$$(L^p(\Omega))^\star = L^q(\Omega), \quad 1/p + 1/q = 1.$$

At the same time the spaces $L^1(\Omega)$ and $L^\infty(\Omega)$ are nonreflexive: a space dual of $L^1(\Omega)$ coincides with $L^\infty(\Omega)$, but a dual one of $L^\infty(\Omega)$ is wider as compared with $L^1(\Omega)$. The notion of $\star$–weak convergence will be used for nonreflexive spaces. Namely, the sequence of elements $u_n^\star \in V^\star$ is called *$\star$–weakly convergent* to the element $u^\star \in V^\star$ if for every fixed $u \in V$

$$u_n^\star(u) \to u^\star(u), \quad n \to \infty.$$

The compactness properties are closely connected with the reflexivity of spaces. On that score we formulate two theorems widely used in this book (Vainberg, 1972).

Theorem 1.5. *A bounded set of reflexive Banach spaces is weakly compact.*

Theorem 1.6. *A bounded set of the space dual of the separable normed one is $\star$–weakly compact.*

Using the term *weakly compact* we mean only that every bounded sequence contains a weakly converging subsequence. The same is related to the term $\star$ - *weakly compact.*

1.2.4 Semicontinuous functionals

A functional $J : V \to R$ is called *weakly lower semicontinuous* at the point u if the condition $u_n \to u$ weakly in V implies

$$\liminf J(u_n) \geq J(u).$$

The weak convergence can be compared with the strong one (convergence in the norm). It is easily seen that if $u_n \to u$ strongly in V then $u_n \to u$ weakly in V. We have to indicate also that a continuity of the functional does not imply, in general, its weak lower semicontinuity. Of course, the weak lower semicontinuity does not imply a continuity. At the same time a weak lower semicontinuous functional is strongly lower semicontinuous, i.e. the condition $u_n \to u$ strongly in V implies

$$\liminf J(u_n) \geq J(u).$$

This property is obvious in so far as the strongly converging sequence u_n is weakly converging.

Now let us prove two theorems containing sufficient conditions of weak lower semicontinuity of the functionals.

Theorem 1.7. *Let the inequality*

$$J(u) - J(u_0) \geq J'_{u_0}(u - u_0) \quad \forall u \in V \tag{1.78}$$

be fulfilled at the point $u_0 \in V$. Then the functional J is weakly lower semicontinuous at the point u_0.

PROOF. Let $u_n \to u_0$ weakly in V. By virtue of the inclusion $J'_{u_0} \in V^\star$ we have

$$J'_{u_0}(u_n - u_0) \to 0.$$

Then it follows from (1.78) that

$$\liminf (J(u_n) - J(u_0)) \geq \lim J'_{u_0}(u_n - u_0) = 0.$$

Hence

$$\liminf J(u_n) \geq J(u_0).$$

The theorem is proved.

The second assertion is formulated as follows.

Theorem 1.8. *Let the functional $J : V \to R$ satisfy the condition*

$$\left(J'_u - J'_{u_0}\right)(u - u_0) \geq 0 \quad \forall u, u_0 \in V. \tag{1.79}$$

Then the functional J is weakly lower semicontinuous on V.

PROOF. It was proved that the inequality (1.79) is equivalent to the convexity of J. On the other hand, the convexity is equivalent to (1.78). Hence the result follows from the previous theorem.

We formulate one more theorem concerning the weak lower semicontinuity.

Theorem 1.9. *Let V be a Banach space, and $J : V \to R$ be a convex functional. Then J is weakly lower semicontinuous on V.*

1.2.5 Existence of solution to minimization problems

A functional $J : V \to R$ is called a *coercive* if

$$J(u) \to +\infty \quad \text{as} \quad ||u|| \to \infty.$$

A set $K \subset V$ is called *weakly closed* if the conditions $u_n \to u$ weakly in V, $u_n \in K$, imply $u \in K$. The following statement is valid.

Theorem 1.10. *A closed convex set of a reflexive Banach space is weakly closed.*

Now we can prove the statement of solution existence to minimization problems.

Theorem 1.11. *Let V be a reflexive Banach space, and $K \subset V$ be a closed convex set. Assume that $J : V \to R$ is a coercive and weakly lower semicontinuous functional. Then the problem*

$$\inf_{u \in K} J(u) \tag{1.80}$$

has a solution.

PROOF. Let us take a minimizing sequence u_n, i.e. a sequence possessing the property

$$J(u_n) \to \inf_{u \in K} J(u).$$

The functional J is coercive, hence the sequence u_n is bounded,

$$||u_n|| \leq c,$$

where the constant c is independent of n. If the converse is valid, the chosen sequence can not be a minimizing one. By the reflexivity of V, one can choose a subsequence u_i from the sequence u_n such that $u_i \to u$ weakly in V. We have $u_i \in K$, whence by Theorem 1.10, the inclusion $u \in K$ follows. Let us denote

$$j = \inf_{u \in K} J(u).$$

Then

$$j = \liminf J(u_n) = \liminf J(u_i) \geq J(u).$$

The inequality used here follows from the weak lower semicontinuity of the functional J. Thus, the element u is found such that

$$j = J(u), \quad u \in K.$$

This means that u is the solution of problem (1.80). The proof of Theorem 1.11 is completed.

Let V be a reflexive Banach space. We consider a bilinear continuous functional $B : V \times V \to R$ such that $B(u, u_1) = B(u_1, u)$, $B(u, u) \geq 0$ for all $u, u_1 \in V$. Let also $F : V \to R$ be a linear continuous functional, and $K \subset V$ be a closed convex set. We define the functional

$$J(u) = \frac{1}{2}B(u, u) - F(u)$$

and investigate the minimization problem

$$\inf_{u \in K} J(u). \tag{1.81}$$

Theorem 1.12. *The solution of the problem* (1.81) *exists if and only if there exists a solution of the variational inequality*

$$u \in K : \quad B(u, \bar{u} - u) \geq F(\bar{u} - u) \quad \forall \bar{u} \in K. \tag{1.82}$$

PROOF. First of all we find the derivative J'_u of the functional J. It is easily seen that

$$\lim_{\lambda \to 0} \frac{J(u + \lambda \bar{u}) - J(u)}{\lambda} = B(u, \bar{u}) - F(\bar{u}).$$

We have used here the symmetry of the functional B. Thus

$$J'_u(\bar{u}) = B(u, \bar{u}) - F(\bar{u}).$$

Hence the mapping $u \to J'_u$ is monotonous. Consequently, the functional J is convex. As it was proved in Theorem 1.1, the problem (1.81) is equivalent to

$$u \in K : \quad J'_u(\bar{u} - u) \geq 0 \quad \forall \bar{u} \in K. \tag{1.83}$$

Taking into account the obtained formula for J'_u, we complete the proof of Theorem 1.12.

Let us formulate sufficient conditions of solvability to the problem (1.82). We additionally assume that there exists a constant $c > 0$ such that

$$B(u, u) \geq c\|u\|^2 \quad \forall u \in V.$$

It will be proved that all conditions of Theorem 1.11 are fulfilled. It provides the solvability of the problem (1.81) and therefore of the problem (1.82).

We first show that the functional J is weakly lower semicontinuous. The function $B(u, u_0)$ is linear and continuous over u_0 for each fixed $u \in V$. Now let $u_n \to u$ weakly. Then

$$B(u_n, u_n) = B(u, u) + 2B(u, u_n - u) + B(u_n - u, u_n - u).$$

The second term of the right-hand side of this relation converges to zero in view of the weak convergence of u_n; the third term is nonnegative. Hence

$$\liminf B(u_n, u_n) \geq B(u, u).$$

The linear functional F obviously possesses the needful property of the weak continuity. Thus we obtain the weak lower semicontinuity of the functional J.

By the inequality $|F(u)| \leq \|F\|_\star \|u\|$, the functional J is also coercive. Indeed,

$$J(u) \geq \frac{c}{2}\|u\|^2 - \|F\|_\star \|u\| \ \to \ +\infty$$

as $\|u\| \to \infty$. Hence we obtain the following statement.

Theorem 1.13. *Let the above assumptions be fulfilled. Then there exists a unique solution to the problem* (1.81).

The set K in Theorem 1.11 may coincide with the space V. For a differentiable functional J it guarantees the solvability of the Euler equation

$$J'_u(\bar{u}) = 0 \quad \forall \bar{u} \in V. \tag{1.84}$$

1.2.6 Existence of solutions to operator equations and inequalities

Let us formulate assertions related to a solvability of problems which are not variational ones in general (Lions, 1969).

Firstly, consider the *operator equation*

$$Au = u^\star \tag{1.85}$$

for which solvability means $Au(v) = u^\star(v)$ for all $v \in V$, for the fixed element $u^\star \in V^\star$. An operator $A : V \to V^\star$ is called a *semicontinuous operator* if the function $\phi(\lambda) = A(u + \lambda v)(w)$ of the variable λ is continuous from R to R for all fixed $u, v, w \in V$. The following theorem holds.

Theorem 1.14. *Let V be a reflexive separable Banach space. Assume that an operator $A : V \to V^\star$ possesses the following properties:*

1. A is bounded and semicontinuous;

2. A is monotonous;

3. A is coercive, i.e. there exists $u_0 \in V$ such that

$$\frac{1}{\|u\|} A(u)(u - u_0) \ \to \ +\infty \quad \text{as} \quad \|u\| \to \infty.$$

Then the equation (1.85) *has at least one solution $u \in V$ for every fixed $u^\star \in V^\star$. This solution is unique if A is strictly monotonous.*

Let K be a closed convex subset of V. We consider the *operator inequality*

$$u \in K : \quad Au(v-u) \geq u^\star(v-u) \quad \forall v \in K \tag{1.86}$$

for fixed $u^\star \in V^\star$, where $A : V \to V^\star$. Note that (1.86) coincides with variational inequality (1.59) if $Au = J'_u$, i.e. A is a potential operator. But it does not always hold and therefore (1.86) is a more general inequality as compared to (1.59). For inequalities the following existence theorem takes place.

Theorem 1.15. *Let V be a reflexive separable Banach space, and K be a closed convex subset in V. Assume that an operator $A : V \to V^\star$ is bounded, semicontinuous, monotonous, and A is coercive or K is bounded. Then the inequality* (1.86) *has at least one solution $u \in K$ for every fixed $u^\star \in V^\star$. This solution is unique if A is strictly monotonous.*

We will prove this theorem in Section 1.3 by a penalty method.

An operator $A : V \to V^\star$ is called *pseudomonotonous* if A is bounded and the conditions $u_n \to u_0$ weakly in V, $\limsup Au_n(u_n - u_0) \leq 0$, imply that

$$\liminf Au_n(u_n - v) \geq Au_0(u_0 - v) \quad \forall v \in V. \tag{1.87}$$

In general, a pseudomonotonous operator is not monotonous but is continuous. Indeed, let $u_n \to u_0$ strongly in V. By the boundedness of A, we have $Au_n \to f$ weakly in $V^\star$. Then $\limsup Au_n(u_n - u_0) = 0$. Hence (1.87) gives

$$\liminf Au_n(u_n - v) = f(u_0 - v) \geq Au_0(u_0 - v) \quad \forall v \in V,$$

i.e. $Au_0 = f$ means that the operator A is a continuous mapping of the space V with the strong topology onto the space $V^\star$ with the weak topology.

Following (Lions, 1969), the next assertion can be proved.

Theorem 1.16. *Let V be a reflexive separable Banach space, and K be a closed convex subset in V. Assume that an operator $A : V \to V^\star$ is pseudomonotonous, and A is coercive or K is bounded. Then the inequality* (1.86) *has a solution.*

The proof of Theorem 1.16 is based on the following property of pseudomonotonous operators. If $u_n \to u_0$ weakly in V, $Au_n \to f$ weakly in $V^\star$ and $\limsup Au_n(u_n) \leq f(u_0)$, then $\limsup Au_n(u_n - u_0) \leq 0$, and (1.87) gives

$$Au_0(u_0 - v) \leq \limsup Au_n(u_n - v) \leq f(u_0 - v) \quad \forall v \in V.$$

Hence $f = Au_0$.

This section is concluded with some remarks related to a connection between normed spaces. Let V, W be two normed spaces such that $V \subset W$, V is dense in W and

$$\|u\|_W \leq c\|u\|_V \quad \forall u \in V, \quad c = \text{const} > 0.$$

Then every element $w^\star \in W^\star$ considered only on V defines some linear and continuous functional on V, i.e. $w^\star_V \in V^\star$. It is clear that the correspondence $w^\star \to w^\star_V$ is one-to-one, since due to the aforementioned density the functional $w^\star$ is uniquely defined by its values on V. Hence the space $W^\star$ can be identified with some subspace of $V^\star$. Moreover,

$$w^\star(v) \le \|w^\star\|_{W^\star}\|v\|_W \le c\|w^\star\|_{W^\star}\|v\|_V \quad \forall v \in V,$$

that is $\|w^\star\|_{V^\star} \le c\|w^\star\|_{W^\star}$, and the injection $W^\star \subset V^\star$ is fulfilled. It can also be shown that for reflexive spaces V the injection $W^\star \subset V^\star$ is dense.

Now let V be a reflexive Banach space, and H be a Hilbert space. Assume that $V \subset H$, V is dense in H and

$$\|u\|_H \le c\|u\|_V \quad \forall u \in V, \quad c = \text{const} > 0.$$

According to the above arguments, the dual space $H^\star$ can be considered as a subspace of $V^\star$. By the reflexivity of V, the space $H^\star$ is dense in $V^\star$ and, moreover,

$$\|u^\star\|_{V^\star} \le c\|u^\star\|_{H^\star} \quad \forall u^\star \in H^\star.$$

Taking into account the Riesz theorem for Hilbert spaces, the spaces H and $H^\star$ could be identified. Thus

$$V \subset H \subset V^\star,$$

and H is continuously and densely injected in $V^\star$.

1.3 Approximation methods

In this section we analyse some approximation methods for variational inequalities considered in Section 1.2. We discuss the penalty and the projection methods and their consequences. As for numerical methods, we refer the reader to (Glowinski et al., 1976).

1.3.1 Duality mapping. Projection

Let V be a reflexive Banach space, and $V^\star$ be a space dual of V. We assume that the functionals $u \to \|u\|$, $u^\star \to \|u^\star\|_\star$ defined on V, $V^\star$, respectively, are strictly convex. In this case the spaces V, $V^\star$ are called strictly convex. A value of a functional $u^\star \in V^\star$ on elements $u \in V$ is denoted by $\langle u^\star, u\rangle$. An operator $I: V \to V^\star$ is called a *duality mapping* if the following conditions hold (Gajewski et al., 1974):

$$\langle Iu, u\rangle = \|Iu\|_\star\|u\|, \quad \|Iu\|_\star = \|u\| \quad \forall u \in V. \tag{1.88}$$

We note that Iu is a linear and continuous functional on V; meanwhile the operator I is not linear, in general.

The existence of a unique duality mapping could easily be proved. Indeed, let $E = \{u \in V \mid \|u\| = 1\}$ be a unit sphere in V. According to the Hahn–Banach theorem, for every fixed $u \in E$ there exists a unique element $u^\star \in V^\star$ such that $\|u^\star\|_\star = 1$, $\langle u^\star, u\rangle = 1$ due to the strict convexity of $V^\star$. Let us define

$$Iu = \|u\| \, (u/\|u\|)^\star ,$$

where $(u/\|u\|)^\star$ is found according to the above theorem. Then it is clear that the properties (1.88) are fulfilled.

A unique *inverse duality mapping* $I^{-1} : V^\star \to V$ exists such that $I(I^{-1}) = (I^{-1})I = 1$, where 1 is a unit operator. By substituting $u = I^{-1}u^\star$ in (1.88), for any element $u^\star \in V^\star$, one has

$$\langle u^\star, I^{-1}u^\star\rangle = \|u^\star\|_\star \|I^{-1}u^\star\|, \quad \|I^{-1}u^\star\| = \|u^\star\|_\star \quad \forall u^\star \in V^\star. \tag{1.89}$$

Now let $V = H$ be a Hilbert space with a scalar product $(\cdot\,,\cdot)$. Fix any element $u \in H$. By Riesz's theorem, for $Iu \in H^\star$, there exists an element $j \in H$ such that

$$\langle Iu, v\rangle = (j, v) \quad \forall v \in H, \quad \|Iu\|_\star = \|j\|. \tag{1.90}$$

Therefore, applying (1.88), (1.90), one can write

$$\|u\|^2 = \|Iu\|_\star \|u\| = \langle Iu, u\rangle = (j, u) \le \|j\|\|u\| = \|Iu\|_\star \|u\|,$$

which implies $(j, u) = \|u\|^2 = \|j\|^2$. Hence, we have

$$\|j - u\|^2 = \|j\|^2 + \|u\|^2 - 2(j, u) = 0,$$

i.e. $j = u$. Substituting $j = u$ into (1.90), we have the formulae

$$\langle Iu, v\rangle = (u, v) \quad \forall v \in H, \quad \|Iu\|_\star = \|u\|, \tag{1.91}$$

for arbitrary $u \in H$. On the other hand, from (1.89) and (1.91) a similar property of the inverse duality mapping $I^{-1} : H^\star \to H$ follows:

$$\langle u^\star, v\rangle = (I^{-1}u^\star, v) \quad \forall v \in V, \quad \|I^{-1}u^\star\| = \|u^\star\|_\star. \tag{1.92}$$

Note that relations (1.91) and (1.92) mean linearity of the duality mapping I and its inverse I^{-1} in Hilbert spaces due to the linearity of the scalar product.

Let $K \subset V$ be a convex closed set. We assume that V is a strictly convex reflexive Banach space. For given $u \subset V$ an element $Pu \subset K$ is called a *projection* of u onto the set K if

$$\|u - Pu\| \le \|u - v\| \quad \forall v \in K. \tag{1.93}$$

Then $P : V \to K$ is called a projection operator. Thus finding the projection (1.93) is equivalent to solving the minimization problem

$$\inf_{v \in K} \|v - u\|^2. \tag{1.94}$$

It is easily seen that Theorem 1.11 guarantees the solvability of problem (1.94). This solution is unique due to the strict convexity of V.

Let $V = H$ be a Hilbert space, and $K \subset H$ be its convex closed subset. Then $||u - v||^2 = (u - v, u - v)$ and we can use Theorem 1.12 since (1.94) is equivalent to the following inequality:

$$(u - Pu, Pu - v) \geq 0 \quad \forall v \in K. \tag{1.95}$$

Lemma 1.2. *P is a Lipschitz continuous operator, i.e.*

$$||Pu_1 - Pu_2|| \leq ||u_1 - u_2|| \quad \forall u_1, u_2 \in H. \tag{1.96}$$

Indeed, the inequality (1.95) for fixed $u_1, u_2 \in H$ gives

$$(u_1 - Pu_1, Pu_1 - v) \geq 0, \quad (u_2 - Pu_2, Pu_2 - w) \geq 0 \quad \forall v, w \in K.$$

Let us take here $v = Pu_2$, $w = Pu_1$ and sum these inequalities; then

$$(u_1 - u_2 - Pu_1 + Pu_2, Pu_1 - Pu_2) \geq 0. \tag{1.97}$$

The consideration of the norm $||Pu_1 - Pu_2||$ gives from (1.97) that

$$||Pu_1 - Pu_2||^2 = -(u_1 - u_2 - Pu_1 + Pu_2, Pu_1 - Pu_2)$$

$$+(u_1 - u_2, Pu_1 - Pu_2) \leq (u_1 - u_2, Pu_1 - Pu_2) \leq ||u_1 - u_2|| ||Pu_1 - Pu_2||.$$

This estimate proves the lemma.

Lemma 1.3. *P is a monotonous operator.*
Indeed, we can write

$$(Pu_1 - Pu_2, u_1 - u_2)$$

$$= (Pu_1 - Pu_2, u_1 - u_2 - Pu_1 + Pu_2) + ||Pu_1 - Pu_2||^2.$$

By (1.97), the right-hand side is nonnegative, which proves the assertion.

Lemma 1.4. *The following estimate takes place:*

$$||u_1 - u_2 - (Pu_1 - Pu_2)|| \leq ||u_1 - u_2|| \quad \forall u_1, u_2 \in H. \tag{1.98}$$

Indeed, by (1.97) and Lemma 1.3,

$$||u_1 - u_2 - (Pu_1 - Pu_2)||^2 = -(u_1 - u_2 - Pu_1 + Pu_2, Pu_1 - Pu_2)$$

$$-(Pu_1 - Pu_2, u_1 - u_2) + ||u_1 - u_2||^2 \leq ||u_1 - u_2||^2.$$

1.3.2 Penalty operators

Let $K \subset V$ be a convex closed subset of a reflexive Banach space V, I be a duality mapping, and P be a projection operator of V onto K. We are in a position to give a definition of a penalty operator. An operator $\beta : V \to V^*$ is called a *penalty operator* connected with the set K if the following conditions are fulfilled. Firstly, β is a monotonous bounded semicontinuous operator. Secondly, a kernel of β coincides with K, i.e.

$$K = \{v \in V \mid \beta(v) = 0\}.$$

The following existence theorem takes place.

Theorem 1.17. *Operator $\beta(v) = I(v - Pv)$ is a penalty operator connected with the set K.*

The proof of this theorem is based on the following lemmas.

Lemma 1.5. *The inequality*

$$\langle \beta(u), Pu - v \rangle \geq 0 \quad \forall v \in K \tag{1.99}$$

holds for arbitrary $u \in V$.

PROOF. We denote $\psi(t) = t^2/2$. Then

$$\psi(\|v\|) - \psi(\|u\|) = \int_{\|u\|}^{\|v\|} \tau \, d\tau \geq \|u\|(\|v\| - \|u\|)$$

$$= \|Iu\|_\star \|v\| - \|Iu\|_\star \|u\| \geq \langle Iu, v \rangle - \langle Iu, u \rangle = \langle Iu, v - u \rangle,$$

i.e.

$$\psi(\|v\|) - \psi(\|u\|) \geq \langle Iu, v - u \rangle \quad \forall u, v \in V. \tag{1.100}$$

It follows from the definition of a projection that

$$\|u - Pu\| \leq \|u - w\| \quad \forall w \in K.$$

Choosing here $w = (1 - \lambda)Pu + \lambda v$, $v \in K$, $\lambda \in (0, 1)$, by the convexity of K, we find

$$\|u - Pu\| \leq \|u - Pu - \lambda(v - Pu)\|.$$

The monotonicity of $\psi(t)$ yields

$$\psi(\|u - Pu\|) \leq \psi(\|u - Pu - \lambda(v - Pu)\|) \quad \forall v \in K.$$

Consequently, taking into account (1.100), we obtain

$$0 \geq \psi(\|u - Pu\|) - \psi(\|u - Pu - \lambda(v - Pu)\|)$$

$$\geq \langle I(u - Pu - \lambda(v - Pu)), \lambda(v - Pu) \rangle.$$

Therefore the inequality

$$\langle I(u - Pu - \lambda(v - Pu)), v - Pu \rangle \leq 0 \quad \forall v \in K, \quad \lambda > 0,$$

holds. Now let $\lambda \to 0$. Due to the semicontinuity of J the inequality

$$\langle I(u - Pu), v - Pu \rangle \leq 0 \quad \forall v \in K$$

takes place for arbitrary $u \in K$. The lemma is proved. The converse assertion is also valid.

Lemma 1.6. *Let $\langle I(u - w), w - v \rangle \geq 0$ for all $v \in K$. Then $w = Pu$.*

PROOF. By (1.100), we have

$$\psi(\|u - v\|) - \psi(\|u - w\|) \geq \langle I(u - w), w - v \rangle \quad \forall v \in K.$$

The right-hand side is nonnegative here, hence in view of the monotony of $\psi(t)$ the relation

$$\|u - v\| \geq \|u - w\| \quad \forall w \in K$$

holds, whence $w = Pu$. This completes the proof.

Lemma 1.7. *The following inequality:*

$$\langle \beta(u) - \beta(v), Pu - Pv \rangle \geq 0 \quad \forall u, v \in V \tag{1.101}$$

is valid.

PROOF. In view of (1.99) we have

$$\langle \beta(u), Pu - w \rangle \geq 0, \quad \langle \beta(v), Pv - w \rangle \geq 0 \quad \forall w \in K.$$

Putting $w = Pv$ in the first inequality, $w = Pu$ in the second one and summing we obtain (1.101).

Lemma 1.8. *The penalty operator β is monotonous.*

PROOF. We can write

$$\langle \beta(u) - \beta(v), u - v \rangle$$

$$= \langle \beta(u) - \beta(v), Pu - Pv \rangle + \langle \beta(u) - \beta(v), (u - Pu) - (v - Pv) \rangle.$$

The first term of the right-hand side is nonnegative here in view of (1.101). The second term can be written in the form

$$\langle I(u - Pu) - I(v - Pv), (u - Pu) - (v - Pv) \rangle.$$

It is nonnegative due to the monotonicity of I. The lemma is proved.

A boundedness and semicontinuity of the operator β follow from the same properties of I and P. The verification of the equivalence of conditions

$\beta(u) = 0$ and $u \in K$ is obvious. Compiling these remarks with the above lemmas we get the proof of Theorem 1.17.

As we can see, penalty operators can be built easier in Hilbert spaces. In applications H is often a Hilbert space such that

$$V \subset H \subset V^\star,$$

where all injections are continuous and dense. Also, let K be a convex closed set in $V \cap H$. In this case, an identity operator can be chosen as the duality operator $I : H \to H$. Thus $\beta(u) = u - Pu$ is the penalty operator connected with the set K. Here P projects H onto K.

In what follows we give applications of the penalty and projection operators to variational inequalities (see Kovtunenko, 1994b, 1994c).

1.3.3 Iteration penalty method

We consider penalized operator equations approximating variational inequalities. For equations with strongly monotonous operators we construct an iterative method, prove convergence of solutions, and obtain error estimates.

Let K be a closed convex subset in a reflexive Banach space V; let an operator A act from V into $V^\star$ and let $f \in V^\star$ be given. Consider the variational inequality

$$u \in K, \quad \langle Au, v-u\rangle \geq \langle f, v-u\rangle \quad \forall v \in K, \tag{1.102}$$

and assume that there exists a Hilbert space H such that

$$V \subset H \subset V^\star.$$

The imbeddings are continuous and dense, and there is a constant $c > 0$ such that

$$\|u\|_H \leq c\|u\| \quad \forall u \in V. \tag{1.103}$$

We have denoted by $\|u\|$ the norm of u in V, and by $\|u\|_H$ the norm of u in H, $(u,u) = \|u\|_H^2$. Suppose that K is a closed convex set in H. Let P be the projection of H onto K. Introduce the penalty operator $\beta : H \to H$ by the formula $\beta(u) = u - Pu$ for $u \in H$. The operator β is monotonous by Lemma 1.3 and continuous by Lemma 1.2.

Given a small parameter $\varepsilon > 0$, we write down the penalized equation

$$Au^\varepsilon + \varepsilon^{-1}\beta(u^\varepsilon) = f, \tag{1.104}$$

which is to be understood as follows:

$$\langle Au^\varepsilon, v\rangle + \varepsilon^{-1}\left(\beta(u^\varepsilon), v\right) = \langle f, v\rangle \quad \forall v \in V.$$

Fix ε and construct the sequence of equations

$$Au^{\varepsilon,n+1} + \varepsilon^{-1}u^{\varepsilon,n+1} = f + \varepsilon^{-1}Pu^{\varepsilon,n}, \tag{1.105}$$

where $n = 0, 1, 2, \ldots$, and $u^{\varepsilon,0} \in V$ is an arbitrary function. Suppose that the operator A is bounded semicontinuous and *strongly monotonous*, i.e. there is a constant $M > 0$ such that

$$\langle Au - Av, u - v\rangle \geq M\|u - v\|^2 \quad \forall u, v \in V.$$

This property obviously implies coercivity and strict monotonicity of A. The right-hand side of (1.105) belongs to $V^\star$ since $H \subset V^\star$. Then, by Theorem 1.14, there exists a unique solution $u^{\varepsilon,n+1} \in V$, $n = 0, 1, 2, \ldots$, to problem (1.105).

Theorem 1.18. *Under the above assumptions, there exists a unique solution* $u^\varepsilon \in V$ *of the problem* (1.104) *and*

$$u^{\varepsilon,n+1} \to u^\varepsilon \quad \textit{strongly in } V \quad \textit{as } n \to \infty,$$

where $u^{\varepsilon,n+1}$ *is a solution to the problem* (1.105). *Moreover, the following estimate holds:*

$$\|u^\varepsilon - u^{\varepsilon,n+1}\| \leq \frac{c\rho^{n/2}}{\varepsilon M^2}\|f - Au^{\varepsilon,0} - \varepsilon^{-1}\beta(u^{\varepsilon,0})\|_\star, \quad \rho = \frac{c^2}{c^2 + 2\varepsilon M} < 1.$$

PROOF. Consider (1.105) at the preceding step in n,

$$Au^{\varepsilon,n} + \varepsilon^{-1}u^{\varepsilon,n} = f + \varepsilon^{-1}Pu^{\varepsilon,n-1},$$

and subtract this equality from (1.105); this gives

$$\langle Au^{\varepsilon,n+1} - Au^{\varepsilon,n}, u^{\varepsilon,n+1} - u^{\varepsilon,n}\rangle + \varepsilon^{-1}\|u^{\varepsilon,n+1} - u^{\varepsilon,n}\|_H^2$$
$$= \varepsilon^{-1}(Pu^{\varepsilon,n} - Pu^{\varepsilon,n-1}, u^{\varepsilon,n+1} - u^{\varepsilon,n}).$$

By making use of the strong monotonicity of A, Holder's inequality, and Lemma 1.2, we obtain the estimate

$$2\|u^{\varepsilon,n+1} - u^{\varepsilon,n}\|^2 + (\varepsilon M)^{-1}\|u^{\varepsilon,n+1} - u^{\varepsilon,n}\|_H^2 \tag{1.106}$$
$$\leq (\varepsilon M)^{-1}\|u^{\varepsilon,n} - u^{\varepsilon,n-1}\|_H^2.$$

Introduce the equivalent norm in V as follows:

$$[u]^2 = \|u\|^2 + (\varepsilon M)^{-1}\|u\|_H^2, \quad u \in V.$$

By (1.103), from (1.106) we obtain

$$[u^{\varepsilon,n+1} - u^{\varepsilon,n}]^2 \leq \rho[u^{\varepsilon,n} - u^{\varepsilon,n-1}]^2 \leq \rho^n[u^{\varepsilon,1} - u^{\varepsilon,0}]^2. \tag{1.107}$$

Thus,

$$u^{\varepsilon,n+1} - u^{\varepsilon,n} \to 0 \quad \text{strongly in } V \quad \text{as } n \to \infty. \tag{1.108}$$

It follows that there exists an element $u_0 \in V$ such that

$$u^{\varepsilon,n+1} \to u_0 \quad \text{strongly in } V \quad \text{as } n \to \infty. \tag{1.109}$$

The mentioned properties of the operator A imply that A acts continuously from V with the strong topology into $V^\star$ with the weak topology; then

$$Au^{\varepsilon,n+1} \to Au_0 \quad \text{weakly in } V \quad \text{as } n \to \infty. \tag{1.110}$$

From the continuity of the penalty operator it follows that

$$\beta(u^{\varepsilon,n}) \to \beta(u_0) \quad \text{strongly in } H \quad \text{as } n \to \infty. \tag{1.111}$$

Representing (1.105) as

$$Au^{\varepsilon,n+1} + \varepsilon^{-1}\beta(u^{\varepsilon,n}) + \varepsilon^{-1}(u^{\varepsilon,n+1} - u^{\varepsilon,n}) = f,$$

passing to the limit as $n \to \infty$, and using (1.108)–(1.111), we obtain (1.104). Hence $u_0 = u^\varepsilon$.

Now we estimate the error. It follows from (1.105) that

$$Au^{\varepsilon,1} - Au^{\varepsilon,0} + \varepsilon^{-1}(u^{\varepsilon,1} - u^{\varepsilon,0}) = f - Au^{\varepsilon,0} - \varepsilon^{-1}\beta(u^{\varepsilon,0}).$$

Multiply this equation by $u^{\varepsilon,1} - u^{\varepsilon,0}$. Then Holder's inequality implies the estimate

$$[u^{\varepsilon,1} - u^{\varepsilon,0}] \leq M^{-1}\|f - Au^{\varepsilon,0} - \varepsilon^{-1}\beta(u^{\varepsilon,0})\|_\star. \tag{1.112}$$

Now write (1.105) in the form

$$Au^{\varepsilon,n+1} + \varepsilon^{-1}\beta(u^{\varepsilon,n+1}) = f + \varepsilon^{-1}(Pu^{\varepsilon,n} - Pu^{\varepsilon,n+1}),$$

subtract it from (1.104), and multiply the difference by $u^\varepsilon - u^{\varepsilon,n+1}$. One easily derives

$$\langle Au^\varepsilon - Au^{\varepsilon,n+1}, u^\varepsilon - u^{\varepsilon,n+1}\rangle + \varepsilon^{-1}\left(\beta(u^\varepsilon) - \beta(u^{\varepsilon,n+1}), u^\varepsilon - u^{\varepsilon,n+1}\right)$$

$$= \varepsilon^{-1}(Pu^{\varepsilon,n+1} - Pu^{\varepsilon,n}, u^\varepsilon - u^{\varepsilon,n+1}).$$

Applying Holder's inequality and using the strong monotonicity of A, the monotonicity of β, Lemma 1.2 and the estimates for the norms, we obtain

$$\|u^\varepsilon - u^{\varepsilon,n+1}\| \leq c(\varepsilon M)^{-1}[u^{\varepsilon,n+1} - u^{\varepsilon,n}]. \tag{1.113}$$

Combining (1.107), (1.112) and (1.113), we arrive at the required estimate. The theorem is proved.

The approximate method developed is constructive in the following sense. If A is a linear operator, then the equation (1.105) is linear too and, therefore, it can be solved by standard numerical methods.

Now let us pass to the limit in (1.104) as $\varepsilon \to 0$ to obtain (1.102). The operator $u \to Au + \varepsilon^{-1}\beta(u)$ is coercive because A is coercive and

$$\langle Au, u - u_0\rangle + \varepsilon^{-1}(\beta(u), u - u_0)$$
$$= \langle Au, u - u_0\rangle + \varepsilon^{-1}(\beta(u) - \beta(u_0), u - u_0) \geq \langle Au, u - u_0\rangle$$

for any $u_0 \in K$ (which implies $\beta(u_0) = 0$). Therefore, we can deduce that

$$\|u^\varepsilon\| \leq c_1$$

uniformly in ε. Operator A is bounded, hence

$$\|Au^\varepsilon\| \leq c_2$$

uniformly in ε. By the reflexivity of V, choosing a subsequence still denoted by u^ε, we obtain

$$u^\varepsilon \to u \quad \text{weakly in } V, \quad \text{strongly in } H \tag{1.114}$$

as $\varepsilon \to 0$ because of the dense imbedding $V \subset H$. Rewriting (1.104) as

$$(\beta(u^\varepsilon), v) = \varepsilon\langle f - Au^\varepsilon, v\rangle \quad \forall v \in H \subset V^\star,$$

by the boundedness of $Au^\varepsilon \in V^\star$, we conclude

$$\beta(u^\varepsilon) \to 0 \quad \text{strongly in } H \quad \text{as } \varepsilon \to 0,$$

i.e. the convergence $u^\varepsilon - Pu^\varepsilon \to 0$ together with (1.114) provide $Pu^\varepsilon \to u$. Hence $u \in K$.

Let us substitute $u^\varepsilon - v$ as a test function in (1.104). In view of the monotonicity of the penalty operator β, it gives

$$\langle Au^\varepsilon, u^\varepsilon - v\rangle \leq \langle Au^\varepsilon, u^\varepsilon - v\rangle + \varepsilon^{-1}(\beta(u^\varepsilon) - \beta(v), u^\varepsilon - v) \leq \langle f, u^\varepsilon - v\rangle$$

for all $v \in K$, i.e.

$$\langle Au^\varepsilon - f, u^\varepsilon - v\rangle \leq 0 \quad \forall v \in K. \tag{1.115}$$

Since $u \in K$, we can substitute $v = u$ in (1.115) and use the monotonicity of A. This provides

$$\langle Au - f, u^\varepsilon - u\rangle \leq \langle Au^\varepsilon - Au, u^\varepsilon - u\rangle + \langle Au - f, u^\varepsilon - u\rangle = \langle Au^\varepsilon - f, u^\varepsilon - u\rangle \leq 0.$$

Note that $\langle Au, u^\varepsilon - u\rangle \to 0$ and $\langle f, u^\varepsilon - u\rangle \to 0$ as $\varepsilon \to 0$ due to the weak convergence (1.114). Therefore, the last inequality implies

$$\langle Au^\varepsilon, u^\varepsilon - u\rangle \to 0. \tag{1.116}$$

At the next step we consider the monotony condition

$$\langle Au^\varepsilon - Aw, u^\varepsilon - w\rangle \geq 0 \quad \forall w \in V$$

for the operator A and substitute here $w = u + \lambda(v-u)$, $\lambda > 0$, $v \in V$. Then the inequality

$$\lambda\langle Au^\varepsilon, u - v\rangle \geq \lambda\langle Aw, u-v\rangle + \langle Aw, u^\varepsilon - u\rangle - \langle Au^\varepsilon, u^\varepsilon - u\rangle$$

follows. In view of the convergencies (1.114), (1.116), we obtain

$$\lambda \liminf \langle Au^\varepsilon, u - v\rangle \geq \lambda\langle Aw, u - v\rangle \quad \forall v \in V.$$

Let us divide this inequality by λ and pass to the limit as $\lambda \to 0$. By the semicontinuity of A, we have $Aw \to Au$ as $\lambda \to 0$. Consequently

$$\liminf \langle Au^\varepsilon, u - v\rangle \geq \langle Au, u - v\rangle \quad \forall v \in V.$$

Combining this estimate with (1.115), one deduces, with the help of (1.116),

$$\langle Au - f, u - v\rangle \leq \liminf \langle Au^\varepsilon - f, u - v\rangle = \liminf \langle Au^\varepsilon - f, u - v\rangle$$

$$+ \liminf \langle Au^\varepsilon - f, u^\varepsilon - u\rangle = \liminf \langle Au^\varepsilon - f, u^\varepsilon - v\rangle \leq 0 \quad \forall v \in K.$$

Thus, we have obtained the inequality (1.102), u is its solution and $u^\varepsilon \to u$ weakly in V. Uniqueness of the solution follows from the strict monotonicity of A. Moreover, we can show the strong convergence of u^ε. Indeed, due to (1.114), (1.116),

$$M\|u^\varepsilon - u\|^2 \leq \langle Au^\varepsilon - Au, u^\varepsilon - u\rangle = \langle Au^\varepsilon, u^\varepsilon - u\rangle - \langle Au, u^\varepsilon - u\rangle \to 0.$$

Therefore, the following assertion is proved.

Theorem 1.19. *There exists a unique solution $u \in K$ of the inequality* (1.102), *and the convergence*

$$u^\varepsilon \to u \quad \text{strongly in } V \quad \text{as } \varepsilon \to 0$$

holds for the solutions u^ε to the penalty equation (1.104).

1.3.4 Iteration penalty method in Hilbert spaces

Let V be a Hilbert space, and $V^\star$ be its dual. Denote by $\langle\cdot,\cdot\rangle$, $(\cdot,\cdot)$, $\|\cdot\|$ and $\|\cdot\|_\star$ a duality pairing between V and $V^\star$, a scalar product in V, norms in V and $V^\star$, respectively. Recall that the duality mapping $I : V \to V^\star$ satisfies the relation

$$\langle Iu, v\rangle = (u, v) \quad \forall u, v \in V,$$

and the inverse duality mapping $I^{-1} : V^\star \to V$ satisfies (1.92).

Let K be a closed convex subset of V. An element $f \in V^\star$ and an operator $A : V \to V^\star$ are given. We assume that A is a bounded semicontinuous and strongly monotonous operator, i.e.

$$\langle Au - Av, u - v\rangle \geq M\|u - v\|^2 \quad \forall u, v \in V, \quad M > 0.$$

In the sequel, the following variational inequality is analysed: to find $u \in K$ such that

$$\langle Au, v-u\rangle \geq \langle f, v-u\rangle \quad \forall v \in K. \tag{1.117}$$

By Theorem 1.15, there exists a unique solution u of (1.117).

Let $P : V \to K$ be the projection operator. By Lemma 1.2, P is Lipschitz continuous, i.e.

$$\|Pv - Pw\| \leq \|v - w\| \quad \forall v, w \in V. \tag{1.118}$$

Let us construct the standard penalty operator $\beta(v) = I(v - Pv)$ and define the penalty problem depending on a small positive parameter ε,

$$Au^\varepsilon + \varepsilon^{-1}\beta(u^\varepsilon) = f. \tag{1.119}$$

Repeating the proof of Theorem 1.19 for this case, one deduces that equation (1.119) has a unique solution $u^\varepsilon \in V$ which satisfies

$$u^\varepsilon \to u \quad \text{strongly in } V \quad \text{as } \varepsilon \to 0.$$

To linearize the penalty operator in (1.119) we use the following iteration scheme similar to (1.105),

$$Au^{\varepsilon,n} + \varepsilon^{-1}Iu^{\varepsilon,n} = f + \varepsilon^{-1}IPu^{\varepsilon,n-1}, \tag{1.120}$$

for $n = 1, 2, 3, \ldots$, where $u^{\varepsilon,0} \in V$ is an arbitrary element. It follows from Theorem 1.14 that there exists a unique solution $u^{\varepsilon,n+1} \in V$ of (1.120).

Lemma 1.9. *The following estimates hold:*

$$\|u^{\varepsilon,n} - u\|^2 \leq \rho_\varepsilon^n\|u^{\varepsilon,0} - u\|^2 + \delta_\varepsilon(1 - \rho_\varepsilon^n)\|f - Au\|_\star^2, \tag{1.121}$$

$$\|u^{\varepsilon,n} - u^\varepsilon\|^2 \leq \rho_\varepsilon^n\|u^{\varepsilon,0} - u^\varepsilon\|^2, \tag{1.122}$$

for the solutions $u, u^\varepsilon, u^{\varepsilon,n}$ *of the problems* (1.117), (1.119), (1.120), *respectively. Here* $\rho_\varepsilon = (1 + M\varepsilon)^{-2} < 1$, $\delta_\varepsilon = \varepsilon M^{-1}(2 + M\varepsilon)^{-1}$.

PROOF. Let us rewrite (1.120) by adding $(-Au - \varepsilon^{-1}Iu)$ to both parts. Since I is linear, we have

$$Au^{\varepsilon,n} - Au + \varepsilon^{-1}I(u^{\varepsilon,n} - u) = f - Au + \varepsilon^{-1}I(Pu^{\varepsilon,n-1} - Pu).$$

Here $u = Pu$ due to $u \in K$. Application of the linear mapping I^{-1} to this equation gives

$$I^{-1}(Au^{\varepsilon,n} - Au) + \varepsilon^{-1}(u^{\varepsilon,n} - u) = I^{-1}(f - Au) + \varepsilon^{-1}(Pu^{\varepsilon,n-1} - Pu).$$

Squaring the above equality, we obtain

$$\|Au^{\varepsilon,n} - Au\|_\star^2 + 2\varepsilon^{-1}\langle Au^{\varepsilon,n} - Au, u^{\varepsilon,n} - u\rangle + \varepsilon^{-2}\|u^{\varepsilon,n} - u\|^2 \tag{1.123}$$

$$= \|f - Au\|_\star^2 + 2\varepsilon^{-1}\langle f - Au, Pu^{\varepsilon,n-1} - u\rangle + \varepsilon^{-2}\|Pu^{\varepsilon,n-1} - Pu\|^2.$$

According to the strong monotonicity of A, the left-hand side of (1.123) is bounded from below by $(M + \varepsilon^{-1})^2\|u^{\varepsilon,n} - u\|^2$. The second term in the right-hand side of (1.123) is negative due to (1.117). Further, using the inequality (1.118), we have

$$\|u^{\varepsilon,n} - u\|^2 \le \rho_\varepsilon \left(\|u^{\varepsilon,n-1} - u\|^2 + \varepsilon^2\|f - Au\|_\star^2\right).$$

Continuing this estimate as n tends to 1, we have

$$\|u^{\varepsilon,n} - u\|^2 \le \rho_\varepsilon^n \left(\|u^{\varepsilon,0} - u\|^2 + \varepsilon^2 \sum_{i=0}^{n-1} \rho_\varepsilon^i \|f - Au\|_\star^2\right). \tag{1.124}$$

With the sum of the geometrical series

$$\sum_{i=0}^{n-1} \rho_\varepsilon^i = (1 - \rho_\varepsilon^n)(1 - \rho_\varepsilon)^{-1},$$

the estimate (1.124) gives the estimate (1.121). Let us next subtract (1.119) from (1.120); then

$$Au^{\varepsilon,n} - Au^\varepsilon + \varepsilon^{-1}I(u^{\varepsilon,n} - u^\varepsilon) = \varepsilon^{-1}I(Pu^{\varepsilon,n-1} - Pu^\varepsilon).$$

Applying the operator I^{-1} to both sides and squaring, we obtain

$$\|u^{\varepsilon,n} - u^\varepsilon\|^2 \le (1 + M\varepsilon)^{-2}\|u^{\varepsilon,n-1} - u^\varepsilon\|^2.$$

This inequality reduces to the estimate (1.122) and completes the proof.

From Lemma 1.9 the following assertion is immediately deduced.

Theorem 1.20. *The following convergencies take place:*

$$u^{\varepsilon,n} \to u^\varepsilon \quad \textit{strongly in } V \quad \textit{as } n \to \infty,$$

for fixed ε, *and* (1.122) *holds;* $u^\varepsilon \to u$ *strongly in* V *as* $\varepsilon \to 0$ *and*

$$\|u^\varepsilon - u\|^2 \le \delta_\varepsilon \|f - Au\|_\star^2 \quad ,$$

where $u^{\varepsilon,n}$, u^ε *and* u *are the solutions of* (1.120), (1.119) *and* (1.117), *respectively.*

1.3.5 Projection methods

We keep the notations of the Hilbert space V, its dual space $V^\star$, with a duality pairing $\langle\cdot\,,\cdot\rangle$, a scalar product $(\cdot\,,\cdot)$ and a norm $\|\cdot\|^2 = (\cdot\,,\cdot)$ in V. Let the duality mapping $I : V \to V^\star$ and the inverse duality mapping

$I^{-1}: V^\star \to V$ be as before. We assume that K is a closed convex set in V, and $P: V \to K$ is a projection operator, i.e.

$$(u - Pu, Pu - v) \geq 0 \quad \forall v \in K. \tag{1.125}$$

An element $f \in V^\star$ and an operator $A: V \to V^\star$ are given. Consider the variational inequality

$$u \in K, \quad \langle Au, v - u\rangle \geq \langle f, v - u\rangle \quad \forall v \in K. \tag{1.126}$$

Lemma 1.10. *The variational inequality* (1.126) *is equivalent to the equation*

$$u = P\left(u + \theta I^{-1}(f - Au)\right) \tag{1.127}$$

for arbitrary constant $\theta > 0$.

PROOF. In view of (1.92) and the linearity of I^{-1}, we can rewrite (1.126) in the form

$$(I^{-1}Au, v - u) \geq (I^{-1}f, v - u) \quad \forall v \in K,$$

or

$$\left(I^{-1}(f - Au), u - v\right) \geq 0 \quad \forall v \in K.$$

Multiplying by $\theta > 0$ and adding $\pm u$, this inequality takes the form

$$\left((u + \theta I^{-1}(f - Au)) - u, u - v\right) \geq 0 \quad \forall v \in K.$$

The comparison of this relation with (1.125) leads to (1.127) due to the uniqueness of the projection. The function u belongs to K because $P: V \to K$. The lemma is proved.

Thus we give the presentation of variational inequalities as projection equations. It is utilized to construct approximate solutions.

We will also use the theorem on contraction mappings. A mapping $S: V \to V$ is called a *contraction mapping* if it is Lipschitz continuous,

$$||Su - Sv|| \leq \rho||u - v|| \quad \forall u, v \in V, \tag{1.128}$$

and $0 < \rho < 1$. The following generalized Banach theorem is valid (see Baiocchi, Capelo, 1984).

Theorem 1.21. *If S is a contraction mapping in a Hilbert space V then there exists* a fixed point *u such that $Su = u$ and solutions u^n of the equation*

$$u^n = Su^{n-1}, \quad n = 1, 2, ..., \quad u^0 \in V$$

converge strongly to u in V as $n \to \infty$. Moreover the estimate

$$||u^n - u|| \leq c(u, u^0)\rho^n$$

holds with the constant ρ taken from (1.128).

It is easy to see that $c(u,u^0) = ||u - u^0||$.

Lemma 1.11. *If the operator $A: V \to V^\star$ is strongly monotonous, i.e.*

$$\exists m > 0: \quad m||u - v||^2 \le \langle Au - Av, u - v\rangle \quad \forall u, v \in V \tag{1.129}$$

and Lipschitz continuous, i.e.

$$\exists M > 0: \quad ||Au - Av||_\star \le M||u - v|| \quad \forall u, v \in V, \tag{1.130}$$

then (1.127) *is a contraction mapping for any $0 < \theta < 2m/M^2$ with $\rho = 1 - 2m\theta + M^2\theta^2$. A minimum value $\rho = \sqrt{1 - m^2/M^2}$ is reached as $\theta = m/M^2$.*

PROOF. Applying the mapping (1.127) to any u, v, we have

$$u - v = P\left(u + \theta I^{-1}(f - Au)\right) - P\left(v + \theta I^{-1}(f - Av)\right).$$

Take the norm of both parts of this equality and use the Lipschitz continuity of P (see Lemma 1.2). By the linearity of I^{-1}, it provides

$$||u - v||^2 = ||P\left(u + \theta I^{-1}(f - Au)\right) - P\left(v + \theta I^{-1}(f - Av)\right)||^2$$

$$\le ||u + \theta I^{-1}(f - Au) - v + \theta I^{-1}(f - Ab)||^2 = ||u - v - \theta I^{-1}(Au - Av)||^2.$$

By (1.129), (1.130), we can obtain an additional estimate,

$$||u - v||^2 \le ||u - v||^2 - 2\theta\left(I^{-1}(Au - Av), u - v\right) + \theta^2||I^{-1}(Au - Av)||^2$$

$$= ||u - v||^2 - 2\theta\langle Au - Av, u - v\rangle + \theta^2||Au - Av||_\star^2$$

$$\le (1 - 2m\theta + M^2\theta^2)||u - v||^2,$$

which yields the desired condition (1.128). The lemma is proved.

Theorem 1.21 and Lemmas 1.10, 1.11 provide the following assertion.

Theorem 1.22. *For strongly monotonous and Lipschitz continuous operator A satisfying* (1.129), (1.130), *there exists a unique solution $u \in K$ of the variational inequality* (1.126) *and*

$$u^n \to u \quad \text{strongly in } V \quad \text{as } n \to \infty$$

with the estimate

$$||u^n - u|| \le \rho^n ||u^0 - u||, \quad \rho = \sqrt{1 - m^2/M^2},$$

where $u^n \in K$ are solutions of the iteration equation

$$u^n = P\left(u^{n-1} + \frac{m}{M^2} I^{-1}(f - Au^{n-1})\right), \quad n = 1, 2, \ldots, \tag{1.131}$$

for arbitrary $u^0 \in V$.

Now let us consider the second presentation of the variational inequality (1.126) by means of the projection operators. Suppose that A is a linear operator such that

$$m\|u\|^2 \leq \langle Au, u\rangle, \quad \|Au\|_\star \leq M\|u\|. \tag{1.132}$$

These conditions obviously provide the fulfilment of (1.129), (1.130).

On the other hand, the duality mapping I is defined by the scalar product in the Hilbert space V. Assume that the operator A is self-conjugate. Then we can define the scalar product in V as follows:

$$(u, v)_A = \langle Au, v\rangle = \langle Av, u\rangle.$$

In view of the properties (1.132), the corresponding norm in V is $\|u\|_A^2 = (u, u)_A$, namely

$$m\|u\|^2 \leq \|u\|_A^2 = \langle Au, u\rangle \leq \|Au\|_\star \|u\| \leq M\|u\|^2.$$

This means that A is the duality mapping connected with the introduced scalar product $(\cdot,\cdot)_A$. Then the variational inequality (1.126) can be rewritten in the form

$$(u, v-u)_A \geq (A^{-1}f, v-u)_A \quad \forall v \in K,$$

or

$$(A^{-1}f - u, u - v)_A \geq 0 \quad \forall v \in K$$

which together with (1.125) provide

$$u = P(A^{-1}f). \tag{1.133}$$

Let us denote $A^{-1}f = u_0$. We understand this equality as $Au_0 = f$, i.e. $u_0 \in V$ is a solution of the equation

$$\langle Au_0, v\rangle = \langle f, v\rangle \quad \forall v \in V. \tag{1.134}$$

Then (1.133) yields the following theorem.

Theorem 1.23. *If $A : V \to V^\star$ is a linear, self-conjugate, strongly monotonous and Lipschitz continuous operator in a Hilbert space V, then there exists a unique solution $u \in K$ of the variational inequality* (1.126) *given by the formula*

$$u = Pu_0$$

for the solution $u^0 = A^{-1}f \in V$ of (1.134).

To verify this theorem, it suffices to note that a unique solution $u_0 \in V$ of (1.134) always exists due to the mentioned properties of the operator A and Theorem 1.14.

Utilizing this approach, we construct the analytical solutions for a few one-dimensional unilateral boundary value problems considered in Chapter 2.

1.4 Problems with singular boundaries

In this section we define trace spaces at boundaries and consider Green's formulae. The statements formulated are applied to boundary value problems for solids with cracks provided that inequality type boundary conditions hold at the crack faces.

1.4.1 Smoothness of a boundary

Let $\Omega \subset R^m$ be a bounded domain with a boundary Γ, $\overline{\Omega} = \Omega \cup \Gamma$, $m = 2, 3$. The boundary Γ *belongs to the class* $C^{k,1}$ if there exist two real numbers $b > 0$, $h > 0$, p coordinate systems

$$(y^j, y^j_m), \quad y^j = (y^j_1, ..., y^j_{m-1}), \quad j = 1, ..., p, \tag{1.135}$$

and p functions θ^j, $j = 1, ..., p$, such that in the cubes

$$\Delta^j = \{y^j \in R^{m-1} \mid \quad |y^j_i| < b, \quad i = 1, ..., m-1\}$$

the functions θ^j belong to $C^{k,1}(\overline{\Delta}^j)$, and for

$$\Sigma^j = \{(y^j, y^j_m) \in R^m \mid \quad y^j \in \Delta^j, \quad y^j_m = \theta^j(y^j)\},$$

$$\Omega^j_+ = \{(y^j, y^j_m) \in R^m \mid \quad y^j \in \Delta^j, \quad \theta^j(y^j) < y^j_m < \theta^j(y^j) + h\},$$

$$\Omega^j_- = \{(y^j, y^j_m) \in R^m \mid \quad y^j \in \Delta^j, \quad \theta^j(y^j) - h < y^j_m < \theta^j(y^j)\},$$

the following conditions hold:

$$\Gamma = \bigcup_{j=1}^{p} \Sigma^j, \quad \Omega^j_+ \subset \Omega, \quad \Omega^j_- \subset R^m \setminus \overline{\Omega}, \quad j = 1, ..., p.$$

Here $C^{k,1}(\overline{\Delta}^j)$ is the space of functions having k Lipschitz continuous derivatives in $\overline{\Delta}^j$, where $k \geq 0$ is an integer. In other words, this definition implies that the boundary Γ can be presented as a union of local graphs Σ^j of the functions θ^j from $C^{k,1}(\overline{\Delta}^j)$, $j = 1, ..., p$, and the set Ω is locally the epigraph for these functions.

Now consider a domain Ω containing a surface Γ_c, whose properties are described in Section 1.1.7. Denote $\Sigma_c = \Gamma_c \setminus \partial\Gamma_c$, $\Omega_c = \Omega \setminus \Gamma_c$. Introduce the unit normal ν to Γ_c and define the opposite faces $\Gamma^{\pm}_c$ of the surface Γ_c. The signs $\pm$ fit positive and negative directions of ν, respectively. Then we denote the boundary of Ω_c by $\partial\Omega_c = \Gamma \cup \Gamma^{\pm}_c$. We assume that there exists a closed extension Σ of Γ_c dividing the domain Ω into two subdomains Ω_1, Ω_2 with boundaries $\partial\Omega_1, \partial\Omega_2$ and such that $\Gamma_c \subset \Sigma$. It is assumed that $\partial\Omega_1 = \Sigma^-$, $\partial\Omega_2 = \Sigma^+ \cup \Gamma$. We say that the boundary $\partial\Omega_c$ belongs to the class $C^{k,1}$ if $\partial\Omega_1, \partial\Omega_2$ belong to $C^{k,1}$.

1.4.2 Trace spaces at the boundary

Introduce the Sobolev spaces (Adams, 1975; Maz'ya, 1985)

$$H^k(\Omega) = \{u \mid \; D^\alpha u \in L^2(\Omega),\ 0 \le |\alpha| \le k\}, \quad H^0(\Omega) = L^2(\Omega),$$

equipped with the norm and the scalar product

$$\|u\|^2_{k,\Omega} = \sum_{|\alpha|=0}^{k} \|D^\alpha u\|^2_{0,\Omega}, \quad (u,v)_{k,\Omega} = \sum_{|\alpha|=0}^{k} \int_\Omega D^\alpha u \, D^\alpha v,$$

where D^α denotes derivatives of the order $|\alpha|$, $\alpha = (\alpha_1, ..., \alpha_m)$, $|\alpha| = \sum_{i=1}^{m} \alpha_i$, k is an integer, and $\|\cdot\|_{0,\Omega}$ is the norm in $L^2(\Omega)$.

The spaces $H^k(\Gamma)$, where $k \ge 1$ is an integer, at the boundary Γ can be introduced in the local coordinates (1.135) as follows (Lions, Magenes, 1968). Let Γ belong to the class $C^{k,1}$. For a given function $s(x)$, $x \in \Gamma$, the functions

$$s^j(y^j) = s(y^j_1, ..., y^j_{m-1}, \theta^j(y^j)), \quad y^j = (y^j_1, ..., y^j_{m-1}) \in \Delta^j, \quad j = 1, ..., p,$$

can be considered in the cubes Δ^j. Then we define

$$H^k(\Gamma) = \{s \in L^2(\Gamma) \mid \; s^j \in H^k(\Delta^j),\ j = 1, ..., p\}$$

with the norm

$$\|s\|^2_{k,\Gamma} = \sum_{j=1}^{p} \|s^j\|^2_{k,\Delta^j}.$$

Introduce also the spaces $H^{k-1/2}(\Gamma)$, $k \ge 1$, equipped with the norm

$$\|s\|^2_{k-1/2,\Gamma} = \|s\|^2_{k-1,\Gamma} \tag{1.136}$$

$$+ \sum_{|\alpha|=0}^{k-1} \sum_{j=1}^{p} \int_{\Delta^j} \int_{\Delta^j} \frac{|D^\alpha s^j(y^j) - D^\alpha s^j(\eta^j)|^2}{|y^j - \eta^j|^m} \, dy^j \, d\eta^j.$$

Let n be a unit outer normal to the boundary Γ. Denote by $\partial^i/\partial n^i$ the ith order normal derivative at Γ. We formulate the general trace theorem (Baiocchi, Capelo, 1984).

Theorem 1.24. *Let the boundary Γ belong to the class $C^{k,1}$, and a function u belong to the space $H^k(\Omega)$. Then there exists a linear continuous operator $\gamma : H^k(\Omega) \to \prod_{i=0}^{k-1} H^{k-i-1/2}(\Gamma)$, which uniquely defines the traces $\gamma u = (\gamma_0 u, ..., \gamma_{k-1} u)$ of u at Γ,*

$$\gamma_i u \in H^{k-i-1/2}(\Gamma), \quad 0 \le i \le k-1.$$

For smooth functions u defined in $\overline{\Omega}$,

$$\gamma_i u = \frac{\partial^i u}{\partial n^i}, \quad 0 \le i \le k-1, \qquad on\ \Gamma.$$

Conversely, there exists a linear continuous operator $\prod_{i=0}^{k-1} H^{k-i-1/2}(\Gamma) \to H^k(\Omega)$ *such that for any given* $\phi_i \in H^{k-i-1/2}(\Gamma)$, $0 \le i \le k-1$, *a function* $u \in H^k(\Omega)$ *can be found such that*

$$\gamma_i u = \phi_i, \quad 0 \le i \le k-1, \qquad on\ \Gamma.$$

For $k=1$, *the smoothness class of the boundary* Γ *can be reduced to* $C^{0,1}$.

In what follows, we use the notation $\partial^i u/\partial n^i$ for $\gamma_i u$.

Consider the domain Ω_c with the boundary $\partial\Omega_c$ described in the previous subsection. Let a function $u \in H^k(\Omega_c)$ be given. We assume that $\partial\Omega_c$ belongs to the class $C^{k,1}$, i.e. Ω can be divided into two domains Ω_1, Ω_2 by the closed surface Σ such that $\Sigma_c \subset \Sigma$, and $\partial\Omega_1, \partial\Omega_2$ belong to the class $C^{k,1}$. For every Ω_i, $i=1,2$, we have $u \in H^k(\Omega_i)$ and, consequently, one can apply Theorem 1.24 and define the normal derivatives at $\partial\Omega_i$. The boundaries $\partial\Omega_1, \partial\Omega_2$ consist of Σ^-, $\Gamma \cup \Sigma^+$, respectively, and $\Sigma_c^\pm$ are the corresponding parts of $\Sigma^\pm$. This provides the following statement.

Lemma 1.12. *If the boundary* $\partial\Omega_c$ *belongs to the class* $C^{k,1}$, *and a function* $u \in H^k(\Omega_c)$ *is given, then the normal derivatives at the boundary* $\partial\Omega_c$ *are uniquely defined,*

$$\frac{\partial^i u}{\partial n^i} \in H^{k-i-1/2}(\Gamma), \quad \frac{\partial^i u^\pm}{\partial \nu^i} \in H^{k-i-1/2}(\Sigma_c), \quad 0 \le i \le k-1.$$

By Lemma 1.12, we define the jumps of the function $u \in H^k(\Omega_c)$ at Σ_c,

$$\left[\frac{\partial^i u}{\partial \nu^i}\right] = \frac{\partial^i u^+}{\partial \nu^i} - \frac{\partial^i u^-}{\partial \nu^i} \in H^{k-i-1/2}(\Sigma_c), \quad 0 \le i \le k-1.$$

The same notations are used for the closed extension Σ, $\Sigma_c \subset \Sigma$, namely

$$\left[\frac{\partial^i u}{\partial \nu^i}\right] = \frac{\partial^i u^+}{\partial \nu^i} - \frac{\partial^i u^-}{\partial \nu^i} \in H^{k-i-1/2}(\Sigma), \quad 0 \le i \le k-1.$$

Note that, by $u \in H^k(\Omega_c)$, the uniqueness of the traces implies

$$\frac{\partial^i u^+}{\partial \nu^i} = \frac{\partial^i u^-}{\partial \nu^i} \quad \text{on } \Sigma \setminus \Sigma_c, \quad 0 \le i \le k-1,$$

or

$$\left[\frac{\partial^i u}{\partial \nu^i}\right] = 0 \quad \text{on } \Sigma \setminus \Sigma_c, \quad 0 \le i \le k-1. \tag{1.137}$$

Condition (1.137) gives an additional property of the traces at Σ_c which is used in studying the space $H_{00}^{k-1/2}(\Sigma_c)$ below.

As before, Σ means the closed extension of Σ_c belonging the class $C^{k,1}$, where $k \geq 1$ is an integer. Let the space $H_0^{k-1/2}(\Sigma_c)$ be a completion in the $H^{k-1/2}(\Sigma_c)$-norm of functions from $C^{k,1}(\Sigma_c)$ having compact supports. Introduce the Hilbert spaces

$$H_{00}^{k-1/2}(\Sigma_c) = \{s \in H_0^{k-1/2}(\Sigma_c) \mid \rho^{-1/2} D^\alpha s \in L^2(\Sigma_c), \quad 0 \leq |\alpha| \leq k-1\},$$

equipped with the norm

$$||s||^2_{k-1/2,00,\Sigma_c} = ||s||^2_{k-1/2,\Sigma_c} + \sum_{|\alpha|=0}^{k-1} ||\rho^{-1/2} D^\alpha s||^2_{0,\Sigma_c},$$

where the function ρ satisfies the properties $\rho \in C^{k,1}(\overline{\Sigma}_c)$, $\rho > 0$ in Σ_c, $\rho = 0$ on $\partial\Sigma_c$, $\lim_{x \to x_0} \rho(x)/d(x, \partial\Sigma_c) = d \neq 0$ for all $x_0 \in \partial\Sigma_c$. Here $d(x, \partial\Sigma_c)$ denotes the distance between the point $x \in \Sigma_c$ and the boundary $\partial\Sigma_c$.

We prove the statement characterizing the functions from $H_{00}^{k-1/2}(\Sigma_c)$.

Lemma 1.13. *The following equivalence takes place:*

$$s \in H_{00}^{k-1/2}(\Sigma_c) \iff \bar{s} = \begin{cases} s & , \quad in \ \Sigma_c \\ 0 & , \quad in \ \Sigma \setminus \Sigma_c \end{cases} \in H^{k-1/2}(\Sigma).$$

PROOF. By utilizing the local coordinate systems (1.135), the assertion of Lemma 1.13 reduces to the case

$$\Sigma = R^{m-1} = \{x = (x_1, ..., x_{m-1})\},$$

$$\Sigma_c = \Delta = \{x \in R^{m-1} \mid \ |x_i| < b, \ 1 \leq i \leq m-1\}.$$

Let $k = 1$. Denote

$$\Delta_i = \{x_j \mid \ |x_j| < b, \quad 1 \leq j \leq m-1, \quad j \neq i\}.$$

Instead of (1.136), we can use the following equivalent norms (see Lions, Magenes, 1968):

$$||\bar{s}||^2_{1/2,R^{m-1}} = ||\bar{s}||^2_{0,R^{m-1}}$$

$$+ \int_R \int_R |t-\tau|^{-2} \sum_{i=1}^{m-1} \int_{\Delta_i} |\bar{s}(x|_{x_i=t}) - \bar{s}(x|_{x_i=\tau})|^2 \, d\Delta_i dt d\tau,$$

$$||s||^2_{1/2,\Delta} = ||s||^2_{0,\Delta} + \int_{-b}^{b} \int_{-b}^{b} |t-\tau|^{-2} \sum_{i=1}^{m-1} \int_{\Delta_i} |s(x|_{x_i=t}) - s(x|_{x_i=\tau})|^2 \, d\Delta_i dt d\tau.$$

Since $\bar{s}(x) = 0$ for $|x_i| \geq b$, $1 \leq i \leq m-1$, we obtain

$$||\bar{s}||^2_{1/2,R^{m-1}} = ||s||^2_{0,\Delta}$$

$$+\int_{-b}^{b}\int_{-b}^{b}|t-\tau|^{-2}\sum_{i=1}^{m-1}\int_{\Delta_i}|s(x|_{x_i=t})-s(x|_{x_i=\tau})|^2\,d\Delta_i dt d\tau$$

$$+2\int_{-b}^{b}\left(\int_{-\infty}^{-b}+\int_{b}^{\infty}\right)|t-\tau|^{-2}\sum_{i=1}^{m-1}\int_{\Delta_i}|s(x|_{x_i=\tau})|^2\,d\Delta_i dt d\tau,$$

which implies

$$||\bar{s}||^2_{1/2,R^{m-1}}=||s||^2_{1/2,\Delta}+2\int_{-b}^{b}\sum_{i=1}^{m-1}\int_{\Delta_i}|s(x|_{x_i=\tau})|^2\left(\int_{-\infty}^{-b}+\int_{b}^{\infty}\right)|t-\tau|^{-2}.$$

The integral in t can be calculated here for $\tau\in(-b,b)$,

$$\left(\int_{-\infty}^{-b}+\int_{b}^{\infty}\right)|t-\tau|^{-2}\,dt=\int_{-\infty}^{-b}(\tau-t)^{-2}\,dt+\int_{b}^{\infty}(t-\tau)^{-2}\,dt=\frac{1}{b+\tau}+\frac{1}{b-\tau}.$$

Thus, we have

$$||\bar{s}||^2_{1/2,R^{m-1}}=||s||^2_{1/2,\Delta}+\sum_{i=1}^{m-1}\int_{\Delta_i}\int_{-b}^{b}\left(\frac{(b-\tau)(b+\tau)}{4b}\right)^{-1}|s(x|_{x_i=\tau})|^2\,d\tau d\Delta_i.$$

Changing the variable τ by x_i in each Δ_i, $1\le i\le m-1$, and denoting

$$\rho^{-1}(x)=\sum_{i=1}^{m-1}\left(\frac{(b-x_i)(b+x_i)}{4b}\right)^{-1},$$

we obtain the equality

$$||\bar{s}||^2_{1/2,R^{m-1}}=||s||^2_{1/2,\Delta}+\int_{\Delta}\rho^{-1}(x)|s(x)|^2\,dx \tag{1.138}$$

which proves the assertion of Lemma 1.13 for $k=1$.

Let $k>1$. The above reasonings applied to $D^\alpha s$, $|\alpha|=0,...,k-1$, provide the equalities similar to (1.138),

$$||D^\alpha\bar{s}||^2_{1/2,R^{m-1}}=||D^\alpha s||^2_{1/2,\Delta}+\int_{\Delta}\rho^{-1}(x)|D^\alpha s(x)|^2\,dx. \tag{1.139}$$

Summing (1.139) from $|\alpha|=0$ up to $|\alpha|=k-1$, one obtains

$$||\bar{s}||^2_{k-1/2,R^{m-1}}=||s||^2_{k-1/2,\Delta}+\sum_{|\alpha|=0}^{k-1}||\rho^{-1/2}D^\alpha s||^2_{0,\Delta}=||s||^2_{k-1/2,00,\Delta}.$$

Lemma 1.13 is proved.

We have to note that $H_{00}^{k-1/2}(\Sigma_c)$ is imbedded in $H_0^{k-1/2}(\Sigma_c)$, and extensions of functions from $H_0^{k-1/2}(\Sigma_c)$ on Σ by zero do not belong to $H^{k-1/2}(\Sigma)$, in general (Lions, Magenes, 1968).

By Lemmas 1.12, 1.13 and property (1.137), from Theorem 1.24 the next statement follows.

Theorem 1.25. *Let the boundary $\partial\Omega_c$ belong to the class $C^{k,1}$, and a function u belong to the space $H^k(\Omega_c)$. Then there exists a linear continuous operator, which uniquely defines at $\partial\Omega_c$ the values*

$$\frac{\partial^i u}{\partial n^i} \in H^{k-i-1/2}(\Gamma), \quad \frac{\partial^i u^\pm}{\partial \nu^i} \in H^{k-i-1/2}(\Sigma_c), \quad \left[\frac{\partial^i u}{\partial \nu^i}\right] \in H_{00}^{k-i-1/2}(\Sigma_c)$$

for $0 \le i \le k-1$. Conversely, there exists a linear continuous operator such that for any given

$$\phi_i^\pm \in H^{k-i-1/2}(\Sigma_c), \quad [\phi_i] \in H_{00}^{k-i-1/2}(\Sigma_c), \quad 0 \le i \le k-1,$$

a function $u \in H^k(\Omega_c)$ can be found such that

$$\frac{\partial^i u^\pm}{\partial \nu^i} = \phi_i^\pm, \quad 0 \le i \le k-1, \qquad on\ \Sigma_c.$$

For $k = 1$, the smoothness class of the boundary $\partial\Omega_c$ can be reduced to $C^{0,1}$.

PROOF. Assume that Σ is the closed extension of Σ_c from the class $C^{k,1}$ dividing Ω into two domains Ω_1, Ω_2 as before. The boundaries $\partial\Omega_1, \partial\Omega_2$ consist of Σ^-, $\Gamma \cup \Sigma^+$, respectively. By Theorem 1.24, for $u \in H^k(\Omega_c)$ we have

$$\frac{\partial^i u^\pm}{\partial \nu^i} \in H^{k-i-1/2}(\Sigma), \quad 0 \le i \le k-1.$$

In view of the property (1.137), one can write

$$\left[\frac{\partial^i u}{\partial \nu^i}\right] = 0 \text{ on } \Sigma \setminus \Sigma_c, \quad \left[\frac{\partial^i u}{\partial \nu^i}\right] \in H^{k-i-1/2}(\Sigma), \quad 0 \le i \le k-1.$$

By Lemma 1.13, this means that

$$\left[\frac{\partial^i u}{\partial \nu^i}\right] \in H_{00}^{k-i-1/2}(\Sigma_c), \quad 0 \le i \le k-1,$$

which proves the first assertion of Theorem 1.25.

Now we prove the converse assertion formulated in Theorem 1.25. Let $\phi_i^\pm \in H^{k-i-1/2}(\Sigma_c)$ be given, $[\phi_i] \in H_{00}^{k-i-1/2}(\Sigma_c)$, $0 \le i \le k-1$. One can construct an arbitrary smooth extension of ϕ_i^- on Σ^- such that

$$\tilde{\phi}_i^- = \begin{cases} \phi_i^- & , \text{ on } \Sigma_c^- \\ \psi_i & , \text{ on } \Sigma^- \setminus \Sigma_c^- \end{cases} \in H^{k-i-1/2}(\Sigma), \quad 0 \le i \le k-1.$$

Let us define on Σ^+ the function

$$\tilde{\phi}_i^+ = \begin{cases} \phi_i^+ & , \text{ on } \Sigma_c^+ \\ \psi_i & , \text{ on } \Sigma^+ \setminus \Sigma_c^+ \end{cases}, \quad 0 \leq i \leq k-1.$$

Since $[\phi_i] \in H_{00}^{k-i-1/2}(\Sigma_c)$ and $[\tilde{\phi}_i] = 0$ at $\Sigma \setminus \Sigma_c$, then, by Lemma 1.13, we obtain $[\tilde{\phi}_i] \in H^{k-i-1/2}(\Sigma)$. In particular, this implies that $\tilde{\phi}_i^+ = [\tilde{\phi}_i] + \tilde{\phi}_i^- \in H^{k-i-1/2}(\Sigma)$. Hence, by Theorem 1.24, there exist functions $u_j \in H^k(\Omega_j)$, $j = 1, 2$, such that $\partial^i u_j / \partial \nu^i$ coincide with $\tilde{\phi}_i^\pm$ on $\Sigma^\pm$. In Ω_c, define the function

$$u = \begin{cases} u_1 & , \text{ in } \Omega_1, \\ u_2 & , \text{ in } \Omega_2. \end{cases}$$

By the property

$$0 = [\tilde{\phi}_i] = \left[\frac{\partial^i u}{\partial \nu^i}\right] \quad \text{on } \Sigma \setminus \Sigma_c, \quad 0 \leq i \leq k-1,$$

we obtain $u \in H^k(\Omega_c)$. Theorem 1.25 is proved.

1.4.3 Green's formulae

Firstly, let us formulate an auxiliary statement concerning boundary values for the vector-functions having square integrable divergence (Baiocchi, Capelo, 1984; Temam, 1979). Consider a bounded domain $\Omega \subset R^m$. Introduce the Hilbert space

$$L^2_{\mathrm{div}}(\Omega) = \{u = (u_1, ..., u_m) \in L^2(\Omega) \mid \ \mathrm{div}\, u \in L^2(\Omega)\}$$

with the scalar product

$$(u, v) = \int_\Omega (uv + \mathrm{div}\, u \cdot \mathrm{div}\, v)\,.$$

Denote by $H^{-s}(\Gamma)$ the space dual of $H^s(\Gamma)$ with a duality pairing $\langle \cdot, \cdot \rangle_{s,\Gamma}$, $s > 0$.

Theorem 1.26. *Let the boundary Γ of the domain Ω belong to the class $C^{0,1}$, n be a unit outward normal to Γ, and $u \in L^2_{\mathrm{div}}(\Omega)$. There exists a linear continuous operator $\Lambda : L^2_{\mathrm{div}}(\Omega) \to H^{-1/2}(\Gamma)$ which uniquely defines at Γ the value $\Lambda u \subset H^{-1/2}(\Gamma)$, and the generalized Green formula holds:*

$$\int_\Omega u \cdot \nabla v = -\int_\Omega \mathrm{div}\, u \cdot v + \langle \Lambda u, v \rangle_{1/2,\Gamma} \quad \forall v \in H^1(\Omega). \tag{1.140}$$

For smooth functions u defined in $\overline{\Omega}$,

$$\Lambda u = u \cdot n \quad \textit{on } \Gamma.$$

Conversely, there exists a linear continuous operator $H^{-1/2}(\Gamma) \to L^2_{\text{div}}(\Omega)$ *such that, for any given* $\lambda \in H^{-1/2}(\Gamma)$, *a function* $u \in L^2_{\text{div}}(\Omega)$ *can be found such that* $\Lambda u = \lambda$ *at* Γ.

Note that, if the function $u \in L^2_{\text{div}}(\Omega)$ allows the presentation

$$u = \nabla w, \quad w \in H^1(\Omega),$$

then (1.140) takes the form

$$\int_\Omega \nabla w \cdot \nabla v = -\int_\Omega v \Delta u + \langle \frac{\partial w}{\partial n}, v\rangle_{1/2,\Gamma} \quad \forall v \in H^1(\Omega),$$

where $\partial w/\partial n$ denotes the element from $H^{-1/2}(\Gamma)$ which coincides with the usual normal derivative for smooth functions w defined in $\overline{\Omega}$.

Using Theorem 1.26, we can consider the *Green formulae* in domains with regular boundaries, which are useful in the sequel.

Let $\Omega \subset R^3$ be a bounded domain with a smooth boundary Γ, and $n = (n_1, n_2, n_3)$ be a unit outward normal vector to Γ. Introduce the stress and strain tensors of linear elasticity (see Section 1.1.1),

$$\sigma_{ij}(u) = a_{ijkl}\varepsilon_{kl}(u), \quad \varepsilon_{ij}(u) = 1/2\,(u_{i,j} + u_{j,i}), \quad i,j = 1,2,3,$$

where $u = (u_1, u_2, u_3)$ are the displacements defined in Ω.

By the symmetry $\sigma_{ij}(u) = \sigma_{ji}(u)$, we can integrate by parts:

$$\int_\Omega \sigma_{ij}(u)\varepsilon_{ij}(v) = -\int_\Omega \sigma_{ij,j}(u)v_i + \int_\Gamma \sigma_{ij}(u)n_j v_i.$$

Decompose the vectors $(\sigma_{1j}(u)n_j, \sigma_{2j}(u)n_j, \sigma_{3j}(u)n_j)$, $v = (v_1, v_2, v_3)$ into normal and tangential components at the boundary as follows:

$$\sigma_{ij}(u)n_j = \sigma_n(u)n_i + \sigma_{Ti}(u),\ i = 1,2,3, \quad \sigma_n(u) = \sigma_{ij}(u)n_j n_i; \qquad (1.141)$$

$$v_i = v_n n_i + v_{Ti}, \quad i = 1,2,3, \quad v_n = v_i n_i.$$

Since $\sigma_{Ti}(u)n_i = \sigma_{ij}(u)n_j n_i - \sigma_n(u) = 0$, $v_{Ti}n_i = v_i n_i - v_n = 0$, one has

$$\sigma_{ij}(u)n_j v_i = (\sigma_n(u)n_i + \sigma_{Ti}(u))(v_n n_i + v_{Ti}) = \sigma_n(u)v_n + \sigma_{Ti}(u)v_{Ti}.$$

Thus, for smooth functions u, v, we obtain the following Green formula:

$$\int_\Omega \sigma_{ij}(u)\varepsilon_{ij}(v) = -\int_\Omega \sigma_{ij,j}(u)v_i + \int_\Gamma (\sigma_n(u)v_n + \sigma_{Ti}(u)v_{Ti}).$$

Introduce the space

$$H^1_\sigma(\Omega) = \{u = (u_1, u_2, u_3) \in H^1(\Omega) \mid \ \sigma_{ij,j}(u) \in L^2(\Omega), \quad i = 1,2,3\}$$

equipped with the norm

$$\|u\|^2_{H^1_\sigma(\Omega)} = \sum_{i=1}^{3} \left(\|u_i\|^2_{1,\Omega} + \|\sigma_{ij,j}(u)\|^2_{0,\Omega} \right).$$

The following result holds true.

Theorem 1.27. *Let the boundary Γ belong to the class $C^{1,1}$, and a function u belong to the space $H^1_\sigma(\Omega)$. There exists a linear continuous operator $H^1_\sigma(\Omega) \to [H^{-1/2}(\Gamma)]^3$ which uniquely defines at the boundary Γ the values*

$$\sigma_n(u), \sigma_{\tau i}(u) \in H^{-1/2}(\Gamma),\ i = 1,2,3, \quad \sigma_{\tau i}(u) n_i = 0,$$

and for all $v \in [H^1(\Omega)]^3$ the generalized Green formula holds:

$$\int_\Omega \sigma_{ij}(u)\varepsilon_{ij}(v) = -\int_\Omega \sigma_{ij,j}(u) v_i \tag{1.142}$$

$$+ \langle \sigma_n(u), v_n \rangle_{1/2,\Gamma} + \langle \sigma_{\tau i}(u), v_{\tau i} \rangle_{1/2,\Gamma}.$$

For smooth functions u defined in $\overline{\Omega}$, formula (1.141) *arises. Conversely, there exists a linear continuous operator $[H^{-1/2}(\Gamma)]^3 \to H^1_\sigma(\Omega)$ such that for any given $\lambda_n, \lambda_{\tau i} \in H^{-1/2}(\Gamma)$, $i = 1,2,3$, $\lambda_{\tau i} n_i = 0$, a function $u \in H^1_\sigma(\Omega)$ can be found such that*

$$\sigma_n(u) = \lambda_n, \quad \sigma_{\tau i}(u) = \lambda_{\tau i},\ i = 1,2,3, \quad \text{on } \Gamma.$$

Now we consider a two-dimensional solid occupying a bounded domain $\Omega \subset R^2$ with a smooth boundary Γ. Let the bilinear form B be introduced by the formula

$$B(u,v) = \int_\Omega \big(u_{,11}v_{,11} + u_{,22}v_{,22} + \kappa(u_{,11}v_{,22} + u_{,22}v_{,11})$$

$$+ 2(1-\kappa) u_{,12} v_{,12} \big),$$

where κ is a constant, $0 < \kappa < 1/2$. Denote by $n = (n_1, n_2)$ a unit outward vector to Γ. We define at the boundary Γ the values

$$m(u) = \kappa \Delta u + (1-\kappa)\frac{\partial^2 u}{\partial n^2}, \quad t(u) = \frac{\partial}{\partial n}\left(\Delta u + (1-\kappa)\frac{\partial^2 u}{\partial \tau^2} \right). \tag{1.143}$$

Here $\tau = (-n_2, n_1)$ is the tangential unit vector at Γ. Integrating by parts, one can obtain the Green formula (Temam, 1983; Khludnev, Sokolowski, 1997)

$$B(u,v) = \int_\Omega v \Delta^2 u + \int_\Gamma m(u)\frac{\partial v}{\partial n} - \int_\Gamma t(u) v.$$

Introduce the space

$$H^2_{\Delta^2}(\Omega) = \{u \in H^2(\Omega) \mid \quad \Delta^2 u \in L^2(\Omega)\}$$

equipped with the norm

$$||u||^2 = ||u||^2_{2,\Omega} + ||\Delta^2 u||^2_{0,\Omega}.$$

The following statement holds.

Theorem 1.28. *Let the boundary Γ belong to the class $C^{2,1}$, and a function u belong to the space $H^2_{\Delta^2}(\Omega)$. There exists a linear continuous operator $\Lambda : H^2_{\Delta^2}(\Omega) \to H^{-1/2}(\Gamma) \times H^{-3/2}(\Gamma)$ which uniquely defines at the boundary Γ the values*

$$\Lambda u = (m(u), t(u)), \quad m(u) \in H^{-1/2}(\Gamma), \quad t(u) \in H^{-3/2}(\Gamma),$$

and the generalized Green formula holds:

$$B(u,v) = \int_\Omega v\Delta^2 u + \langle m(u), \frac{\partial v}{\partial n}\rangle_{1/2,\Gamma} - \langle t(u), v\rangle_{3/2,\Gamma}, \quad v \in H^2(\Omega). \quad (1.144)$$

For smooth functions u defined in $\overline{\Omega}$, formula (1.143) takes place. Conversely, there exists a linear continuous operator $H^{-1/2}(\Gamma) \times H^{-3/2}(\Gamma) \to H^2_{\Delta^2}(\Omega)$ such that, for any given $m \in H^{-1/2}(\Gamma)$, $t \in H^{-3/2}(\Gamma)$, a function $u \in H^2_{\Delta^2}(\Omega)$ can be found such that $\Lambda u = (m,t)$ on Γ.

In the two-dimensional theory of solids, the following theorem is also useful. For vector-functions $M = \{M_{ij}\}^2_{i,j=1}$, introduce the space

$$W(\Omega) = \{M_{ij} = M_{ji} \in L^2(\Omega),\ i,j = 1,2 \mid \quad M_{ij,ij} \in L^2(\Omega)\}$$

equipped with the norm

$$||M||^2_{W(\Omega)} = \sum_{i,j=1}^{2} ||M_{ij}||^2_{0,\Omega} + ||M_{ij,ij}||^2_{0,\Omega}.$$

Theorem 1.29. *Let the boundary Γ belong to the class $C^{2,1}$, and a function M belong to the space $W(\Omega)$. There exists a linear continuous operator $\Lambda : W(\Omega) \to H^{-1/2}(\Gamma) \times H^{-3/2}(\Gamma)$ which uniquely defines at the boundary Γ the values*

$$\Lambda M = (m,t), \quad m \in H^{-1/2}(\Gamma), \quad t \in H^{-3/2}(\Gamma),$$

and the generalized Green formula holds:

$$-\int_\Omega M_{ij} v_{,ij} = -\int_\Omega M_{ij,ij} v + \langle m, \frac{\partial v}{\partial n}\rangle_{1/2,\Gamma} - \langle t, v\rangle_{3/2,\Gamma} \quad (1.145)$$

$$\forall v \in H^2(\Omega).$$

For smooth functions M defined in $\overline{\Omega}$,

$$m = -M_{ij}n_jn_i, \quad t = -M_{ij,k}\tau_k\tau_jn_i - M_{ij,j}n_i \quad \text{on } \Gamma. \tag{1.146}$$

Conversely, there exists a linear continuous operator Λ^{-1}: $H^{-1/2}(\Gamma) \times H^{-3/2}(\Gamma) \to W(\Omega)$ such that, for any given $\lambda \in H^{-1/2}(\Gamma)$, $\mu \in H^{-3/2}(\Gamma)$, a function $M \in W(\Omega)$ can be found such that $\Lambda M = (\lambda, \mu)$.

PROOF. On the space $H^{3/2}(\Gamma) \times H^{1/2}(\Gamma)$, consider the linear functional

$$L(\phi_0, \phi_1) = \int_\Omega (-M_{ij}v_{,ij} + M_{ij,ij}v), \tag{1.147}$$

$$\phi_0 \in H^{3/2}(\Gamma), \quad \phi_1 \in H^{1/2}(\Gamma),$$

where the function $v \in H^2(\Omega)$ is such that

$$v = \phi_0, \quad \partial v/\partial n = \phi_1 \quad \text{on } \Gamma. \tag{1.148}$$

It is clear that the functional L is well defined. By Theorem 1.24, one can obtain the estimate

$$|L(\phi_0, \phi_1)| \le c\|M\|^2_{W(\Omega)} \left(\|\phi_0\|^2_{3/2,\Gamma} + \|\phi_1\|^2_{1/2,\Gamma}\right).$$

Therefore, there exist $m \in H^{-1/2}(\Gamma)$, $t \in H^{-3/2}(\Gamma)$ such that

$$L(\phi_0, \phi_1) = \langle m, \phi_1\rangle_{1/2,\Gamma} - \langle t, \phi_0\rangle_{3/2,\Gamma}.$$

By (1.147), (1.148), this equality provides (1.145). For the smooth function u defined in $\overline{\Omega}$, integrating by parts, one can easily deduce (1.146).

Conversely, assume that $\lambda \in H^{-1/2}(\Gamma)$, $\mu \in H^{-3/2}(\Gamma)$ are given. For a constant $c > 0$, one can solve the problem

$$B(u, v) + c\int_\Omega uv = \langle \lambda, \frac{\partial v}{\partial n}\rangle_{1/2,\Gamma} - \langle \mu, v\rangle_{3/2,\Gamma} \quad \forall v \in H^2(\Omega).$$

By the Green formula (1.144), it is equivalent to the following problem:

$$\Delta^2 u + cu = 0 \text{ in } \Omega, \quad m(u) = \lambda, \quad t(u) = \mu \text{ on } \Gamma.$$

Then the solution u belongs to $H^2_{\Delta^2}(\Omega)$. Let us define

$$M_{11} = -(u_{,11} + \kappa u_{,22}), \quad M_{22} = -(u_{,22} + \kappa u_{,11}), \quad M_{12} = -(1 - \kappa)u_{,12}.$$

These functions satisfy the conditions

$$M_{ij} = M_{ji} \in L^2(\Omega), \quad -M_{ij,ij} = \Delta^2 u \in L^2(\Omega).$$

By the Green formula (1.145), $m = \lambda$, $t = \mu$ on Γ. Theorem 1.29 is proved.
Note that for isotropic plates we have the relations

$$M_{11} = -(u_{,11} + \kappa u_{,22}), \quad M_{22} = -(u_{,22} + \kappa u_{,11}), \quad M_{12} = -(1-\kappa)u_{,12},$$

where u is a displacement. Then

$$-\int_{\Omega} M_{ij} v_{,ij} = B(u,v), \quad -M_{ij,ij} = \Delta^2 u, \quad m = m(u), \quad t = t(u),$$

and formula (1.145) coincides with (1.144). In this case, for smooth displacements u defined in $\overline{\Omega}$, we also have at Γ

$$-M_{ij}n_j n_i = \kappa \Delta u + (1-\kappa)\frac{\partial^2 u}{\partial n^2},$$

$$-M_{ij,k}\tau_k \tau_j n_i - M_{ij,j}n_i = \frac{\partial}{\partial n}\left(\Delta u + (1-\kappa)\frac{\partial^2 u}{\partial \tau^2}\right), \quad \tau = (-n_2, n_1).$$

1.4.4 Solid with a crack

Let a solid occupy the domain $\Omega_c \subset R^3$ with the crack Σ_c such that its boundary $\partial\Omega_c$ belongs to the class $C^{1,1}$ in accord with Section 1.4.1. Introduce the space

$$H^{1,0}(\Omega_c) = \{u = (u_1, u_2, u_3) \in H^1(\Omega_c) \mid \quad u = 0 \quad \text{on } \Gamma\}$$

equipped with the norm

$$\|u\|_1^2 = \sum_{i=1}^{3} \|u_i\|_{0,\Omega_c}^2 + \sum_{i,j=1}^{3} \|u_{i,j}\|_{0,\Omega_c}^2.$$

This space corresponds to the solid clamped at the boundary,

$$u = 0 \quad \text{on } \Gamma.$$

The nonpenetration condition of the crack surfaces has the form

$$[u_\nu] = [u_i]\nu_i \geq 0 \quad \text{on } \Sigma_c.$$

Introduce the admissible displacements set

$$K = \{u \in H^{1,0}(\Omega_c) \mid \quad [u_\nu] \geq 0 \quad \text{on } \Sigma_c\}$$

which is convex and closed. Here $\nu = (\nu_1, \nu_2, \nu_3)$ corresponds to a unit normal vector at Σ_c, and $n = (n_1, n_2, n_3)$ is a unit outward normal vector to Γ.

For a given force $f = (f_1, f_2, f_3) \in L^2(\Omega_c)$, consider the potential energy functional

$$\Pi(u) = \frac{1}{2}\int\limits_{\Omega_c} \sigma_{ij}(u)\varepsilon_{ij}(u) - \int\limits_{\Omega_c} f_i u_i.$$

The stress and strain tensors $\sigma_{ij}(u), \varepsilon_{ij}(u)$ are defined by the Hooke and Cauchy laws

$$\sigma_{ij}(u) = a_{ijkl}\varepsilon_{kl}(u), \quad \varepsilon_{ij}(u) = 1/2\,(u_{i,j} + u_{j,i}), \quad i,j = 1,2,3,$$

$$a_{ijkl} = a_{jikl} = a_{klij}, \quad c_1\xi_{ij}\xi_{ij} \le a_{ijkl}\xi_{kl}\xi_{ij} \le c_2\xi_{ij}\xi_{ij}, \quad c_1, c_2 > 0.$$

By Theorem 1.12, the equilibrium problem

$$\Pi(u) = \inf_{\bar{u}\in K} \Pi(\bar{u}) \tag{1.149}$$

is equivalent to the variational inequality

$$u \in K, \quad \Pi'_u(v-u) \ge 0 \quad \forall v \in K. \tag{1.150}$$

Note that the functional Π is convex and continuous, and consequently, it is weakly lower semicontinuous.

Extend Σ_c up to the boundary Γ such that Ω is divided into two domains with Lipschitz boundaries $\partial\Omega_1, \partial\Omega_2$. Assume that *meas* $(\Gamma \cap \partial\Omega_i) > 0$, $i = 1,2$. In each of these domains, for $u \in H^{1,0}(\Omega_c)$, the second Korn inequality (see Reshetnyak, 1982; Hlavaček, Nečas, 1970)

$$\int\limits_{\Omega_i} \varepsilon_{ij}(u)\varepsilon_{ij}(u) \ge c\|u\|^2_{1,\Omega_i}, \; i = 1,2, \quad u = (u_1, u_2, u_3),$$

is fulfilled since $u = 0$ at $\Gamma \cap \partial\Omega_i$, $i = 1,2$. Consequently, we have the following estimate in Ω_c,

$$\int\limits_{\Omega_c} \varepsilon_{ij}(u)\varepsilon_{ij}(u) \ge c\|u\|^2_1.$$

This estimate provides the coercivity of the functional Π,

$$\Pi(u) \ge c\|u\|^2_1 - \|f_i\|_{0,\Omega_c}\|u_i\|_{0,\Omega_c} \to +\infty, \quad \|u\|_1 \to \infty.$$

Thus, all conditions of Theorem 1.11 are fulfilled, hence there exists a unique solution $u \in K$ of the problem (1.149). One can calculate the derivative

$$\Pi'_u(v) = \int\limits_{\Omega_c} (\sigma_{ij}(u)\varepsilon_{ij}(v) - f_i v_i)$$

and substitute in (1.150). This implies the following variational inequality

$$\int_{\Omega_c} \sigma_{ij}(u)\varepsilon_{ij}(v-u) \geq \int_{\Omega_c} f_i(v_i - u_i) \quad \forall v \in K. \tag{1.151}$$

Denote next by $H_{00}^{1/2}(\Sigma_c)^\star$ the space dual of $H_{00}^{1/2}(\Sigma_c)$.

Theorem 1.30. *There exists a unique solution $u \in K$ to the problem* (1.151) *such that*

$$-\sigma_{ij,j}(u) = f_i, \quad i = 1,2,3, \quad \text{in } \Omega_c,$$

$$\sigma_\tau(u) = 0, \quad [u_\nu] \geq 0, \quad [\sigma_\nu(u)] = 0, \quad \sigma_\nu(u) \leq 0, \quad \sigma_\nu(u)[u_\nu] = 0 \ \text{ on } \Sigma_c.$$

PROOF. The existence and uniqueness of $u \in K$ was already proved. The substitution of $v = u \pm \phi$, $\phi \in [C_0^\infty(\Omega_c)]^3$ in (1.151) as a test function provides the identity

$$\int_{\Omega_c} \sigma_{ij}(u)\varepsilon_{ij}(\phi) = \int_{\Omega_c} f_i\phi_i \quad \forall \phi \in [C_0^\infty(\Omega_c)]^3. \tag{1.152}$$

This means that the equilibrium equations

$$-\sigma_{ij,j}(u) = f_i, \quad i = 1,2,3,$$

hold in the sense of distributions. Since $f_i \in L^2(\Omega_c)$, we have $\sigma_{ij,j}(u) \in L^2(\Omega_c)$, $i = 1,2,3$, i.e. $u \in H_\sigma^1(\Omega_c)$ (see notations of Section 1.4.3).

Let us extend Σ_c up to the closed surface Σ, $\Sigma_c \subset \Sigma$, dividing Ω into two domains Ω_1, Ω_2 with boundaries $\partial\Omega_1, \partial\Omega_2$ of the class $C^{1,1}$ (see Section 1.4.1). In each Ω_k, $k = 1,2$, we have $u \in H_\sigma^1(\Omega_k)$. Hence, we can apply Theorem 1.27 which provides the existence of $\sigma_\nu^\pm(u), \sigma_{\tau i}^\pm(u) \in H^{-1/2}(\Sigma)$, $i = 1,2,3$, and obtain the Green formula

$$\int_{\Omega_c} \sigma_{ij}(u)\varepsilon_{ij}(v) = -\int_{\Omega_c} \sigma_{ij,j}(u)v_i \tag{1.153}$$

$$- \left[\langle \sigma_\nu(u), v_\nu \rangle_{1/2,\Sigma}\right] - \left[\langle \sigma_{\tau i}(u), v_{\tau i} \rangle_{1/2,\Sigma}\right].$$

Here the signs $\pm$ correspond to the faces $\Sigma^\pm$ of the surface Σ, and $\langle \cdot, \cdot \rangle_{1/2,\Sigma}$ means the duality pairing between $H^{1/2}(\Sigma)$ and $H^{-1/2}(\Sigma)$. By (1.152), (1.153), from (1.151) one deduces

$$\left[\langle \sigma_\nu(u), v_\nu - u_\nu \rangle_{1/2,\Sigma}\right] + \left[\langle \sigma_{\tau i}(u), v_{\tau i} - u_{\tau i} \rangle_{1/2,\Sigma}\right] \leq 0 \quad \forall v \in K. \tag{1.154}$$

For $\phi \in [H_0^1(\Omega)]^3$ we have $\phi^\pm \in H^{1/2}(\Sigma)$, $[\phi] = 0$ and $v = u \pm \phi \in K$. Substituting $v = u \pm \phi$ in (1.154) as a test function, we obtain

$$\langle \sigma_\nu^+(u) - \sigma_\nu^-(u), \phi_\nu \rangle_{1/2,\Sigma} + \langle \sigma_{\tau i}^+(u) - \sigma_{\tau i}^-(u), \phi_{\tau i} \rangle_{1/2,\Sigma} = 0 \quad \forall \phi \in [H_0^1(\Omega)]^3.$$

The independence between $\phi_\nu, \phi_{\tau i} \in H^{1/2}(\Sigma)$ provides

$$\langle \sigma_\nu^+(u) - \sigma_\nu^-(u), \psi \rangle_{1/2,\Sigma} = \langle \sigma_{\tau i}^+(u) - \sigma_{\tau i}^-(u), \psi \rangle_{1/2,\Sigma} = 0, \tag{1.155}$$

$$i = 1, 2, 3, \quad \forall \psi \in H^{1/2}(\Sigma).$$

These identities mean that

$$[\sigma_\nu(u)] = 0, \quad [\sigma_\tau(u)] = 0 \quad \text{on } \Sigma.$$

Using (1.155) and Lemma 1.13, let us define the functionals $\sigma_\nu(u), \sigma_{\tau i}(u) \in H_{00}^{1/2}(\Sigma_c)^\star$ by the formulae

$$\langle \sigma_\nu(u), \psi \rangle_{1/2,\Sigma_c} = \langle \sigma_\nu^\pm(u), \bar{\psi} \rangle_{1/2,\Sigma},$$

$$\langle \sigma_{\tau i}(u), \psi \rangle_{1/2,\Sigma_c} = \langle \sigma_{\tau i}^\pm(u), \bar{\psi} \rangle_{1/2,\Sigma}, \quad i = 1, 2, 3,$$

$$\forall \psi \in H_{00}^{1/2}(\Sigma_c), \quad \bar{\psi} = \psi \text{ in } \Sigma_c, \quad \bar{\psi} = 0 \text{ in } \Sigma \setminus \Sigma_c, \quad \bar{\psi} \in H^{1/2}(\Sigma).$$

Here $\langle \cdot, \cdot \rangle_{1/2,\Sigma_c}$ means the duality pairing between the spaces $H_{00}^{1/2}(\Sigma_c)$ and $H_{00}^{1/2}(\Sigma_c)^\star$. This allows us to rewrite (1.154) in the form

$$\langle \sigma_\nu(u), [v_\nu - u_\nu] \rangle_{1/2,\Sigma_c} + \langle \sigma_{\tau i}(u), [v_{\tau i} - u_{\tau i}] \rangle_{1/2,\Sigma_c} \leq 0 \quad \forall v \in K. \tag{1.156}$$

Take here $v = u \pm \phi$, $\phi \in H^{1,0}(\Omega_c)$, $\phi_\nu = 0$ at Σ_c. Then $v \in K$, and we have

$$\langle \sigma_{\tau i}(u), \psi_i \rangle_{1/2,\Sigma_c} = 0 \quad \forall \psi \in [H_{00}^{1/2}(\Sigma_c)]^3, \ \psi_i \nu_i = 0. \tag{1.157}$$

By utilizing (1.157), the inequality (1.156) provides

$$\langle \sigma_\nu(u), [v_\nu] \rangle_{1/2,\Sigma_c} \leq \langle \sigma_\nu(u), [u_\nu] \rangle_{1/2,\Sigma_c} \quad \forall v \in K.$$

Substituting here $v = 0$, $v = 2u$, one obtains

$$\langle \sigma_\nu(u), [u_\nu] \rangle_{1/2,\Sigma_c} = 0. \tag{1.158}$$

Consequently, $\langle \sigma_\nu(u), [v_\nu] \rangle_{1/2,\Sigma_c} \leq 0$ for all $v \in H^{1,0}(\Omega_c)$, $[v_\nu] \geq 0$. This implies the inequality

$$\langle \sigma_\nu(u), \psi \rangle_{1/2,\Sigma_c} \leq 0 \quad \forall \psi \in H_{00}^{1/2}(\Sigma_c), \ \psi \geq 0. \tag{1.159}$$

The system (1.152), (1.155), (1.157)–(1.159) gives the exact meaning of the relations formulated in Theorem 1.30. The theorem is proved.

At this point we have to mention different approaches to the crack problem with equality type boundary conditions (Osadchuk, 1985; Panasyuk et al., 1977; Duduchava, Wendland, 1995).

1.4.5 Solid with a crack and friction

We keep the notation of Section 1.4.4. Let us prove an auxiliary statement.

Lemma 1.14. *For $s \in H_{00}^{1/2}(\Sigma_c)$, if $r \in C^{0,1}(\overline{\Sigma}_c)$, then $rs \in H_{00}^{1/2}(\Sigma_c)$.*

PROOF. This assertion is a consequence of the norm definition given in Section 1.4.2. Indeed, we can write

$$||rs||^2_{1/2,00,\Sigma_c} = ||rs||^2_{1/2,\Sigma_c} + ||\rho^{-1/2} rs||^2_{0,\Sigma_c} = ||rs||^2_{0,\Sigma_c} + ||\rho^{-1/2} rs||^2_{0,\Sigma_c}$$

$$+ \sum_{j=1}^{p} \sum_{i=1}^{m-1} \int_{-b}^{b} \int_{\Delta^j} \frac{|r^j(y^j|_{y_i^j=t}) s^j(y^j|_{y_i^j=t}) - r^j(y^j) s^j(y^j)|^2}{|t - y_i^j|^2} dy^j dt.$$

The following equality takes place:

$$r^j(y^j|_{y_i^j=t}) s^j(y^j|_{y_i^j=t}) - r^j(y^j) s^j(y^j)$$

$$= r^j(y^j|_{y_i^j=t}) \left(s^j(y^j|_{y_i^j=t}) - s^j(y^j) \right) + s^j(y^j) \left(r^j(y^j|_{y_i^j=t}) - r^j(y^j) \right).$$

By the Lipschitz continuity of r, r^j in $\overline{\Sigma}_c, \overline{\Delta}^j$, $j = 1, ..., p$, respectively, we can evaluate the terms

$$||rs||^2_{0,\Sigma_c} \le \sup_{x \in \overline{\Sigma}_c} |r(x)|^2 \, ||s||^2_{0,\Sigma_c},$$

$$||\rho^{-1/2} rs||^2_{0,\Sigma_c} \le \sup_{x \in \overline{\Sigma}_c} |r(x)|^2 \, ||\rho^{-1/2} s||^2_{0,\Sigma_c},$$

$$\int_{-b}^{b} \int_{\Delta^j} |r^j(y^j|_{y_i^j=t})|^2 \frac{|s^j(y^j|_{y_i^j=t}) - s^j(y^j)|^2}{|t - y_i^j|^2} dy^j dt$$

$$\le \sup_{y^j \in \overline{\Delta}^j} |r^j(y^j)|^2 \int_{-b}^{b} \int_{\Delta^j} \frac{|s^j(y^j|_{y_i^j=t}) - s^j(y^j)|^2}{|t - y_i^j|^2} dy^j dt,$$

$$\int_{\Delta^j} |s^j(y^j)|^2 \int_{-b}^{b} \frac{|r^j(y^j|_{y_i^j=t}) - r^j(y^j)|^2}{|t - y_i^j|^2} dt dy^j$$

$$\le \sup_{y^j, \eta^j \in \overline{\Delta}^j} \left(\frac{|r^j(y^j) - r^j(\eta^j)|}{|y^j - \eta^j|} \right)^2 \cdot 2b \cdot ||s^j||^2_{0,\Delta^j}.$$

Hence, we obtain the estimate

$$||rs||_{1/2,00,\Sigma_c} \le c||r||_{C^{0,1}(\overline{\Sigma}_c)} ||s||_{1/2,00,\Sigma_c}$$

which proves the lemma.

Lemma 1.15. *For* $s \in H_{00}^{1/2}(\Sigma_c)^\star$, *if* $r \in C^{0,1}(\overline{\Sigma}_c)$, *then* $rs \in H_{00}^{1/2}(\Sigma_c)^\star$.

Indeed, by Lemma 1.14, we define rs from the formula

$$\langle rs, \phi\rangle_{1/2,\Sigma_c} = \langle s, r\phi\rangle_{1/2,\Sigma_c} \quad \forall \phi \in H_{00}^{1/2}(\Sigma_c)$$

which proves Lemma 1.15.

Let $\mathcal{F} \in C^{0,1}(\overline{\Sigma}_c)$ be a given friction coefficient, and $F \in H_{00}^{1/2}(\Sigma_c)^\star$ be a given friction force between the crack faces, $\mathcal{F} \geq 0$. Assume that $F \geq 0$ in the following sense:

$$\langle F, \phi\rangle_{1/2,\Sigma_c} \geq 0 \quad \forall \phi \in H_{00}^{1/2}(\Sigma_c),\ \phi \geq 0.$$

For a given external force $f = (f_1, f_2, f_3) \in L^2(\Omega_c)$, we introduce the potential energy functional

$$P(u) = \Pi(u) + \langle \mathcal{F}F, |[u_\tau]|\rangle_{1/2,\Sigma_c}, \quad \Pi(u) = \frac{1}{2}\int\limits_{\Omega_c} \sigma_{ij}(u)\varepsilon_{ij}(u) - \int\limits_{\Omega_c} f_i u_i.$$

The properties of $\Pi(u)$ were discussed in Section 1.4.4. The functional

$$I(u) = \langle \mathcal{F}F, |[u_\tau]|\rangle_{1/2,\Sigma_c}$$

is positive since $\mathcal{F}, F$ are positive; it is continuous by Theorem 1.25, and convex. Thus, P is a coercive, strictly convex, weakly lower semicontinuous functional on $H^{1,0}(\Omega_c)$.

We recall the admissible displacements set

$$K = \{u \in H^{1,0}(\Omega_c) \mid \quad [u_\nu] \geq 0 \quad \text{on } \Sigma_c\}.$$

By Theorem 1.4, the equilibrium problem

$$P(u) = \inf_{\bar{u}\in K} P(\bar{u}) \tag{1.160}$$

is equivalent to the variational inequality

$$u \in K, \quad \Pi'_u(v-u) + I(v) - I(u) \geq 0 \quad \forall v \in K,$$

which has the form

$$\int\limits_{\Omega_c} \sigma_{ij}(u)\varepsilon_{ij}(v-u) + \langle \mathcal{F}F, |[v_\tau]| - |[u_\tau]|\rangle_{1/2,\Sigma_c} \tag{1.161}$$

$$\geq \int\limits_{\Omega_c} f_i(v_i - u_i) \quad \forall v \in K.$$

By the properties of P, there exists a unique solution $u \in K$ to the problems (1.160), (1.161).

Theorem 1.31. *There exists a unique solution* $u \in K$ *to the problem* (1.161) *such that*

$$-\sigma_{ij,j}(u) = f_i, \quad i = 1,2,3, \quad \text{in } \Omega_c,$$

$$[u_\nu] \geq 0, \quad [\sigma_\nu(u)] = 0, \quad \sigma_\nu(u) \leq 0, \quad \sigma_\nu(u)[u_\nu] = 0 \quad \text{on } \Sigma_c,$$

$$[\sigma_\tau(u)] = 0, \quad |\sigma_\tau(u)| \leq \mathcal{F}F, \quad \sigma_{\tau i}(u)[u_{\tau i}] - \mathcal{F}F|[u_\tau]| = 0 \quad \text{on } \Sigma_c.$$

PROOF. Substituting $v = u \pm \phi$, $\phi \in [C_0^\infty(\Omega_c)]^3$ in (1.161) as a test function, one obtains

$$\int\limits_{\Omega_c} \sigma_{ij}(u)\varepsilon_{ij}(\phi) = \int\limits_{\Omega_c} f_i\phi_i \quad \forall \phi \in [C_0^\infty(\Omega_c)]^3. \tag{1.162}$$

This means that the following equations hold:

$$-\sigma_{ij,j}(u) = f_i, \quad i = 1,2,3, \quad \text{a.e. in } \Omega_c,$$

and $\sigma_{ij,j}(u) \in L^2(\Omega_c)$, $i = 1,2,3$. By (1.162) and the Green formula

$$\int\limits_{\Omega_c} \sigma_{ij}(u)\varepsilon_{ij}(v-u) = -\int\limits_{\Omega_c} \sigma_{ij,j}(u)(v_i - u_i) \tag{1.163}$$

$$-\left[\langle \sigma_\nu(u), v_\nu - u_\nu \rangle_{1/2,\Sigma}\right] - \left[\langle \sigma_{\tau i}(u), v_{\tau i} - u_{\tau i} \rangle_{1/2,\Sigma}\right],$$

from (1.161) one can deduce

$$\langle \mathcal{F}F, |[v_\tau]| - |[u_\tau]| \rangle_{1/2,\Sigma_c} - \left[\langle \sigma_\nu(u), v_\nu - u_\nu \rangle_{1/2,\Sigma}\right] \tag{1.164}$$

$$-\left[\langle \sigma_{\tau i}(u), v_{\tau i} - u_{\tau i} \rangle_{1/2,\Sigma}\right] \geq 0,$$

where $v \in K$. For $\phi \in [H_0^1(\Omega)]^3$ we have $[\phi] = 0$ at Σ and, therefore, we can substitute $v = u \pm \phi \in K$ in (1.164) as a test function. This gives

$$\langle \sigma_\nu^+(u) - \sigma_\nu^-(u), \phi_\nu \rangle_{1/2,\Sigma} + \langle \sigma_{\tau i}^+(u) - \sigma_{\tau i}^-(u), \phi_{\tau i} \rangle_{1/2,\Sigma} = 0 \quad \forall \phi \in [H_0^1(\Omega)]^3.$$

Hence $[\sigma_\nu(u)] = [\sigma_{\tau i}(u)] = 0$, $i = 1,2,3$. Using Lemma 1.13, we introduce $\sigma_\nu(u), \sigma_{\tau i}(u) \in H_{00}^{1/2}(\Sigma_c)^\star$ by the formulae

$$\langle \sigma_\nu(u), \psi \rangle_{1/2,\Sigma_c} = \langle \sigma_\nu^\pm(u), \bar\psi \rangle_{1/2,\Sigma},$$

$$\langle \sigma_{\tau i}(u), \psi \rangle_{1/2,\Sigma_c} = \langle \sigma_{\tau i}^\pm(u), \bar\psi \rangle_{1/2,\Sigma}, \quad i = 1,2,3,$$

$$\forall \psi \in H_{00}^{1/2}(\Sigma_c), \quad \bar\psi = \psi \text{ in } \Sigma_c, \quad \bar\psi = 0 \text{ in } \Sigma \setminus \Sigma_c, \quad \bar\psi \in H^{1/2}(\Sigma).$$

Utilizing these notations, by the independence between v_ν, v_τ, we rewrite (1.164) in the form

$$\langle \sigma_\nu(u), [v_\nu] \rangle_{1/2,\Sigma_c} \le \langle \sigma_\nu(u), [u_\nu] \rangle_{1/2,\Sigma_c} \quad \forall v \in K, \tag{1.165}$$

$$\langle \mathcal{F}F, |[v_\tau]| \rangle_{1/2,\Sigma_c} - \langle \sigma_{\tau i}(u), [v_{\tau i}] \rangle_{1/2,\Sigma_c} \tag{1.166}$$
$$\ge \langle \mathcal{F}F, |[u_\tau]| \rangle_{1/2,\Sigma_c} - \langle \sigma_{\tau i}(u), [u_{\tau i}] \rangle_{1/2,\Sigma_c} \quad \forall v \in K.$$

The arguments of Section 1.4.4 applied to the inequality (1.165) provide the conditions

$$\langle \sigma_\nu(u), [u_\nu] \rangle_{1/2,\Sigma_c} = 0, \quad \langle \sigma_\nu(u), \psi \rangle_{1/2,\Sigma_c} \le 0 \tag{1.167}$$
$$\forall \psi \in H^{1/2}_{00}(\Sigma_c), \quad \psi \ge 0$$

which imply the first line of boundary conditions formulated in Theorem 1.31.

Consider the inequality (1.166). We can replace v_τ by $\pm\lambda v_\tau$ in (1.166), where $\lambda \ge 0$ is a constant, which gives

$$\lambda \left(\langle \mathcal{F}F, |[v_\tau]| \rangle_{1/2,\Sigma_c} \mp \langle \sigma_{\tau i}(u), [v_{\tau i}] \rangle_{1/2,\Sigma_c} \right)$$
$$\ge \langle \mathcal{F}F, |[u_\tau]| \rangle_{1/2,\Sigma_c} - \langle \sigma_{\tau i}(u), [u_{\tau i}] \rangle_{1/2,\Sigma_c}.$$

By the arbitrariness of λ, this inequality means that

$$\langle \mathcal{F}F, |[u_\tau]| \rangle_{1/2,\Sigma_c} - \langle \sigma_{\tau i}(u), [u_{\tau i}] \rangle_{1/2,\Sigma_c} = 0, \tag{1.168}$$
$$\langle \mathcal{F}F, |[v_\tau]| \rangle_{1/2,\Sigma_c} \mp \langle \sigma_{\tau i}(u), [v_{\tau i}] \rangle_{1/2,\Sigma_c} \ge 0 \quad \forall v \in K.$$

The last relation implies

$$\langle \mathcal{F}F, |\psi| \rangle_{1/2,\Sigma_c} \mp \langle \sigma_{\tau i}(u), \psi_i \rangle_{1/2,\Sigma_c} \ge 0 \quad \forall \psi \in [H^{1/2}_{00}(\Sigma_c)]^3, \ \psi_i \nu_i = 0,$$

i.e.

$$\left| \langle \sigma_{\tau i}(u), \psi_i \rangle_{1/2,\Sigma_c} \right| \le \langle \mathcal{F}F, |\psi| \rangle_{1/2,\Sigma_c} \tag{1.169}$$
$$\forall \psi \in [H^{1/2}_{00}(\Sigma_c)]^3, \quad \psi_i \nu_i = 0.$$

Equations and inequalities (1.162), (1.155), (1.167)–(1.169) give the exact meaning of the relations formulated in Theorem 1.31. The theorem is proved.

Chapter 2

Cracks in plates and shells

In this chapter we analyse a wide class of equilibrium problems with cracks. It is well known that the classical approach to the crack problem is characterized by the equality type boundary conditions considered at the crack faces, in particular, the crack faces are considered to be stress-free (Cherepanov, 1979, 1983; Kachanov, 1974; Morozov, 1984). This means that displacements found as solutions of these boundary value problems do not satisfy nonpenetration conditions. There are practical examples showing that interpenetration of crack faces may occur in these cases. An essential feature of our consideration is that restrictions of Signorini type are considered at the crack faces which do not allow the opposite crack faces to penetrate each other. The restrictions can be written as inequalities for the displacement vector. As a result a complete set of boundary conditions at crack faces is written as a system of equations and inequalities. The presence of inequality type boundary conditions implies the boundary problems to be nonlinear, which requires the investigation of corresponding boundary value problems. In the chapter, plates and shells with cracks are considered. Properties of solutions are established: existence of solutions, regularity up to the crack faces, convergence of solutions as parameters of a system are varying and so on. We analyse different constitutive laws: elastic, viscoelastic.

We start with contact problems for plates. The contact problems with nonpenetration conditions can be viewed as a specific type of crack problem. On the other hand, the analysis of solution properties when the contact occurs is useful in the sequel.

2.1 Viscoelastic contact problem for a plate

The viscoelastic contact problem for a plate with the constitutive law (see Section 1.1.4)

$$M_{ij} = -b_{ijkl}w_{,kl} - \frac{\partial}{\partial t}\bar{b}_{ijkl}w_{,kl}, \quad i,j = 1,2, \tag{2.1}$$

is considered in this section. Here M_{ij}, w are bending moments and vertical displacements, respectively. As we know the equilibrium relations for a plate in contact with a punch have the following form (see Section 1.1.5):

$$w - \psi \geq 0, \quad -M_{ij,ij} - f \geq 0, \quad (w - \psi)(M_{ij,ij} + f) = 0, \tag{2.2}$$

where $z = \psi(x)$ is the equation of the punch shape (see Fig.2.1). Assuming that the plate is isotropic, the relations (2.1), (2.2) allow one to formulate the following boundary value problem.

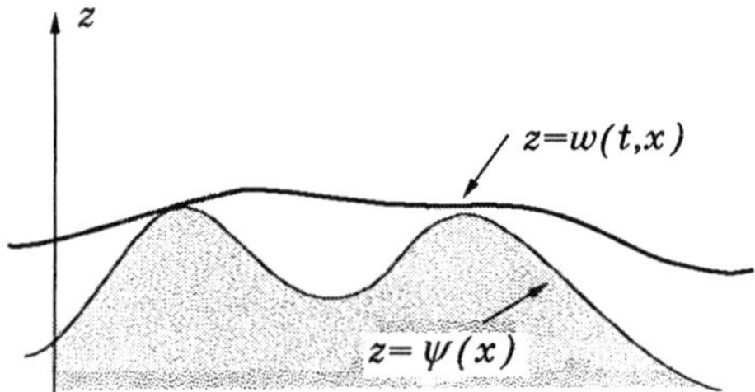

Fig.2.1. Contact of a plate with a punch

In the domain $Q = \Omega \times (0,T)$ it is required to find a function w satisfying the inequalities

$$w - \psi \geq 0, \tag{2.3}$$

$$(\Delta^2 w + \Delta^2 w_t - f)(\bar{w} - w) \geq 0 \quad \forall \bar{w} \geq \psi. \tag{2.4}$$

We consider the following initial and boundary conditions:

$$w = w_0 \quad \text{at} \quad t = 0, \tag{2.5}$$

$$w = \frac{\partial w}{\partial \nu} = 0 \quad \text{on} \quad \Gamma \times (0,T). \tag{2.6}$$

Here $\Omega \subset R^2$ is a bounded domain with boundary Γ, and ν is the unit exterior normal to Γ. The results of this section can be found in (Khludneva, 1990a).

2.1.1 Solution existence

Let $\psi \in H^2(\Omega)$. For simplicity the assumption $\psi < 0$ in Ω is used below. This assumption does not restrict the generality. We introduce the closed and convex set

$$K = \{w \in L^2(0,T; H_0^2(\Omega)) \mid \quad w(t,x) \geq \psi(x) \quad \text{a.e. in } Q\}.$$

The brackets $(\cdot,\cdot)$ denote the scalar product in $L^2(Q)$. The aim of further reasonings is a proof of the following statement.

Theorem 2.1. *Assume $f \in H^1(Q)$, $w_0 \in H_0^2(Q)$, $w_0(x) \geq \psi(x)$, $x \in \Omega$. Then, there exists a function w satisfying the initial condition* (2.5) *and the relations*

$$w_t \in L^\infty(0,T; H_0^2(\Omega)), \quad w \in K, \tag{2.7}$$

$$(\Delta^2 w + \Delta^2 w_t, \bar{w} - w) \geq (f, \bar{w} - w) \quad \forall \bar{w} \in K. \tag{2.8}$$

PROOF. We introduce the penalty operator $p(w) = -(w-\psi)^-$ and consider the auxiliary boundary value problem with the positive parameter $\varepsilon > 0$,

$$\Delta^2 w + \Delta^2 w_t + \varepsilon^{-1} p(w) = f, \tag{2.9}$$

$$w = w_0 \quad \text{as} \quad t = 0, \tag{2.10}$$

$$w = \frac{\partial w}{\partial \nu} = 0 \quad \text{on } \Gamma \times (0,T). \tag{2.11}$$

First of all, the solvability of (2.9)–(2.11) is stated. To obtain an a priori estimate we multiply (2.9) by w and integrate over $\Omega \times (0,T)$. This provides

$$\int_0^T \|\Delta w\|^2 \, d\tau + \frac{1}{2}\|\Delta w(t)\|^2 - \frac{1}{2}\|\Delta^2 w_0\|^2 \tag{2.12}$$

$$+ \varepsilon^{-1} \int_0^T \langle p(w), w \rangle \, d\tau = \int_0^T \langle f, w \rangle \, d\tau.$$

As usual, $\langle \cdot,\cdot \rangle$ denotes the scalar product in $L^2(\Omega)$. The norm in $H^s(\Omega)$ is denoted by $\|\cdot\|_s$, $\|\cdot\|_0 = \|\cdot\|$. Since $0 \in K$ the penalty term of the left-hand side of (2.12) is nonnegative. Hence, from (2.12) we obtain

$$\max_{0 \leq t \leq T} \|w(t)\|_2 \leq c. \tag{2.13}$$

The constant c depends only on T, f, w_0. When $t = 0$, from (2.9) we obtain the equation

$$\Delta^2 w_0 + \Delta^2 w_t(0) + \varepsilon^{-1} p(w_0) = f(0).$$

Note that, by the imbedding theorems, there exists a constant $c > 0$ such that

$$\|f(0)\| \le c\|f\|_{H^1(Q)}.$$

Taking into account the property $p(w_0) = 0$, we see that

$$\Delta^2 w_t(0) \in H^{-2}(\Omega)$$

and hence

$$\|w_t(0)\|_2 \le c.$$

One can differentiate the equation (2.9) with respect to t and multiply by w_t. By the inequality (see Lions, 1969)

$$\langle p(w)_t, w_t(t)\rangle \ge 0,$$

valid a.e. on $(0, T)$, we arrive at the estimate

$$\max_{0\le t\le T} \|w_t(t)\|_2 \le c. \tag{2.14}$$

We use the Galerkin approach to prove the existence of the solution to the boundary value problem (2.9)–(2.11). It is well known that the eigenvalue functions

$$\Delta^2 \phi_i = \lambda_i \phi_i, \quad \phi_i \in H_0^2(\Omega),$$

form the basis in the space $H_0^2(\Omega)$. The approximate solution of the problem (2.9)–(2.11) is sought in the form

$$w^n(t) = \sum_{i=1}^{n} c_{in}(t)\phi_i,$$

where the functions $c_{in}(t)$ are found from the following system of ordinary differential equations:

$$\langle \Delta w_t^n + \Delta w^n, \Delta \phi_j\rangle + \varepsilon^{-1}\langle p(w^n), \phi_j\rangle = \langle f, \phi_j\rangle, \quad j = 1, 2, ..., n. \tag{2.15}$$

The initial data for c_{in} are defined from the representation

$$w^n(0) = \sum_{i=1}^{n} b_{in}\phi_i,$$

where $\sum_{i=1}^{n} b_{in}\phi_i \to w_0$ strongly in $H_0^2(\Omega)$. A priori estimates (2.13), (2.14) for w^n can be reproduced in the usual way. To do this one has to multiply (2.15) by c_{jn} and to sum over j from 1 to n. Then, we differentiate these equations with respect to t, multiply by c'_{jn} and sum over j. As a result, the estimate

$$\max_{0\le t\le T} (\|w_t^n\|_2 + \|w^n(t)\|_2) \le c \tag{2.16}$$

follows. It is of importance that the constant c does not depend on ε, n. The choice of the special basis allows one to write (2.15) in the normal form. Moreover, the estimate (2.16) guarantees the solvability of (2.15) on the interval $(0,T)$. Hence, the existence of $w^n(t)$ is proved. Taking into account (2.16), one can choose a subsequence w^n with the previous notation such that as $n \to \infty$

$$w^n, w_t^n \to w, w_t \quad \star\text{- weakly in } L^\infty(0,T;H_0^2(\Omega)),$$

$$w^n \to w \quad \text{strongly in } L^2(Q).$$

The second line is the corollary of the imbedding $H^1(Q)$ in $L^2(Q)$ which is compact. Hence, we have

$$p(w^n) \to p(w) \quad \text{strongly in } L^2(Q).$$

Using the above convergence, from (2.15) we get

$$(\Delta w_t + \Delta w, \Delta\phi) + \varepsilon^{-1}(p(w),\phi) = (f,\phi) \quad \forall\phi \in L^2(0,T;H_0^2(\Omega)). \quad (2.17)$$

This means that equation (2.9) is fulfilled in the sense of distributions. Moreover, in view of (2.13), (2.14) one has

$$w^n(0) \to w(0) = w_0 \quad \text{weakly in } H_0^2(\Omega).$$

Consequently, the initial condition (2.10) holds.

Now we have to justify the passage to the limit as $\varepsilon \to 0$. As it was mentioned the estimates (2.13), (2.14) are uniform in ε. This means that the constructed solutions denoted by w^ε satisfy the estimate

$$||w^\varepsilon||_{L^\infty(0,T;H_0^2(\Omega))} + ||w_t^\varepsilon||_{L^\infty(0,T;H_0^2(\Omega))} \le c.$$

Choosing a subsequence, if necessary, we assume that as $\varepsilon \to 0$

$$w^\varepsilon, w_t^\varepsilon \to w, w_t \quad \star\text{- weakly in } L^\infty(0,T;H_0^2(\Omega)).$$

The functions w^ε satisfy the identity (2.17). Let $\bar{w} \in K$. We can substitute $\bar{w} - w^\varepsilon$ in (2.17) as a test function. Taking into account the monotonicity of p, we arrive at the inequality

$$(\Delta w_t^\varepsilon + \Delta w^\varepsilon, \Delta\bar{w} - \Delta w^\varepsilon) \ge (f, \bar{w} - w^\varepsilon)$$

which can be written in the form

$$(\Delta w_t^\varepsilon + \Delta w^\varepsilon, \Delta\bar{w}) \ge ||\Delta w^\varepsilon||^2_{L^2(Q)} \quad (2.18)$$

$$+\frac{1}{2}||\Delta w^\varepsilon(T)||^2 - \frac{1}{2}||\Delta w_0||^2 + (f, \bar{w} - w^\varepsilon).$$

In addition to the above convergence of w^ε, we suppose that as $\varepsilon \to 0$

$$w^\varepsilon(T) \to w(T) \quad \text{weakly in } H_0^2(\Omega).$$

Hence

$$\liminf \left(\|\Delta w^\varepsilon\|_{L^2(Q)}^2 + \|\Delta w^\varepsilon(T)\|^2 \right) \geq \|\Delta w\|_{L^2(Q)}^2 + \|\Delta w(T)\|^2.$$

This allows us to pass to the lower limit in (2.18) as $\varepsilon \to 0$, and we have

$$(\Delta w_t + \Delta w, \Delta \bar{w}) \geq \|\Delta w\|_{L^2(Q)}^2 + \frac{1}{2}\|\Delta w(T)\|^2$$

$$-\frac{1}{2}\|\Delta w_0\|^2 + (f, \bar{w} - w) \quad \forall \bar{w} \in K.$$

This inequality coincides with (2.8). To conclude the proof we have to state the inclusion $w \in K$. From (2.9) it follows that as $\varepsilon \to 0$

$$p(w^\varepsilon) \to 0 \quad \text{in } L^\infty(0, T; H^{-2}(\Omega)).$$

Meanwhile we know that $p(w^\varepsilon) \to p(w)$ strongly in $L^2(Q)$. Hence $p(w) = 0$. This means $w \in K$. Theorem 2.1 is proved.

It is easy to see that the function w in (2.7)–(2.8) is uniquely defined.

2.1.2 Optimal control of exterior forces

Let $F \subset H^1(Q)$ be a bounded, closed and convex set. As it was proved in the previous subsection, for every $f \in F$ there exists a solution $w \equiv w_f$ of the problem (2.7)–(2.8). We consider the cost functional

$$J(f) = \|w_f(T) - w_\star\|,$$

where $w_\star \in L^2(\Omega)$ is given. The optimal control problem to be considered here is formulated as follows:

$$\inf_{f \in F} J(f). \tag{2.19}$$

Theorem 2.2. *There exists a solution of the problem* (2.19).

PROOF. Let $f^n \in F$ be a minimizing sequence. It is bounded in $H^1(Q)$. Without loss of generality, we can assume that

$$f^n \to f \quad \text{weakly in } H^1(Q), \quad \text{strongly in } L^2(Q), \quad f \in F.$$

For every $f^n \in F$, the solution $w^n \equiv w_{f^n}$ can be found such that

$$w^n \in K, \quad w_t^n \in L^\infty(0, T; H_0^2(\Omega)),$$

$$(\Delta w^n + \Delta w_t^n, \Delta \bar{w} - \Delta w^n) \geq (f^n, \bar{w} - w^n) \quad \forall \bar{w} \in K,$$

$$w^n = w_0 \quad \text{at} \quad t = 0.$$

By the boundedness of f^n in $H^1(Q)$, the estimate

$$\|w^n\|_{L^\infty(0,T;H_0^2(\Omega))} + \|w_t^n\|_{L^\infty(0,T;H_0^2(\Omega))} \leq c$$

takes place being uniform in n. Assume that a subsequence with the same notation possesses the property

$$w^n, w_t^n \to w, w_t \quad \star\text{- weakly in } L^\infty(0,T;H_0^2(\Omega)),$$

$$w^n(T) \to w(T) \quad \text{weakly in } H_0^2(\Omega).$$

The inequality for w^n can be written in the form

$$(\Delta w^n + \Delta w_t^n, \Delta \bar{w}) \geq \|\Delta w^n\|_{L^2(Q)}^2 + \frac{1}{2}\|\Delta w^n(T)\|^2 - \frac{1}{2}\|\Delta w_0\|^2 + (f^n, \bar{w} - w^n).$$

The above convergence of w^n, f^n allows us to pass to the lower limit in the last inequality. The resulting relation can be written as follows:

$$(\Delta w + \Delta w_t, \Delta \bar{w} - \Delta w) \geq (f, \bar{w} - w) \quad \forall \bar{w} \in K.$$

Since $w \in K$ we therefore have $w = w_f$. Now it is easy to complete the proof. Indeed,

$$\liminf \|w^n(T) - w_\star\| \geq \|w(T) - w_\star\| = J(f).$$

Thus, the function f solves the optimal control problem (2.19). Theorem 2.2 is proved.

Different optimal control problems can be found in the monographs and papers (Khludnev, Sokolowski, 1997; Banichuk, 1980; Barbu, 1984; Céa, 1971; Lions, 1968a, 1968b; Litvinov, 1987; Mignot, 1976; Puel, 1987; Bock, Lovišek, 1987, Haslinger et al., 1986).

2.1.3 Optimal control in the regularized problem

As we know the vertical displacements of the plate defined from (2.7), (2.8) can be found as a limit of solutions to the problem (2.9)–(2.11). Two questions arise in this case. The first one is the following. Is it possible to solve an optimal control problem like (2.19) when $w \equiv w_f$ is defined from (2.9)–(2.11)? The second question concerns relationships between solutions of (2.19) and those of the regularized optimal control problem. Our goal in this subsection is to answer these questions.

First of all let us formulate the regularized optimal control problem. If the set F is introduced in a similar way and $w_f \equiv w$ is found from the equation

$$\Delta^2 w + \Delta^2 w_t + \varepsilon^{-1} p(w) = f, \tag{2.20}$$

$$w = \frac{\partial w}{\partial \nu} = 0 \quad \text{on } \Gamma \times (0, T). \tag{2.21}$$

$$w = w_0 \quad \text{at} \quad t = 0, \tag{2.22}$$

we arrive at the problem

$$\inf_{f \in F} J_\varepsilon(f), \tag{2.23}$$

where $J_\varepsilon(f) = \|w(T) - w_\star\|$. At this step ε is assumed to be fixed. Now we can formulate the following statement.

Theorem 2.3. *Let the above hypotheses be fulfilled. Then a solution of the problem* (2.23) *exists.*

PROOF. We briefly indicate the line of the proof in so far as it follows the arguments of Theorem 2.2. It can be assumed for a minimizing sequence f^n that as $n \to \infty$

$$f^n \to f \quad \text{weakly in } H^1(Q).$$

The solution w^n of (2.20)–(2.22) can be found for all $f^n \in F$. Moreover the estimate

$$\|w^n\|_{L^\infty(0,T;H_0^2(\Omega))} + \|w_t^n\|_{L^\infty(0,T;H_0^2(\Omega))} \le c$$

takes place being uniform in n. Choosing a subsequence, if necessary, one can assume that as $n \to \infty$

$$w^n, w_t^n \to w, w_t \quad \star\text{- weakly in } L^\infty(0, T; H_0^2(\Omega)),$$

$$w^n(T) \to w(T) \quad \text{weakly in } H_0^2(\Omega),$$

$$w^n \to w \quad \text{strongly in } L^2(Q).$$

This convergence allows one to pass to the limit in the identity for w^n, as $n \to \infty$, and to obtain

$$(\Delta w + \Delta w_t, \Delta\phi) + \varepsilon^{-1}(p(w), \phi) = (f, \phi) \quad \forall \phi \in L^2(0, T; H_0^2(\Omega)).$$

This means $w = w_f$ and consequently

$$\liminf J_\varepsilon(f^n) = \liminf \|w^n(T) - w_\star\| \ge J_\varepsilon(f)$$

which guarantees that f is the solution of the problem (2.23). The proof of Theorem 2.3 is completed.

Henceforth the solution of the problem (2.23) is denoted by f^ε. Accordingly, w^ε is the solution of (2.20)–(2.22) for $f = f^\varepsilon$. Denoting by f the solution of the optimal control problem (2.19) we put

$$j = J(f), \qquad j^\varepsilon = J_\varepsilon(f^\varepsilon).$$

Now we are in a position to study a connection between the solutions of (2.23) and (2.19).

Theorem 2.4. *From the sequence f^ε, w^ε, j^ε, one can choose a subsequence such that as $\varepsilon \to 0$ (the notation for the subsequence is the same)*

$$f^\varepsilon \to \tilde{f} \quad \textit{weakly in } H^1(Q), \quad j^\varepsilon \to j,$$

$$w^\varepsilon, w_t^\varepsilon \to w, w_t \quad \star\textit{- weakly in } L^\infty(0,T;H_0^2(\Omega)),$$

where $w \equiv w_{\tilde{f}}$ and $\tilde{f}$ is the solution of the problem (2.19).

PROOF. We consider the equations

$$\Delta^2 w^\varepsilon + \Delta^2 w_t^\varepsilon + \varepsilon^{-1} p(w^\varepsilon) = f^\varepsilon. \tag{2.24}$$

The boundedness of f^ε in $H^1(Q)$ provides the validity of the estimate

$$\|w^\varepsilon\|_{L^\infty(0,T;H_0^2(\Omega))} + \|w_t^\varepsilon\|_{L^\infty(0,T;H_0^2(\Omega))} \le c$$

uniformly in ε. Choosing a subsequence we assume that, as $\varepsilon \to 0$,

$$w^\varepsilon, w_t^\varepsilon \to w, w_t \quad \star\text{- weakly in } L^\infty(0,T;H_0^2(\Omega)),$$

$$w^\varepsilon \to w \quad \text{strongly in } L^2(Q),$$

$$w^\varepsilon(T) \to w(T) \quad \text{weakly in } H_0^2(\Omega).$$

Besides, one can suppose that

$$f^\varepsilon \to \tilde{f} \quad \text{weakly in } H^1(Q), \quad \tilde{f} \in F.$$

Let $\bar{w} \in K$ be an arbitrary function. The multiplication of (2.24) by $\bar{w} - w^\varepsilon$ implies

$$(\Delta w^\varepsilon + \Delta w_t^\varepsilon, \Delta\bar{w} - \Delta w^\varepsilon) \ge (f^\varepsilon, \bar{w} - w^\varepsilon).$$

This inequality can be rewritten in the form

$$(\Delta w^\varepsilon + \Delta w_t^\varepsilon, \Delta\bar{w}) \ge \|\Delta w^\varepsilon\|_{L^2(Q)}^2 + \frac{1}{2}\|\Delta w^\varepsilon(T)\|^2 - \frac{1}{2}\|\Delta w_0\|^2 + (f^\varepsilon, \bar{w} - w^\varepsilon).$$

After the passage to the lower limit in both sides of this relation we arrive at the inequality

$$(\Delta w + \Delta w_t, \Delta\bar{w} - \Delta w) \ge (\tilde{f}, \bar{w} - w) \quad \forall \bar{w} \in K.$$

Moreover it is easily seen that $w \in K$. Indeed, we have $p(w^\varepsilon) \to p(w)$ strongly in $L^2(Q)$. On the other hand, the equation (2.24) gives $p(w^\varepsilon) \to 0$ weakly in $L^\infty(0,T;H^{-2}(\Omega))$. Hence $p(w) = 0$, that is $w \subset K$. Thus we conclude $w = w_{\tilde{f}}$ and consequently

$$\liminf j^\varepsilon \ge J(\tilde{f}). \tag{2.25}$$

We know that for a fixed f the solutions w_f^ε of the equation (2.20) converge to the solution w_f of the inequality (2.7)–(2.8) as $\varepsilon \to 0$ so that, in particular,

$$w_f^\varepsilon(T) \to w_f(T) \quad \text{strongly in } L^2(\Omega).$$

This convergence means that

$$J_\varepsilon(f) \equiv \|w_f^\varepsilon(T) - w_\star\| \to \|w_f(T) - w_\star\|.$$

So we have proved the convergence

$$J_\varepsilon(f) \to J(f)$$

for every fixed f. Now let f be the solution of the problem (2.19). Then

$$j^\varepsilon = J_\varepsilon(f^\varepsilon) \leq J_\varepsilon(f).$$

By the above arguments,

$$\limsup j^\varepsilon \leq J(f). \tag{2.26}$$

The relations (2.25), (2.26) guarantee that $\tilde{f}$ is the solution of (2.19) and, besides, $j^\varepsilon \to j$. The equality $w = w_{\tilde{f}}$ has already been proved. The proof is completed.

2.1.4 Other cost functional

Let $w = w_f$ be the solution of the inequality

$$w \in K, \quad w_t \in L^\infty(0,T; H_0^2(\Omega)), \tag{2.27}$$

$$(\Delta w + \Delta w_t, \Delta \bar{w} - \Delta w) \geq (f, \bar{w} - w) \quad \forall \bar{w} \in K, \tag{2.28}$$

$$w = w_0 \quad \text{at} \quad t = 0, \tag{2.29}$$

and $w_\star^0 \in L^2(Q)$ be a given function. We consider the cost functional

$$J(f) = \|w_f - w_\star^0\|_{L^2(Q)}$$

and formulate the optimal control problem

$$\inf_{f \in F} J(f). \tag{2.30}$$

In addition to this we put

$$J_\varepsilon(f) = \|w_f^\varepsilon - w_\star^0\|_{L^2(Q)},$$

where w_f^ε is the solution of (2.20)–(2.22), and introduce one more optimal control problem

$$\inf_{f \in F} J_\varepsilon(f). \tag{2.31}$$

Now we can formulate the assertion.

Theorem 2.5. *The following statements are valid:*

1. *The solution of the problem* (2.27)–(2.30) *exists.*
2. *The solution of the problem* (2.31), (2.20)–(2.22) *also exists.*

The proof follows the lines of those of Theorems 2.2, 2.3 and therefore is omitted. We can also prove the statement on convergence like Theorem 2.4 concerning the optimal control problems (2.27)–(2.30) and (2.31), (2.20)–(2.22).

2.2 A plate under creep conditions

2.2.1 Existence of solutions

Let $\Omega \subset R^2$ be a bounded domain with smooth boundary Γ, $Q = \Omega \times (0,T)$. Our object is to study a contact problem for a plate under creep conditions (see Khludneva, 1990b). The formulation of the problem is as follows. In the domain Q, it is required to find functions w, M_{ij}, $i,j = 1,2$, satisfying the relations

$$w - \psi \geq 0, \quad -M_{ij,ij} - f \geq 0, \quad (w-\psi)(M_{ij,ij} + f) = 0, \tag{2.32}$$

$$-w_{,ij} = a_{ijkl}M_{kl} + \int_0^t b_{ijkl}M_{kl}(\tau)\,d\tau, \quad i,j = 1,2, \tag{2.33}$$

$$w = \frac{\partial w}{\partial \nu} = 0 \quad \text{on } \Gamma \times (0,T). \tag{2.34}$$

Here $f \in L^2(Q)$ is a given function, and the equation $z = \psi(x)$ describes a punch shape, $\psi \in H^2(\Omega)$, $\psi < 0$ on Γ. It is assumed that $a_{ijkl}, b_{ijkl} \in L^\infty(\Omega)$ depend only on x and possess the usual properties of symmetry and positive definiteness. As a matter of convenience we choose arbitrary fixed functions $w^0 \in H_0^2(\Omega)$ and $M_{ij}^0 \in L^2(Q)$ satisfying the conditions

$$w^0(x) \geq \psi(x), \quad x \in Q, \quad -M_{ij,ij}^0 = f \quad \text{in } Q.$$

The equation for M^0 holds in the distribution sense. We are in a position to formulate the existence result. The set of functions $w \in L^2(0,T; H_0^2(\Omega))$ satisfying the inequality $w \geq \psi$ a.e. in Q is denoted by K.

Theorem 2.6. *Under the above hypotheses there exist unique functions w, $M = \{M_{ij}\}$ such that:*

$$w \in K, \quad M \in L^2(Q), \tag{2.35}$$

$$a_{ijkl}M_{kl} + \int_0^t b_{ijkl}M_{kl}\,d\tau + w_{,ij} = 0, \quad i,j = 1,2, \tag{2.36}$$

$$-(M_{ij}, \bar{w}_{,ij} - w_{,ij}) \geq (f, \bar{w} - w) \quad \forall \bar{w} \in K. \tag{2.37}$$

PROOF. Let ε, δ be positive parameters, and p be the penalty operator introduced in the previous section. We consider the auxiliary problem

$$\varepsilon \Delta^2 w - M_{ij,ij} + \delta^{-1} p(w) = f, \tag{2.38}$$

$$a_{ijkl} M_{kl} + \int_0^t b_{ijkl} M_{kl}\, d\tau + w_{,ij} = 0, \quad i,j = 1,2, \tag{2.39}$$

$$w = \frac{\partial w}{\partial \nu} = 0 \quad \text{on } \Gamma \times (0,T). \tag{2.40}$$

First of all the solvability of the problem (2.38)–(2.40) is established for the fixed parameter ε, δ. In doing so the dependence of w, M_{ij} on these parameters is not indicated. To state an a priori estimate, we multiply (2.38), (2.39) by $w - w^0$, $M_{ij} - M_{ij}^0$, respectively, integrate over Q and sum the obtained relations. This provides the inequality

$$\varepsilon \int_0^T \|\Delta w(\tau)\|^2\, d\tau + \int_0^T \|M(\tau)\|^2\, d\tau + \left(\int_0^t b_{ijkl} M_{kl}\, d\tau, M_{ij} \right)$$

$$\leq \left(\int_0^t b_{ijkl} M_{kl}\, d\tau, M_{ij}^0 \right) + \varepsilon(\Delta w, \Delta w^0) + (a_{ijkl} M_{kl}, M_{ij}^0) \tag{2.41}$$

$$- (M_{ij}, w_{,ij}^0) - (f, w^0).$$

Here we have used the notation

$$\|M\|^2 = \int_\Omega a_{ijkl} M_{kl} M_{ij}\, dx$$

which is correct by the properties of a_{ijkl}. The nonnegative term containing the penalty operator has been neglected. The third term on the left-hand side of (2.41) is nonnegative since it is equal to

$$\frac{1}{2} \left\langle \int_0^T b_{ijkl} M_{kl}\, d\tau, \int_0^T M_{ij}\, d\tau \right\rangle.$$

Thus we see that the estimate

$$\varepsilon \int_0^T \|\Delta w(\tau)\|^2\, d\tau + \int_0^T \|M(\tau)\|^2\, d\tau \leq c \tag{2.42}$$

follows from (2.41) with the constant c independent of $\varepsilon \le \varepsilon_0$, δ. This means that the operator could be constructed acting from $L^2(0,T; H_0^2(\Omega)) \times L^2(Q)$ into the dual space, which puts to the pair (w, M_{ij}) the left-hand side of (2.38)–(2.39). This operator satisfies all conditions of Theorem 1.14. Its coercivity follows from the above reasonings used to obtain the estimate (2.42). The monotonicity of the operator can be verified by the following way. We take two points (w^1, M_{ij}^1), (w^2, M_{ij}^2) and calculate the value of the operator at these points. Denoting $w = w^1 - w^2$, $M = M^1 - M^2$ we obtain

$$\varepsilon \int_0^T \|\Delta w(\tau)\|^2 \, d\tau + \int_0^T \|M(\tau)\|^2 \, d\tau + \frac{1}{2} \left\langle \int_0^T b_{ijkl} M_{kl} \, d\tau, \int_0^T M_{ij} \, d\tau \right\rangle$$

$$+ \delta^{-1} \int_0^T \langle p(w^1) - p(w^2), w^1 - w^2 \rangle \, d\tau \ge 0$$

which proves the assertion. The boundedness and the semicontinuity of the operator are obvious. Thus the solution of the problem (2.38)–(2.40) exists such that

$$w \in L^2(0,T; H_0^2(\Omega)), \quad M \in L^2(Q).$$

In what follows the passage to the limit $\varepsilon \to 0$, $\delta \to 0$ is justified. First of all we note that, in addition to (2.42), the estimate

$$\int_0^T \|w(\tau)\|_2^2 \, d\tau \le c \tag{2.43}$$

pertains, as a consequence of (2.42) and (2.39). Denoting the solution of (2.38)–(2.40) by $w^\varepsilon, M^\varepsilon$, we can assume that for a subsequence, as $\varepsilon \to 0$

$$w^\varepsilon \to w^\delta \quad \text{weakly in } L^2(0,T; H_0^2(\Omega)),$$

$$M^\varepsilon \to M^\delta \quad \text{weakly in } L^2(Q).$$

After the passage to the limit the following identities are obtained:

$$-(M_{ij}^\delta, \bar{w}_{,ij}) + \delta^{-1}(p(w^\delta), \bar{w}) = (f, \bar{w}) \quad \forall \bar{w} \in L^2(0,T; H_0^2(\Omega)), \tag{2.44}$$

$$\left(a_{ijkl} M_{kl}^\delta + \int_0^t b_{ijkl} M_{kl}^\delta \, d\tau + w_{,ij}^\delta, \bar{M}_{ij} \right) = 0 \quad \forall \bar{M} \in L^2(Q). \tag{2.45}$$

The functions w^δ, M^δ satisfy the same a priori estimate as $w^\varepsilon, M^\varepsilon$. Hence, choosing a subsequence, if necessary, we can assume that as $\delta \to 0$

$$w^\delta \to w \quad \text{weakly in } L^2(0,T; H_0^2(\Omega)),$$

$$M^\delta \to M \quad \text{weakly in } L^2(Q).$$

In this case the identity (2.45) implies

$$\left(a_{ijkl}M_{kl} + \int_0^t b_{ijkl}M_{kl}\,d\tau + w_{,ij}, \bar{M}_{ij} \right) = 0 \quad \forall \bar{M} \in L^2(Q). \tag{2.46}$$

Let $\bar{w} \in K$. We substitute $\bar{w} - w^\delta$ in (2.44) as a test function and derive the inequality

$$-(M_{ij}^\delta, \bar{w}_{,ij} - w_{ij}^\delta) \geq (f, \bar{w} - w^\delta). \tag{2.47}$$

Meantime the identity (2.45) provides the equalities

$$w_{,ij}^\delta = -a_{ijkl}M_{kl}^\delta - \int_0^t b_{ijkl}M_{kl}^\delta\,d\tau, \quad i,j = 1,2.$$

Hence (2.47) can be written in the form

$$-(M_{ij}^\delta, \bar{w}_{,ij}) \geq \int_0^T \|M^\delta(\tau)\|^2\,d\tau + \frac{1}{2}\left\langle \int_0^T b_{ijkl}M_{kl}^\delta\,d\tau, \int_0^T M_{ij}^\delta\,d\tau \right\rangle + (f, \bar{w} - w^\delta).$$

Passing to the lower limit on both sides of the last relation we arrive at the inequality

$$-(M_{ij} + a_{ijkl}M_{kl} + \int_0^t b_{ijkl}M_{kl}\,d\tau, \bar{w}_{ij}) \geq (f, \bar{w} - w).$$

In view of (2.46) this means that

$$-(M_{ij}, \bar{w}_{,ij} - w_{ij}) \geq (f, \bar{w} - w) \quad \forall \bar{w} \in K. \tag{2.48}$$

The inclusion $w \in K$ follows from (2.44) in a standard way for variational inequalities. Thus the existence of w, M satisfying (2.35)–(2.37) is proved.

To verify the uniqueness we assume the existence of two solutions w^1, M^1 and w^2, M^2. From (2.46), (2.48) we have

$$\int_0^T \|M(\tau)\|^2\,d\tau + \frac{1}{2}\left\langle \int_0^T b_{ijkl}M_{kl}\,d\tau, \int_0^T M_{ij}\,d\tau \right\rangle \leq 0,$$

where $M = M^1 - M^2$. Hence $M \equiv 0$. Taking into account the equalities

$$-w_{,ij} = a_{ijkl}M_{kl} + \int_0^t b_{ijkl}M_{kl}(\tau)\,d\tau, \quad i,j = 1,2,$$

valid for $w = w^1 - w^2$, we get $w \equiv 0$. Theorem 2.6 is completely proved.

2.2.2 Optimal control of exterior forces

We proceed with an investigation of the contact problem for a plate under creep conditions. We know that for every fixed $f \in L^2(Q)$ there exists a unique solution w, M satisfying (2.35)–(2.37). Let $w^0 \in L^2(Q)$ be a given element and $F \subset H^1(Q)$ be a closed convex and bounded set. We introduce the cost functional

$$J(f) = \|w - w^0\|_{L^2(Q)},$$

where $w \equiv w_f$ is defined from (2.35)–(2.37), and consider the optimal control problem

$$\inf_{f \in F} J(f). \tag{2.49}$$

The main result of this subsection can be formulated as follows.

Theorem 2.7. *Under the above conditions, there exists a solution of the optimal control problem* (2.49).

PROOF. First of all we note that the solution of (2.35)–(2.37) satisfies the estimate

$$\int_0^T \|M(\tau)\|^2 \, d\tau + \int_0^T \|w(\tau)\|_2^2 \, d\tau \le c \tag{2.50}$$

with the constant c depending on $\|f\|_{L^2(Q)}$. Let $f^n \in F$ be a minimizing sequence and M^n, w^n be the solutions of (2.35)–(2.37) corresponding to f^n. By the boundedness of f^n in $H^1(Q)$, the constant c in the inequality like (2.50), written for M^n, w^n, is bounded uniformly in n. This means that the solutions of the problems

$$w^n \in K, \quad M^n \in L^2(Q),$$

$$-(M_{ij}^n, \bar{w}_{,ij} - w_{,ij}^n) \ge (f^n, \bar{w} - w^n) \quad \forall \bar{w} \in K, \tag{2.51}$$

$$-w_{,ij}^n = a_{ijkl} M_{kl}^n + \int_0^t b_{ijkl} M_{kl}^n \, d\tau, \quad i, j = 1, 2, \tag{2.52}$$

possess the properties: w^n are bounded in $L^2(0, T; H_0^2(\Omega))$; M^n are bounded in $L^2(Q)$. Hence, it can be assumed that for a subsequence having the same notation as $n \to \infty$

$$w^n \to w \quad \text{weakly in } L^2(0, T; H_0^2(\Omega)),$$

$$M^n \to M \quad \text{weakly in } L^2(Q),$$

$$f^n \to f \quad \text{weakly in } H^1(Q), \quad \text{strongly in } L^2(Q).$$

On the basis of this convergence, the passage to the limit as $n \to \infty$ can be justified in (2.51), (2.52). Indeed, in view of (2.52) the inequality (2.51) can be written as follows:

$$-(M_{ij}^n, \bar{w}_{,ij}) \ge (a_{ijkl} M_{kl}^n, M_{ij}^n) \tag{2.53}$$

$$+\left(\int_0^t b_{ijkl}M_{kl}^n\,d\tau, M_{ij}^n\right)+(f^n,\bar{w}-w^n).$$

The second term on the right-hand side of (2.53) is equal to

$$\frac{1}{2}\left\langle\int_0^T b_{ijkl}M_{kl}^n\,d\tau, \int_0^T M_{ij}^n\,d\tau\right\rangle.$$

Hence the passage to the lower limit in (2.53) implies

$$-(M_{ij},\bar{w}_{,ij})\geq(a_{ijkl}M_{kl},M_{ij})+\left(\int_0^t b_{ijkl}M_{kl}\,d\tau, M_{ij}\right)+(f,\bar{w}-w).$$

It is easily seen that the last relation can be written in the form

$$-(M_{ij},\bar{w}_{,ij}-w_{ij})\geq(f,\bar{w}-w)\quad\forall\bar{w}\in K$$

provided that we use the limit equations obtained from (2.52),

$$-w_{,ij}=a_{ijkl}M_{kl}+\int_0^t b_{ijkl}M_{kl}(\tau)\,d\tau,\quad i,j=1,2.$$

Consequently $w=w_f$. Now let

$$j=\inf_{f\in F}J(f).$$

We see that

$$\liminf\|w^n-w_0\|_{L^2(Q)}\geq J(f)$$

which means that f is the solution of the optimal control problem (2.49). Theorem 2.7 is proved.

2.2.3 Optimal control in the penalty problem

As we know, for every fixed f the solution of (2.35)–(2.37) can be approximated by the solution of the following problem:

$$-M_{ij,ij}+\delta^{-1}p(w)=f,\tag{2.54}$$

$$-w_{,ij}=a_{ijkl}M_{kl}+\int_0^t b_{ijkl}M_{kl}\,d\tau,\quad i,j=1,2,\tag{2.55}$$

$$w=\frac{\partial w}{\partial\nu}=0\quad\text{on }\Gamma\times(0,T),\tag{2.56}$$

as $\delta \to 0$. Let us define the cost functional

$$J_\delta(f) = \|w - w^0\|_{L^2(Q)},$$

where $w \equiv w_f$ is the solution of (2.54)–(2.56), and consider the optimal control problem

$$\inf_{f \in F} J_\delta(f). \tag{2.57}$$

The goal of this subsection is to establish a relationship between the solutions of (2.54)–(2.57) and (2.49), (2.35)–(2.37). At the first stage of our consideration we intend to prove the existence of a solution of (2.54)–(2.57). Let f^n be a minimizing sequence. In view of the boundedness of f^n in $H^1(Q)$, without loss of generality, we assume that as $n \to \infty$

$$f^n \to f \quad \text{weakly in } H^1(Q).$$

For every f^n, there exists a solution w^n, M^n of the problem (2.54)–(2.56). This solution satisfies the identities

$$-(M_{ij}^n, \bar{w}_{,ij}) + \delta^{-1}(p(w^n), \bar{w}) = (f^n, \bar{w}) \quad \forall \bar{w} \in L^2(0,T; H_0^2(\Omega)), \tag{2.58}$$

$$\left(a_{ijkl} M_{kl}^n + \int_0^t b_{ijkl} M_{kl}^n \, d\tau + w_{,ij}^n, \bar{M}_{ij} \right) = 0 \quad \forall \bar{M} \in L^2(Q). \tag{2.59}$$

Moreover we know that the estimate

$$\int_0^T \|M^n(\tau)\|^2 \, d\tau + \int_0^T \|w^n(\tau)\|_2^2 \, d\tau \le c$$

holds being uniform in n. Choosing a subsequence, if necessary, we assume that as $n \to \infty$

$$w^n \to w^\delta \quad \text{weakly in } L^2(0,T; H_0^2(\Omega)),$$

$$M^n \to M^\delta \quad \text{weakly in } L^2(Q).$$

The above convergence of f^n, w^n, M^n allows us to pass to the limit in (2.58), (2.59) and to get $w = w_{f^\delta}$. Hence

$$\inf_{\bar{f} \in F} J_\delta(\bar{f}) = \liminf_{n \to \infty} J_\delta(f^n) \ge J_\delta(f^\delta);$$

that is, f^δ solves the optimal control problem (2.54)–(2.57). We denote $j^\delta = J_\delta(f^\delta)$ and formulate the assertion characterizing a relationship between optimal control problems under consideration.

Theorem 2.8. *There exists a subsequence $w^\delta, M^\delta, f^\delta$ for which we use the previous notation such that as $\delta \to 0$*

$$w^\delta \to w \quad \textit{weakly in } L^2(0,T; H_0^2(\Omega)),$$

$$M^\delta \to M \quad \textit{weakly in} \;\; L^2(Q),$$

$$f^\delta \to f \quad \textit{weakly in} \;\; H^1(Q), \qquad \liminf j^\delta \geq j,$$

where w, M *correspond to* f *and can be defined from* (2.35)–(2.37).

PROOF. We have the equations

$$-M^\delta_{ij,ij} + \delta^{-1} p(w^\delta) = f^\delta, \tag{2.60}$$

$$-w^\delta_{,ij} = a_{ijkl} M^\delta_{kl} + \int_0^t b_{ijkl} M^\delta_{kl} \, d\tau, \quad i,j = 1,2. \tag{2.61}$$

Since f^δ are bounded in $H^1(Q)$, the estimate

$$\int_0^T \|M^\delta(\tau)\|^2 \, d\tau + \int_0^T \|w^\delta(\tau)\|_2^2 \, d\tau \leq c$$

is valid being uniform in δ. We can suppose that a subsequence with the previous notation possesses the following properties as $\delta \to 0$:

$$w^\delta \to w \quad \text{weakly in} \;\; L^2(0,T; H_0^2(\Omega)),$$

$$M^\delta \to M \quad \text{weakly in} \;\; L^2(Q),$$

$$f^\delta \to f \quad \text{weakly in} \;\; H^1(Q), \quad \text{strongly in} \;\; L^2(Q).$$

Equation (2.60) is fulfilled in the sense of the identity

$$-(M^\delta_{ij}, \bar{w}_{,ij}) + \delta^{-1}(p(w^\delta), \bar{w}) = (f^\delta, \bar{w}),$$

valid for all test functions $\bar{w} \in L^2(0,T; H_0^2(\Omega))$. Substituting $\bar{w} - w^\delta$ as a test function, where $\bar{w} \in K$, we obtain

$$-(M^\delta_{ij}, \bar{w}_{,ij}) \geq (a_{ijkl} M^\delta_{kl}, M^\delta_{ij}) \tag{2.62}$$

$$+\frac{1}{2}\left\langle \int_0^T b_{ijkl} M^\delta_{kl} \, d\tau, \int_0^T M^\delta_{ij} \, d\tau \right\rangle + (f^\delta, \bar{w} - w^\delta).$$

It is easily seen that the above convergence provides the passage to the lower limit in (2.62), i.e.

$$-(M_{ij}, \bar{w}_{,ij}) \geq (a_{ijkl} M_{kl}, M_{ij}) + \frac{1}{2}\left\langle \int_0^T b_{ijkl} M_{kl} \, d\tau, \int_0^T M_{ij} \, d\tau \right\rangle \tag{2.63}$$

$$+(f, \bar{w} - w) = \left(a_{ijkl} M_{kl} + \int_0^t b_{ijkl} M_{kl}(\tau) \, d\tau, M_{ij} \right) + (f, \bar{w} - w).$$

On the other hand, after the passage to the limit, from (2.61) one has

$$-w_{,ij} = a_{ijkl}M_{kl} + \int_0^t b_{ijkl}M_{kl}(\tau)\,d\tau, \quad i,j = 1,2. \tag{2.64}$$

Consequently, the substitution of these values w, M in (2.63) provides the fulfilment of the inequality

$$-(M_{ij}, \bar{w}_{,ij} - w_{ij}) \geq (f, \bar{w} - w) \quad \forall \bar{w} \in K. \tag{2.65}$$

By (2.64), (2.65), one concludes that $w = w_f$, $M = M_f$, since the inclusion $w \in K$ is verified by standard arguments. To complete the proof we notice

$$\liminf \|w^\delta - w_0\|_{L^2(Q)} \geq \|w - w_0\|_{L^2(Q)},$$

that is, $\liminf j^\delta \geq J(f) \geq j$. Theorem 2.8 is proved.

2.2.4 Other cost functional

In this subsection we consider the other cost functional which describes a deflection of the moments M_{ij} from given functions $M_{ij}^0 \in L^2(Q)$. Namely, let

$$J(f) = \|M_f - M^0\|_{L^2(Q)}$$

be the cost functional, where $M_f = M$ is the solution of the problem

$$w \in K, \quad M \in L^2(Q),$$

$$-(M_{ij}, \bar{w}_{,ij} - w_{,ij}) \geq (f, \bar{w} - w) \quad \forall \bar{w} \in K,$$

$$-w_{,ij} = a_{ijkl}M_{kl} + \int_0^t b_{ijkl}M_{kl}\,d\tau, \quad i,j = 1,2.$$

We shall state the existence of a solution of the problem

$$\inf_{f \in F} J(f). \tag{2.66}$$

Theorem 2.9. *There exists a solution of the problem* (2.66).

PROOF. Let f^n be a minimizing sequence. Without loss of generality we assume that

$$f^n \to f \quad \text{weakly in } H^1(Q), \quad \text{strongly in } L^2(Q).$$

As it was mentioned in the previous subsection the solution w^n, M^n of the problem

$$w^n \in K, \quad M^n \in L^2(Q),$$

$$-(M_{ij}^n, \bar{w}_{,ij} - w_{,ij}^n) \geq (f^n, \bar{w} - w^n) \quad \forall \bar{w} \in K, \tag{2.67}$$

$$-w_{,ij}^n = a_{ijkl} M_{kl}^n + \int_0^t b_{ijkl} M_{kl}^n \, d\tau, \quad i, j = 1, 2, \tag{2.68}$$

possesses the property

$$\|M^n\|_{L^2(Q)}^2 + \|w^n\|_{L^2(0,T;H_0^2(\Omega))}^2 \leq c$$

with a constant c independent of n. Hence one can assume that as $n \to \infty$

$$w^n \to w \quad \text{weakly in } L^2(0, T; H_0^2(\Omega)),$$

$$M^n \to M^\delta \quad \text{weakly in } L^2(Q).$$

The convergence of f^n, w^n, M^n allows us to pass to the limit in (2.67), (2.68). It can be done as in Theorem 2.7. As a consequence we obtain $w = w_f$, $M = M_f$, that is

$$-(M_{ij}, \bar{w}_{,ij} - w_{,ij}) \geq (f, \bar{w} - w) \quad \forall \bar{w} \in K,$$

$$-w_{,ij} = a_{ijkl} M_{kl} + \int_0^t b_{ijkl} M_{kl} \, d\tau, \quad i, j = 1, 2.$$

The inclusion $w \in K$ follows from the weak closeness of the set K. Hence

$$\liminf \|M^n - M^0\|_{L^2(Q)} \geq \|M - M^0\|_{L^2(Q)}.$$

This means

$$\inf_{\bar{f} \in F} J(\bar{f}) = \liminf J(f^n) \geq J(f)$$

which proves that the element f solves the problem (2.66). Theorem 2.9 is proved.

2.3 A plate with vertical and horizontal displacements

We continue the investigation of the contact problem for a plate under creep conditions. In this section the case of both normal and tangential displacements of the plate is considered.

2.3.1 Existence of the solution

First of all we formulate the equilibrium problem. Let $\Omega \subset R^2$ be a bounded domain with smooth boundary Γ, $Q = \Omega \times (0,T)$. Denote by w_1, w_2, w the horizontal and vertical displacements of the plate, $\chi = (w_1, w_2, w)$. It is assumed that the equation $z = \psi(x)$ describes a punch shape, $x \in \Omega$. The nonpenetration condition can be written in the form

$$w - W\nabla\psi \geq \psi \quad \text{in } \Omega, \tag{2.69}$$

where $W = (w_1, w_2)$. In the domain Q, we wish to find the functions w_1, w_2, w, σ_{ij}, M_{ij}, $i,j = 1,2$, satisfying the relation (2.69) and

$$(M_{ij,ij} + f)(\bar{w} - w) + (\sigma_{ij,j} + f_i)(\bar{w}_i - w_i) \leq 0 \tag{2.70}$$

$$\forall(\bar{w}_1, \bar{w}_2, \bar{w}), \quad \bar{w} - \bar{W}\nabla\psi \geq \psi,$$

$$-w_{,ij} = a_{ijkl}M_{kl} + \int_0^t a^0_{ijkl}M_{kl}\,d\tau, \quad i,j = 1,2, \tag{2.71}$$

$$\varepsilon_{ij}(W) = b_{ijkl}\sigma_{kl} + \int_0^t b^0_{ijkl}\sigma_{kl}\,d\tau, \quad i,j = 1,2, \tag{2.72}$$

as well as the boundary conditions

$$w_1 = w_2 = w = \frac{\partial w}{\partial \nu} = 0 \quad \text{on } \Gamma \times (0,T). \tag{2.73}$$

Here $\varepsilon_{ij}(W) = (w_{i,j} + w_{j,i})/2$; $f_1, f_2, f \in L^2(Q)$ are given functions; the coefficients a^0_{ijkl}, b^0_{ijkl}, as well as a_{ijkl}, b_{ijkl} depend on x and possess the usual properties of symmetry and positive definiteness. It is assumed that these coefficients belong to $L^\infty(\Omega)$. Let $H(\Omega) = H^1_0(\Omega) \times H^1_0(\Omega) \times H^2_0(\Omega)$. We denote by K the set of all functions $\chi = (w_1, w_2, w)$ from $L^2(0,T; H(\Omega))$ satisfying the inequality (2.69) and consider the penalty operator p connected with K and acting from the space $L^2(0,T; H(\Omega))$ into its dual.

For convenience we introduce the functions $M^0_{ij}, \sigma^0_{ij} \in L^2(Q)$ satisfying the equations

$$-M^0_{ij,ij} = f, \quad -\sigma^0_{ij,j} = f_i, \quad i = 1,2, \quad \text{in } Q.$$

In addition to this, a function $\chi^0 = (w^0_1, w^0_2, w^0) \in H(\Omega)$ is assumed to be chosen satisfying the inequality (2.69). It can be done, in particular, when $\psi \in H^2(\Omega)$, $\psi < 0$ on Γ. Now we are in a position to formulate the existence theorem.

Theorem 2.10. *Let the above assumptions be fulfilled. Then there exist functions* $\chi = (w_1, w_2, w)$, $M = \{M_{ij}\}$, $\sigma = \{\sigma_{ij}\}$ *such that the equations* (2.71), (2.72) *hold and*

$$\chi \in K, \quad M, \sigma \in L^2(Q),$$

$$-(M_{ij}, \bar{w}_{,ij} - w_{,ij}) + \big(\sigma_{ij}, \varepsilon_{ij}(\bar{W}) - \varepsilon_{ij}(W)\big) \geq (f, \bar{w} - w) + (f_i, \bar{w}_i - w_i)$$
$$\forall \bar{\chi} \in K.$$

PROOF. We put $p = (p_1, p_2, p_0)$. Let ε, δ be positive parameters. The following auxiliary boundary value problem is analysed at the first stage:

$$\varepsilon \Delta^2 w - M_{ij,ij} + \delta^{-1} p_0(\chi) = f, \tag{2.74}$$

$$-\varepsilon \Delta w_i - \sigma_{ij,j} + \delta^{-1} p_i(\chi) = f_i, \quad i = 1, 2, \tag{2.75}$$

$$a_{ijkl} M_{kl} + w_{,ij} + \int_0^t a^0_{ijkl} M_{kl}\, d\tau = 0, \quad i, j = 1, 2, \tag{2.76}$$

$$b_{ijkl} \sigma_{kl} - \varepsilon_{ij}(W) + \int_0^t b^0_{ijkl} \sigma_{kl}\, d\tau = 0, \quad i, j = 1, 2, \tag{2.77}$$

$$w_1 = w_2 = w = \frac{\partial w}{\partial \nu} = 0 \quad \text{on } \Gamma \times (0, T). \tag{2.78}$$

At the first step of our reasonings the existence of a solution to the problem (2.74)–(2.78) is established. To obtain an a priori estimate, we multiply (2.74)–(2.77) by $w - w^0$, $w_i - w_i^0$, $M_{ij} - M_{ij}^0$, $\sigma_{ij} - \sigma_{ij}^0$, respectively, integrate over Q and sum. In doing so we have to note that

$$\left(\int_0^t a^0_{ijkl} M_{kl}\, d\tau, M_{ij} \right) = \frac{1}{2} \left\langle \int_0^T a^0_{ijkl} M_{kl}\, d\tau, \int_0^T M_{ij}\, d\tau \right\rangle \geq 0.$$

Analogously,

$$\left(\int_0^t b^0_{ijkl} \sigma_{kl}\, d\tau, \sigma_{ij} \right) \geq 0.$$

Thus, after some simple calculations one has

$$\varepsilon \|\Delta w\|^2_{L^2(Q)} + \varepsilon \|\nabla W\|^2_{L^2(Q)} + \|M\|^2_{L^2(Q)} + \|\sigma\|^2_{L^2(Q)}$$

$$\leq \frac{\varepsilon}{2} \left(\|\Delta w\|^2_{L^2(Q)} + \|\nabla W\|^2_{L^2(Q)} \right) + \frac{1}{2} \left(\|M\|^2_{L^2(Q)} + \|\sigma\|^2_{L^2(Q)} \right) + c.$$

The constant c does not depend on w, W, M, σ and is uniform in ε, δ, $\varepsilon \leq \varepsilon_0$. Whence

$$\varepsilon \|\Delta w\|^2_{L^2(Q)} + \varepsilon \|\nabla W\|^2_{L^2(Q)} + \|M\|^2_{L^2(Q)} + \|\sigma\|^2_{L^2(Q)} \leq c. \tag{2.79}$$

It is easy to prove the solvability of the problem (2.74)–(2.78) for fixed parameters ε, δ on the basis of (2.79). To do this, we consider the operator acting from the space $L^2(0, T; H(\Omega)) \times L^2(Q)$ into the dual one, which puts a correspondence between every element (χ, M, σ) and the left-hand side of (2.74)–(2.77). The way we have used to get the estimate (2.79) actually provides the coercivity of the operator. It is easily seen that the operator is monotonous, bounded and semicontinuous. Hence Theorem 1.14 is applicable which provides the existence of the solution to the problem (2.74)–(2.78) such that

$$w \in L^2(0,T;H_0^2(\Omega)), \quad w_i \in L^2(0,T;H_0^1(\Omega)), \quad i = 1,2,$$

$$M_{ij} \in L^2(Q), \quad \sigma_{ij} \in L^2(Q), \quad i,j = 1,2.$$

To indicate the dependence of the solution on the parameter ε we write w^ε, w_i^ε, M^ε, σ^ε. To justify the passage to the limit as $\varepsilon \to 0$ we should get one more a priori estimate. First of all let us note that the solutions of (2.74)–(2.78) satisfy the following relations:

$$\varepsilon(\Delta w^\varepsilon, \Delta \bar{w}) - (M_{ij}^\varepsilon, \bar{w}_{,ij}) + \delta^{-1}(p_0(\chi^\varepsilon), \bar{w}) = (f, \bar{w}) \tag{2.80}$$

$$\forall \bar{w} \in L^2(0,T;H_0^2(\Omega)),$$

$$\varepsilon(\nabla w_i^\varepsilon, \nabla \bar{w}_i) + (\sigma_{ij,j}^\varepsilon, \varepsilon_{ij}(\bar{W})) + \delta^{-1}(p_i(\chi^\varepsilon, \bar{w}_i)) = (f_i, \bar{w}_i) \tag{2.81}$$

$$\forall \bar{W} \in L^2(0,T;H_0^1(\Omega)),$$

$$a_{ijkl}M_{kl}^\varepsilon + w_{,ij}^\varepsilon + \int_0^t a_{ijkl}^0 M_{kl}^\varepsilon \, d\tau = 0, \quad i,j = 1,2, \tag{2.82}$$

$$b_{ijkl}\sigma_{kl}^\varepsilon - \varepsilon_{ij}(W^\varepsilon) + \int_0^t b_{ijkl}^0 \sigma_{kl}^\varepsilon \, d\tau = 0, \quad i,j = 1,2. \tag{2.83}$$

The brackets $(\cdot,\cdot)$ mean here both the scalar product in $L^2(Q)$ and the duality pairing between $L^2(0,T;H_0^k(\Omega))$ and $L^2(0,T;H^{-k}(\Omega))$, $k = 1,2$. Let us recall the boundary conditions for w^ε, W^ε. By (2.82), (2.83), we see that $\varepsilon_{ij}(W^\varepsilon)$, w_{ij}^ε are bounded in $L^2(Q)$ uniformly in $\varepsilon \le \varepsilon_0$, δ. Hence

$$\|W^\varepsilon\|_{L^2(0,T;H_0^1(\Omega))}^2 + \|w^\varepsilon\|_{L^2(0,T;H_0^2(\Omega))}^2 \le c. \tag{2.84}$$

It follows from (2.79), (2.84) that a subsequence can be chosen such that as $\varepsilon \to 0$

$$w^\varepsilon \to w^\delta \quad \text{weakly in } L^2(0,T;H_0^2(\Omega)),$$

$$W^\varepsilon \to W^\delta \quad \text{weakly in } L^2(0,T;H_0^1(\Omega)),$$

$$M^\varepsilon, \sigma^\varepsilon \to M^\delta, \sigma^\delta \quad \text{weakly in } L^2(Q).$$

For convenience we keep the same notation for the subsequence. In doing so we can fulfil the passage to the limit as $\varepsilon \to 0$ in (2.80)–(2.83) and obtain

$$-(M_{ij}^\delta, \bar{w}_{,ij}) + \delta^{-1}(p_0(\chi^\delta), \bar{w}) = (f, \bar{w}) \quad \forall \bar{w} \in L^2(0,T;H_0^2(\Omega)), \tag{2.85}$$

$$(\sigma^\delta_{ij,j}, \varepsilon_{ij}(\bar{W})) + \delta^{-1}(p_i(\chi^\delta, \bar{w}_i)) = (f_i, \bar{w}_i) \quad \forall \bar{W} \in L^2(0,T; H^1_0(\Omega)), \quad (2.86)$$

$$a_{ijkl}M^\delta_{kl} + w^\delta_{,ij} + \int\limits_0^t a^0_{ijkl}M^\delta_{kl}\, d\tau = 0, \quad i,j = 1,2, \quad (2.87)$$

$$b_{ijkl}\sigma^\delta_{kl} - \varepsilon_{ij}(W^\delta) + \int\limits_0^t b^0_{ijkl}\sigma^\delta_{kl}\, d\tau = 0, \quad i,j = 1,2. \quad (2.88)$$

The weak convergence of $p(\chi^\varepsilon)$ to $p(\chi^\delta)$ can be justified by the standard way on the basis of the monotonicity. As it was mentioned the estimates (2.79), (2.84) are uniform in $\varepsilon \le \varepsilon_0$, δ. This provides the boundedness of the solutions w^δ, W^δ, M^δ, σ^δ in the same spaces. In particular, choosing a subsequence, if necessary, one can assume that as $\delta \to 0$

$$w^\delta \to w \quad \text{weakly in } L^2(0,T; H^2_0(\Omega)),$$

$$W^\delta \to W \quad \text{weakly in } L^2(0,T; H^1_0(\Omega)),$$

$$M^\delta, \sigma^\delta \to M, \sigma \quad \text{weakly in } L^2(Q).$$

The passage to the limit in (2.87), (2.88) is obvious. We shall give the explanations related to (2.85), (2.86). Let $\bar{\chi} \in K$ be an arbitrary fixed element. Substituting $\bar{w} - w^\delta$, $\bar{W} - W^\delta$ in (2.85), (2.86) as the test functions and taking into account the monotonicity of p, we arrive at the inequality

$$-(M^\delta_{ij}, \bar{w}_{,ij} - w^\delta_{,ij}) + (\sigma^\delta_{ij}, \varepsilon_{ij}(\bar{W}) - \varepsilon_{ij}(W^\delta)) \ge (f, \bar{w} - w^\delta) + (f_i, \bar{w}_i - w^\delta_i).$$

We can substitute in this relation the values w^δ_{ij}, $\varepsilon_{ij}(W^\delta)$ taken from (2.87), (2.88) and obtain

$$-(M^\delta_{ij}, \bar{w}_{,ij}) + (\sigma^\delta_{ij}, \varepsilon_{ij}(\bar{W})) \ge (a_{ijkl}M^\delta_{kl}, M^\delta_{ij})$$

$$+(b_{ijkl}\sigma^\delta_{kl}, \sigma^\delta_{ij}) + \frac{1}{2}\left\langle \int\limits_0^T a^0_{ijkl}M^\delta_{kl}\, d\tau, \int\limits_0^T M^\delta_{ij}\, d\tau \right\rangle \quad (2.89)$$

$$+\frac{1}{2}\left\langle \int\limits_0^T b^0_{ijkl}\sigma^\delta_{kl}\, d\tau, \int\limits_0^T \sigma^\delta_{ij}\, d\tau \right\rangle + (f, \bar{w} - w^\delta) + (f_i, \bar{w}_i - w^\delta_i).$$

The convergence, we have for w^δ, W^δ, M^δ, σ^δ, and the properties of the coefficients a_{ijkl}, a^0_{ijkl}, b_{ijkl}, b^0_{ijkl} provide an opportunity to pass to the lower limit in (2.89) as $\delta \to 0$. The limiting relation for the functions w, W, M, σ has the form exactly like (2.89). At the same time we know that the passage to the limit in (2.87), (2.88) implies (2.71), (2.72). Hence the values

$$a_{ijkl}M_{kl} + \int\limits_0^t a^0_{ijkl}M_{kl}\, d\tau, \quad b_{ijkl}\sigma_{kl} + \int\limits_0^t b^0_{ijkl}\sigma_{kl}\, d\tau$$

taken from (2.71), (2.72) can be substituted in the above limiting relation. As a result we arrive at the inequality

$$-(M_{ij}, \bar{w}_{,ij} - w_{,ij}) + (\sigma_{ij}, \varepsilon_{ij}(\bar{W}) - \varepsilon_{ij}(W)) \tag{2.90}$$
$$\geq (f, \bar{w} - w) + (f_i, \bar{w}_i - w_i) \quad \forall \bar{\chi} \in K.$$

The inclusion $\chi = (w_1, w_2, w) \in K$ is proved by the standard arguments. Theorem 2.10 is proved.

To conclude this subsection we notice that the solution χ, M, σ is unique.

2.3.2 Optimal control of exterior forces

We intend to prove an existence theorem for the optimal control problem of exterior forces. Let $F = (f_1, f_2, f) \in G$, where $G \subset H^1(Q)$ is a convex closed and bounded set. As it was proved for every $F \in G$ there exists a solution $\chi \in K$, $M, \sigma \in L^2(Q)$ satisfying (2.71), (2.72), (2.90). We take $\chi_0 \in L^2(Q)$ and consider the cost functional

$$J(F) = \|\chi - \chi_0\|_{L^2(Q)}.$$

Our goal is to solve the optimal control problem

$$\inf_{F \in G} J(F). \tag{2.91}$$

The main assertion to be proved is as follows.

Theorem 2.11. *There exists a solution of the problem* (2.91).

PROOF. We choose a minimizing sequence F^n. Due to its boundedness one can assume that as $n \to \infty$

$$F^n \to F \quad \text{weakly in } H^1(Q), \quad \text{strongly in } L^2(Q).$$

Of course, we have $F \in G$. For every F^n, the solution of the following problem can be found:

$$\chi^n \in K, \quad M^n, \sigma^n \in L^2(Q),$$

$$-(M^n_{ij}, \bar{w}_{,ij} - w^n_{,ij}) + (\sigma^n_{ij}, \varepsilon_{ij}(\bar{W}) - \varepsilon_{ij}(W^n)) \tag{2.92}$$
$$\geq (F^n, \bar{\chi} - \chi^n) \quad \forall \bar{\chi} \in K,$$

$$a_{ijkl} M^n_{kl} + w^n_{,ij} + \int_0^t a^0_{ijkl} M^n_{kl} \, d\tau = 0, \quad i, j = 1, 2, \tag{2.93}$$

$$b_{ijkl} \sigma^n_{kl} - \varepsilon_{ij}(W^n) + \int_0^t b^0_{ijkl} \sigma^n_{kl} \, d\tau = 0, \quad i, j = 1, 2. \tag{2.94}$$

Moreover, the arguments we have used to prove the existence of the solutions validate the estimate

$$\|w^n\|^2_{L^2(0,T;H_0^2(\Omega))} + \|W^n\|^2_{L^2(0,T;H_0^1(\Omega))} + \|M^n\|^2_{L^2(Q)} + \|\sigma^n\|^2_{L^2(Q)} \le c.$$

The constant c can be chosen uniform in n since it depends on $\|F^n\|_{L^2(Q)}$. This means that there exists a subsequence such that as $n \to \infty$

$$w^n \to w \quad \text{weakly in} \quad L^2(0,T;H_0^2(\Omega)),$$

$$W^n \to W \quad \text{weakly in} \quad L^2(0,T;H_0^1(\Omega)),$$

$$M^n, \sigma^n \to M, \sigma \quad \text{weakly in} \quad L^2(Q).$$

The convergence of $F^n, \chi^n, M^n, \sigma^n$ allows us to pass to the limit in (2.92)–(2.94). First of all we note that (2.93)–(2.94) imply

$$a_{ijkl}M_{kl} + w_{,ij} + \int_0^t a^0_{ijkl}M_{kl}\,d\tau = 0, \quad i,j=1,2, \tag{2.95}$$

$$b_{ijkl}\sigma_{kl} + \varepsilon_{ij}(W) + \int_0^t b^0_{ijkl}\sigma_{kl}\,d\tau = 0, \quad i,j=1,2. \tag{2.96}$$

To justify the passage to the limit in (2.92) we substitute $w^n_{,ij}$, $\varepsilon_{ij}(W^n)$ in the inequality in accordance with (2.93), (2.94). This leads to the following relation:

$$-(M^n_{ij}, \bar{w}_{,ij}) + (\sigma^n_{ij}, \varepsilon_{ij}(\bar{W})) \ge (a_{ijkl}M^n_{kl}, M^n_{ij}) + (b_{ijkl}\sigma^n_{kl}, \sigma^n_{ij})$$

$$+\frac{1}{2}\left\langle \int_0^T a^0_{ijkl}M^n_{kl}\,d\tau, \int_0^T M^n_{ij}\,d\tau \right\rangle + \frac{1}{2}\left\langle \int_0^T b^0_{ijkl}\sigma^n_{kl}\,d\tau, \int_0^T \sigma^n_{ij}\,d\tau \right\rangle$$

$$+(F^n, \bar{\chi} - \chi^n).$$

We see that after the passage to the lower limit on both sides of this relation the following inequality holds:

$$-(M_{ij}, \bar{w}_{,ij}) + (\sigma_{ij}, \varepsilon_{ij}(\bar{W})) \ge \left(a_{ijkl}M_{kl} + \int_0^T a^0_{ijkl}M_{kl}\,d\tau, M_{ij}\right)$$

$$+\left(b_{ijkl}\sigma_{kl} + \int_0^T b^0_{ijkl}\sigma_{kl}\,d\tau, \sigma_{ij}\right) + (F, \bar{\chi} - \chi^n).$$

Taking into account (2.95), (2.96) it can be written in the following form:

$$-(M_{ij}, \bar{w}_{,ij} - w_{,ij}) + (\sigma_{ij}, \varepsilon_{ij}(\bar{W}) - \varepsilon_{ij}(W)) \ge (F, \bar{\chi} - \chi) \quad \forall \bar{\chi} \in K. \tag{2.97}$$

Therefore, we have established that $\chi = \chi_F$. This allows us to complete the proof. Indeed,

$$\liminf \|\chi^n - \chi_0\|_{L^2(Q)} \geq \|\chi - \chi_0\|_{L^2(Q)},$$

and hence

$$\inf_{\bar{F}\in G} J(\bar{F}) = \liminf J(F^n) \geq J(F),$$

which proves that F is the solution of the optimal control problem (2.91). Theorem 2.11 is proved.

2.4 Contact problem for a plate having a crack

A contact problem for a plate having a vertical crack is considered. The solution satisfies two restrictions of the inequality type. The first restriction is imposed in the domain and represents the mutual nonpenetration condition in the plate–punch system; the second one is put on the crack faces and corresponds to the nonpenetration of these faces. The corresponding variational inequality describing the equilibrium of the plate has its fourth order along the normal to the plate and its second order in the horizontal direction. The regularity of the solution is analysed. Boundary conditions having a natural physical interpretation are found on the crack faces. The existence of extreme crack shapes is also investigated. Specifically, the cost functional is defined on the feasible set of functions describing the crack shapes. The functional characterizes the deviation of the displacement vector from a given function. The problem consists in maximizing this functional. The existence of solutions of the formulated problem is proved. This section follows (Khludnev, 1995a).

The results on contact problems for plates without cracks can be found in (Caffarelli, Friedman, 1979; Caffarelli et al., 1982). Properties of solutions to elliptic problems with thin obstacles were analysed in (Frehse, 1975; Schild, 1984; Nečas, 1975; Kovtunenko, 1994a). Problems with boundary conditions of equality type at the crack faces are investigated in (Friedman, Lin, 1996).

2.4.1 Problem formulation

The model of the plate considered in this section actually corresponds to a shallow shell having zeroth curvatures. The gradient of the punch surface is assumed to be rather small, so that the nonpenetration condition imposed in the domain is the same as in the usual case for a plate. Meanwhile, the restriction imposed on the crack faces contains three components of the displacement vector.

Let $\Omega \subset R^2$ be a bounded domain with a smooth boundary Γ, and the equation $y = \psi(x)$, $x \in [0,1]$, describe a crack shape on the plane x, y. The graph of the function $y = \psi(x)$ is denoted by Γ_ψ, $\psi \in H_0^3(0,1)$, $\Omega_\psi = \Omega \backslash \Gamma_\psi$. Denote next by $\chi = (W, w)$ a displacement vector of the mid-surface points of the plate, where $W = (w^1, w^2)$ is horizontal displacements and w is a vertical one. Let $\varepsilon_{ij} = \varepsilon_{ij}(W)$ be the strain tensor of the mid-surface points, and $\sigma_{ij} = \sigma_{ij}(W)$ be the integrated stresses,

$$\varepsilon_{ij} = \frac{1}{2}\left(\frac{\partial w^i}{\partial x_j} + \frac{\partial w^j}{\partial x_i}\right), \quad i,j = 1,2, \quad x_1 = x,\ x_2 = y,$$

$$\sigma_{11} = \varepsilon_{11} + \kappa\varepsilon_{22}, \quad \sigma_{22} = \varepsilon_{22} + \kappa\varepsilon_{11}, \quad \sigma_{12} = (1-\kappa)\varepsilon_{12},$$
$$0 < \kappa < 1/2, \quad \kappa = \text{const}.$$

Introduce the energy functional of the plate,

$$\Pi_\psi(\chi) = \frac{1}{2}B_\psi(w,w) + \frac{1}{2}\langle \sigma_{ij}(W), \varepsilon_{ij}(W)\rangle_\psi - \langle f, \chi\rangle_\psi .$$

Herein

$$f = (f_1, f_2, f_3) \in L^2(\Omega), \quad \langle p, q\rangle_\psi = \int\limits_{\Omega_\psi} pq\, d\Omega_\psi$$

and the bilinear form $B_\psi(\cdot,\cdot)$ is as follows:

$$B_\psi(w,\bar w) = \int\limits_{\Omega_\psi} (w_{xx}\bar w_{xx} + w_{yy}\bar w_{yy} + \kappa w_{xx}\bar w_{yy} + \kappa w_{yy}\bar w_{xx} + 2(1-\kappa)w_{xy}\bar w_{xy}).$$

Assume that the equation $z = \Phi(x,y)$ describes a punch shape, $(x,y) \in \Omega$, $\Phi \in C^1(\overline{\Omega})$. A nonpenetration condition for the plate–punch system can be written as

$$w \geq \Phi \quad \text{in} \quad \Omega_\psi \tag{2.98}$$

provided $\nabla\Phi$ is small enough (see Section 1.1.5). Denote next by

$$\nu = (-\psi_x, 1)/\sqrt{1+\psi_x^2}$$

the normal vector to the curve $y = \psi(x)$ and by $2h$ the thickness of the plate, $\nu = (\nu_1, \nu_2)$. Taking into account the linear dependence of the horizontal displacements $W(z) = (w^1(z), w^2(z))$ on the distance z from the mid-plane (see Vol'mir, 1972),

$$w^i(z) = w^i - z w_{x_i}, \quad i = 1,2, \quad |z| \leq h,$$

the nonpenetration condition of crack faces takes the form (see Section 1.1.7)

$$[W - z\nabla w]\nu \geq 0 \quad \text{on } \Gamma_\psi, \quad |z| \leq h, \tag{2.99}$$

where $[U] = U^+ - U^-$ is the jump of U on Γ_ψ and $U^\pm$ correspond to the positive and negative directions of ν. For simplicity we put $h = 1$. The following boundary conditions are assumed to be given at the external boundary:

$$w = \frac{\partial w}{\partial n} = W = 0 \quad \text{on} \quad \Gamma.$$

Here n is the unit external vector to Γ. Let the subspace $H^{1,0}(\Omega_\psi) \subset H^1(\Omega_\psi)$ consist of the functions equal to zero on Γ, and the subspace $H^{2,0}(\Omega_\psi) \subset H^2(\Omega_\psi)$ consist of functions equal to zero on Γ with the first derivatives, $H(\Omega_\psi) = H^{1,0}(\Omega_\psi) \times H^{1,0}(\Omega_\psi) \times H^{2,0}(\Omega_\psi)$. Consider the convex and closed set

$$K_\psi(\Omega_\psi) = \{(W, w) \in H(\Omega_\psi) \mid \quad (W, w) \quad \text{satisfy} \quad (2.98), (2.99)\}$$

assuming that the boundary value Φ provides the nonemptiness of $K_\psi(\Omega_\psi)$. The equilibrium problem for the plate contacting with the punch $z = \Phi(x, y)$ and having the crack shape $y = \psi(x)$ can be formulated as variational,

$$\inf_{\chi \in K_\psi(\Omega_\psi)} \Pi_\psi(\chi).$$

In view of the convexity and the differentiability of Π_ψ this problem is equivalent to the next one: find the function $\chi = (W, w) \in K_\psi(\Omega_\psi)$ satisfying the inequality

$$B_\psi(w, \bar{w} - w) + \langle \sigma_{ij}(W), \varepsilon_{ij}(\bar{W} - W)\rangle_\psi \geq \langle f, \bar{\chi} - \chi\rangle_\psi \tag{2.100}$$

$$\forall \bar{\chi} \in K_\psi(\Omega_\psi).$$

Note that the following inequalities hold:

$$B_\psi(w, w) \geq c\|w\|^2_{2,\Omega_\psi} \quad \forall w \in H^{2,0}(\Omega_\psi), \tag{2.101}$$

$$\langle \sigma_{ij}(W), \varepsilon_{ij}(W)\rangle_\psi \geq c\|W\|^2_{1,\Omega_\psi} \quad \forall W \in H^{1,0}(\Omega_\psi) \tag{2.102}$$

with the constants independent of w and W, respectively. The inequalities (2.101), (2.102) provide the coercivity and the weak lower semicontinuity of the functional Π_ψ on $H(\Omega_\psi)$, and hence the problem (2.100) has a unique solution (see Theorems 1.11, 1.12).

2.4.2 Boundary conditions at the crack faces

Let us elucidate the boundary conditions on Γ_ψ for the solution (W, w) of (2.100) assuming that $w > \Phi$ in some neighbourhood $\mathcal{W}$ of the graph Γ_ψ. To this end, we first note that the equation

$$\Delta^2 w = f_3 \tag{2.103}$$

holds in $\mathcal{W} \setminus \Gamma_\psi$ in the sense of distributions (Schwartz, 1967; Vladimirov, 1981). Indeed, to verify this equation, the test elements of the form $(W, w)+(0, \varepsilon\varphi)$ are substituted in (2.100), where φ is a smooth function having a compact support in $\mathcal{W} \setminus \Gamma_\psi$ and ε is a small parameter. Moreover, the following equations hold in Ω_ψ:

$$-\frac{\partial \sigma_{ij}}{\partial x_j} = f_i, \quad i = 1, 2, \tag{2.104}$$

in the sense of distributions. We next denote $F = (f_1, f_2)$ and assume that the solution (W, w) is quite regular. This assumption means that the arguments given below are formal. The restriction (2.99) can be written in the equivalent form as

$$[W]\nu \geq \left|\left[\frac{\partial w}{\partial \nu}\right]\right| \quad \text{on } \Gamma_\psi. \tag{2.105}$$

Let us put functions of the form $(\bar{W}, w)$ as test ones in (2.100), where w is the third component of the solution (W, w). This yields

$$\langle \sigma_{ij}(W), \varepsilon_{ij}(\bar{W} - W)\rangle_\psi \geq \langle F, \bar{W} - W\rangle_\psi. \tag{2.106}$$

In so doing, the test functions $\bar{W}$ should satisfy the inequality

$$[\bar{W}]\nu \geq \left|\left[\frac{\partial w}{\partial \nu}\right]\right| \quad \text{on } \Gamma_\psi, \quad \bar{W} \in H^{1,0}(\Omega_\psi).$$

One can represent the vector $\{\sigma_{ij}\nu_j\}$ on Γ_ψ^- as a sum of the normal and tangential components

$$\{\sigma_{ij}\nu_j\} = \sigma_\nu \nu + \sigma_s s, \quad s = (-\nu_2, \nu_1).$$

A similar formula can be written on Γ_ψ^+. Choosing the functions $\tilde{W}$ having the property $[\tilde{W}]\nu \geq 0$ on Γ_ψ, the test elements $\bar{W} = W + \tilde{W}$ can be substituted in (2.106). Since the boundary $\partial\Omega_\psi$ of domain Ω_ψ is a combination of the sets Γ, Γ_ψ^+, Γ_ψ^-, the integration by parts is easily carried out. This implies

$$\sigma_s = 0, \quad \sigma_\nu \leq 0, \quad \text{on } \Gamma_\psi. \tag{2.107}$$

On the other hand, let the functions $\bar{\chi} = (W, \bar{w})$ be chosen as the test ones in (2.100). This leads to the relation

$$B_\psi(w, \bar{w} - w) \geq \langle f_3, \bar{w} - w\rangle_\psi \tag{2.108}$$

satisfied for test functions $\bar{w}$ such that

$$[W]\nu \geq \left|\left[\frac{\partial \bar{w}}{\partial \nu}\right]\right| \quad \text{on } \Gamma_\psi, \quad \bar{w} \in H^{2,0}(\Omega_\psi). \tag{2.109}$$

Consider the following boundary operators on Γ_ψ:

$$m(u) = \kappa \Delta u + (1-\kappa)\frac{\partial^2 u}{\partial \nu^2},$$

$$t(u) = \frac{\partial}{\partial \nu}\Delta u + (1-\kappa)\frac{\partial^3 u}{\partial \nu \partial s^2}, \quad s = (-\nu_2, \nu_1).$$

Making use of the Green formula

$$B_\psi(u,v) = \langle m(u), \frac{\partial v}{\partial \nu}\rangle_{\Gamma_\psi} - \langle t(u), v\rangle_{\Gamma_\psi} + \langle \Delta^2 u, v\rangle_\psi$$

the relations (2.108), (2.109) imply

$$t(w) = 0, \quad m(w)\left[\frac{\partial w}{\partial \nu}\right] + \sigma_\nu [W]\nu = 0 \quad \text{on} \quad \Gamma_\psi. \tag{2.110}$$

In particular, the strict inequality in (2.105) provides $m(w) = 0$. We have to note at this point that the boundary conditions (2.107), (2.110) hold on $\Gamma_\psi^\pm$ and

$$[\sigma_\nu] = 0, \quad [m(w)] = 0.$$

Besides, (2.107) holds good irrespective of the inequality $w > \Phi$ in $\mathcal{W}$, i.e. this condition takes place in the general case $w \geq \Phi$. At the same time, to derive (2.110), we make use of the equation (2.103) in $\mathcal{W} \setminus \Gamma_\psi$, which takes place provided that $w > \Phi$ in $\mathcal{W}$. Moreover, the inequality $w > \Phi$ in $\mathcal{W}$ provides one more relation

$$|m(w)| \leq -\sigma_\nu.$$

As a result, we obtained a complete system of boundary conditions on Γ_ψ, provided that $w > \Phi$ in $\mathcal{W}$:

$$[W]\nu \geq \left|\left[\frac{\partial w}{\partial \nu}\right]\right|, \quad \sigma_s = 0, \quad \sigma_\nu \leq 0,$$

$$t(w) = 0, \quad m(w)\left[\frac{\partial w}{\partial \nu}\right] + \sigma_\nu [W]\nu = 0,$$

$$[\sigma_\nu] = 0, \quad [m(w)] = 0, \quad |m(w)| \leq -\sigma_\nu.$$

The completeness of this system of boundary conditions and its detailed derivation and discussion will be presented later on, in Sections 3.1, 3.3, 3.4, where more complicated constitutive laws are considered.

2.4.3 Solution regularity

Let $x^0 \in \Gamma_\psi \setminus \partial\Gamma_\psi$ be any fixed point. Assume that $\Phi = c$ in some neighbourhood $\mathcal{O}(x^0)$ of the point x^0, $c = \text{const}$, and $\Gamma_\psi \cap \mathcal{O}(x^0)$ be a segment parallel to the x-axis (see Fig.2.2).

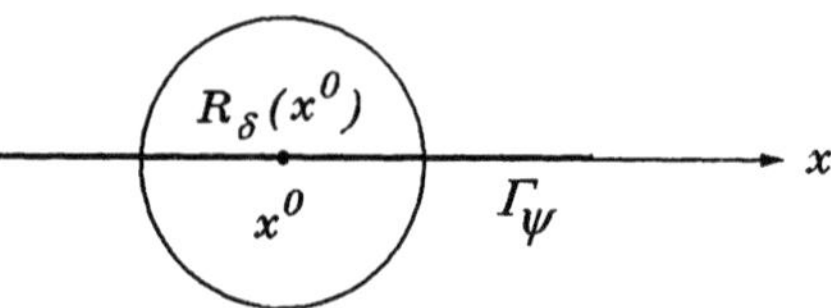

Fig.2.2. Smoothness of the solution

Denote next by $R_\delta(x^0)$ the ball of the radius δ centred at the point x^0. The following assertion holds.

Theorem 2.12. *Let the above hypotheses be fulfilled. Then the inclusions*

$$W \in H^2(R_\delta(x^0) \cap \Omega_\psi), \quad \frac{\partial w}{\partial x} \in H^2(R_\delta(x^0) \cap \Omega_\psi)$$

take place for δ small enough.

PROOF. Choose a smooth function φ such that $\varphi \equiv 1$ in $R_\delta(x^0)$, $\varphi \equiv 0$ outside $R_{3\delta/2}(x^0)$, $0 \le \varphi \le 1$ everywhere, $\partial\varphi/\partial y = 0$ on $\Gamma_\psi \cap \mathcal{O}(x_0)$. The inclusion $R_{2\delta}(x^0) \subset \mathcal{O}(x^0)$ is assumed to be valid. Introduce the notations

$$d_{\pm\tau} p(\bar{x}) = \tau^{-1}(p(\bar{x} \pm \tau e) - p(\bar{x})), \quad \Delta_\tau = -d_{-\tau} d_\tau,$$

where e is a unit vector of the axis x, $0 < |\tau| < \delta/2$. In this case the functions

$$w_\tau^i = w^i + \frac{\tau^2}{2}\varphi^2 \Delta_\tau w^i, \quad i = 1, 2, \quad w_\tau = w + \frac{\tau^2}{2}\varphi^2 \Delta_\tau w$$

can be considered in Ω_ψ. By virtue of the assumptions, the normal ν has the coordinates $(0, 1)$ near x^0, hence the nonpenetration condition (2.105) on $\Gamma_\psi \cap \mathcal{O}(x^0)$ is of the form

$$[w^2] \ge \left| \left[\frac{\partial w}{\partial y} \right] \right|. \tag{2.111}$$

Let us notice the following. Assuming that a function $p \ge 0$ on $\Gamma_\psi \cap \mathcal{O}(x^0)$, it is easy to check that with the above function φ the relation

$$p + \frac{\tau^2}{2}\varphi^2 \Delta_\tau p \ge 0 \quad \text{on} \quad \Gamma_\psi \cap \mathcal{O}(x^0)$$

holds. In fact, one has for $\bar{x} \in \Gamma_\psi \cap \mathcal{O}(x^0)$

$$p(\bar{x}) + \frac{\tau^2}{2}\varphi^2(\bar{x})\Delta_\tau p(\bar{x})$$

$$= (1 - \varphi^2(\bar{x}))p(\bar{x}) + \frac{\varphi^2(\bar{x})}{2}[p(\bar{x} - \tau e) + p(\bar{x} + \tau e)] \geq 0.$$

Bearing in mind this fact the vector $\chi_\tau = (w_\tau^1, w_\tau^2, w_\tau)$ is easily proved to satisfy the restriction (2.111); that is,

$$[w_\tau^2] \geq \left| \left[\frac{\partial w_\tau}{\partial y} \right] \right| \quad \text{on } \Gamma_\psi \cap \mathcal{O}(x^0).$$

Consequently

$$[W_\tau]\nu \geq \left| \left[\frac{\partial w_\tau}{\partial \nu} \right] \right| \quad \text{on } \Gamma_\psi.$$

Moreover, $w_\tau \geq \Phi$ in Ω_ψ, since $\Phi = c$ in $\mathcal{O}(x^0)$. To state this, we first notice that $w_\tau = w$ outside $R_{2\delta}(x^0)$, so that $w_\tau \geq \Phi$ in $\Omega_\psi \setminus R_{2\delta}(x^0)$. On the other hand, one has in $R_{2\delta}(x^0)$

$$w - c + \frac{\tau^2}{2}\varphi^2\Delta_\tau w = (w - c) + \frac{\tau^2}{2}\varphi^2\Delta_\tau(w - c) \geq 0.$$

The above arguments show that $\chi_\tau \in K_\psi(\Omega_\psi)$. Substitute χ_τ in (2.100) as a test function. In this case we easily arrive at the inequality

$$B_\psi(w, \varphi^2\Delta_\tau w) + \langle \sigma_{ij}(W), \varepsilon_{ij}(\varphi^2\Delta_\tau W)\rangle_\psi \geq 2\tau^{-2}\langle f, \chi_\tau - \chi\rangle_\psi. \quad (2.112)$$

It can be verified that the difference between the terms $B_\psi(w, \varphi^2\Delta_\tau w)$ and $-B_\psi(d_\tau(\varphi w), d_\tau(\varphi w))$ can be estimated from above by the value being in the right-hand side of the inequality (2.113) below. Analogously, the difference between $\langle \sigma_{ij}(W), \varepsilon_{ij}(\varphi^2\Delta_\tau W)\rangle_\psi$ and $-\langle \sigma_{ij}(d_\tau\varphi W), \varepsilon_{ij}(d_\tau\varphi W)\rangle_\psi$ can be estimated from above by the same quantity. Thus, the relation (2.112) implies

$$B_\psi(d_\tau(\varphi w), d_\tau(\varphi w)) + \langle \sigma_{ij}(d_\tau(\varphi W)), \varepsilon_{ij}(d_\tau(\varphi W))\rangle_\psi \quad (2.113)$$

$$\leq c\left(\|\chi\|^2_{H(\Omega_\psi)} + \|d_\tau(\varphi\chi)\|_{H(\Omega_\psi)}(\|\chi\|_{H(\Omega_\psi)} + \|f\|_{0,\Omega_\psi})\right).$$

In view of (2.101), (2.102) the estimate

$$\|d_\tau(\varphi\chi)\|_{H(\Omega_\psi)} \leq c$$

follows, being uniform in τ. It clearly yields

$$\frac{\partial}{\partial x}(\varphi\chi) \in H(\Omega_\psi).$$

So, the assertion of Theorem 2.12 related to w is proved. Meanwhile, equations (2.104) can be written down as

$$W_{yy} = G.$$

The function G depends on f_1, f_2, W_{xy}, W_{xx} linearly, so that in view of the above result, we have $G \in L^2(R_\delta(x^0) \cap \Omega_\psi)$. Hence, all derivatives of W up to the second order belong to $L^2(R_\delta(x^0) \cap \Omega_\psi)$. Theorem 2.12 is completely proved.

In what follows we prove the solution regularity in a neighbourhood of points belonging to the crack faces and not having contact with the punch. Let $x^0 \in \Gamma_\psi \setminus \partial\Gamma_\psi$ be any fixed point such that $w^\pm(x^0) > \Phi(x^0)$; moreover, a neighbourhood $\mathcal{O}(x^0)$ of the point x^0 is assumed to be chosen such that $\Gamma_\psi \cap \mathcal{O}(x^0)$ is a segment parallel to the axis x. The following statement is valid.

Theorem 2.13. *Let the above hypotheses be fulfilled. Then the inclusions*

$$W \in H^2(R_\delta(x^0) \cap \Omega_\psi), \quad w \in H^3(R_\delta(x^0) \cap \Omega_\psi)$$

hold provided δ is small enough.

PROOF. The condition $w^\pm(x^0) > \Phi(x^0)$ implies the fulfilment of the equation

$$\Delta^2 w = f_3 \tag{2.114}$$

in $R_{2\delta}(x^0) \cap \Omega_\psi$ for small δ. Take the function φ and construct the vector $\chi_\tau = (w_\tau^1, w_\tau^2, w_\tau)$ as in Theorem 2.12, $0 < |\tau| < \delta/2$. The parameter δ is supposed to be fixed such that $R_{2\delta}(x^0) \subset \mathcal{O}(x^0)$ and $w_\tau \geq \Phi$ in $R_{3\delta/2}(x^0)$. In this case it is seen that $w_\tau \geq \Phi$ in Ω_ψ. Moreover, it has been proved that χ_τ satisfies the restriction (2.105). Hence, the inclusion $\chi_\tau \in K_\psi(\Omega_\psi)$ holds. Substituting χ_τ in (2.100) as a test function results in the relation like (2.112). Further arguments are those of Theorem 2.12, so that

$$W \in H^2(R_\delta(x^0) \cap \Omega_\psi), \quad \frac{\partial w}{\partial x} \in H^2(R_\delta(x^0) \cap \Omega_\psi). \tag{2.115}$$

Meantime, equation (2.114) can be written as

$$w_{yyyy} = Q.$$

According to (2.115) the inclusion $Q \in H^{-1}(R_\delta(x^0) \cap \Omega_\psi)$ holds. Whence, taking into account the relations $w_{yyy}, w_{yyyx} \in H^{-1}(R_\delta(x^0) \cap \Omega_\psi)$ and the results of (Duvaut, Lions, 1972, Ch.3, Sect.3, Th.3.2) we arrive at the desired conclusion:

$$w_{yyy} \in L^2(R_\delta(x^0) \cap \Omega_\psi).$$

Theorem 2.13 is proved.

REMARK. Seemingly, the hypothesis relating to $\Gamma_\psi \cap \mathcal{O}(x^0)$ in Theorem 2.12 and Theorem 2.13 may be omitted, but it is not proved.

2.4.4 Extreme crack shapes

Suppose that the crack shape is described by the equation $y = \delta\psi(x)$ with a parameter δ (see Fig.2.3). The space $H(\Omega_\delta)$ and the set $K_\delta(\Omega_\delta)$ are introduced analogously to $H(\Omega_\psi)$ and $K_\psi(\Omega_\psi)$, respectively.

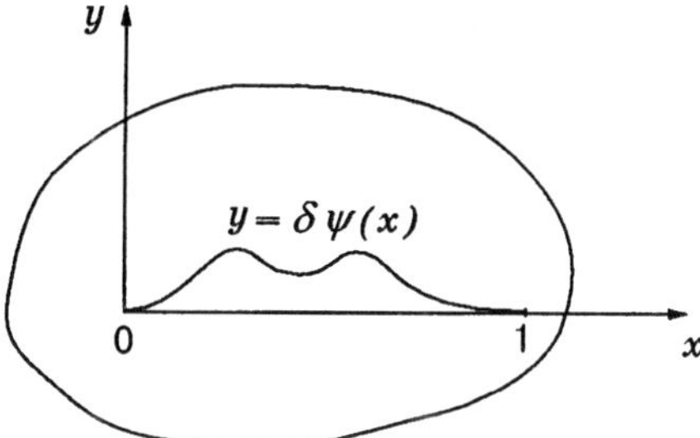

Fig.2.3. Perturbations of the crack shape

In the following we analyse the behaviour of the solution as $\delta \to 0$. It will enable us in the sequel to prove the existence of extreme crack shapes. The formulation of this problem is given below. So, for every fixed δ there exists a solution $\chi^\delta = (W^\delta, w^\delta)$ of the problem

$$B_\delta(w^\delta, \bar{w} - w^\delta) + \langle\sigma_{ij}(W^\delta), \varepsilon_{ij}(\bar{W} - W^\delta)\rangle_\delta \geq \langle f, \bar{\chi} - \chi^\delta\rangle_\delta, \qquad (2.116)$$

$$\chi^\delta \in K_\delta(\Omega_\delta), \quad \forall\bar{\chi} \in K_\delta(\Omega_\delta).$$

In order to study the solution convergence as $\delta \to 0$ we carry out the mapping of Ω_δ onto Ω_0. Of course, the graphs $y = \delta\psi(x)$ are assumed to belong to Ω for all $0 \leq \delta \leq \delta_0$. Extend the function ψ beyond $[0, 1]$ by zero, then choose domains Ω_1, Ω_2 such that $\overline{\Omega}_1 \subset \Omega_2$, $\overline{\Omega}_2 \subset \Omega$, $\Gamma_\delta \subset \Omega_1$ for all δ small enough and a function ξ possessing the properties: $\xi \equiv 1$ in Ω_1, $\xi \equiv 0$ in $\Omega \setminus \Omega_2$. The following transformation of the independent variables can be considered:

$$\tilde{x} = x, \quad \tilde{y} = y - \delta\xi\psi. \qquad (2.117)$$

It is clear that the Jacobian $q_\delta = 1 - \delta\psi\xi_y$ of this transformation converges uniformly to the unit on Ω as $\delta \to 0$. Introduce the notations

$$U^\delta(\tilde{x}, \tilde{y}) = W^\delta(x, y), \quad u^\delta(\tilde{x}, \tilde{y}) = w^\delta(x, y), \quad \omega^\delta = (U^\delta, u^\delta).$$

A substitution of a fixed test function $\bar{\chi}$ in (2.116) results in the relation

$$B_\delta(w^\delta, w^\delta) + \langle\sigma_{ij}(W^\delta), \varepsilon_{ij}(W^\delta)\rangle_\delta$$

$$\leq B_\delta(w^\delta, \bar{w}) + \langle\sigma_{ij}(W^\delta), \varepsilon_{ij}(\bar{W})\rangle_\delta + \langle f, \chi^\delta - \bar{\chi}\rangle_\delta.$$

Omitting the sign δ in the functions it is easy to rewrite this inequality in the new variables

$$\int\limits_{\Omega_0} \left(u_{\tilde{x}\tilde{x}}^2 + u_{\tilde{y}\tilde{y}}^2 + 2\kappa u_{\tilde{x}\tilde{x}} u_{\tilde{y}\tilde{y}} + 2(1-\kappa) u_{\tilde{x}\tilde{y}}^2\right) q_\delta^{-1}\, d\Omega_0$$

$$+ \langle \sigma_{ij}(U), \varepsilon_{ij}(U) q_\delta^{-1} \rangle_0 - \langle f^\delta, (\omega - \bar{\omega}) q_\delta^{-1} \rangle_0 \tag{2.118}$$

$$+ \delta \int\limits_{\Omega_0} g(\tilde{x}, \tilde{y}, \delta, D^\alpha u, D^\alpha \bar{u}, D^\beta U, D^\beta \bar{U})\, d\Omega_0 \le 0.$$

Herein $f^\delta(\tilde{x}, \tilde{y}) = f(x, y)$, $|\alpha| \le 2$, $|\beta| \le 1$. A dependence of the function g on its arguments is fully determined by the transformation (2.117). It is of importance that this function has quadratic growth in the higher order derivatives. In view of the inequality $q_\delta^{-1} > 1/2$ holding for small δ, from (2.118) we conclude that

$$\|\omega^\delta\|_{H(\Omega_0)} \le c$$

uniformly in $\delta \le \delta_0$. Choosing a subsequence, if necessary, one can assume that as $\delta \to 0$

$$\omega^\delta \to \omega \quad \text{weakly in } H(\Omega_0). \tag{2.119}$$

The solution (U^δ, u^δ) satisfies the inequalities (2.98), (2.99) written in the new variables. To be more precise, we denote $\Phi^\delta(\tilde{x}, \tilde{y}) = \Phi(x, y)$. Then the inequality (2.98) takes the form

$$u^\delta \ge \Phi^\delta \quad \text{in} \quad \Omega_0, \tag{2.120}$$

and the inequality (2.99) can be written as

$$\left[U^\delta - z({u^\delta}_{\tilde{x}} - \delta\psi_x {u^\delta}_{\tilde{y}}, {u^\delta}_{\tilde{y}})\right](-\delta\psi_x, 1) \ge 0 \quad \text{on} \quad \Gamma_0, \quad |z| \le 1. \tag{2.121}$$

Let the set of all functions (U, u) from the space $H(\Omega_0)$ satisfying (2.120), (2.121) be denoted by $K_\delta(\Omega_0)$. The following statement is useful for further consideration.

Lemma 2.1. *For any fixed* $(\bar{U}, \bar{u}) \in K_0(\Omega_0)$ *there exists a sequence* $(\bar{U}^\delta, \bar{u}^\delta) \in K_\delta(\Omega_0)$ *such that as* $\delta \to 0$

$$(\bar{U}^\delta, \bar{u}^\delta) \to (\bar{U}, \bar{u}) \quad \textit{strongly in } H(\Omega_0). \tag{2.122}$$

PROOF. We make use of Lemma 4.4. Namely, for any fixed function $(\bar{U}, \bar{u}) \in H(\Omega_0)$ satisfying the inequality

$$\left[\bar{U} - z\nabla\bar{u}\right]\nu \ge 0 \quad \text{on} \quad \Gamma_0, \quad |z| \le 1, \quad \nu = (0, 1),$$

a sequence $(\bar{U}^\delta, \bar{u}^\delta) \in H(\Omega_0)$ can be constructed such that $(\bar{U}^\delta, \bar{u}^\delta) \to (\bar{U}, \bar{u})$ strongly in $H(\Omega_0)$ and, moreover, the inequality (2.121) holds for $(\bar{U}^\delta, \bar{u}^\delta)$, and $\bar{u}^\delta = \bar{u}$ for all δ. Let us take a fixed element $(\bar{U}, \bar{u}) \in K_0(\Omega_0)$ and bearing in mind the above arguments, construct a sequence $\bar{\chi}^\delta = (\bar{U}^\delta, \bar{u})$ having the above properties. We show that the appropriate substitution of the third component of $\bar{\chi}^\delta$ by $\bar{u}^\delta$ will imply that the sequence $(\bar{U}^\delta, \bar{u}^\delta)$ is needed, that is $(\bar{U}^\delta, \bar{u}^\delta) \in K_\delta(\Omega_0)$, and (2.122) arises. Since $\Phi^\delta \to \Phi$ uniformly on Ω and $\Phi^\delta = \Phi$ near Γ there exists a function θ^δ such that

$$\theta^\delta \geq |\Phi^\delta - \Phi| \quad \text{in } \Omega, \qquad \theta^\delta \to 0 \quad \text{strongly in } H^2(\Omega).$$

We should remark at this step that $\bar{u} \geq \Phi$ in Ω_0. Putting $\bar{u}^\delta = \bar{u} + \theta^\delta$ it is easily verified that the sequence $(\bar{U}^\delta, \bar{u}^\delta)$ satisfies all conditions. Indeed, the restriction (2.120) for $\bar{u}^\delta$ holds by the construction of θ^δ. Since the jump $\left[(\theta^\delta_{\tilde{x}} - \delta\psi_x\theta^\delta_{\tilde{y}}, \theta^\delta_{\tilde{y}})\right]$ is equal to zero on Γ_0 the restriction (2.121) for $(\bar{U}^\delta, \bar{u}^\delta)$ also holds. The convergence (2.122) is evident. Lemma 2.1 is proved.

Let us now rewrite (2.116) in the new variables $\tilde{x}, \tilde{y}$. The convergence (2.119) and Lemma 2.1 allow us to carry out the limiting procedure as $\delta \to 0$. Moreover, the limiting function $\omega = (U, u)$ is a solution of the variational inequality

$$B_0(u, \bar{u} - u) + \langle \sigma_{ij}(U), \varepsilon_{ij}(\bar{U} - U)\rangle_0 \geq \langle f, \bar{\omega} - \omega\rangle_0, \tag{2.123}$$

$$\omega \in K_0(\Omega_0), \quad \forall \bar{\omega} \in K_0(\Omega_0).$$

So, the following statement is proved.

Theorem 2.14. *From the sequence $\chi^\delta = \omega^\delta$ of solutions of the problem* (2.116) *one can choose a subsequence, still denoted by ω^δ, such that as $\delta \to 0$ the convergence* (2.119) *takes place and, moreover, the limiting function satisfies* (2.123).

This result enables us to investigate the extreme crack shape problem. The formulation of the last one is as follows. Let $\Psi \subset H_0^3(0,1)$ be a convex, closed and bounded set. Assume that for every $\psi \in \Psi$ the graph $y = \psi(x)$ describes the crack shape. Consequently, for a given $\psi \in \Psi$ there exists a unique solution of the problem

$$B_\psi(w, \bar{w} - w) + \langle \sigma_{ij}(W), \varepsilon_{ij}(\bar{W} - W)\rangle_\psi \geq \langle f, \bar{\chi} - \chi\rangle_\psi, \tag{2.124}$$

$$\chi = (W, w) \in K_\psi(\Omega_\psi), \quad \forall \bar{\chi} \in K_\psi(\Omega_\psi).$$

Consider the cost functional

$$J(\psi) = \|\chi - \chi_0\|_{0,\Omega_\psi}$$

where $\chi_0 \in L^2(\Omega)$ is a prescribed element. We have to find a solution of the maximization problem

$$\sup_{\psi \in \Psi} J(\psi). \tag{2.125}$$

The following assertion holds.

Theorem 2.15. *Let the above hypotheses be fulfilled. Then, there exists a solution of the problem* (2.125).

We shall confine ourselves to short remarks. A maximizing sequence $\psi^n \in \Psi$ is evidently bounded in $H_0^3(0,1)$. Hence, without any loss, one can assume that as $n \to \infty$

$$\psi^n \to \psi \quad \text{weakly in } H_0^3(0,1), \quad \psi \in \Psi, \tag{2.126}$$

$$|\psi^n_{xx} - \psi_{xx}| < 1/n \quad \text{on} \quad [0,1].$$

The second line here follows from the imbedding theorem. For any n, there exists a solution (W^n, w^n) of the problem

$$B_{\psi^n}(w^n, \bar{w} - w^n) + \langle \sigma_{ij}(W^n), \varepsilon_{ij}(\bar{W} - W^n)\rangle_{\psi^n} \tag{2.127}$$

$$\geq \langle f, \bar{\chi} - \chi^n \rangle_{\psi^n} \quad \forall \bar{\chi} \in K_{\psi^n}(\Omega_{\psi^n}).$$

The domains Ω_1, Ω_2 and the function ξ can be chosen as in the proof of Theorem 2.14. The transformation of the independent variables is of the form

$$\tilde{x} = x, \quad \tilde{y} = y + (\psi - \psi^n)\xi.$$

We prove that the solution $U^\delta(\tilde{x}, \tilde{y}) = W^\delta(x, y)$, $u^\delta(\tilde{x}, \tilde{y}) = w^\delta(x, y)$ satisfies the following estimate:

$$\|U^n\|_{1,\Omega_\psi} + \|u^n\|_{2,\Omega_\psi} \leq c.$$

Without loss of generality, one can suppose that as $n \to \infty$

$$(U^n, u^n) \to (U, u) \quad \text{weakly in } H(\Omega_0), \quad \text{strongly in } L^2(\Omega_0). \tag{2.128}$$

To justify the passage to the limit in the relations obtained from (2.127) by a change of variables, we use the convergence (2.128) and the statement analogous to Lemma 2.1. The limiting function $\chi = (U, u)$ is a solution of the variational inequality (2.100) with the function ψ from (2.126), that is $\chi = \chi_\psi$. Finally, it is easy to verify that

$$J(\psi) = \sup_{\bar{\psi} \in \Psi} J(\bar{\psi}).$$

This precisely means that the limiting function ψ is a solution of the extreme crack shape problem (2.125).

As for approximate methods of finding crack shapes we refer the reader to (Banichuk, 1970). Qualitative properties of solutions to boundary value problems in nonsmooth domains are in (Oleinik et al., 1981; Nazarov, 1989; Nazarov, Plamenevslii, 1991; Nicaise, 1992; Maz'ya, Nazarov, 1987; Grisvard, 1985, 1991; Kondrat'ev et al., 1982; Kondrat'ev, Oleinik, 1983; Dauge, 1988; Costabel, Dauge, 1994; Sändig et al., 1989; Movchan A.B., Movchan N.V., 1995).

2.5 Cracks of minimal opening in plates

This section concerns the equilibrium problem for a plate contacting a rigid punch and having a vertical crack. Two conditions of inequality type are assumed to be imposed on the solution. These conditions describe a mutual nonpenetration in the plate–punch system and a nonpenetration of crack faces. The first one is of the form $w \geq \varphi$, where w is the vertical displacement of the plate, and φ corresponds to the punch shape. The second one can be written as $[W]\nu \geq \varepsilon|[\partial w/\partial \nu]|$, where $W = (w^1, w^2)$ is the horizontal displacement, ν is the normal to the crack shape curve, 2ε is the thickness of the plate, and $[\cdot]$ is the jump of a function at crack faces. The aim of the section is to study the solution properties of the optimal control problem of the punch shape φ. The existence theorem is proved as providing the minimal jump of the displacement $\chi = (W, w)$. The solution regularity up to the interior crack points is analysed. In particular, the inclusion $\chi \in C^\infty$ is stated to be valid for the crack points having a zeroth jump. The convergence of solution is investigated as $\varepsilon \to 0$. The results of this section can be found in (Khludnev, 1996a).

2.5.1 Formulation of the problem

Our aim is to analyze the solution properties of the variational inequality describing the equilibrium state of the elastic plate. The plate is assumed to have a vertical crack and, simultaneously, to contact with a rigid punch.

Considering the crack, we impose the nonpenetration condition of the inequality type at the crack faces. The nonpenetration condition for the plate–punch system also is the inequality type. It is well known that, in general, solutions of problems having restrictions of inequality type are not smooth. In this section, we establish existence and regularity results related to the problem considered. Namely, the following questions are under consideration:

1. The existence of punch shape which provides the minimal opening of the crack.

2. The regularity of solutions in the case of minimal opening of the crack.

3. The solutions properties related to the case where the thickness of the plate tends to zero.

We consider the Kirchhoff–Love model of the plate for which both vertical and horizontal displacements of the mid-surface points are to be found.

Let us introduce the notations and give the appropriate formulae of the Kirchhoff–Love model which can be found, for instance, in (Vol'mir, 1972). Denote a bounded domain with a smooth boundary Γ by $\Omega \subset R^2$, and $y = \psi(x)$ signifies the function describing a crack face, $x \in [0, 1]$, $(x, y) \in \Omega$. Let Γ_ψ be the graph of the function $y = \psi(x)$, and $\Omega_\psi = \Omega \backslash \Gamma_\psi$. The domain Ω_ψ is identified with thc mid-surface of the plate in its nondeformable state.

The displacement vector of the mid-surface points is denoted by $\chi = (W, w)$, where $W = (w^1, w^2)$ is the horizontal displacement and w is the vertical one.

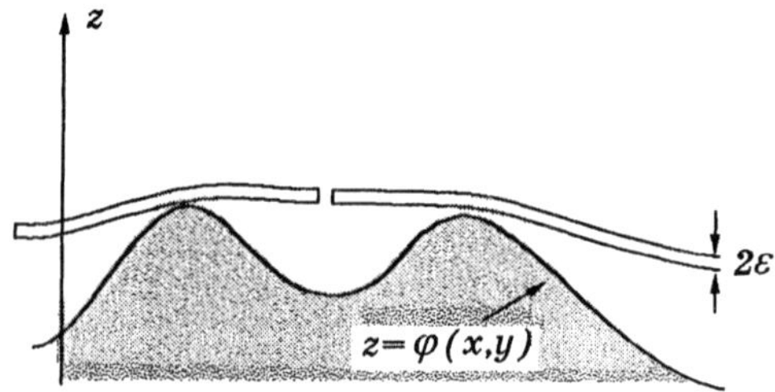

Fig.2.4. Cracked plate in contact with the punch

We next assume that the graph $z = \varphi(x, y)$ corresponds to the punch shape, $(x, y) \in \Omega$ (see Fig.2.4). Then the nonpenetration condition for the plate–punch system can be written as

$$w \geq \varphi \quad \text{in} \quad \Omega_\psi. \tag{2.129}$$

The Kirchhoff–Love model of the plate is characterized by the linear dependence of the horizontal displacements on the distance from the mid-surface, that is

$$W(z) = W - z\nabla w, \quad -\varepsilon \leq z \leq \varepsilon,$$

where $z = 0$ corresponds to the mid-surface, and the axis z is orthogonal to the (x, y)-plane, and 2ε is the thickness of the plate. Denote the normal to the graph Γ_ψ by $\nu = (-\psi_x, 1)/\sqrt{1 + \psi_x^2}$, $\nu = (\nu_1, \nu_2)$. In this case, the nonpenetration condition of Signorini type imposed at the crack faces is as follows:

$$[W - z\nabla w]\nu \geq 0 \quad \text{on } \Gamma_\psi, \quad |z| \leq \varepsilon,$$

where $[V] = V^+ - V^-$ is the jump of V, and $V^\pm$ correspond to the positive and negative directions of ν, respectively. As evident from the above, the nonpenetration condition can be rewritten in the equivalent form

$$[W]\nu \geq \varepsilon \left| \left[\frac{\partial w}{\partial \nu} \right] \right| \quad \text{on } \Gamma_\psi. \tag{2.130}$$

Thus, we see that there is no penetration for all points of the crack faces since condition (2.130) is independent of $z \in [-\varepsilon, \varepsilon]$.

The strain and integrated stress tensors are denoted by $e_{ij} = e_{ij}(W)$, $\sigma_{ij} = \sigma_{ij}(W)$, respectively:

$$e_{ij} = \frac{1}{2}\left(\frac{\partial w^i}{\partial x_j} + \frac{\partial w^j}{\partial x_i} \right), \quad i, j = 1, 2, \quad x_1 = x, \ x_2 = y,$$

$$\sigma_{11} = e_{11} + \kappa e_{22}, \quad \sigma_{22} = e_{22} + \kappa e_{11}, \quad \sigma_{12} = (1-\kappa)e_{12}.$$

Here $\kappa = $ const is Poisson's ratio, $0 < \kappa < 1/2$.

The following boundary conditions are assumed to be fulfilled at the external boundary

$$w = \frac{\partial w}{\partial n} = W = 0 \quad \text{on } \Gamma.$$

Let the subspace $H^{1,0}(\Omega_\psi)$ of the Sobolev space $H^1(\Omega_\psi)$ consist of functions equal to zero on Γ. Analogously, the functions of $H^{2,0}(\Omega_\psi)$ are equal to zero on Γ together with the first derivatives, $H^{2,0}(\Omega_\psi) \subset H^2(\Omega_\psi)$. Define the space $H(\Omega_\psi) = H^{1,0}(\Omega_\psi) \times H^{1,0}(\Omega_\psi) \times H^{2,0}(\Omega_\psi)$ and consider the energy functional of the plate

$$\Pi(\chi) = \frac{1}{2}B(w,w) + \frac{1}{2}\langle \sigma_{ij}(W), e_{ij}(W)\rangle - \langle f, \chi\rangle.$$

Here $f = (f_1, f_2, f_3) \in L^2(\Omega)$ is the given vector of exterior forces, the brackets $\langle \cdot, \cdot \rangle$ mean the integration over Ω_ψ,

$$B(u,v) = \int\limits_{\Omega_\psi} (u_{xx}v_{xx} + u_{yy}v_{yy} + \kappa u_{xx}v_{yy} + \kappa u_{yy}v_{xx} + 2(1-\kappa)u_{xy}v_{xy})\, d\Omega_\psi.$$

The above formula for $\Pi(\chi)$ contains three different terms which correspond to the bending energy of the plate, to the deformation energy of the mid-surface, and to the work of the exterior force f, respectively. Also, we introduce the set of admissible displacements

$$K_\varepsilon^\varphi = \{(W,w) \in H(\Omega_\psi) \mid \quad (W,w) \quad \text{satisfy} \ (2.129), (2.130)\}.$$

The equilibrium problem for the plate can be formulated as variational, namely, it corresponds to the minimum of the functional Π over the set of admissible displacements. To minimize the functional Π over the set K_ε^φ we can consider the variational inequality

$$B(w, \bar{w} - w) + \langle \sigma_{ij}(W), e_{ij}(\bar{W} - W)\rangle \geq \langle f, \bar{\chi} - \chi\rangle, \tag{2.131}$$

$$\chi = (W,w) \in K_\varepsilon^\varphi, \quad \forall\, \bar{\chi} \in K_\varepsilon^\varphi.$$

Nonemptiness of the set K_ε^φ depends on the values of the function φ on Γ. For further consideration we should note that the following inequality holds in Ω_ψ

$$B(w,w) \geq c\|w\|^2_{2,\Omega_\psi} \quad \forall\, w \in H^{2,0}(\Omega_\psi), \tag{2.132}$$

and the first Korn inequality takes place,

$$\langle \sigma_{ij}(W), e_{ij}(W)\rangle \geq c\|W\|^2_{1,\Omega_\psi} \quad \forall\, W \in H^{1,0}(\Omega_\psi), \tag{2.133}$$

with the constants independent of w, W, respectively. The relations (2.132), (2.133) provide the coercivity of the functional Π on $H(\Omega_\psi)$. Thus, considering the weak lower semicontinuity of Π, one concludes that there exists a solution of (2.131). Moreover, the solution is unique.

In the sequel we shall study an optimal control problem. Let $\Phi \subset H^2(\Omega)$ be a convex, bounded and closed set. Assume that $\varphi < 0$ on Γ for each $\varphi \in \Phi$. In particular, this condition provides nonemptiness for K_ε^φ. Denote the solution of (2.131) by $\chi = \chi(\varphi)$, and introduce the cost functional which characterizes the opening of the crack (Goldshtein, Entov, 1989),

$$J_\varepsilon(\varphi) = \int_{\Gamma_\psi} \|[\chi]\| \, d\Gamma_\psi.$$

The problem of finding an obstacle providing the minimal opening of the crack can be formulated as follows:

$$\inf_{\varphi \in \Phi} J_\varepsilon(\varphi). \tag{2.134}$$

The crack shape is defined by the function ψ. This function is assumed to be fixed. It is noteworthy that the problems of choice of the so-called extreme crack shapes were considered in (Khludnev, 1994; Khludnev, Sokolowski, 1997). We also address this problem in Sections 2.4 and 4.9. The solution regularity for biharmonic variational inequalities was analysed in (Frehse, 1973; Caffarelli et al., 1979; Schild, 1984). The last paper also contains the results on the solution smoothness in the case of thin obstacles. As for general solution properties for the equilibrium problem of the plates having cracks, one may refer to (Morozov, 1984). Referring to this book, the boundary conditions imposed on crack faces have the equality type. In this case there is no interaction between the crack faces.

In the next two subsections the parameter ε is supposed to be fixed. The convergence of solutions of the optimal control problem (2.134) as $\varepsilon \to 0$ will be analysed in Section 2.5.4. For this reason the ε-dependence of the cost functional is indicated.

2.5.2 Existence of minimal opening cracks

Let ε be fixed. Before proving the theorem an auxiliary statement is to be established. It is formulated as a lemma.

Lemma 2.2. *Let a sequence $\varphi_m \in \Phi$ possess the properties*

$$\varphi_m \to \varphi \quad \textit{weakly in } H^2(\Omega), \quad \textit{uniformly in } \overline{\Omega}. \tag{2.135}$$

Then for any fixed function $\bar{\chi} = (\bar{W}, \bar{w}) \in K_\varepsilon^\varphi$ there exists a sequence $\bar{\chi}_m = (\bar{W}_m, \bar{w}_m)$ from $K_\varepsilon^{\varphi_m}$ such that

$$\bar{\chi}_m \to \bar{\chi} \quad \textit{strongly in } H(\Omega_\psi). \tag{2.136}$$

PROOF. Without loss, the following inequality,

$$|\varphi_m - \varphi| < \frac{1}{m} \quad \text{in } \overline{\Omega},$$

is supposed to be held. Set $\tilde{w}_m = \bar{w} + 1/m$. In this case $\tilde{w}_m \geq \varphi_m$ in Ω_ψ. There exists a neighbourhood $\mathcal{O}$ of the boundary Γ such that $\varphi < -\delta < 0$ in $\mathcal{O} \cap \Omega$ with a constant $\delta > 0$. Let $\Gamma_\psi \cap \mathcal{O} = \emptyset$. In view of the uniform convergence of φ_m the estimate $\varphi_m < -\delta/2$ holds in $\mathcal{O} \cap \Omega$. It is easy to construct a sequence $\xi_m \in C^\infty$ such that the supports of ξ_m belong to $\mathcal{O}$ and

$$\xi_m = \frac{1}{m}, \quad \nabla \xi_m = 0 \quad \text{on } \Gamma,$$

$$|D^\alpha \xi_m| \leq \frac{c}{m} \quad \text{in } \mathcal{O}, \quad |\alpha| \leq 2,$$

with a constant c independent of m. Now, we can define

$$\bar{w}_m = \tilde{w}_m - \xi_m.$$

It is clear that $\bar{w}_m \geq \varphi_m$ in Ω_ψ and $[\partial \bar{w}_m/\partial \nu] = [\partial \bar{w}/\partial \nu]$ on Γ_ψ. Thus the functions $\bar{\chi}_m = (\bar{W}, \bar{w}_m)$ belong to $K_\varepsilon^{\varphi_m}$ for all m. Moreover, the convergence (2.136) takes place. The proof is completed. Note that in (Mosco, 1969) general results on convergence of convex sets can be found.

We now are in a position to establish the solvability of the optimal control problem (2.134), (2.131).

Theorem 2.16. *There exists a solution of the problem* (2.134), (2.131).

PROOF Let $\varphi_m \in \Phi$ be a minimizing sequence. It is bounded in $H^2(\Omega)$, and hence the convergence (2.135) can be assumed. For every m, the solution of the following variational inequality can be found:

$$B(w_m, \bar{w}_m - w_m) + \langle \sigma_{ij}(W_m), e_{ij}(\bar{W}_m - W_m) \rangle \geq \langle f, \bar{\chi}_m - \chi_m \rangle, \quad (2.137)$$

$$\chi_m = (W_m, w_m) \in K_\varepsilon^{\varphi_m}, \quad \forall\, \bar{\chi}_m \in K_\varepsilon^{\varphi_m}.$$

By virtue of the uniform convergence of φ_m there exists a function $\bar{\chi}$ such that $\bar{\chi} \in K_\varepsilon^{\varphi_m}$ for all m. Substituting this function in (2.137) as $\bar{\chi}_m$ implies

$$\|\chi_m\|_{H(\Omega_\psi)} \leq c$$

uniformly in m. Deriving this estimate we make use of the inequalities (2.132), (2.133). Hence choosing a subsequence, if necessary, we assume as $m \to \infty$

$$\chi_m \to \chi \quad \text{weakly in } H(\Omega_\psi). \quad (2.138)$$

Let $\bar{\chi} \in K_\varepsilon^{\varphi}$ be any fixed element where φ is the function from (2.135). Lemma 2.2 provides an existence of a sequence $\bar{\chi}_m \in K_\varepsilon^{\varphi_m}$ strongly converging to $\bar{\chi}$ in $H(\Omega_\psi)$. Bearing in mind (2.138), this allows us to carry out the limiting procedure in (2.137). The resulting relation precisely coincides with (2.131), i.e. $\chi = \chi(\varphi)$. An additional assumption,

$$\chi_m^\pm \to \chi^\pm \quad \text{weakly in } L^1(\Gamma_\psi),$$

easily yields the relations

$$\inf_{\bar\varphi\in\Phi} J_\varepsilon(\bar\varphi) = \liminf \int_{\Gamma_\psi} |[\chi_m]|\, d\Gamma_\psi \geq \int_{\Gamma_\psi} |[\chi]| d\Gamma_\psi \geq \inf_{\bar\varphi\in\Phi} J_\varepsilon(\bar\varphi).$$

This means that the function φ is a solution of the problem (2.134), (2.131), which completes the proof.

2.5.3 Solution regularity

Let $Q \subset R^2$ be a bounded domain with a smooth boundary γ. An external normal to γ is denoted by $n = (n_1, n_2)$. Introduce the following operators defined at γ by

$$m(u) = \kappa\Delta u + (1-\kappa)\frac{\partial^2 u}{\partial n^2},$$

$$t(u) = \frac{\partial}{\partial n}\Delta u + (1-\kappa)\frac{\partial^3 u}{\partial n \partial s^2}, \qquad s = (-n_2, n_1).$$

We know (see Section 1.4.3) that for any fixed $u \in H^2(Q)$, $\Delta^2 u \in L^2(Q)$ the values $m(u), t(u)$ can be considered as elements of $H^{-1/2}(\gamma)$ and $H^{-3/2}(\gamma)$, respectively. Moreover, the Green formula

$$B_Q(u,v) = \left\langle m(u), \frac{\partial v}{\partial n}\right\rangle_{1/2,\gamma} - \langle t(u), v\rangle_{3/2,\gamma} + \langle \Delta^2 u, v\rangle_Q \tag{2.139}$$

takes place for all $v \in H^2(Q)$. Here, the integration is carried out over Q, and $\langle\cdot,\cdot\rangle_{p,\gamma}$ means a duality pairing between $H^{-p}(\gamma)$ and $H^p(\gamma)$. Besides, once more Green's formula holds good (see Section 1.4.3). Namely, for any $U \in H^1(Q)$, $\sigma_{ij} = \sigma_{ij}(U)$, $\partial\sigma_{ij}/\partial x_j \in L^2(Q)$, $i = 1,2$, one has $\sigma_{ij}n_j \in H^{-1/2}(\gamma)$ and

$$\left\langle \frac{\partial\sigma_{ij}}{\partial x_j}, v\right\rangle_Q = \langle \sigma_{ij}n_j, v\rangle_{\frac{1}{2},\gamma} - \left\langle \sigma_{ij}, \frac{\partial v}{\partial x_j}\right\rangle_Q \tag{2.140}$$

$$\forall v \in H^1(Q), \quad i = 1,2.$$

Assume next that

$$w > \varphi \quad \text{in } \mathcal{W}, \tag{2.141}$$

where $\mathcal{W}$ is a neighbourhood of the graph Γ_ψ. In this case the inequality (2.131) implies that the following equations are satisfied in the sense of distributions in $\mathcal{W}\setminus\Gamma_\psi$:

$$\Delta^2 w = f_3, \tag{2.142}$$

$$-\frac{\partial \sigma_{ij}}{\partial x_j} = f_i, \quad i = 1, 2, \tag{2.143}$$

where $\sigma_{ij} = \sigma_{ij}(W)$. The relation (2.141) means, in particular, that the inequality $w > \varphi$ holds at both crack faces. This last fact will be written as $w > \varphi$ on $\Gamma_\psi^\pm$.

Consider a closed curve such that it confines the bounded simply connected domain $Q \subset \mathcal{W} \setminus \Gamma_\psi$ and contains Γ_ψ as a part of its boundary. According to the above the equations (2.142), (2.143) hold in Q, and hence

$$m(w) \in H^{-1/2}(\gamma),\ t(w) \in H^{-3/2}(\gamma),\ \sigma_{ij} n_j \in H^{-1/2}(\gamma),\ i = 1, 2. \tag{2.144}$$

Obviously, the domain Q can be constructed in different ways. Nevertheless, in any case one of the inclusions $\Gamma_\psi^+ \subset \gamma$, $\Gamma_\psi^- \subset \gamma$ will be valid, and (2.144) will take place. The exact form of the boundary conditions on Γ_ψ was obtained in Section 2.4. We omit the derivation of these conditions here. All we want to do is to discuss briefly their general form in connection with the subsequent regularity result. These conditions are as follows. Let

$$\{\sigma_{ij}\nu_j\} = \sigma_\nu \nu + \sigma_s s, \quad s = (-\nu_2, \nu_1),$$

be a decomposition of the vector $\{\sigma_{ij}\nu_j\}$, $i = 1, 2$, into the sum of normal and tangential components on Γ_ψ^-. Then, assuming $\varepsilon = 1$ on account of the reasons shown at the beginning of Section 2.5.4, we have on Γ_ψ:

$$\sigma_s = 0, \quad t(w) = 0, \quad [m(w)] = 0, \quad [\sigma_\nu] = 0, \quad \sigma_\nu \le 0,$$

$$|m(w)| \le -\sigma_\nu, \quad m(w)\left[\frac{\partial w}{\partial \nu}\right] + \sigma_\nu [W]\nu = 0.$$

Here $t(w) = 0$ means that for any smooth function θ in Ω with a compact trace on $\Gamma_\psi \setminus \partial\Gamma_\psi$ the relation

$$\langle t(w), \theta \rangle_{3/2,\gamma^+} = 0 \tag{2.145}$$

holds, where a domain Q^+ is chosen in such a way that $\Gamma_\psi^+ \subset \gamma^+$. A similar relation takes place in the case $\Gamma_\psi^- \subset \gamma^-$. The zeroth jumps of $m(w)$, $\sigma_{ij}\nu_j$ on Γ_ψ mean that

$$\left\langle m(w), \frac{\partial \theta}{\partial \nu} \right\rangle_{1/2,\gamma^+} = \left\langle m(w), \frac{\partial \theta}{\partial \nu} \right\rangle_{\frac{1}{2},\gamma^-},$$

$$\langle \sigma_{ij}\nu_j, \theta \rangle_{1/2,\gamma^+} = \langle \sigma_{ij}\nu_j, \theta \rangle_{1/2,\gamma^-}, \quad i = 1, 2.$$

In general, the above boundary conditions hold provided that (2.141) is fulfilled and the solution is quite regular. In fact, some part of the boundary conditions can be considered as holding in the strong sense without any additional assumptions on regularity. In particular, as proved in Section 2.4, if $x^0 \in \Gamma_\psi \setminus \partial\Gamma_\psi$ and $\mathcal{O}(x^0)$ is a neighbourhood of x^0 such that $\mathcal{O}(x^0) \subset \mathcal{W}$ and $\Gamma_\psi \cap \mathcal{O}(x^0)$ is a segment of a straight line, the following inclusions take place:

$$W \in H^2(\mathcal{O}(x^0) \cap \Omega_\psi), \quad w \in H^3(\mathcal{O}(x^0) \cap \Omega_\psi).$$

Hence

$$m(w),\ \sigma_{ij}\nu_j \in L^2(\Gamma_\psi^\pm \cap \mathcal{O}(x^0)), \quad i = 1, 2.$$

The condition $[\chi] = 0$ on Γ_ψ implies that the cost functional of the problem (2.134), (2.131) is equal to zero, i.e.

$$J_\varepsilon(\varphi) = \int_{\Gamma_\psi} |[\chi]| d\Gamma_\psi = 0.$$

In this case the crack is said to have a zeroth opening. The cracks of a zeroth opening prove to possess a remarkable property which is the main result of the present section. Namely, the solution χ is infinitely differentiable in a vicinity of $\Gamma_\psi \setminus \partial\Gamma_\psi$ provided that f is infinitely differentiable. This statement is interpreted as a removable singularity property. In what follows this assertion is proved. Let $x^0 \in \Gamma_\psi \setminus \partial\Gamma_\psi$ and $w > \varphi$ in $\mathcal{O}(x^0)$, where $\mathcal{O}(x^0)$ is a neighbourhood of x^0. For convenience, the boundary of the domain $\mathcal{O}(x^0)$ ia assumed to be smooth.

Theorem 2.17. *Let* $f \in C^\infty(\mathcal{O}(x^0))$ *and* $[\chi] = 0$ *on* $\mathcal{O}(x^0) \cap \Gamma_\psi$. *Then* $\chi = (W, w) \in C^\infty(\mathcal{O}(x^0))$.

PROOF. In view of (2.130) the hypotheses of the theorem imply that $[\partial w/\partial\nu] = 0$ on $\mathcal{O}(x^0) \cap \Gamma_\psi$. Consequently (see Mikhailov, 1976),

$$W \in H^1(\mathcal{O}(x^0)), \quad w \in H^2(\mathcal{O}(x^0)). \tag{2.146}$$

Incidentally, the equations (2.142), (2.143) hold in $\mathcal{O}(x^0) \cap \Omega_\psi$ in the sense of distributions, and hence

$$\Delta^2 w \in L^2(\mathcal{O}(x^0) \cap \Omega_\psi), \quad \frac{\partial \sigma_{ij}}{\partial x_j} \in L^2(\mathcal{O}(x^0) \cap \Omega_\psi), \quad i = 1, 2.$$

Let us show that the equation

$$\Delta^2 w = f_3 \tag{2.147}$$

holds in $\mathcal{O}(x^0)$. The brackets $(\cdot, \theta)$ will mean the value of a distribution evaluated at the point θ. The inclusions (2.146) are essential in our further reasoning. Let $\theta \in C_0^\infty(\mathcal{O}(x^0))$ be any fixed function. In view of the Green formula (2.139) one has

$$(\Delta^2 w - f_3, \theta) = B(w, \theta) - (f_3, \theta) = B_+(w, \theta) + B_-(w, \theta) - (f_3, \theta) \tag{2.148}$$

$$= \left\langle m(w), \frac{\partial\theta}{\partial\nu} \right\rangle_{1/2}^- - \left\langle m(w), \frac{\partial\theta}{\partial\nu} \right\rangle_{1/2}^+ - \langle t(w), \theta\rangle_{\frac{3}{2}}^- + \langle t(w), \theta\rangle_{\frac{3}{2}}^+$$

$$+\langle \Delta^2 w - f_3, \theta \rangle_\pm .$$

The signs $+, -$ mean that the formulae are concerned with the domains $\mathcal{O}^+(x^0)$, $\mathcal{O}^-(x^0)$, respectively, where

$$\mathcal{O}^+(x^0) = \mathcal{O}(x^0) \cap \{y > \psi(x)\}$$

and $\mathcal{O}^-(x^0)$ is defined in a similar way. The presence of two corner points at the boundaries $\partial\mathcal{O}^\pm(x^0)$ is not essential since θ has a compact support in $\mathcal{O}(x^0)$. In view of (2.145) the boundary terms of (2.148) containing $t(w)$ are equal to zero. Besides, the equation (2.142) holds in $\mathcal{O}^\pm(x^0)$ so that the two last terms of (2.148) are equal to zero. Lastly, the condition $[m(w)] = 0$ on $\mathcal{O}(x^0) \cap \Gamma_\psi$ provides two vanishing terms of (2.148) containing $m(w)$. Thus, (2.148) yields

$$(\Delta^2 w - f_3, \theta) = 0.$$

The proof of (2.147) is completed.

Analogously, the Green formula (2.140) and the first relation of (2.146) imply

$$\left(-\frac{\partial \sigma_{ij}}{\partial x_j} - f_i, \theta\right) = \left(\sigma_{ij}, \frac{\partial \theta}{\partial x_j}\right) - (f_i, \theta) = \left\langle \sigma_{ij}, \frac{\partial \theta}{\partial x_j}\right\rangle_\pm - (f_i, \theta)$$

$$= \langle \sigma_{ij}\nu_j, \theta\rangle^-_{1/2} - \langle \sigma_{ij}\nu_j, \theta\rangle^+_{1/2} - (f_i, \theta) - \left\langle \frac{\partial \sigma_{ij}}{\partial x_j}, \theta\right\rangle_\pm$$

$$= \left\langle -\frac{\partial \sigma_{ij}}{\partial x_j} - f_i, \theta\right\rangle_\pm = 0$$

for $i = 1, 2$. In so doing the equations (2.143) are used as holding good in $\mathcal{O}^\pm(x^0)$. The equations $[\sigma_{ij}\nu_j] = 0$ on $\mathcal{O}(x^0) \cap \Gamma_\psi$ are also used. Consequently, the following equations,

$$-\frac{\partial \sigma_{ij}}{\partial x_j} = f_i, \quad i = 1, 2, \tag{2.149}$$

hold in $\mathcal{O}(x^0)$ in the sense of distributions. The results on the internal solution regularity of (2.147), (2.149) (see Fichera, 1972; Lions, Magenes, 1968) provide the validity of the theorem assertion. The proof is completed.

2.5.4 Convergence of solutions

Consider the limit case corresponding to $c = 0$ in (2.130). The restriction obtained in such a way describes approximately a mutual nonpenetration of the crack faces. Note that in reality a complete account of the thickness implies the dependence of the energy functional on ε. This dependence is as follows (Vol'mir, 1972):

$$\Pi(\chi) = \frac{\varepsilon^3}{2} B(w, w) + \frac{\varepsilon}{2}\langle \sigma_{ij}(W), e_{ij}(W)\rangle - \langle f, \chi\rangle. \tag{2.150}$$

Moreover, in this case $m(w), t(w), e_{ij}(W)$ should also depend on ε. In spite of this, in this section the parameter ε is equal to 1 in the formula (2.150) just as in the previous subsections. Thus, the case $\varepsilon = 0$ in (2.130), in fact, means both the approximate description of the nonpenetration condition and a fixed thickness. Hence, in the case under consideration a solution should satisfy the following restriction:

$$w \geq \varphi \quad \text{in } \Omega_\psi, \tag{2.151}$$

$$[W]\nu \geq 0 \quad \text{on } \Gamma_\psi. \tag{2.152}$$

As a result the set of feasible displacements is as follows:

$$K_0^\varphi = \{(W, w) \in H(\Omega_\psi) \mid \ (W, w) \quad \text{satisfy} \ (2.151), (2.152)\}.$$

Herewith the problem of minimizing Π over the set K_0^φ is equivalent to the variational inequality

$$B(w, \bar{w} - w) + \langle \sigma_{ij}(W), e_{ij}(\bar{W} - W)\rangle \geq \langle f, \bar{\chi} - \chi\rangle, \tag{2.153}$$

$$\chi = (W, w) \in K_0^\varphi, \quad \forall\, \bar{\chi} \in K_0^\varphi.$$

Let the set Φ be the same as in Section 2.5.2. Consider the optimal control problem

$$\inf_{\varphi \in \Phi} J_\varepsilon(\varphi) = \inf_{\varphi \in \Phi} \int\limits_{\Gamma_\psi} |[\chi]|\, d\Gamma_\psi. \tag{2.154}$$

There exists a solution of (2.154), (2.153). We omit the arguments.

So, instead of precise nonpenetration condition (2.130) we consider the approximate condition (2.152) in this subsection. In application this approach is interesting since it is easier to find the solutions of (2.153) as compared to (2.131). In particular, it is possible to find solutions of (2.153) by using the penalty operator relative to the restriction (2.152). The displacements W and w are uncoupled in (2.153), and one can write down two variational inequalities for finding W and w, respectively. Meanwhile, when the optimal control problem (2.154) is solved, the solution φ depends on the pair (W, w), which actually means the coupling of W and w. The problem is to prove the solution proximity of (2.134), (2.131) and (2.154), (2.153), as $\varepsilon \to 0$.

A relationship between the solutions of (2.134), (2.131) and (2.154), (2.153) is characterized by the following statement. Introduce the notation

$$j_0 = \inf_{\varphi \in \Phi} J_0(\varphi), \quad j_\varepsilon = \inf_{\varphi \in \Phi} J_\varepsilon(\varphi).$$

Let φ_ε be the solution of (2.134), (2.131), and χ_ε correspond to φ_ε.

Theorem 2.18. *From the sequence $\varphi_\varepsilon, \chi_\varepsilon$ one can choose a subsequence such that*

$$\varphi_\varepsilon \to \varphi \ \textit{weakly in}\ H^2(\Omega), \quad \chi_\varepsilon \to \chi \ \textit{weakly in}\ H(\Omega_\psi), \quad j_\varepsilon \to j_0$$

as $\varepsilon \to 0$*, where* φ *is a solution of* (2.154), (2.153) *and* $\chi = \chi(\varphi)$ *is defined from* (2.153).

PROOF. Consider first any fixed element $\varphi \in \Phi$ and prove that

$$J_\varepsilon(\varphi) \to J_0(\varphi), \quad \varepsilon \to 0. \tag{2.155}$$

Let $\chi_\varepsilon(\varphi)$ be the solution of (2.131). There exists an element $\bar{\chi}$ such that $\bar{\chi} \in K_\varepsilon^\varphi$ for all ε. Substituting $\bar{\chi}$ in (2.131) as a test function implies

$$\|\chi_\varepsilon(\varphi)\|_{H(\Omega_\psi)} \le c$$

with a constant c independent of ε. Without loss of generality as $\varepsilon \to 0$ we assume that

$$\chi_\varepsilon(\varphi) \to \tilde{\chi} \text{ weakly in } H(\Omega_\psi), \tag{2.156}$$
$$\chi_\varepsilon^\pm(\varphi) \to \tilde{\chi}^\pm \text{ strongly in } L^1(\Gamma_\psi).$$

Moreover, the pair $(\varphi, \tilde{\chi})$ is a solution of the variational inequality

$$B(\tilde{w}, \bar{w} - \tilde{w}) + \langle \sigma_{ij}(\tilde{W}), e_{ij}(\bar{W} - \tilde{W})\rangle \ge \langle f, \bar{\chi} - \tilde{\chi}\rangle,$$
$$\tilde{\chi} = (\tilde{W}, \tilde{w}) \in K_0^\varphi, \quad \forall\, \bar{\chi} \in K_0^\varphi.$$

To verify this it suffices to fulfil the limiting transition in (2.131) as $\varepsilon \to 0$. Thus, $\tilde{\chi} = \chi(\varphi)$. In view of (2.156) we arrive at the desired convergence (2.155).

Let φ be a solution of the optimal control problem (2.154), (2.153). The above arguments imply

$$j_\varepsilon \le J_\varepsilon(\varphi) \to J_0(\varphi) = j_0.$$

Whence

$$\limsup j_\varepsilon \le J_0(\varphi) = j_0. \tag{2.157}$$

On the other hand, the boundedness of the set Φ provides the estimate

$$\|\varphi_\varepsilon\|_{2,\Omega} \le c \tag{2.158}$$

which is uniform in ε. Consequently, the inequality

$$B(w_\varepsilon, \bar{w} - w_\varepsilon) + \langle \sigma_{ij}(W_\varepsilon), e_{ij}(\bar{W} - W_\varepsilon)\rangle \ge \langle f, \bar{\chi} - \chi_\varepsilon\rangle, \tag{2.159}$$
$$\chi_\varepsilon = (W_\varepsilon, w_\varepsilon) \in K_\varepsilon^{\varphi_\varepsilon}, \quad \forall\, \bar{\chi} \in K_\varepsilon^{\varphi_\varepsilon}$$

enables us to derive the following estimate,

$$\|\chi_\varepsilon\|_{H(\Omega_\psi)} \le c, \tag{2.160}$$

being uniform in ε. Choosing subsequences, still denoted by $\varphi_\varepsilon, \chi_\varepsilon$, we assume that

$$\varphi_\varepsilon \to \tilde{\varphi} \quad \text{weakly in } H^2(\Omega), \quad \text{uniformly in } \overline{\Omega},$$

$$\chi_\varepsilon \to \tilde{\chi} \quad \text{weakly in } H(\Omega_\psi), \quad \varepsilon \to 0.$$

Moreover, it can be proved that for every fixed $\bar{\chi} \in K_0^{\tilde{\varphi}}$ there exists a sequence $\bar{\chi}_\varepsilon \in K_\varepsilon^{\varphi_\varepsilon}$ such that

$$\bar{\chi}_\varepsilon \to \bar{\chi} \quad \text{strongly in } H(\Omega_\psi).$$

Combining this convergence and Lemma 2.2, one can pass on to the limit in (2.159). Hence the following relation is obtained:

$$B(\tilde{w}, \bar{w} - \tilde{w}) + \langle \sigma_{ij}(\tilde{W}), e_{ij}(\bar{W} - \tilde{W}) \rangle \geq \langle f, \bar{\chi} - \tilde{\chi} \rangle,$$

$$\tilde{\chi} = (\tilde{W}, \tilde{w}) \in K_0^{\tilde{\varphi}}, \quad \forall\, \bar{\chi} \in K_0^{\tilde{\varphi}},$$

that is $\tilde{\chi} = \chi(\tilde{\varphi})$. Furthermore, just as in the proof of (2.155), the convergence $J_\varepsilon(\varphi_\varepsilon) \to J_0(\tilde{\varphi})$ holds. Hence

$$\liminf j_\varepsilon \geq J_\varepsilon(\tilde{\varphi}). \tag{2.161}$$

A comparison of (2.157) and (2.161) results in the conclusion that $\tilde{\varphi}$ is a solution of (2.154), (2.153) and $j_\varepsilon \to j_0$. As noted above, $\tilde{\chi} = \chi(\tilde{\varphi})$. Theorem 2.18 is proved.

The condition $[\chi] = 0$ is shown to provide the infinite differentiability of the solution only for $\varepsilon > 0$. For the problem (2.153), corresponding to $\varepsilon = 0$, one cannot state that $w \in H^2(\mathcal{O}(x^0))$ provided that $[\chi] = 0$ on $\mathcal{O}(x^0) \cap \Gamma_\psi$, since, in general, in this case $\partial w / \partial \nu \neq 0$ on $\mathcal{O}(x^0) \cap \Gamma_\psi$. The result of Theorem 2.17 on C^∞-regularity actually shows that the condition $[\chi] = 0$ provides the disappearance of singularity which takes place in view of the presence of a crack. It means that under the condition mentioned, we can 'forget' about the crack since the behaviour of the plate is the same as that without the crack.

2.6 Solving methods for plates with cracks

In this section we proceed to study the plate model with the crack described in Sections 2.4, 2.5. The corresponding variational inequality is analysed provided that the nonpenetration condition holds. By the principles of Section 1.3, we propose approximate equations in the two-dimensional case and analytical solutions in the one-dimensional case (see Kovtunenko, 1996b, 1997b).

2.6.1 Iteration penalty method

Let a plate occupy a bounded domain $\Omega \subset R^2$ with smooth boundary Γ. Inside Ω there is a graph Γ_c of a sufficiently smooth function. The graph Γ_c corresponds to the crack in the plate (see Section 1.1.7). A unit vector $\nu = (\nu^1, \nu^2)$ being normal to Γ_c defines the surfaces of the crack, positive Γ_c^+ and negative Γ_c^-. Denote $\Omega_c = \Omega\backslash\Gamma_c$. The unknown vector $\chi = (W, w)$ of horizontal $W = (w^1, w^2)$ and vertical w displacements of the mid-surface points of the plate meets the following boundary conditions: first, the clamping condition at the outer boundary

$$w = \frac{\partial w}{\partial n} = w^1 = w^2 = 0 \quad \text{on } \Gamma, \tag{2.162}$$

with n standing for the normal to Γ; second, the nonpenetration condition at the crack surfaces

$$[W]\nu \geq \left|\left[\frac{\partial w}{\partial \nu}\right]\right| \quad \text{on } \Gamma_c, \tag{2.163}$$

with $[s]$ denoting the jump of a function s across Γ_c, i.e. $[s] = s|_{\Gamma_c^+} - s|_{\Gamma_c^-}$. Introduce the linear functions

$$\varphi(\chi) = [W]\nu + \left[\frac{\partial w}{\partial \nu}\right], \quad \psi(\chi) = [W]\nu - \left[\frac{\partial w}{\partial \nu}\right].$$

Then condition (2.163) is equivalent to the following inequalities: $\varphi(\chi) \geq 0$ and $\psi(\chi) \geq 0$. Define the Hilbert space

$$H(\Omega_c) = \{\chi = (W, w) \in H^1(\Omega_c) \times H^1(\Omega_c) \times H^2(\Omega_c) \mid \quad \chi \text{ satisfies } (2.162)\}$$

and the closed convex set

$$K = \{\chi \in H(\Omega_c) \mid \quad \varphi(\chi) \geq 0, \quad \psi(\chi) \geq 0\}.$$

Denote by $H(\Omega_c)^\star$ the space dual of $H(\Omega_c)$. Let us recall the following bilinear forms from Sections 2.4, 2.5:

$$a(\chi, \bar{\chi}) = B(W, \bar{W}) + b(w, \bar{w}), \quad \bar{\chi} = (\bar{W}, \bar{w}), \quad \bar{W} = (\bar{w}^1, \bar{w}^2),$$

$$B(W, \bar{W}) = \int_{\Omega_c} \sigma_{ij}(W)\varepsilon_{ij}(\bar{W})\, d\Omega_c,$$

$$b(w, \bar{w}) = \int_{\Omega_c} (w_{xx}\bar{w}_{xx} + w_{yy}\bar{w}_{yy} + \kappa(w_{xx}\bar{w}_{yy} + w_{yy}\bar{w}_{xx})$$

$$+2(1-\kappa)w_{xy}\bar{w}_{xy})\, d\Omega_c.$$

The Poisson ratio $0 < \kappa < 1/2$ is given. By the first Korn inequality, we have

$$B(W, W) \geq c_1\|W\|_1^2, \quad c_1 > 0.$$

We also know that

$$b(w, w) \geq c_2\|w\|_2^2, \quad c_2 > 0,$$

with $\|\cdot\|_k$ to be the norm in $H^k(\Omega_c)$, $k=1,2$. Therefore, we have the inequality

$$a(\chi,\chi)\geq c_3\|\chi\|^2,\quad c_3>0, \tag{2.164}$$

with $\|\cdot\|$ to be the norm in $H(\Omega_c)$. So we can introduce the scalar product

$$(\chi,\bar{\chi})=a(\chi,\bar{\chi})+\int\limits_{\Gamma_c}\Big(\varphi(\chi)\varphi(\bar{\chi})+\psi(\chi)\psi(\bar{\chi})\Big)\,d\Gamma_c$$

and the equivalent norm $\|\chi\|^2=(\chi,\chi)$ in $H(\Omega_c)$. Let $f=(f^1,f^2,f^3)\in L^2(\Omega_c)$ be some given functions of the external forces. The equilibrium problem for a plate with a crack is formulated as the following variational inequality:

$$\chi\in K,\quad a(\chi,\bar{\chi}-\chi)\geq\langle f,\bar{\chi}-\chi\rangle\quad\forall\bar{\chi}\in K. \tag{2.165}$$

The angular brackets $\langle\cdot,\cdot\rangle$ denote the integration over Ω_c. In virtue of the linearity, boundedness, and coercivity of the form $a(\cdot,\cdot)$, there exists a unique solution to (2.165).

We shall use the Green formula

$$a(\chi,\bar{\chi})=\langle A\chi,\bar{\chi}\rangle-\int\limits_{\Gamma_c}\left[\sigma_\nu(W)\bar{W}\nu+\sigma_\tau(W)\bar{W}\tau+m(w)\frac{\partial\bar{w}}{\partial\nu}-t(w)\bar{w}\right]d\Gamma_c,$$

where

$$A\chi=\left(-\sigma_{1j,j}(W),\ -\sigma_{2j,j}(W),\ \Delta^2 w\right),$$

$$\sigma_\nu(W)=\sigma_{ij}(W)\nu^j\nu^i,\quad \sigma_{\tau i}(W)=\sigma_{ij}(W)\nu^j-\sigma_\nu(W)\nu^i,\quad i=1,2,$$

$$m(w)=\kappa\Delta w+(1-\kappa)\frac{\partial^2 w}{\partial\nu^2},\quad t(w)=\frac{\partial}{\partial\nu}\left(\Delta w+(1-\kappa)\frac{\partial^2 w}{\partial\tau^2}\right).$$

Here $\tau=(-\nu^2,\nu^1)$ is the unit tangent vector to Γ_c. Let a solution χ possess a sufficient smoothness. Then we can rewrite (2.165) as

$$\langle A\chi-f,\bar{\chi}-\chi\rangle-\int\limits_{\Gamma_c}\left([\sigma_\nu(W)(\bar{W}-W)\nu]+\left[m(w)\frac{\partial(\bar{w}-w)}{\partial\nu}\right]\right.$$

$$\left.+[\sigma_\tau(W)(\bar{W}-W)\tau]-[t(w)(\bar{w}-w)]\right)d\Gamma_c\geq 0\quad\forall\bar{\chi}\in K.$$

Further, we vary the test functions $\bar{\chi}$ and use the relation

$$\sigma_\nu(W)[\bar{W}-W]\nu+m(w)\left[\frac{\partial(\bar{w}-w)}{\partial\nu}\right]$$

$$=\frac{1}{2}\left(\sigma_\nu(W)+m(w)\right)\varphi(\bar{\chi}-\chi)+\frac{1}{2}\left(\sigma_\nu(W)-m(w)\right)\psi(\bar{\chi}-\chi).$$

Then it follows that the solution of the variational inequality (2.165) is characterized by the equilibrium equation

$$A\chi = f \quad \text{in } \Omega_c,$$

and the boundary conditions at Γ_c

$$[\sigma_\nu(W)] = 0, \quad [m(w)] = 0, \quad \sigma_\tau(W) = 0,$$

$$t(w) = 0, \quad [W]\nu \geq \left| \left[\frac{\partial w}{\partial \nu} \right] \right|, \quad -\sigma_\nu(W) \geq |m(w)|,$$

$$\Big(\sigma_\nu(W) + m(w)\Big)\varphi(\chi) = 0, \quad \Big(\sigma_\nu(W) - m(w)\Big)\psi(\chi) = 0.$$

Note that, summing the two last equalities, we obtain the condition

$$\sigma_\nu(W)[W]\nu + m(w)\left[\frac{\partial w}{\partial \nu}\right] = 0,$$

which has the same form as in Section 2.5.

We now construct a penalized problem. To this end, define the penalty operator $\beta : H(\Omega_c) \to H(\Omega_c)^\star$ by the formula

$$\langle \beta(\chi), \bar{\chi} \rangle = -\int_{\Gamma_c} \Big(\varphi^-(\chi)\varphi(\bar{\chi}) + \psi^-(\chi)\psi(\bar{\chi})\Big)\, d\Gamma_c.$$

Here the angular brackets denote the duality pairing between $H(\Omega_c)$ and $H(\Omega_c)^\star$; the superscript '-' signifies the negative part of a function, i.e. $s = s^+ - s^-$, $s^+, s^- \geq 0$, $s^+ s^- = 0$. It is seen that β is a monotonous and semicontinuous operator. Denote by $\chi^\varepsilon \in H(\Omega_c)$ a solution of the equation

$$a(\chi^\varepsilon, \bar{\chi}) + \varepsilon^{-1}\langle \beta(\chi^\varepsilon), \bar{\chi} \rangle = \langle f, \bar{\chi} \rangle \quad \forall \bar{\chi} \in H(\Omega_c) \tag{2.166}$$

depending on a small parameter $\varepsilon > 0$. Supposing the solution χ^ε to be sufficiently smooth, penalized equation (2.166) is equivalent to the following boundary value problem:

$$A\chi^\varepsilon = f \quad \text{in } \Omega_c,$$

$$[\sigma_\nu(W^\varepsilon)] = 0, \quad [m(w^\varepsilon)] = 0, \quad \sigma_\tau(W^\varepsilon) = 0, \quad t(w^\varepsilon) = 0,$$

$$\sigma_\nu(W^\varepsilon) = -\varepsilon^{-1}(\varphi^-(\chi^\varepsilon) + \psi^-(\chi^\varepsilon)), \quad m(w^\varepsilon) = -\varepsilon^{-1}(\varphi^-(\chi^\varepsilon) - \psi^-(\chi^\varepsilon)).$$

Fix the parameter ε. To linearize the left-hand side of (2.166), for arbitrary $\chi^{\varepsilon,0} \in H(\Omega_c)$, we construct the following iterative procedure for $n = 0, 1, 2, \ldots$,

$$a(\chi^{\varepsilon,n+1}, \bar{\chi}) + \varepsilon^{-1}(\chi^{\varepsilon,n+1}, \bar{\chi}) \tag{2.167}$$
$$= \langle f, \bar{\chi} \rangle + \varepsilon^{-1}(\chi^{\varepsilon,n}, \bar{\chi}) - \varepsilon^{-1}\langle \beta(\chi^{\varepsilon,n}), \bar{\chi} \rangle.$$

By the above properties for the operator of (2.167), there exists a unique solution $\chi^{\varepsilon,n+1} \in H(\Omega_c)$ to the problem (2.167). Then, for $\chi^{\varepsilon,n+1}$ sufficiently smooth, the corresponding boundary value problem takes the form

$$A\left(\chi^{\varepsilon,n+1} + \frac{1}{\varepsilon}(\chi^{\varepsilon,n+1} - \chi^{\varepsilon,n})\right) = f \quad \text{in } \Omega_c,$$

$$\left[\sigma_\nu\left(W^{\varepsilon,n+1} + \frac{1}{\varepsilon}(W^{\varepsilon,n+1} - W^{\varepsilon,n})\right)\right] = 0,$$

$$\left[m\left(w^{\varepsilon,n+1} + \frac{1}{\varepsilon}(w^{\varepsilon,n+1} - w^{\varepsilon,n})\right)\right] = 0,$$

$$\sigma_\tau\left(W^{\varepsilon,n+1} + \frac{1}{\varepsilon}(W^{\varepsilon,n+1} - W^{\varepsilon,n})\right) = 0,$$

$$t\left(w^{\varepsilon,n+1} + \frac{1}{\varepsilon}(w^{\varepsilon,n+1} - w^{\varepsilon,n})\right) = 0,$$

$$\sigma_\nu\left(W^{\varepsilon,n+1} + \frac{1}{\varepsilon}(W^{\varepsilon,n+1} - W^{\varepsilon,n})\right) - \frac{2}{\varepsilon}[W^{\varepsilon,n+1} - W^{\varepsilon,n}]$$

$$= -\frac{1}{\varepsilon}(\varphi^-(\chi^{\varepsilon,n}) + \psi^-(\chi^{\varepsilon,n})),$$

$$m\left(w^{\varepsilon,n+1} + \frac{1}{\varepsilon}(w^{\varepsilon,n+1} - w^{\varepsilon,n})\right) - \frac{2}{\varepsilon}\left[\frac{\partial(w^{\varepsilon,n+1} - w^{\varepsilon,n})}{\partial\nu}\right]$$

$$= -\frac{1}{\varepsilon}(\varphi^-(\chi^{\varepsilon,n}) - \psi^-(\chi^{\varepsilon,n})).$$

Theorem 2.19. *We have* $\chi^{\varepsilon,n+1} \to \chi^\varepsilon$ *strongly in* $H(\Omega_c)$ *as* $n \to \infty$. *Moreover,*

$$\|\chi^{\varepsilon,n+1} - \chi^\varepsilon\|^2 \le (1 + 2c_3\varepsilon)^{-(n+1)}\|\chi^{\varepsilon,0} - \chi^\varepsilon\|^2 \tag{2.168}$$

and $\chi^\varepsilon \to \chi$ *strongly in* $H(\Omega_c)$ *as* $\varepsilon \to 0$, *where* $\chi^{\varepsilon,n+1}$, χ^ε, χ *are the solutions of* (2.167), (2.166), (2.165), *respectively.*

PROOF. Subtract (2.166) from (2.167) and add the term $-\varepsilon^{-1}(\chi^\varepsilon, \bar\chi)$ to both sides of the obtained inequality. It gives

$$a(\chi^{\varepsilon,n+1} - \chi^\varepsilon, \bar\chi) + \varepsilon^{-1}(\chi^{\varepsilon,n+1} - \chi^\varepsilon, \bar\chi) = \varepsilon^{-1}a(\chi^{\varepsilon,n} - \chi^\varepsilon, \bar\chi)$$

$$+\varepsilon^{-1}\int\limits_{\Gamma_c}\Big(\left(\varphi(\chi^{\varepsilon,n}) - \varphi(\chi^\varepsilon) + \varphi^-(\chi^{\varepsilon,n}) - \varphi^-(\chi^\varepsilon)\right)\varphi(\bar\chi) \tag{2.169}$$

$$+\left(\psi(\chi^{\varepsilon,n}) - \psi(\chi^\varepsilon) + \psi^-(\chi^{\varepsilon,n}) - \psi^-(\chi^\varepsilon)\right)\psi(\bar\chi)\Big)\,d\Gamma_c.$$

Since $s - t + s^- - t^- = s^+ - t^+ \le |s - t|$, by using the Holder inequality, we can estimate the right-hand side of (2.169) from above by the value

$$\frac{1}{2\varepsilon}\Big(a(\chi^{\varepsilon,n} - \chi^{\varepsilon}, \chi^{\varepsilon,n} - \chi^{\varepsilon}) + a(\bar{\chi},\bar{\chi}) + \int\limits_{\Gamma_c} (\varphi^2(\chi^{\varepsilon,n} - \chi^{\varepsilon}) + \varphi^2(\bar{\chi})$$

$$+ \psi^2(\chi^{\varepsilon,n} - \chi^{\varepsilon}) + \psi^2(\bar{\chi}))\, d\Gamma_c\Big) = \frac{1}{2\varepsilon}\left(\|\chi^{\varepsilon,n} - \chi^{\varepsilon}\|^2 + \|\bar{\chi}\|^2\right).$$

Consider this relation at the point $\bar{\chi} = \chi^{\varepsilon,n+1} - \chi^{\varepsilon}$. By (2.164), from (2.169) we have the inequality

$$(c_3 + \varepsilon^{-1})\|\chi^{\varepsilon,n+1} - \chi^{\varepsilon}\|^2 \le (2\varepsilon)^{-1}(\|\chi^{\varepsilon,n} - \chi^{\varepsilon}\|^2 + \|\chi^{\varepsilon,n+1} - \chi^{\varepsilon}\|^2).$$

Thus,

$$\|\chi^{\varepsilon,n+1} - \chi^{\varepsilon}\|^2 \le (1 + 2c_3\varepsilon)^{-1}\|\chi^{\varepsilon,n} - \chi^{\varepsilon}\|^2.$$

Repeating this estimate for n tending to 0, we obtain (2.168) and the first assertion of Theorem 2.19 on strong convergence.

In a standard way (see Section 1.3), the properties of the operators $a(\cdot\,,\cdot)$ and $\beta(\cdot)$ imply that

$$\chi^{\varepsilon} \to \chi \quad \text{weakly in } H(\Omega_c) \quad \text{as } \varepsilon \to 0. \tag{2.170}$$

Subtract $a(\chi, \bar{\chi})$ from (2.166) and consider the obtained equation at the element $\bar{\chi} = \chi^{\varepsilon} - \chi$. It gives

$$a(\chi^{\varepsilon} - \chi, \chi^{\varepsilon} - \chi) - \varepsilon^{-1}\int\limits_{\Gamma_c} (\varphi^-(\chi^{\varepsilon})\varphi(\chi^{\varepsilon} - \chi) + \psi^-(\chi^{\varepsilon})\psi(\chi^{\varepsilon} - \chi))\, d\Gamma_c$$

$$= \langle f, \chi^{\varepsilon} - \chi\rangle - a(\chi, \chi^{\varepsilon} - \chi).$$

We estimate the left-hand side of this equality from below which provides

$$c_3\|\chi^{\varepsilon} - \chi\|^2 + \varepsilon^{-1}\int\limits_{\Gamma_c} ((\varphi^-(\chi^{\varepsilon}))^2 + (\psi^-(\chi^{\varepsilon}))^2)\, d\Gamma_c \le \langle f, \chi^{\varepsilon} - \chi\rangle - a(\chi, \chi^{\varepsilon} - \chi).$$

Therefore, (2.170) implies the second assertion of Theorem 2.19 on strong convergence. The theorem is proved.

Unlike (2.166), the constructed iterative equation (2.167) is linear, which allows us to apply the standard numerical methods to solve it.

2.6.2 A bar with a cut

We consider here a one-dimensional case corresponding to a bar with a cut. Let the mid-line of a bar coincide with the segment $[0, 1]$, and the bar have a vertical cut at the fixed point y, $0 < y < 1$. Denote $\Omega_y = (0, y) \cup (y, 1)$. We have to find the vector $\chi = (W, w)$ of horizontal displacements $W = W(x)$ and vertical displacements $w = w(x)$ of the bar points $x \in \Omega_y$ under the action of the external forces $f = (f^1, f^2)$.

The clamping conditions

$$W = w = w_x = 0 \quad \text{at} \;\; x = 0, 1 \tag{2.171}$$

are assumed to hold. In accordance with the Kirchhoff–Love kinematic hypothesis, the displacements field along the thickness $z \in [-\varepsilon, \varepsilon]$ of the bar is given by the following formulae:

$$W(x, z) = W(x) - z w_x(x), \quad w(x, z) = w(x).$$

The condition providing the nonpenetration of the cut faces along the cut thickness is as follows:

$$W(y + 0, z) - W(y - 0, z) \geq 0 \quad \forall z, \; |z| \leq \varepsilon.$$

Substituting here the function values, one gets

$$[W] \geq z[w_x] \quad \forall z, \; |z| \leq \varepsilon,$$

where $[s]$ denotes the jump of the function $s(x)$ at the point y, i.e. $[s] = s(y + 0) - s(y - 0)$. Obviously, the last inequality is equivalent to

$$[W] \geq \varepsilon \Big| [w_x] \Big|.$$

Thus, we obtain the nonpenetration condition of the cut faces, the same as for a plate with a crack. Later on, for simplicity we consider the case $\varepsilon = 1$, i.e.

$$[W] \geq \Big| [w_x] \Big|. \tag{2.172}$$

Consider the linear functions $\phi(\chi) = [W] + [w_x]$, $\psi(\chi) = [W] - [w_x]$. In this case (2.172) is equivalent to the inequalities

$$\phi(\chi) \geq 0, \quad \psi(\chi) \geq 0. \tag{2.173}$$

Define next the Hilbert space

$$H(\Omega_y) = \{\chi = (W, w) \in H^1(\Omega_y) \times H^2(\Omega_y) \mid \;\; \chi \;\text{ satisfies }\; (2.171)\},$$

its dual space $H(\Omega_y)^\star$ and a closed convex subset

$$K = \{\chi \in H(\Omega_y) \mid \;\; \chi \;\text{ satisfies }\; (2.173)\}.$$

Also, we introduce the scalar product in $H(\Omega_y)$ as follows,

$$(\chi, \bar{\chi}) = \int\limits_{\Omega_y} W_x \bar{W}_x \, dx + \int\limits_{\Omega_y} w_{xx} \bar{w}_{xx} \, dx, \quad \bar{\chi} = (\bar{W}, \bar{w}),$$

and the corresponding norm $(\chi, \chi) = ||\chi||^2$ in agreement with the estimates

$$\int_{\Omega_y} W^2\,dx \le \int_{\Omega_y} W_x^2\,dx, \quad \int_{\Omega_y} w^2\,dx \le \int_{\Omega_y} w_x^2\,dx \le \int_{\Omega_y} w_{xx}^2\,dx.$$

Let $f \in H(\Omega_y)^\star$. The equilibrium problem for the clamped elastic bar with the cut is formulated as the following variational inequality:

$$\chi \in K, \quad (\chi, \bar{\chi} - \chi) \ge \langle f, \bar{\chi} - \chi\rangle \quad \forall \bar{\chi} \in K. \tag{2.174}$$

Here the brackets $\langle \cdot, \cdot \rangle$ denote the duality pairing between $H(\Omega_y)$ and $H(\Omega_y)^\star$. It is easy to see that there exists a unique solution to (2.174).

2.6.3 Construction of analytical solutions

Let $I^{-1} : H(\Omega_y)^\star \to H(\Omega_y)$ be the inverse duality mapping; then

$$(I^{-1}f, \bar{\chi}) = \langle f, \bar{\chi}\rangle \quad \forall \bar{\chi} \in H(\Omega_y). \tag{2.175}$$

We denote $\chi^0 = I^{-1}f$. Let $f = (f^1, f^2) \in L^2(\Omega_y)$ be given. Integrating by parts, we see that (2.175) is equivalent to the following boundary value problem,

$$\begin{gathered} -W_{xx}^0 = f^1, \quad w_{xxxx}^0 = f^2, \qquad \text{in } \Omega_y, \\ W_x^0(y) = w_{xx}^0(y) = w_{xxx}^0(y) = 0, \\ W^0(0) = w_x^0(0) = w_{xx}^0 = W^0(1) = w_x^0(1) = w_{xx}^0(1) = 0, \end{gathered} \tag{2.176}$$

and there exists a unique solution $\chi^0 = (W^0, w^0) \in H(\Omega_y)$, $\chi^0 \in H^2(\Omega_y) \times H^4(\Omega_y)$.

By the above notation, the variational inequality (2.174) is equivalent to

$$\chi \in K, \quad (\chi^0 - \chi, \chi - \bar{\chi}) \ge 0 \quad \forall \bar{\chi} \in K. \tag{2.177}$$

Let P be the projection operator of $H(\Omega_y)$ onto K, i.e. for any $s \in H(\Omega_y)$ the unique projection $Ps \in K$ exists such that

$$(s - Ps, Ps - \bar{\chi}) \ge 0 \quad \forall \bar{\chi} \in K. \tag{2.178}$$

Comparing (2.177) and (2.178), it is clear that (2.177) is equivalent to the following equation (see Section 1.3):

$$\chi = P\chi^0. \tag{2.179}$$

To construct this projection, we introduce the function $\alpha \in H(\Omega_y) \cap C^\infty(\Omega_y)$ by the formula

$$\alpha(x) = \frac{1}{2}\begin{cases} x^2 & , x \in [0, y) \\ (x-1)^2 & , x \in (y, 1] \end{cases}$$

and introduce the function $\theta(\chi^0) = (\theta^1(\chi^0), \theta^2(\chi^0))$ from $H(\Omega_y) \cap C^\infty(\Omega_y)$ as follows:

$$\theta^1(\chi^0) = \frac{1}{2}\Big(\phi^-(\chi^0) + \psi^-(\chi^0)\Big)\alpha_x, \quad \theta^2(\chi^0) = \frac{1}{2}\Big(\phi^-(\chi^0) - \psi^-(\chi^0)\Big)\alpha.$$

Here the superscript '-' means the negative part of a number, i.e. $a = a^+ - a^-$; $a^+, a^- \geq 0$; $a^+a^- = 0$. We indicate the following properties of the constructed functions:

$$\theta^1(\chi^0)_{xx} = \theta^2(\chi^0)_{xxx} = 0, \qquad \text{in } \Omega_y, \tag{2.180}$$

$$\theta^1(\chi^0)_x(y) = \frac{1}{2}\Big(\phi^-(\chi^0) + \psi^-(\chi^0)\Big), \tag{2.181}$$

$$\theta^2(\chi^0)_{xx}(y) = \frac{1}{2}\Big(\phi^-(\chi^0) - \psi^-(\chi^0)\Big), \tag{2.182}$$

$$\phi(\theta(\chi^0)) = -\phi^-(\chi^0), \quad \psi(\theta(\chi^0)) = -\psi^-(\chi^0). \tag{2.183}$$

Theorem 2.20. *The function*

$$\chi = \chi^0 - \theta(\chi^0) \tag{2.184}$$

is the unique solution of the variational inequality (2.174), *where* χ^0 *is the solution of* (2.176).

Proof. Taking into account (2.179), we have to prove that

$$P\chi^0 = \chi^0 - \theta(\chi^0).$$

Note that $\chi^0 - \theta(\chi^0)$ belongs to K. Actually, in view of the linearity for ϕ and ψ, (2.183) provides

$$\phi(\chi^0 - \theta(\chi^0)) = \phi(\chi^0) - \phi(\theta(\chi^0)) = \phi^+(\chi^0) - \phi^-(\chi^0) + \phi^-(\chi^0) = \phi^+(\chi^0) \geq 0$$

and, similarly, $\psi(\chi^0 - \theta(\chi^0)) = \psi^+(\chi^0) \geq 0$.

We next verify (2.178), i.e.

$$(\theta(\chi^0), \chi^0 - \theta(\chi^0) - \bar{\chi}) \geq 0, \quad \forall \bar{\chi} \in K.$$

By the smoothness of $\theta(\chi^0)$, the following formula holds for every $\bar{\chi} = (\bar{W}, \bar{w}) \in H(\Omega_y)$:

$$(\theta(\chi^0), \bar{\chi}) = -\int\limits_{\Omega_y} (\theta^1(\chi^0)_{xx}\bar{W} + \theta^2(\chi^0)_{xxx}\bar{w}_x)\,dx$$

$$- \left[\theta^1(\chi^0)_x\bar{W} + \theta^2(\chi^0)_{xx}\bar{w}_x\right].$$

Relations (2.180)–(2.182) imply

$$(\theta(\chi^0), \bar{\chi}) = -\frac{1}{2}\Big((\phi^-(\chi^0) + \psi^-(\chi^0))\,[\bar{W}] + (\phi^-(\chi^0) - \psi^-(\chi^0))\,[\bar{w}_x] \Big)$$

$$= -\frac{1}{2}\Big(\phi^-(\chi^0)\phi(\bar{\chi}) + \psi^-(\chi^0)\psi(\bar{\chi}) \Big).$$

Taking into account (2.183), by the inclusion $\bar{\chi} \in K$, we get

$$(\theta(\chi^0), \chi^0 - \theta(\chi^0) - \bar{\chi}) = -\frac{1}{2}\Big(\phi^-(\chi^0)\,(\phi^+(\chi^0) - \phi(\bar{\chi}))$$

$$+ \psi^-(\chi^0)\,(\psi^+(\chi^0) - \psi(\bar{\chi}))\Big) = \frac{1}{2}\,(\phi^-(\chi^0)\phi(\bar{\chi}) + \psi^-(\chi^0)\psi(\bar{\chi})) \geq 0.$$

The proof is completed.

It follows from (2.184) and (2.176) that, if $f \in H^n(\Omega_y) \times H^m(\Omega_y)$ for $n, m \geq 0$, then $u \in H^{n+2}(\Omega_y) \times H^{m+4}(\Omega_y)$.

By the above properties of $\chi^0, \theta(\chi^0)$, one can easily verify that the constructed function $\chi = (W, w)$ justifies the following boundary value problem:

$$-W_{xx} = f^1, \quad w_{xxxx} = f^2, \qquad \text{in } \Omega_y,$$

$$W_x(y) = -\frac{1}{2}\,(\phi^-(\chi^0) + \psi^-(\chi^0)),$$

$$w_{xx}(y) = -\frac{1}{2}\,(\phi^-(\chi^0) - \psi^-(\chi^0)), \quad w_{xxx}(y) = 0,$$

$$[W] = \frac{1}{2}\,(\phi^+(\chi^0) + \psi^+(\chi^0)), \quad [w_x] = \frac{1}{2}\,(\phi^+(\chi^0) - \psi^+(\chi^0)).$$

Let some function $\chi = (W, w)$ belong to $H(\Omega_y) \cap \big(H^2(\Omega_y) \times H^4(\Omega_y)\big)$, and the following boundary conditions be fulfilled:

$$[W_x] = [w_{xx}] = 0, \quad w_{xxx}(y) = 0, \quad (W_x(y) + w_{xx}(y))\,\phi(\chi) = 0,$$

$$(W_x(y) - w_{xx}(y))\,\psi(\chi) = 0, \quad \phi(\chi) \geq 0, \quad \psi(\chi) \geq 0, \quad -W_x(y) \geq |w_{xx}(y)|.$$

Then χ is the solution of the variational inequality (2.174) with the right-hand side $f = (-W_{xx}, w_{xxxx})$. For instance, this holds provided that $u \in H(\Omega_y) \cap \big(H_0^2(\Omega_y) \times H_0^4(\Omega_y)\big)$.

We give some examples of exact solutions for given f.

EXAMPLE 1. Let $f^1(x) \equiv a$, $a > 0$, $f^2(x) \equiv 0$; then $w(x) \equiv 0$. There are two cases. If $0 < y \leq 1/2$, then

$$W(x) = -\frac{a}{2}\begin{cases} x^2 - 2yx & , x \in [0, y), \\ (x-1)^2 - 2(y-1)(x-1) & , x \in (y, 1], \end{cases}$$

with $[W] = (1/2 - y)a \geq 0$. If $1/2 \leq y < 1$, then

$$W(x) = -1/2\,ax(x-1), \quad [W] = 0.$$

EXAMPLE 2. Let $y = 1/2$, $f^2(x) \equiv 0$,

$$f^1(x) = \begin{cases} a_1 & , x \in [0, 1/2), \\ a_2 & , x \in (1/2, 1]. \end{cases}$$

Therefore, $w(x) \equiv 0$. If $a_2 \leq a_1$, then

$$W(x) = -\frac{1}{8}\begin{cases} 4a_1x^2 - (3a_1 + a_2)x & , x \in [0, 1/2), \\ 4a_2(x-1)^2 + (3a_2 + a_1)(x-1) & , x \in (1/2, 1], \end{cases}$$

and $[W] = 0$. If $a_2 \geq a_1$, then

$$W(x) = -1/2\,x(x-1)\,f^1(x), \quad [W] = 1/8(a_2 - a_1) \geq 0.$$

EXAMPLE 3. Let $f^1(x) \equiv 0$, $y = 1/2$,

$$f^2(x) = \begin{cases} b_1 & , x \in [0, 1/2), \\ b_2 & , x \in (1/2, 1]. \end{cases}$$

If $b_1 + b_2 \geq 0$, then

$$W(x) = -\frac{b_1 + b_2}{96}\begin{cases} x & , x \in [0, 1/2), \\ x - 1 & , x \in (1/2, 1], \end{cases}$$

$$w(x) = \frac{1}{192}\begin{cases} 8b_1x^4 - 16b_1x^3 + (11b_1 - b_2)x^2, \\ 8b_2(x-1)^4 + 16b_2(x-1)^3 + (11b_2 - b_1)(x-1)^2, \end{cases}$$

and

$$[W] = \frac{1}{96}(b_1 + b_2) \geq 0, \quad [w] = \frac{1}{128}(b_2 - b_1), \quad [w_x] = -\frac{1}{96}(b_1 + b_2) \leq 0.$$

We come to the following conclusions: the presentation (2.184) gives that $f^2(x) \equiv 0$ entails $w(x) \equiv 0$ (Examples 1, 2); $f^1(x) \equiv 0$ does not necessarily entail $W(x) \equiv 0$ (Example 3); $[f^1] = 0$ or $[f^2] = 0$ do not guarantee $[W] = 0$, $[w] = 0$, $[w_x] = 0$ (Examples 1, 3).

2.7 Contact problem for a shell with a crack

The contact problem for a shallow shell containing a vertical crack is considered. The solution of the problem satisfies two inequality restrictions describing the mutual nonpenetration of the shell and a punch, and the condition of nonpenetration for the crack faces. The purpose of this section is to investigate a control problem using external loading with an objective functional describing the crack opening. The regularity of the solution is investigated near the tips of the crack. In particular, for a crack with zero opening the solution is shown to belong to the class C^∞. The convergence of the solutions of the optimal control problems is analysed provided that the parameters of the model are perturbed. The results proved in this section were obtained in (Khludnev, 1995b).

2.7.1 Statement of the problem

Consider a shallow shell whose mid-surface occupies the domain $\Omega_\psi = \Omega \setminus \Gamma_\psi$, where $\Omega \subset R^2$ is a bounded domain with smooth boundary Γ, and Γ_ψ is the graph of the function $y = \psi(x)$, $x \in [0,1]$, $(x,y) \in \Omega$. Let $\chi = (W, w)$ be the displacement vector for points of the mid-surface of the shell, and $W = (w^1, w^2)$. We introduce the following notation for the components of the strain and stress tensors,

$$e_{ij} = \varepsilon_{ij}(W) + k_{ij}w, \quad \varepsilon_{ij}(W) = \frac{1}{2}\left(\frac{\partial w^i}{\partial x_j} + \frac{\partial w^j}{\partial x_i}\right), \quad (x_1 = x,\ x_2 = y)$$

$$\sigma_{11} = e_{11} + \kappa e_{22}, \quad \sigma_{22} = e_{22} + \kappa e_{11}, \quad \sigma_{12} = (1-\kappa)e_{12},$$

where $\kappa = const$, $0 < \kappa < 1/2$. Assume that the curvatures of the shell satisfy the inclusions $k_{ij} \in C^1(\overline{\Omega}_\psi)$. Here and throughout $i,j = 1,2$. The energy functional of the shell can be written in the form

$$\Pi_u(\chi) = \frac{1}{2}B(w,w) + \frac{1}{2}\langle \sigma_{ij}(W), e_{ij}(W)\rangle - \langle u, \chi\rangle,$$

where $u = (u_1, u_2, u_3)$ is the external force vector, the brackets $\langle\cdot,\cdot\rangle$ denote integration over Ω_ψ, and the bilinear form describing the bending properties of the shell has the form

$$B(w,v) = \int\limits_{\Omega_\psi} (w_{xx}v_{xx} + w_{yy}v_{yy} + \kappa w_{xx}v_{yy} + \kappa w_{yy}v_{xx} + 2(1-\kappa)w_{xy}v_{xy})\, d\Omega_\psi.$$

By simplicity we specify the following boundary conditions on the outer boundary:

$$w = \partial w/\partial n = W = 0 \quad \text{on} \quad \Gamma.$$

The model of the shell under consideration is therefore described by the fact that its mid-surface is identified with a plane domain, while at the same time the curvature of the shell is not in general zero (see Section 1.1.3). Let $\psi \in H_0^3(0,1)$, and ν be the normal to the curve $y = \psi(x)$, $x \in (0,1)$. Then the condition of mutual nonpenetration for the crack faces can be written as follows:

$$[W]\nu \geq \delta \left| \left[\frac{\partial w}{\partial \nu} \right] \right| \quad \text{on } \Gamma_\psi. \tag{2.185}$$

We assume that the surface $z = \Phi(x,y)$ describes the shape of the punch, $(x,y) \in \Omega$, $\Phi \in C^1(\overline{\Omega}) \cap C^\infty(\Omega)$. In this case the mutual nonpenetration condition for the shell and the punch, in the linear approximation, has the form (see Section 1.1.5)

$$w - W\nabla\Phi \geq \Phi \quad \text{in} \quad \Omega_\psi. \tag{2.186}$$

Suppose further that the subspace $H^{1,0}(\Omega_\psi)$ of the Sobolev space $H^1(\Omega_\psi)$ consists of elements which vanish on Γ. Elements from $H^{2,0}(\Omega_\psi)$ vanish similarly together with their first derivatives on Γ, $H^{2,0}(\Omega_\psi) \subset H^2(\Omega_\psi)$. We denote by $H(\Omega_\psi)$ the space $H^{1,0}(\Omega_\psi) \times H^{1,0}(\Omega_\psi) \times H^{2,0}(\Omega_\psi)$ and introduce the set of admissible displacements of the shell,

$$K_\delta = \{(W, w) \in H(\Omega_\psi) \mid \quad (W, w) \quad \text{satisfy} \ \ (2.185), (2.186)\}.$$

Here inequalities (2.185), (2.186) are assumed to be satisfied almost everywhere in the Lebesgue sense on Γ_ψ and in Ω_ψ. We assume that $\Phi < 0$ on Γ, so that the set K_δ is nonempty. The equilibrium problem for a shallow shell with a solution satisfying the nonpenetration conditions (2.185), (2.186) can be formulated as follows:

$$\inf_{\chi \in K_\delta} \Pi_u(\chi). \tag{2.187}$$

Because of the convexity and differentiability of the functional Π_u on $H(\Omega_\psi)$, problem (2.187) is equivalent to the variational inequality

$$\Pi_u'(\chi)(\bar{\chi} - \chi) \geq 0, \quad \chi \in K_\delta, \quad \forall\, \bar{\chi} \in K_\delta,$$

where $\Pi_u'(\chi)$ is the derivative of the functional Π_u at the point χ. This inequality has the form

$$B(w, \bar{w} - w) + \langle k_{ij}\sigma_{ij}, \bar{w} - w\rangle + \langle \sigma_{ij}, \varepsilon_{ij}(\bar{W} - W)\rangle - \langle u, \bar{\chi} - \chi\rangle \geq 0 \tag{2.188}$$

$$\chi \in K_\delta, \quad \forall\, \bar{\chi} = (\bar{W}, \bar{w}) \in K_\delta.$$

It can be proved that the functional Π_u is coercive on $H(\Omega_\psi)$. Using the weak lower semicontinuity of this functional, we verify that a solution of the equilibrium problem (2.188) exists. It will be unique.

We shall investigate the problem of controlling the external forces with an objective functional describing the crack opening

$$J_\delta(u) = \int\limits_{\Gamma_\psi} |[\chi]| d\Gamma_\psi,$$

where $\chi = \chi(u)$ is the solution of the variational inequality (2.188).

Let $U \subset L^2(\Omega)^3$ be a convex closed and bounded set. The problem of finding the crack with the least opening can be formulated as follows:

$$\inf_{u \in U} J_\delta(u). \tag{2.189}$$

Here and below we emphasize the dependence of the objective functional on δ, because later we shall investigate the convergence of the solutions of problem (2.189) as $\delta \to 0$.

Suppose that δ is fixed for the time being. We shall prove that a solution of the optimal control problem (2.189), (2.188) exists. We choose a minimizing sequence $u_m \in U$. It is bounded in $L^2(\Omega)$, and so we can assume that

$$u_m \to u \quad \text{weakly in } L^2(\Omega), \quad u \in U. \tag{2.190}$$

For every m one can find a unique solution $\chi_m \in K_\delta$ of the problem

$$\Pi'_{u_m}(\chi_m)(\bar{\chi} - \chi_m) \geq 0 \quad \forall\, \bar{\chi} \in K_\delta. \tag{2.191}$$

Fixing the test function $\bar{\chi}$, we derive the estimate

$$\|\chi_m\|_{H(\Omega_\psi)} \leq c,$$

which is uniform with respect to m. Having by necessity to choose a subsequence, we assume that as $m \to \infty$

$$\chi_m \to \chi \quad \text{weakly in } H(\Omega_\psi), \quad \text{strongly in } L^2(\Omega_\psi). \tag{2.192}$$

The convergence of (2.190) and (2.192) enables us to pass to the limit in (2.191) and thus show that $\chi = \chi(u)$. Moreover, additionally assuming that $\chi_m^\pm \to \chi^\pm$ weakly in $L^1(\Gamma_\psi)$, we obtain

$$\inf_{\bar{u}\in U} J_\delta(\bar{u}) = \liminf_{m\to\infty} J_\delta(u_m) \geq J_\delta(u) \geq \inf_{\bar{u}\in U} J_\delta(\bar{u}).$$

This means that u is a solution of problem (2.189), (2.188). The assertion is proved.

2.7.2 Regularity of solutions up to the crack faces

We note that if the crack opening is zero on Γ_ψ, i.e. $[\chi] = 0$, the value of the objective functional $J_\delta(u)$ is zero. We also assume that near Γ_ψ the punch does not interact with the shell. It turns out that in this case the solution $\chi = (W, w)$ of problem (2.188) is infinitely differentiable in a neighbourhood of points of the crack. This property is local, so that a zero opening of the crack near the fixed point guarantees infinite differentiability of the solution in some neighbourhood of this point. Here it is undoubtedly necessary to require appropriate regularity of the curvatures k_{ij} and the external forces u. The aim of the following discussion is to justify this fact. At this point the external force u is taken to be fixed.

Let $\mathcal{O} \subset R^2$ be a bounded domain with smooth boundary γ and outward normal $n = (n_1, n_2)$. We introduce the following notation for the bending moment and transverse forces on γ:

$$m(w) = \kappa\Delta w + (1-\kappa)\frac{\partial^2 w}{\partial n^2},$$

$$t(w) = \frac{\partial}{\partial n}\Delta w + (1-\kappa)\frac{\partial^3 w}{\partial n \partial s^2}, \quad s = (-n_2, n_1).$$

The quantities $m(w)$ and $t(w)$ can be interpreted as elements from the spaces $H^{-\frac{1}{2}}(\gamma)$ and $H^{-\frac{3}{2}}(\gamma)$, respectively, if $w \in H^2(\mathcal{O})$, $\Delta^2 w \in L^2(\mathcal{O})$. Moreover, the following generalized Green's formula holds:

$$B_{\mathcal{O}}(w,v) = \langle m(w), \frac{\partial v}{\partial n}\rangle_{\frac{1}{2},\gamma} - \langle t(w), v\rangle_{\frac{3}{2},\gamma} + \langle \Delta^2 w, v\rangle_{\mathcal{O}} \tag{2.193}$$

$$\forall v \in H^2(\mathcal{O}).$$

The symbol $\mathcal{O}$ means that the integration is performed over $\mathcal{O}$, while the brackets $\langle \cdot,\cdot\rangle_{p,\gamma}$ denote the duality pairing between $H^{-p}(\gamma)$ and $H^p(\gamma)$. Another Green formula is also needed. Suppose that $\theta = (\theta_1, \theta_2) \in L^2(\mathcal{O})$, $\operatorname{div}\theta \in L^2(\mathcal{O})$. Then the quantity θn is defined on the boundary as an element of $H^{-\frac{1}{2}}(\gamma)$, and we have the formula

$$\langle \operatorname{div}\theta, w\rangle_{\mathcal{O}} = \langle \theta n, w\rangle_{\frac{1}{2},\gamma} - \langle \theta, \nabla w\rangle_{\mathcal{O}} \quad \forall w \in H^1(\mathcal{O}). \tag{2.194}$$

We shall investigate the regularity of the solution in a neighbourhood of the crack tip $x^0 \equiv (1,0)$. Suppose, first, that (W, w) is a solution of the equilibrium problem (2.188). We assume that a neighbourhood $\mathcal{W}$ of the graph Γ_ψ exists such that for any function $\varphi \in C_0^\infty(\mathcal{W})$ there is an $\varepsilon > 0$, for which

$$\varepsilon\varphi + w - W\nabla\Phi \geq \Phi \quad \text{almost everywhere in} \quad \mathcal{W} \setminus \Gamma_\psi. \tag{2.195}$$

Condition (2.195) can be interpreted as the absence of contact between the shell and the punch in $\mathcal{W} \setminus \Gamma_\psi$.

We smoothly continue the function $\psi(x)$ for $x > 1$, keeping the previous notation. We take an arbitrary function $\varphi \in C_0^\infty(R(x^0))$, where $R(x^0)$ is a ball centred at the point x^0 such that $R(x^0) \subset \mathcal{W}$. Then

$$[\partial\varphi/\partial\nu] = 0 \quad \text{on } R(x^0) \cap \Gamma_\psi.$$

From what has been said, for small $\varepsilon > 0$ the function $(W, \varepsilon\varphi + w)$ belongs to the set K_δ. Outside $R(x^0)$ the function φ can be taken to be zero. We now substitute $(W, \varepsilon\varphi + w)$ in (2.188). We arrive at the inequality

$$B_+(w,\varphi) + B_-(w,\varphi) + \langle k_{ij}\sigma_{ij}, \varphi\rangle \geq \langle u_3, \varphi\rangle. \tag{2.196}$$

The plus and minus subscripts denote integration over $\mathcal{O}^+$ and $\mathcal{O}^-$, respectively, where $\mathcal{O}^+ = R(x^0) \cap \{y > \psi(x)\}$, and similarly for $\mathcal{O}^-$. The boundaries of the domains $\mathcal{O}^\pm$ are denoted $\gamma^\pm$. Note that when (2.195) holds, the equation

$$\Delta^2 w + k_{ij}\sigma_{ij} = u_3 \tag{2.197}$$

is satisfied in $\mathcal{W} \setminus \Gamma_\psi$ in the distribution sense.

In order to verify this, it is sufficient to substitute test functions of the form $\chi + c\theta$ into (2.188), where θ is an infinitely differentiable function with support $\mathcal{W}\backslash\Gamma_\psi$ and ε is a small parameter. Thus, applying Green's formula (2.193) to $B_\pm(w,\varphi)$ in (2.196) and using equation (2.197), we obtain

$$\langle m(w), \frac{\partial\varphi}{\partial n^-}\rangle_{1/2,\gamma^-} - \langle t(w),\varphi\rangle_{3/2,\gamma^-} \tag{2.198}$$

$$+\langle m(w), \frac{\partial\varphi}{\partial n^+}\rangle_{1/2,\gamma^+} - \langle t(w),\varphi\rangle_{3/2,\gamma^+} \geq 0.$$

Note that the function $\Delta^2 w + k_{ij}\sigma_{ij} - u_3$ is zero almost everywhere in $\mathcal{W}\backslash\Gamma_\psi$ and so the integral over the domain vanishes.

Below, ν will also denote the normal to the continued graph $\tilde{\Gamma}_\psi$ of the function $\psi(x)$. Using the arbitrariness and the finiteness of φ in $R(x^0)$, from (2.198) we find

$$\langle [m(w)], \partial\varphi/\partial\nu\rangle_{1/2,\gamma} = 0, \quad \langle [t(w)],\varphi\rangle_{3/2,\gamma} = 0 \tag{2.199}$$

$$\forall\varphi \in C_0^\infty(R(x^0)),$$

where γ can be taken to be either γ^+ or γ^-. The proven identities (2.199) mean that

$$[m(w)] = 0, \quad [t(w)] = 0 \quad \text{on } \tilde{\Gamma}_\psi. \tag{2.200}$$

When conditions (2.195) are satisfied we also have the following distribution equations:

$$-\partial\sigma_{ij}/\partial x_j = u_i \quad \text{in } \mathcal{W}\setminus\Gamma_\psi. \tag{2.201}$$

This is proved simultaneously with (2.197).

Suppose that the function $\theta \equiv (\theta_1,\theta_2)$ belongs to $C_0^\infty(\Omega)$ and has support in $R(x^0)$. Then, as before, for small $\varepsilon > 0$ we have $(W+\varepsilon\theta, w) \in K_\delta$. We substitute $(W+\varepsilon\theta, w)$ into (2.188) as a test function. This implies

$$\langle\sigma_{ij},\varepsilon_{ij}(\theta)\rangle_+ + \langle\sigma_{ij},\varepsilon_{ij}(\theta)\rangle_- \geq \langle u_i,\theta_i\rangle.$$

Using Green's formula (2.194), it follows from this that

$$-\langle[\sigma_{ij}\nu_j],\theta_i\rangle_{1/2,\gamma} - \langle\partial\sigma_{ij}/\partial x_j,\theta_i\rangle_+ - \langle\partial\sigma_{ij}/\partial x_j,\theta_i\rangle_- \geq \langle u_i,\theta_i\rangle,$$

where one can take either γ^+ or γ^- to be γ. Bearing in mind equation (2.201), the relation obtained gives

$$\langle[\sigma_{ij}\nu_j],\theta_i\rangle_{1/2,\gamma} = 0, \quad \forall\,\theta \in C_0^\infty(R(x^0)),$$

i.e.

$$[\sigma_{ij}\nu_j] = 0 \quad \text{on} \quad \tilde{\Gamma}_\psi. \tag{2.202}$$

The established properties (2.200) and (2.202) enable us to investigate the regularity of the solution in a neighbourhood of the crack tip x^0 in the case when there is no contact between the shell and the punch near to x^0, and the crack opening is zero.

Theorem 2.21. *Suppose that* $k_{ij}, u \in C^\infty(R(x^0))$, *that condition* (2.195) *is satisfied, and* $[\chi] = 0$ *on* $R(x^0) \cap \Gamma_\psi$. *Then* $\chi \in C^\infty(R(x^0))$.

PROOF. We shall show that equation (2.197) is satisfied in the distribution sense in $R(x^0)$. The condition of the theorem and inequality (2.185) ensure the validity of $[\partial w/\partial\nu] = 0$ on $R(x^0) \cap \Gamma_\psi$. Bearing in mind that $w \in H^2(\mathcal{O}^\pm)$ and that $[w] = 0$ on $R(x^0) \cap \Gamma_\psi$, we conclude that $w \in H^2(R(x^0))$. Note that equation (2.197) is satisfied in $\mathcal{O}^\pm$, and so $\Delta^2 w \in L^2(\mathcal{O}^\pm)$.

Let the brackets $(\cdot, \varphi)$ denote the action of the distribution on the element φ. We choose $\varphi \in C_0^\infty(R(x^0))$. Using formula (2.193) we have

$$(\Delta^2 w, \varphi) = B_+(w, \varphi) + B_-(w, \varphi) = -\langle [m(w)], \partial\varphi/\partial\nu\rangle_{1/2,\gamma}$$

$$+ \langle [t(w)], \varphi\rangle_{3/2,\gamma} + \langle \Delta^2 w, \varphi\rangle_+ + \langle \Delta^2 w, \varphi\rangle_-.$$

The jumps $[m(w)]$, $[t(w)]$ are zero, from which the necessary equation that proves the assertion follows:

$$(\Delta^2 w + k_{ij}\sigma_{ij} - u_3, \varphi) = \langle \Delta^2 w + k_{ij}\sigma_{ij} - u_3, \varphi\rangle_+$$

$$+ \langle \Delta^2 w + k_{ij}\sigma_{ij} - u_3, \varphi\rangle_- = 0, \quad \forall\, \varphi \in C_0^\infty(R(x^0))$$

We shall now show that equations (2.201) are satisfied in $R(x^0)$. Because $[W] = 0$ on $R(x^0) \cap \Gamma_\psi$ and $W \in H^1(\mathcal{O}^\pm)$, we have $W \in H^1(R(x^0))$. Consequently, $\sigma_{ij} \equiv \sigma_{ij}(\chi) \in L^2(R(x^0))$. From the validity of equations (2.201) in $\mathcal{O}^\pm$, we conclude that $\partial\sigma_{ij}/\partial x_j \in L^2(\mathcal{O}^\pm)$. This means that one can apply Green's formula (2.194) to the domain $\mathcal{O}^\pm$. Let $\varphi \in C_0^\infty(R(x^0))$. We have

$$-(\partial\sigma_{ij}/\partial x_j + u_i, \varphi) = \langle \sigma_{ij}, \partial\varphi/\partial x_j\rangle_+ + \langle \sigma_{ij}, \partial\varphi/\partial x_j\rangle_- - (u_i, \varphi) \tag{2.203}$$

$$= -\langle [\sigma_{ij}\nu_j], \varphi\rangle_{1/2,\gamma} - \langle \partial\sigma_{ij}/\partial x_j + u_i, \varphi\rangle_+ - \langle \partial\sigma_{ij}/\partial x_j + u_i, \varphi\rangle_- = 0.$$

However, the jumps $[\sigma_{ij}\nu_j]$ are zero, and equations (2.201) are satisfied in $\mathcal{O}^\pm$. Hence the right-hand side of (2.203) vanishes, which confirms the validity of

$$-\partial\sigma_{ij}/\partial x_j = u_i \quad \text{in } R(x^0) \tag{2.204}$$

in the distribution sense. Equations (2.204) can be written as linear equations in the two-dimensional theory of elasticity,

$$L(W) = F \quad \text{in } R(x^0), \tag{2.205}$$

with right-hand side $F = (f_1, f_2)$, where $f_1 = u_1 + (k_{11}w + \kappa k_{22}w)_x + (k_{12}w)_y$ and f_2 is defined similarly. Moreover, equation (2.197) can be conveniently represented in the form

$$\Delta^2 w = u_3 - k_{ij}\sigma_{ij} \quad \text{in } R(x^0). \tag{2.206}$$

The right-hand side of equation (2.205) belongs to $H^1(R(x^0))$ and the right-hand side (2.206) belongs to $L^2(R(x^0))$. Applying in turn the results on the internal regularity of the solutions of equations (2.205) and (2.206) (Lions, Magenes, 1968; Fichera, 1972), we obtain the necessary inclusion

$$\chi = (W, w) \in C^\infty(R(x^0)). \tag{2.207}$$

The theorem is proved.

We make a number of remarks. For the inclusion (2.207) to be valid it is sufficient only to require that (2.195) is satisfied in $R(x^0) \setminus \Gamma_\psi$ or $\varphi \in C_0^\infty(R(x^0))$.

According to the imbedding theorems the function w is continuous in $\overline{\Omega}_\psi$. Hence if $\nabla\Phi \equiv 0$ in some neighbourhood $\mathcal{W}$ of the graph Γ_ψ and $w > \Phi$ in $\mathcal{W}$ (and, in particular, $w^\pm > \Phi$ on Γ_ψ), then condition (2.195) is obviously satisfied.

If x^0 is an internal point of the crack, i.e. $x^0 \in \Gamma_\psi \setminus \partial\Gamma_\psi$, condition (2.195) is satisfied and $[\chi] = 0$ near x^0, then the corresponding assertion on the infinite diffirentiability of χ can be proved more simply.

2.7.3 Convergence of the solutions

We consider the limiting case corresponding to $\delta = 0$ in (2.185). A restriction obtained in this manner corresponds to the condition of mutual nonpenetration of the crack faces without including the thickness of the shell. We note that in taking full account of the thickness one must bear in mind that the stresses σ_{ij}, the moments $m(w)$ and the transverse forces $t(w)$ depend on δ. Thus $\delta = 0$ in (2.185) carries the implication that the thickness of the shell is taken to be fixed, and the nonpenetration conditions on the crack faces are described approximately. At this point we mention other problems of a passage to limit (Attouch, Picard, 1983; Schuss, 1976; Roubiček, 1997; Oleinik et al., 1992; Moet, 1982; Telega, Lewinski, 1994).

Thus, in the case under consideration the solution satisfies the following restrictions:

$$[W]\nu \geq 0 \quad \text{on } \Gamma_\psi, \qquad w - W\nabla\Phi \geq \Phi \quad \text{in } \Omega_\psi. \tag{2.208}$$

The set of admissible displacements in this case has the form

$$K_0 = \{(W, w) \in H(\Omega_\psi) \mid \ (W, w) \quad \text{satisfy } (2.208)\}.$$

Here the solution of the problem of minimizing the functional Π_u over the set K_0 is equivalent to the following variational inequality:

$$\Pi_u'(\chi)(\bar{\chi} - \chi) \geq 0, \quad \chi \in K_0, \quad \forall\, \bar{\chi} \in K_0. \tag{2.209}$$

Let the set U be chosen as before. We consider the optimal control problem

$$\inf_{u \in U} J_0(u), \quad J_0(u) = \int_{\Gamma_\psi} \|[\chi]\|\, d\Gamma_\psi, \tag{2.210}$$

where χ is defined in (2.209) for given u. A solution of problem (2.210), (2.209) exists (but we will not dwell on the proof).

We introduce the following notation:

$$j_\delta = \inf_{u\in U} J_\delta(u), \quad j_0 = \inf_{u\in U} J_0(u). \tag{2.211}$$

The connection between solutions of the problems (2.210), (2.209) and (2.189), (2.188) is characterized by the theorem given below. Let u_δ be the solution of problem (2.189), (2.188), while χ_δ corresponds to u_δ and is defined by (2.188).

Theorem 2.22. *Let $\nabla\Phi \equiv 0$ in some neighbourhood $\mathcal{W}$ of the graph Γ_ψ. From the sequence (u_δ, χ_δ) one can choose a subsequence such that as $\delta \to 0$*

$$u_\delta \to u_0 \ \text{weakly in } L^2(\Omega), \quad \chi_\delta \to \chi_0 \ \text{weakly in } H(\Omega_\psi), \quad j_\delta \to j_0,$$

where u_0 is a solution of the problem (2.210), (2.209), *and χ_0 corresponds to u_0 and is defined by* (2.209).

PROOF. Let $\chi_\delta(u)$ be a solution of the variational inequality (2.188) with given fixed $u \in U$. We take an arbitrary element $\bar{\chi} \in K_{\delta_0}$. Then $\bar{\chi} \in K_\delta$ for all $\delta \le \delta_0$. We substitute $\bar{\chi}$ into (2.188) as a test element. We arrive at the estimate

$$\|\chi_\delta(u)\|_{H(\Omega_\psi)} \le c$$

which is uniform with respect to $\delta \le \delta_0$. Consequently, one can assume that as $\delta \to 0$

$$\chi_\delta(u) \to \tilde{\chi} \quad \text{weakly in } H(\Omega_\psi), \tag{2.212}$$

$$[\chi_\delta(u)] \to [\tilde{\chi}] \quad \text{strongly in } L^1(\Gamma_\psi), \tag{2.213}$$

$$\delta\,\|[\partial w_\delta(u)/\partial\nu]\| \to 0 \quad \text{strongly in } L^2(\Gamma_\psi). \tag{2.214}$$

We choose an arbitrary element $\bar{\chi} \in K_0$ and construct, in accordance with Lemma 2.3 (see below), a sequence $\bar{\chi}_\delta \in K_\delta$ which strongly converges to $\bar{\chi}$ in $H(\Omega_\psi)$. Substituting the $\bar{\chi}_\delta$ as the test functions into inequality (2.188) and using (2.212) we pass to the limit as $\delta \to 0$. Condition (2.214) ensures the inclusion $\tilde{\chi} \in K_0$. The limiting variational inequality has the form

$$\Pi'_u(\tilde{\chi})(\bar{\chi} - \tilde{\chi}) \ge 0, \quad \tilde{\chi} \in K_0, \quad \forall\, \bar{\chi} \in K_0$$

which means $\tilde{\chi} = \chi(u)$. Here, from (2.213) we obtain

$$J_\delta(u) \to J_0(u), \quad \delta \to 0. \tag{2.215}$$

Suppose that u is now a solution of the optimal control problem (2.210), (2.209). From (2.215) we have $j_\delta \le J_\delta(u) \to J_0(u) = j_0$; hence

$$\limsup j_\delta \le j_0. \tag{2.216}$$

On the other hand, bearing in mind the boundedness of the set U, we can assume

$$\|u_\delta\|_{L^2(\Omega)} \leq c \tag{2.217}$$

uniformly with respect to δ. Then from the variational inequalities

$$\Pi'_{u_\delta}(\chi_\delta)(\bar{\chi} - \chi_\delta) \geq 0, \quad \chi_\delta \in K_\delta, \quad \forall\, \bar{\chi} \in K_\delta \tag{2.218}$$

we derive the estimate

$$\|\chi_\delta\|_{H(\Omega_\psi)} \leq c \tag{2.219}$$

uniform in δ. According to (2.217) and (2.219), we can assume without loss of generality that

$$u_\delta \to u_0 \text{ weakly in } L^2(\Omega), \quad \delta\,|[\partial w_\delta/\partial\nu]| \to 0 \text{ strongly in } L^2(\Gamma_\psi).$$

$$\chi_\delta \to \chi_0 \text{ weakly in } H(\Omega_\psi), \quad \text{strongly in } L^2(\Omega_\psi).$$

This convergence and Lemma 2.3 enable us to pass to the limit in inequality (2.218) and thus obtain

$$\Pi'_{u_0}(\chi_0)(\bar{\chi} - \chi_0) \geq 0, \quad \chi_0 \in K_0, \quad \forall\, \bar{\chi} \in K_0,$$

so that $\chi_0 = \chi(u_0)$. As in the proof of relation (2.215), it can be shown in our case that $J_\delta(u_\delta) \to J_0(u_0)$ and therefore

$$\liminf j_\delta \geq J_0(u_0). \tag{2.220}$$

Comparing (2.216) and (2.220), we conclude that u_0 is a solution of the optimal control problem (2.210), (2.209) and $j_\delta \to j_0$. The theorem is proved.

It remains to establish the assertion used in the proof of Theorem 2.22.

Lemma 2.3. *Let $\nabla\Phi \equiv 0$ in some neighbourhood $\mathcal{W}$ of the graph Γ_ψ. Then for every fixed element $\bar{\chi} = (\bar{W}, \bar{w}) \in K_0$ one can construct a sequence $\bar{\chi}_\delta = (\bar{W}_\delta, \bar{w}_\delta) \in K_\delta$ such that*

$$(\bar{W}_\delta, \bar{w}_\delta) \to (\bar{W}, \bar{w}) \quad \textit{strongly in } H(\Omega_\psi). \tag{2.221}$$

PROOF. We construct a function $\tilde{W}$ from the space $[H^{1,0}(\Omega_\psi)]^2$ equal to zero outside $\mathcal{W}$ and with the property

$$[\tilde{W}]\nu = |[\partial\bar{w}/\partial\nu]| \quad \text{on} \quad \Gamma_\psi.$$

If such a function is constructed, the sequence $(\bar{W}_\delta, \bar{w}_\delta) = (\bar{W} + \delta\tilde{W}, \bar{w})$ will be needed. Indeed, the convergence of (2.221) is obvious, and moreover

$$\bar{w}_\delta - \bar{W}_\delta \nabla\Phi \geq \Phi \quad \text{in } \Omega_\psi, \quad [\bar{W}_\delta]\nu \geq \delta\,|[\partial\bar{w}_\delta/\partial\nu]| \quad \text{on } \Gamma_\psi.$$

We therefore choose a simply connected domain $\mathcal{O}$, $\overline{\mathcal{O}} \subset \Omega$, with smooth boundary γ such that Γ_ψ is a part of γ, and the outward normal $n = (n_1, n_2)$ to γ coincides with ν on Γ_ψ. We put $g = -|[\partial \bar{w}/\partial n]|$. Then $g \in H^{1/2}(\gamma)$, with $g \equiv 0$ outside Γ_ψ. Since the components of the normal n belong to $C^1(\gamma)$, we have $gn \in [H^{1/2}(\gamma)]^2$. Hence a function $W^0 \in [H^1(\mathcal{O})]^2$ exists such that $W^0 = gn$ on γ. We put $W^0 \equiv 0$ outside $\mathcal{O}$. Let φ be an infinitely differentiable function on Ω such that $\varphi = 1$ on Γ_ψ and $\varphi \equiv 0$ outside $\mathcal{W}$. The required function $\tilde{W}$ is obtained as $\tilde{W} = \varphi W^0$. The lemma is proved.

In conclusion we note that the conditions of Theorem 2.21 do not, in general, ensure the validity of the inclusion (2.207) for the solution $\chi = (W, w)$ of problem (2.209). Indeed, in the case of problem (2.209) the jump $[\partial w/\partial \nu]$ is not, in general, zero on $\Gamma_\psi \cap R(x^0)$, and hence for $[\chi] = 0$ one cannot assert that $w \in H^2(R(x^0))$.

2.8 Signorini problem for cracks in shells

We consider an equilibrium problem for a shell with a crack. The faces of the crack are assumed to satisfy a nonpenetration condition, which is an inequality imposed on the horizontal shell displacements. The properties of the solution are analysed – in particular, the smoothness of the stress field in the vicinity of the crack. The character of the contact between the crack faces is described in terms of a suitable nonnegative measure. The stability of the solution is investigated for small perturbations to the crack geometry. The results presented were obtained in (Khludnev, 1996b).

2.8.1 Setting the problem

Consider a shell whose mid-surface occupies a domain $\Omega_\psi = \Omega \setminus \Gamma_\psi$, where $\Omega \subset R^2$ is a bounded domain with a smooth boundary Γ, Γ_ψ is a graph of the function $y = \psi(x)$, $x \in [0,1]$, $(x,y) \in \Omega$. The horizontal displacements of the mid-surface points are denoted by $W = (w^1, w^2)$ and the vertical displacements are denoted by w.

In what follows the Kirchhoff–Love model of the shell is used. We identify the mid-surface with the domain Ω_ψ in R^2. However, the curvatures of the shell are assumed to be small but nonzero. For such a configuration, following (Vol'mir, 1972), we introduce the components of the strain tensor for the mid-surface,

$$e_{ij} = \varepsilon_{ij}(W) + k_{ij} w, \quad i,j = 1,2,$$

where $k_{ij} = k_{ji} \in C^1(\bar{\Omega})$ are the given curvatures of the shell, and

$$\varepsilon_{ij}(W) = \frac{1}{2}\left(\frac{\partial w^i}{\partial x_j} + \frac{\partial w^j}{\partial x_i} \right), \quad x_1 = x, \; x_2 = y.$$

The components of the stresses integrated across the shell can be written as follows:

$$\sigma_{11} = e_{11} + \kappa e_{22}, \quad \sigma_{22} = e_{22} + \kappa e_{11}, \quad \sigma_{12} = (1 - \kappa) e_{12},$$

where $\kappa = \text{const}$ is Poisson's ratio, $0 < \kappa < 1/2$. Let $\chi = (W, w)$. Then the energy functional of the shell is

$$\Pi_\psi(\chi) = \frac{1}{2} B_\psi(w, w) + \frac{1}{2} \langle \sigma_{ij}, e_{ij} \rangle_\psi - \langle f, \chi \rangle_\psi,$$

where $f = (f_1, f_2, f_3) \in L^2(\Omega)$ is a given vector of exterior forces, the brackets $\langle \cdot, \cdot \rangle_\psi$ denote the integral over Ω_ψ, and the bilinear form $B_\psi(\cdot, \cdot)$ is defined by the formula

$$B_\psi(w, \bar{w}) = \int_{\Omega_\psi} (w_{xx}\bar{w}_{xx} + w_{yy}\bar{w}_{yy} + \kappa w_{xx}\bar{w}_{yy} + \kappa w_{yy}\bar{w}_{xx} + 2(1-\kappa) w_{xy}\bar{w}_{xy}).$$

The above representation for the shell energy contains three different terms describing the bending energy of the shell, the deformation energy of the middle surface, and the work done by the exterior force f, respectively.

Let $\nu_\psi = (-\psi_x, 1)/\sqrt{1 + \psi_x^2}$ be the normal vector to the curve $y = \psi(x)$, $\nu_\psi = (\nu_\psi^1, \nu_\psi^2)$. To avoid interpenetration of the crack faces, we consider the following condition of the Signorini type:

$$[W]\nu_\psi \geq 0 \quad \text{on } \Gamma_\psi. \tag{2.222}$$

Here $[W] = W^+ - W^-$, where $W^\pm$ are the quantities of W evaluated at the positive and negative crack faces with respect to ν_ψ. Notice that the function w also has different values at the opposite crack faces of Γ_ψ, in general.

At the external boundary we consider the following conditions, corresponding to clamping the shell:

$$w = \frac{\partial w}{\partial n} = W = 0 \quad \text{on } \Gamma.$$

The equilibrium problem for the shell corresponds to minimization of the energy functional over the set of admissible displacements. To this end, introduce the convex sets

$$K(\Omega_\psi) = \{\chi = (W, w) \in H(\Omega_\psi) \mid \; W \text{ satisfying } (2.222)\},$$

$$K^1(\Omega_\psi) = \{W \in H^{1,0}(\Omega_\psi)^2 \mid \; W \text{ satisfying } (2.222)\},$$

where $H^{1,0}(\Omega_\psi)$ is the space of functions from $H^1(\Omega_\psi)$ which are equal to zero on Γ, $H^{2,0}(\Omega_\psi)$ is introduced analogously, $H(\Omega_\psi) = H^{1,0}(\Omega_\psi) \times$

$H^{1,0}(\Omega_\psi) \times H^{2,0}(\Omega_\psi)$. We assume that $\psi \in H_0^3(0,1)$ and consider the minimization problem

$$\inf_{\chi \in K(\Omega_\psi)} \Pi_\psi(\chi). \tag{2.223}$$

The functional Π_ψ is convex and differentiable, hence the problem (2.223) is equivalent to the variational inequality

$$\chi \in K(\Omega_\psi): \quad \langle \Pi'_\psi(\chi), \bar{\chi} - \chi \rangle \geq 0 \quad \forall \bar{\chi} \in K(\Omega_\psi), \tag{2.224}$$

where $\Pi'_\psi(\chi)$ is a derivative of Π_ψ at the point χ. In view of the coercivity and the weak lower semicontinuity of Π_ψ on the space $H(\Omega_\psi)$, it is easy to prove the solvability of the problem (2.223) or the problem (2.224). Letting $F = (f_1, f_2)$, the inequality (2.224) can be written as

$$B_\psi(w, \bar{w}) + \langle k_{ij}\sigma_{ij} - f_3, \bar{w} \rangle_\psi = 0 \quad \forall \bar{w} \in H^{2,0}(\Omega_\psi), \tag{2.225}$$

$$\langle \sigma_{ij}, \varepsilon_{ij}(\bar{W} - W) \rangle_\psi \geq \langle F, \bar{W} - W \rangle_\psi \quad \forall \bar{W} \in K^1(\Omega_\psi). \tag{2.226}$$

This form of (2.224) is more convenient for further consideration and, moreover, (2.225), (2.226) imply the following equilibrium equations in Ω_ψ in the sense of distributions:

$$\Delta^2 w + k_{ij}\sigma_{ij} = f_3, \tag{2.227}$$

$$-\frac{\partial \sigma_{ij}}{\partial x_j} = f_i, \quad i = 1, 2. \tag{2.228}$$

To prove this, it suffices to substitute $\bar{\chi} = \chi + \chi_0$ in inequality (2.224), where $\chi_0 \in C_0^\infty(\Omega_\psi)$.

The structure of the section is as follows. In Section 2.8.2 we give necessary definitions and construct a Borel measure μ which describes the work of the interaction forces, i.e. for a set $A \subset \Gamma_\psi \setminus \partial\Gamma_\psi$, the value $\mu(A)$ characterizes the forces at the set A. The next step is a proof of smoothness of the solution provided the exterior data are regular. In particular, we prove that horizontal displacements W belong to H^2 in a neighbourhood of the crack faces. Consequently, the components of the strain and stress tensors belong to the space H^1. In this case the measure μ is absolutely continuous with respect to the Lebesgue measure. This confirms the existence of a locally integrable function q called a density of the measure μ such that

$$\mu(A) = \int_A q \, d\Gamma_\psi.$$

Given $W \in H^2$ we have $q = -\sigma_{ij}\nu_\psi^i \nu_\psi^j \geq 0$, and hence, the density q is defined by the normal component of the surface forces at Γ_ψ. At the end, in Section 2.8.3, we establish the stability of solutions with respect to perturbations in the crack shape.

2.8.2 Construction of a measure. Regularity of solutions

We recall some definitions which are useful in the work to follow. The smallest σ-algebra containing all compact sets in $\Gamma_\psi \setminus \partial\Gamma_\psi$ is called the Borel σ-algebra (Landkof, 1966). Any σ-additive real-valued function defined on the Borel σ-algebra which is finite for all compact sets $B \subset \Gamma_\psi \setminus \partial\Gamma_\psi$ is called a measure on $\Gamma_\psi \setminus \partial\Gamma_\psi$. Thus, for a measure μ and a set A, the σ-additivity means

$$\mu(A) = \sum_k \mu(A_k), \quad A = \bigcup_k A_k, \quad A_i \cap A_j = \emptyset, \quad i \neq j.$$

Let $C_0(\Gamma_\psi)$ be the space of continuous functions defined on $\Gamma_\psi \setminus \partial\Gamma_\psi$ and having compact support in $\Gamma_\psi \setminus \partial\Gamma_\psi$. Convergence in this space is introduced in the usual way: we say $\varphi_n \to \varphi$ in $C_0(\Gamma_\psi)$ if the supports of all φ_n belong to a fixed compact set $B \subset \Gamma_\psi \setminus \partial\Gamma_\psi$ and if, in addition, φ_n converge to φ uniformly. Denote by $H^{s,0}(\Omega_\psi) \cap C_0(\Gamma_\psi)$ the space of functions such that each of its elements belongs to $H^{s,0}(\Omega_\psi)$ and is continuous and has compact support at both crack faces $\Gamma_\psi \setminus \partial\Gamma_\psi$.

Let $\chi = (W, w)$ be the solution of (2.225)–(2.226). Then $W + \bar{W} \in K^1(\Omega_\psi)$ for every element $\bar{W} \in [H^{1,0}(\Omega_\psi)]^2$ such that $[\bar{W}]\nu_\psi \geq 0$ on Γ_ψ. Hence it follows from (2.226) that for the above $\bar{W}$ the inequality

$$\langle \sigma_{ij}, \varepsilon_{ij}(\bar{W})\rangle_\psi \geq \langle F, \bar{W}\rangle_\psi \tag{2.229}$$

is satisfied. Now we are in a position to prove the following statement.

Theorem 2.23. *On the σ-algebra of Borel sets of $\Gamma_\psi \setminus \partial\Gamma_\psi$, we can construct a nonnegative measure μ such that, for all $\bar{\chi} = (\bar{W}, \bar{w}) \in H(\Omega_\psi) \cap C_0(\Gamma_\psi)$, a representation*

$$\langle \Pi'_\psi(\chi), \bar{\chi}\rangle = \int_{\Gamma_\psi \setminus \partial\Gamma_\psi} [\bar{W}]\nu_\psi \, d\mu \tag{2.230}$$

holds .

Proof. Let us consider a linear space $\mathcal{W}$ of functions defined on $\Gamma_\psi \setminus \partial\Gamma_\psi$:

$$\mathcal{W} = \{\bar{W}^*\},$$

where $\bar{W}^* = [\bar{W}]\nu_\psi$, $\bar{W} \in [H^{1,0}(\Omega_\psi) \cap C_0(\Gamma_\psi)]^2$. A linear functional can be defined on $\mathcal{W}$ as

$$L(\bar{W}^*) = \langle \sigma_{ij}, \varepsilon_{ij}(\bar{W})\rangle_\psi - \langle F, \bar{W}\rangle_\psi.$$

It is easily seen that the functional L is well-defined. In fact, if $\bar{W}_1^* = \bar{W}_2^*$, then, in view of (2.229), $L(\bar{W}_1^*) = L(\bar{W}_2^*)$.

Let us next show that the space $C_0^1(\Gamma_\psi)$ is included in $\mathcal{W}$. By $C_0^1(\Gamma_\psi)$ we denote the space of continuously differentiable functions defined on $\Gamma_\psi \setminus \partial\Gamma_\psi$

and having compact supports. To this end, we choose an arbitrary function $\lambda \in C_0^1(\Gamma_\psi)$. Extend the function ψ beyond $(0,1)$ by zero and define a smooth function γ along the normal ν_ψ:

$$\gamma(\tilde{x}) = \begin{cases} 1, \text{ if } & \tilde{x} = \bar{x} + \varepsilon\nu_\psi(\bar{x}),\ \bar{x} \in \Gamma_\psi,\ 0 \le \varepsilon \le \varepsilon_0/2, \\ 0, \text{ if } & \varepsilon > \varepsilon_0. \end{cases}$$

The function γ is well-defined since the function ψ is smooth. Now one can construct a function Λ in the domain $y > \psi(x)$ assuming that λ is also extended by zero. Specifically, we put, in the domain $y > \psi(x)$,

$$\Lambda(\tilde{x}) = \lambda(\bar{x})\gamma(\tilde{x})\nu_\psi(\bar{x}), \quad \tilde{x} = \bar{x} + \varepsilon\nu_\psi(\bar{x}), \quad \bar{x} \in \Gamma_\psi, \quad \varepsilon \ge 0.$$

Then

$$\Lambda(\bar{x})\nu_\psi(\bar{x}) = \lambda(\bar{x}), \quad \bar{x} \in \Gamma_\psi.$$

We have to remark that the graph of the extended function $y = \psi(x)$ has intersections with Γ so that the points such as $\tilde{x} = \bar{x} + \varepsilon\nu_\psi(\bar{x})$ may not belong to Ω_ψ, in general. This should present no problems since λ is equal to zero near the ends of the extended graph $y = \psi(x)$. Assuming that the function Λ is identically equal to zero for $y < \psi(x)$, we conclude that $\Lambda \in [H^{1,0}(\Omega_\psi) \cap C_0(\Gamma_\psi)]^2$ and

$$[\Lambda]\nu_\psi = \lambda \quad \text{on } \Gamma_\psi \setminus \partial\Gamma_\psi.$$

Thus, we have $C_0^1(\Gamma_\psi) \subset \mathcal{W}$. This implies that the functional L can be correctly defined on the space $C_0(\Gamma_\psi)$ since it is positive on $C_0^1(\Gamma_\psi)$ (Landkof, 1966). Consequently, there exists a measure $\mu \ge 0$ defined on Borel subsets of $\Gamma_\psi \setminus \partial\Gamma_\psi$ and moreover

$$L(\varphi) = \int\limits_{\Gamma_\psi \setminus \partial\Gamma_\psi} \varphi\, d\mu \quad \forall \varphi \in C_0(\Gamma_\psi).$$

Taking into account the structure of the formula $\langle \Pi'(\chi), \bar{\chi} \rangle$ and the validity of (2.225), we arrive at the conclusion that the representation (2.230) holds for any function $\bar{\chi} = (\bar{W}, \bar{w}) \in H(\Omega_\psi) \cap C_0(\Gamma_\psi)$. The proof is complete.

The properties of our measure μ depend on the regularity of the solution. The inequality (2.226) is actually a Signorini-type problem for finding W provided that the function w is already known. It can be written as follows:

$$\langle n_{ij}, \varepsilon_{ij}(\bar{W} - W)\rangle_\psi + \langle l^1, \bar{w}_x^1 - w_x^1\rangle_\psi + \langle l^2, \bar{w}_y^2 - w_y^2\rangle_\psi \tag{2.231}$$

$$-\langle F, \bar{W} - W\rangle_\psi \ge 0.$$

Here $n_{ij} = n_{ij}(W)$ coincide with σ_{ij} if we put $k_{ij} = 0$, and $l^1 = (k_{11} + \nu k_{22} + (1-\nu)k_{12})w$, l^2 can be obtained by replacing k_{11} on k_{22}.

Let us assume that $k_{ij} = 0$ on Γ_ψ. This enables us to integrate by parts in the second and the third terms of (2.231) and to obtain the inequality

$$\langle n_{ij}, \varepsilon_{ij}(\bar{W} - W)\rangle_\psi \geq \langle \bar{F}, \bar{W} - W\rangle_\psi \tag{2.232}$$

with the function $\bar{F} \in L^2(\Omega_\psi)$. In this case one can use the results (Khludnev, 1983) on the regularity of solutions for problems like (2.232). In particular, for any point $x_0 \in \Gamma_\psi \setminus \partial\Gamma_\psi$ there exists a neighbourhood $\mathcal{O}(x_0)$ such that

$$W \in H^2(\mathcal{O}(x_0) \cap \Omega_\psi).$$

By making use of this result, the density of the measure μ can be found. In view of (2.230), the representation

$$\langle \sigma_{ij}, \varepsilon_{ij}(\bar{W})\rangle_\psi - \langle F, \bar{W}\rangle_\psi = \int\limits_{\Gamma_\psi \setminus \partial\Gamma_\psi} [\bar{W}]\nu_\psi \, d\mu \tag{2.233}$$

holds for all $\bar{W} \in [H^{1,0}(\Omega) \cap C_0(\Gamma_\psi)]^2$. Using equations (2.228), integration by parts in (2.233) implies

$$-\int\limits_{\Gamma_\psi^+} \sigma_{ij}^+ \nu_\psi^j \bar{w}^i \, d\Gamma_\psi + \int\limits_{\Gamma_\psi^-} \sigma_{ij}^- \nu_\psi^j \bar{w}^i \, d\Gamma_\psi = \int\limits_{\Gamma_\psi \setminus \partial\Gamma_\psi} [\bar{W}]\nu_\psi \, d\mu. \tag{2.234}$$

Here the values $\Gamma_\psi^\pm$ correspond to the positive and negative directions of ν_ψ, respectively. Denote by $\sigma_\tau^\pm \equiv 0$ the tangent components of the vectors $\{\sigma_{kj}^\pm \nu_\psi^j\}$ and make use of the formulae

$$\{\sigma_{kj}^\pm \nu_\psi^j\} = (\sigma_{ij}^\pm \nu_\psi^i \nu_\psi^j)\nu_\psi + \sigma_\tau^\pm.$$

Taking into account the equality $\sigma_{ij}^+ \nu_\psi^i \nu_\psi^j = \sigma_{ij}^- \nu_\psi^i \nu_\psi^j$ we derive from (2.234) that

$$-\int\limits_{\Gamma_\psi} (\sigma_{ij}\nu_\psi^i \nu_\psi^j)[\bar{W}]\nu_\psi \, d\Gamma_\psi = \int\limits_{\Gamma_\psi \setminus \partial\Gamma_\psi} [\bar{W}]\nu_\psi \, d\mu.$$

The quantity $\sigma_{ij}^+ \nu_\psi^i \nu_\psi^j$ has been herein denoted by $\sigma_{ij}\nu_\psi^i \nu_\psi^j$. Thus, the above formula implies that the density of the measure μ is equal to $q = -\sigma_{ij}\nu_\psi^i \nu_\psi^j$. Moreover, the regularity of (W, w) provides the inclusion $q \in L^2_{loc}(\Gamma_\psi \setminus \partial\Gamma_\psi)$.

Now we shall prove the H^2-regularity of W up to the points of $\Gamma_\psi \setminus \partial\Gamma_\psi$ without the condition $k_{ij} = 0$ on Γ_ψ. For simplicity, only the case $\psi(x) \equiv 0$, $x \in [0, 1]$, will be considered. Let $x_0 \in \Gamma_\psi \setminus \partial\Gamma_\psi$ be any fixed point and R_δ be the ball of radius δ centred at the point x_0.

Theorem 2.24. *We have $W \in H^2(R_\delta \cap \Omega_\psi)$ provided that δ is small enough.*

Proof. Let $R_{3\delta/2}, R_{2\delta}$ be balls with centre x_0. Choose a smooth function φ such that $\varphi \equiv 1$ in R_δ, $\varphi \equiv 0$ outside $R_{3\delta/2}$, $0 \leq \varphi \leq 1$ everywhere.

Since the normal ν_ψ has the coordinates $(0,1)$, the nonpenetration condition $[W]\nu_\psi \geq 0$ can be written as

$$[w^2] \geq 0, \quad W = (w^1, w^2).$$

Let e be the unit vector in the x-direction,

$$d_{\pm\tau} h(\bar{x}) = [h(\bar{x} \pm \tau e) - h(\bar{x})]\tau^{-1}, \quad \Delta_\tau = -d_{-\tau} d_\tau, \quad 0 < |\tau| < \delta/2.$$

Define the function $W_\tau = (w_\tau^1, w_\tau^2)$ with components

$$w_\tau^i = w^i + \frac{\tau^2}{2} \varphi \Delta_\tau (\varphi w^i), \quad i = 1, 2.$$

It is easily verified that $[w_\tau^2] \geq 0$ on Γ_ψ. Consequently, $W_\tau \in K^1(\Omega_\psi)$. Let us substitute W_τ in (2.231) as a test function. Dividing by $\tau^2/2$ we see that

$$\langle l^2, (\varphi \Delta_\tau \varphi w^2)_y \rangle_\psi = \langle l^2 \varphi_y, \Delta_\tau (\varphi w^2) \rangle_\psi - \langle d_\tau (l^2 \varphi), d_\tau (\varphi w^2)_y \rangle_\psi.$$

The term corresponding to l^1 can be estimated easily since one can directly integrate by parts:

$$\langle l^1, (\varphi \Delta_\tau \varphi w^1)_x \rangle_\psi = -\langle \varphi l_x^1, \Delta_\tau \varphi w^1 \rangle_\psi.$$

Moreover, the difference between the terms

$$\langle n_{ij}(W), \varepsilon_{ij}(\varphi \Delta_\tau (\varphi W)) \rangle_\psi \quad \text{and} \quad - \langle n_{ij}(d_\tau(\varphi W)), \varepsilon_{ij}(d_\tau(\varphi W)) \rangle_\psi$$

can be estimated from the above by the right-hand side of the inequality (2.235) below. As a result we arrive at the inequality

$$\langle n_{ij}(d_\tau(\varphi W)), \varepsilon_{ij}(d_\tau(\varphi W)) \rangle_\psi \tag{2.235}$$

$$\leq c \left(\|f\|_{0,\Omega_\psi}^2 + \|d_\tau(\varphi W)\|_{1,\Omega_\psi} (\|\chi\|_{1,\Omega_\psi} + \|F\|_{0,\Omega_\psi}) \right)$$

with a constant c independent of τ. We can then use the first Korn inequality in Ω_ψ:

$$\|d_\tau(\varphi W)\|_{1,\Omega_\psi}^2 \leq c \langle n_{ij}(d_\tau(\varphi W)), \varepsilon_{ij}(d_\tau(\varphi W)) \rangle_\psi.$$

With (2.235), this implies

$$\|d_\tau(\varphi W)\|_{1,\Omega_\psi} \leq c$$

uniformly in τ. Consequently, $D(\varphi W)$ have first derivatives with respect to x, which belong to $L^2(\Omega_\psi)$. This yields that the second derivatives of φW with the exception of $(\varphi W)_{yy}$ belong to $L^2(\Omega_\psi)$. Meanwhile, it is evident that the equation

$$W_{yy} = G$$

holds in Ω_ψ with $G \in L^2(R_\delta \cap \Omega_\psi)$. The proof is complete.

The internal regularity of the solution follows from (2.227), (2.228). In particular, (2.227) implies $w \in H^4_{loc}(\Omega_\psi)$. It is seen that (2.228) are the equations of two-dimensional elasticity with respect to (w^1, w^2) and the right-hand side $\bar{f}$ from $L^2(\Omega_\psi)$:

$$-\frac{\partial n_{ij}}{\partial x_j} = \bar{f}_i, \quad i = 1, 2,$$

whence $W \in H^2_{loc}(\Omega_\psi)$. Moreover, $(W, w) \in H^2 \times H^4$ near Γ provided that $\Gamma \in C^4$ and $w \in H^4$ in an appropriate neighbourhood of any fixed point $x_0 \in \Gamma_\psi \setminus \partial\Gamma_\psi$ provided $\psi \in C^4(0,1)$.

2.8.3 Stability of the solution

We now aim to study the stability of the solution with respect to the crack shape. Let $y = \delta\psi(x)$ be the crack shape, and δ be a parameter which will subsequently tend to zero.

For any fixed δ the solution of the problem

$$\chi^\delta \in K(\Omega_\delta): \quad \langle \Pi'_\delta(\chi^\delta), \bar{\chi} - \chi^\delta \rangle \geq 0 \quad \forall \bar{\chi} \in K(\Omega_\delta) \tag{2.236}$$

can be found. Here $\Omega_\delta, \Pi_\delta$ are introduced analogously to those of Ω_ψ, Π_ψ. The inequality (2.236) can be written as follows:

$$B_\delta(w^\delta, \bar{w}) + \langle k_{ij}\sigma^\delta_{ij} - f_3, \bar{w} \rangle_\delta = 0 \quad \forall \bar{w} \in H^{2,0}(\Omega_\delta), \tag{2.237}$$

$$\langle \sigma^\delta_{ij}, \varepsilon_{ij}(\bar{W} - W^\delta) \rangle_\delta \geq \langle F, \bar{W} - W^\delta \rangle_\delta \quad \forall \bar{W} \in K^1(\Omega_\delta). \tag{2.238}$$

To derive an estimate for the solution, we put $\bar{w} = -w^\delta$ in (2.237) and $\bar{W} = 0$ in (2.238). Simple reasoning implies

$$2\Pi_\delta(\chi^\delta) + \langle f, \chi^\delta \rangle_\delta \leq 0. \tag{2.239}$$

Let us next transform the independent variables to map Ω_δ onto Ω_0. To this end, we extend the function ψ by zero beyond the interval $(0,1)$. Let Ω_1, Ω_2 be domains such that $\overline{\Omega}_1 \subset \Omega_2$, $\overline{\Omega}_2 \subset \Omega$, $\Gamma_\delta \subset \Omega_1$ for all δ small enough. We can choose a function $\xi \in C_0^\infty(\Omega_2)$ with the property $\xi \equiv 1$ on Ω_1 and consider the transformation

$$\tilde{x} = x, \quad \tilde{y} = y - \delta\psi\xi \tag{2.240}$$

which has a positive Jacobian q_δ for a small δ. This transformation sets up a one-to-one correspondence between Ω_δ and Ω_0. Let us denote $u^\delta(\tilde{x}, \tilde{y}) = W^\delta(x, y)$, $U^\delta(\tilde{x}, \tilde{y}) = W^\delta(x, y)$, $\omega^\delta = (U^\delta, u^\delta)$. For the structure of the following relations to be clear we write down one of the second derivatives of w^δ:

$$w^\delta_{xx} = u^\delta_{\tilde{x}\tilde{x}} - 2\delta u^\delta_{\tilde{x}\tilde{y}}(\psi\xi)_x + \delta^2 u^\delta_{\tilde{y}\tilde{y}}(\psi\xi)^2_x - \delta u^\delta_{\tilde{y}}(\psi\xi)_{xx}.$$

Thus, the inequality (2.239) can be written as

$$\int\limits_{\Omega_0} \left(u_{\tilde{x}\tilde{x}}^2 + u_{\tilde{y}\tilde{y}}^2 + 2\nu u_{\tilde{x}\tilde{x}} u_{\tilde{y}\tilde{y}} + 2(1-\nu) u_{\tilde{x}\tilde{y}}^2\right) q_\delta^{-1}\, d\Omega_0 \tag{2.241}$$

$$+ \langle \sigma_{ij}^\delta, e_{ij}^\delta q_\delta^{-1}\rangle_0 - \langle f^\delta, \omega q_\delta^{-1}\rangle_0 + \delta \int\limits_{\Omega_0} g(\tilde{x}, \tilde{y}, \delta, D^\alpha u, D^\beta U)\, d\Omega_0 \le 0.$$

For simplicity the superscript δ in U^δ, u^δ is omitted; we have also used the notation

$$e_{ij}^\delta = \varepsilon_{ij}(U) + k_{ij}^\delta u, \quad k_{ij}^\delta(\tilde{x}, \tilde{y}) = k_{ij}(x, y), \quad f^\delta(\tilde{x}, \tilde{y}) = f(x, y).$$

The arguments of g are determined by the transformation (2.240). Note that the higher order terms are quadratic in $D^\alpha u$, $D^\beta U$, $|\alpha| \le 2$, $|\beta| \le 1$. Since the inequality $q_\delta^{-1} > 1/2$ holds for small δ, we derive from (2.241) that

$$\Pi_0^\delta(\omega) + \delta \int\limits_{\Omega_0} g(\tilde{x}, \tilde{y}, \delta, D^\alpha u, D^\beta U)\, d\Omega_0 \le 0. \tag{2.242}$$

Here

$$\Pi_0^\delta(\omega) = \frac{1}{2} B_0(u, u) + \frac{1}{2}\langle \sigma_{ij}^\delta, e_{ij}^\delta\rangle_0 - \langle f^\delta, \omega q_\delta^{-1}\rangle_0.$$

Let us prove a complementary statement concerning the coercivity of the functional Π_0^δ (compare Vorovich, Lebedev, 1972). We should remark at this point that the coercivity of Π_ψ has been used to prove the solvability of the problem (2.223).

Lemma 2.4. *The functional* Π_0^δ *is coercive on* $H(\Omega_0)$ *uniformly in* δ, $|\delta| \le \delta_0$, *that is*

$$\Pi_0^\delta(\omega) \to \infty \quad \text{as} \quad \|\omega\|_{H(\Omega_0)} \to \infty, \quad |\delta| \le \delta_0.$$

PROOF. Consider the nonlinear part of Π_0^δ:

$$\pi_\delta(\omega) = \frac{1}{2} B_0(u, u) + \frac{1}{2}\langle \sigma_{ij}^\delta, e_{ij}^\delta\rangle_0, \quad \omega = (U, u).$$

A transformation of the unit sphere S can be defined for $r > 0$ by the formulae

$$\tilde{U} \to \lambda r \tilde{U}, \quad \tilde{u} \to r\tilde{u}, \quad \tilde{U} = (\tilde{u}^1, \tilde{u}^2), \quad (\tilde{U}, \tilde{u}) \in S$$

with the constant $\lambda > 0$ to be defined later on. By virtue of the first Korn inequality, the square norm in $[H^{1,0}(\Omega_0)]^2$ is equal to

$$\|U\|_{1,\Omega_0}^2 = \langle \sigma_{ij}(U), \varepsilon_{ij}(U)\rangle_0.$$

In addition to this we can assume that $\|u\|^2_{2,\Omega_0} = B_0(u,u)$, so that

$$\|\omega\|^2_{H(\Omega_0)} = \|U\|^2_{1,\Omega_0} + \|u\|^2_{2,\Omega_0}.$$

Now let $(\tilde{U},\tilde{u}) \in S$ and $\|\tilde{u}\|_{2,\Omega_0} \geq 1/2$. The image of $(\tilde{U},\tilde{u})$ for the above transformation is denoted by ω. Then $\pi_\delta(\omega) \geq r^2/8$ uniformly in δ. Assuming $\|\tilde{U}\|_{1,\Omega_0} \geq \sqrt{3}/2$, the inequality $\pi_\delta(\omega) \geq 3\lambda^2 r^2/8 - c_0\lambda r^2/2$ easily follows, and is uniform in δ. The constant c_0 bounds the integral

$$2\left\langle u, k^\delta_{11}u^1_{\tilde{x}} + k^\delta_{22}u^2_{\tilde{y}} + \nu(k^2_{11}u^2_{\tilde{y}} + k^\delta_{22}u^1_{\tilde{x}})\right\rangle_0$$

on S uniformly in δ, $|\delta| \leq \delta_0$. Choosing the constant λ from the equation $3\lambda^2/8 - c_0\lambda/2 = 1$, the above arguments provide the estimate $\pi_\delta(\omega) \geq r^2/8$ for all images of the sphere S. This implies

$$\Pi^\delta_0(\omega) \geq \frac{r^2}{8} - cr$$

uniformly in δ, $|\delta| \leq \delta_0$. The proof of Lemma 2.4 is complete.

With Lemma 2.4, we easily conclude from (2.242) that

$$\|\omega^\delta\|_{H(\Omega_0)} \leq c$$

uniformly in δ, $|\delta| \leq \delta_0$. Without any loss of generality, one can assume that there exists a subsequence, still denoted by U^δ, u^δ, such that as $\delta \to 0$

$$U^\delta \to U \ \text{ weakly in } H^{1,0}(\Omega_0), \quad u^\delta \to u \ \text{ weakly in } H^{2,0}(\Omega_0). \quad (2.243)$$

The inequality (2.236) can be rewritten in the variables $\tilde{x}, \tilde{y}$. The structure of the relation obtained in this way is analogous to (2.241), that is the smaller order terms are proportional to δ^i, $i \geq 1$. The convergence (2.243) enables us to pass to the limit in (2.237)–(2.238). To complete the reasoning, we should consider in detail the transformation of $K^1(\Omega_\delta)$. After a change of variables in (2.238), the test functions belong to the set

$$K^1_\delta(\Omega_0) = \{W \in [H^{1,0}(\Omega_0)]^2 \mid \ [W]\nu^\delta \geq 0 \quad \text{on } \Gamma_0\}.$$

Here $\nu^\delta = (-\delta\psi_x, 1)/\sqrt{1+\delta^2\psi^2_x}$ is the normal vector to the curve $y = \delta\psi(x)$. For the passage to the limit to be justified, the following property must be proved: for any fixed $\bar{W} \in K^1(\Omega_0)$, there exists a sequence $\bar{W}^\delta \in K^1_\delta(\Omega_0)$ such that

$$\bar{W}^\delta \to \bar{W} \ \text{ strongly in } H^{1,0}(\Omega_0).$$

Indeed, let $\bar{W} = (\bar{w}^1, \bar{w}^2)$. Choosing ξ just as in (2.240) we put $\bar{W}^\delta = \bar{W} + (0, \delta\psi_x\xi\bar{w}^1)$. In so doing the function $\bar{W}^\delta$ will satisfy all the necessary properties. Really, the condition $\bar{W} \in K^1(\Omega_0)$ means the validity of the inequality $[\bar{w}^2] \geq 0$ on Γ_0. At the same time, for the function $\bar{W}^\delta$ to belong to the set $K_\delta^1(\Omega_0)$, the relation $[\bar{w}_\delta^1](-\delta\psi_x)+[\bar{w}_\delta^2] \geq 0$ on Γ_0 must be satisfied. Obviously, our function satisfies this condition. The strong convergence of $\bar{W}^\delta$ to W^δ is evident. So, the limit of the relations obtained by changing the variables in (2.237), (2.238) leads to the problem

$$U \in K^1(\Omega_0), \quad u \in H^{2,0}(\Omega_0):$$

$$B_0(u,\bar{u}) + \langle k_{ij}\sigma_{ij} - f_3, \bar{u}\rangle_0 = 0 \quad \forall \bar{u} \in H^{2,0}(\Omega_0), \tag{2.244}$$

$$\langle \sigma_{ij}, \varepsilon_{ij}(\bar{U} - U)\rangle_0 \geq \langle F, \bar{U} - U\rangle_0 \quad \forall \bar{U} \in K^1(\Omega_0).$$

This means that the limiting function (U, u) is the solution of the problem corresponding to the crack shape $y = \psi(x) \equiv 0$, $x \in [0,1]$.

In conclusion we formulate the statement which has just been proved.

Theorem 2.25. *From the solution* $\chi^\delta = \omega^\delta$ *of the problem* (2.236) *one can choose a subsequence which weakly converges to* $\omega = (U, u)$ *in* $H(\Omega_0)$. *The function* ω *is the solution of the problem* (2.244).

The nonpenetration condition considered in this section leads to new effects such as the appearance of interaction forces between crack faces. It is of interest to establish the highest regularity of the solution up to the crack faces and thus to analyse the smoothness of the interaction forces. The regularity of the solution stated in this section entails the components of the strain and stress tensors to belong to H^1 in the vicinity of the crack and the interaction forces to belong to L^2. If the crack shape is not regular, i.e. $\psi \notin H_0^3(0,1)$, the interaction forces can be characterized by the nonnegative measure μ defined on the subsets of the crack faces.

2.9 Simplified nonpenetration conditions in contact problems

We consider a problem similar to the one considered in Section 2.8. The nonpenetration condition between crack faces is taken in simplified form. Our aim is to obtain some qualitative properties of solutions for a contact problem for a plate having a crack.

2.9.1 Formulation of the problem

Let $\Omega \subset R^2$ be a bounded domain with a smooth boundary Γ, and $y = \psi(x)$ be a crack shape on the (x,y)-plane, $(x,y) \in \Omega$, $x \in [0,1]$. A displacement vector of the mid-surface points of the plate is denoted by $\chi = (W, w)$, $W = (w^1, w^2)$. Herewith the functions W, w are horizontal and vertical displacements, respectively, $\Omega_\psi = \Omega \backslash \Gamma_\psi$, and Γ_ψ is the graph of the function $y = \psi(x)$. Assume that $\psi \in H_0^3(0,1)$. Let $\varepsilon_{ij} = \varepsilon_{ij}(W)$, $f = (f_1, f_2, f_3) \in$

$L^2(\Omega)$, $\langle p, q\rangle_\psi = \int\limits_{\Omega_\psi} pq \, d\Omega_\psi$,

$$\varepsilon_{ij} = \frac{1}{2}\left(\frac{\partial w^i}{\partial x_j} + \frac{\partial w^j}{\partial x_i}\right), \quad i,j = 1,2, \quad x_1 = x, \ x_2 = y.$$

The energy functional of the plate is as follows:

$$\Pi(\chi) = \frac{1}{2}B(w,w) + \frac{1}{2}\langle\sigma_{ij}, \varepsilon_{ij}\rangle_\psi - \langle f, \chi\rangle_\psi,$$

where

$$\sigma_{11} = \varepsilon_{11} + \kappa\varepsilon_{22}, \quad \sigma_{22} = \varepsilon_{22} + \kappa\varepsilon_{11}, \quad \sigma_{12} = (1-\kappa)\varepsilon_{12},$$

the constant κ such that $0 < \kappa < 1/2$. The bilinear form $B(\cdot,\cdot)$ characterizing the bending properties of the plate is defined by a formula

$$B(w,\bar{w}) = \int\limits_{\Omega_\psi} (w_{xx}\bar{w}_{xx} + w_{yy}\bar{w}_{yy} + \kappa w_{xx}\bar{w}_{yy} + \kappa w_{yy}\bar{w}_{xx} + 2(1-\kappa)w_{xy}\bar{w}_{xy}).$$

The following boundary conditions are supposed to be given at the external boundary

$$w = \frac{\partial w}{\partial n} = W = 0 \quad \text{on } \Gamma. \tag{2.245}$$

The nonpenetration condition is imposed both in the domain Ω_ψ and on Γ_ψ. Thus, let the equation $z = \Phi(x,y)$ describe the punch shape, $(x,y) \in \Omega$, $\Phi \in C^\infty(\bar{\Omega})$. Then the nonpenetration condition for the plate–punch system in a linear approach takes the form

$$w - W\nabla\Phi \geq \Phi \quad \text{in } \Omega_\psi. \tag{2.246}$$

Denote by $\nu = (-\psi_x, 1)/\sqrt{1+\psi_x^2}$ the normal vector to the curve $y = \psi(x)$. Let $[W] = W^+ - W^-$ be the jump of the function W on Γ_ψ. The signs $+, -$ correspond to the positive and negative directions with respect to ν. In this case the simplified nonpenetration condition for the crack faces can be written as

$$[W]\nu \geq 0 \quad \text{on } \Gamma_\psi. \tag{2.247}$$

We next introduce the subspace $H^{1,0}(\Omega_\psi) \subset H^1(\Omega_\psi)$ whose elements are equal to zero on Γ; the subspace $H^{2,0}(\Omega)$ is defined analogously, $H(\Omega_\psi) = H^{1,0}(\Omega_\psi) \times H^{1,0}(\Omega_\psi) \times H^{2,0}(\Omega_\psi)$. In particular, the condition $\chi \in H(\Omega_\psi)$ provides the fulfilment of (2.245). The norm in $H^{s,0}(\Omega_\psi)$ is denoted by $\|\cdot\|_{s,\Omega_\psi}$. Consider the convex and closed set in $H(\Omega_\psi)$:

$$K = \{(W,w) \in H(\Omega_\psi) \mid \ (W,w) \text{ satisfy } (2.246), (2.247)\}.$$

In so doing, the boundary value of Φ on Γ is assumed to provide nonemptiness of the set K. The equilibrium problem for the plate contacting with the punch and having the crack can be formulated as a variational one:

$$\inf_{\chi \in K} \Pi(\chi).$$

In view of the convexity and differentiability of Π this problem has an equivalent form

$$B(w, \bar{w} - w) + \langle \sigma_{ij}, \varepsilon_{ij}(\bar{W} - W) \rangle_\psi \geq \langle f, \bar{\chi} - \chi \rangle_\psi, \quad \chi \in K, \ \forall \bar{\chi} \in K, \tag{2.248}$$

which precisely means the following inequality:

$$\langle \Pi'(\chi), \bar{\chi} - \chi \rangle \geq 0, \quad \chi \in K, \quad \forall \bar{\chi} \in K.$$

Herein $\Pi'(\chi)$ is a derivative of Π evaluated at the point χ.

2.9.2 Construction of the measures

We first note that the coercivity and weak lower semicontinuity of the functional Π imply that the problem (2.248) has a (unique) solution $\chi \in K$. The coercivity is provided by the following two inequalities,

$$B(w, w) \geq c\|w\|^2_{2,\Omega_\psi} \quad \forall\, w \in H^{2,0}(\Omega_\psi), \tag{2.249}$$

$$\langle \sigma_{ij}, \varepsilon_{ij}(W) \rangle_\psi \geq c\|W\|^2_{1,\Omega_\psi} \quad \forall\, W \in H^{1,0}(\Omega_\psi), \tag{2.250}$$

where $\sigma_{ij} = \sigma_{ij}(W)$. Denote next by

$$N = (-\nabla\Phi, 1)/\sqrt{1 + |\nabla\Phi|^2}$$

the normal vector to the surface $z = \Phi(x, y)$. The space of continuous functions in Ω_ψ with compact supports is designated by $C_0(\Omega_\psi)$. The convergence in this space can be introduced as follows: $\phi_n \to \phi$ in $C_0(\Omega_\psi)$, if ϕ_n converges to ϕ uniformly and supports of all ϕ_n belong to a fixed compact $S \subset \Omega_\psi$. The σ-algebra of Borel's subsets of Ω_ψ is denoted hereinafter by $\sigma(\Omega_\psi)$. In what follows we prove the existence of a measure characterizing the interaction forces between the plate and the punch.

Lemma 2.5. *There exists a nonnegative measure $\mu \in \sigma(\Omega_\psi)$ such that for all $\bar{\chi} \in H(\Omega_\psi) \cap C_0(\Omega_\psi)$ the following representation holds:*

$$\langle \Pi'(\chi), \bar{\chi} \rangle = \int_{\Omega_\psi} \bar{\chi} N \, d\mu. \tag{2.251}$$

PROOF. Let us take any fixed element $\bar{\chi} \in H(\Omega_\psi) \cap C_0(\Omega_\psi)$ satisfying the inequality $\bar{\chi} N \geq 0$ in Ω_ψ. In this case the inclusion $\chi + \bar{\chi} \in K$ is evident. Let us substitute the function $\chi + \bar{\chi}$ in (2.248) as a test one. This implies

$$\langle \Pi'(\chi), \bar{\chi} \rangle \geq 0. \tag{2.252}$$

Consider next the variety V of all functions $\bar{\chi}^*$ such that $\bar{\chi}^* = \bar{\chi}N$, $\bar{\chi} \in H(\Omega_\psi) \cap C_0(\Omega_\psi)$, and define a linear and positive functional on V:

$$L(\bar{\chi}^*) = \langle \Pi'(\chi), \bar{\chi} \rangle.$$

By the inequality (2.252), valid for all functions $\bar{\chi}$ possessing the property $\bar{\chi}N \geq 0$ in Ω_ψ, this definition of L is correct. Similar to (Landkof, 1966) the functional L can be extended on $C_0(\Omega_\psi)$. The extended functional is linear and positive, and hence it is continuous. This implies that there exists a nonnegative measure $\mu \in \sigma(\Omega_\psi)$ such that for all $\varphi \in C_0(\Omega_\psi)$ a representation

$$L(\varphi) = \int_{\Omega_\psi} \varphi \, d\mu$$

holds. This formula exactly coincides with (2.251) for $\varphi = \bar{\chi}N$, $\bar{\chi} \in H(\Omega_\psi) \cap C_0(\Omega_\psi)$. Lemma 2.5 is proved.

Now, let us prove the existence of the measure which characterizes an interaction of the crack faces. The space of continuous functions having compact supports in $\Gamma_\psi \setminus \partial\Gamma_\psi$ is denoted by $C_0(\Gamma_\psi)$, $F = (f_1, f_2)$. We next denote by $\sigma(\Gamma_\psi \setminus \partial\Gamma_\psi)$ the Borel σ-algebra of subsets of $\Gamma_\psi \setminus \partial\Gamma_\psi$. If a function p is defined in Ω_ψ and has the traces on Γ_ψ belonging to $C_0(\Gamma_\psi)$ we shall write $p \in C_0(\Gamma_\psi)$. Let $\mathcal{W}_\psi$ be a neighbourhood of the graph Γ_ψ and $\mathcal{W} = \mathcal{W}_\psi \setminus \Gamma_\psi$.

Lemma 2.6. *Assume that $\nabla\Phi = 0$ in $\mathcal{W}$. Then, there exists a nonnegative measure $\gamma \in \sigma(\Gamma_\psi \setminus \partial\Gamma_\psi)$ such that for all $\bar{W} \in (H^{1,0}(\Omega_\psi) \times H^{1,0}(\Omega_\psi)) \cap C_0(\Gamma_\psi)$, $\bar{W} \equiv 0$ outside $\mathcal{W}$, the following representation holds:*

$$\langle \Pi'(\chi), (\bar{W}, 0) \rangle = \int_{\Gamma_\psi \setminus \partial\Gamma_\psi} [\bar{W}]\nu \, d\gamma. \tag{2.253}$$

PROOF. First of all, we note that the function $\chi + (\bar{W}, 0)$ is an element of the set K provided that $\bar{W}$ belongs to the space shown in the Lemma 2.6 formulation and $[\bar{W}]\nu \geq 0$ on Γ_ψ. It should be recalled at this point that $\chi = (W, w) \in K$ is the solution of the problem (2.248). Hence, substituting $\chi + (\bar{W}, 0)$ in (2.248) as a test function yields

$$\langle \sigma_{ij}, \varepsilon_{ij}(\bar{W}) \rangle_\psi - \langle F, \bar{W} \rangle_\psi \geq 0. \tag{2.254}$$

The variety of all functions defined on $\Gamma_\psi \setminus \partial\Gamma_\psi$ by the formula $\bar{W}^* = [\bar{W}]\nu$ is denoted by U. Here $\bar{W} \in (H^{1,0}(\Omega_\psi) \times H^{1,0}(\Omega_\psi)) \cap C_0(\Gamma_\psi)$, $\bar{W} \equiv 0$ outside $\mathcal{W}$. It is easily seen that the functional M defined on U by the formula

$$M(\bar{W}^*) = \langle \sigma_{ij}, \varepsilon_{ij}(\bar{W}) \rangle_\psi - \langle F, \bar{W} \rangle_\psi$$

is positive. To verify this we use the inequality (2.254). The same inequality provides the definition of M to be correct, so that the equality $\bar{W}_1^* = \bar{W}_2^*$

implies $M(\bar{W}_1^*) = M(\bar{W}_2^*)$. Continuously differentiable functions belonging to $C_0(\Gamma_\psi)$ are further denoted by $C_0^1(\Gamma_\psi)$. Let us state that the variety U contains the space $C_0^1(\Gamma_\psi)$. To this end we choose any fixed function $\varphi \in C_0^1(\Gamma_\psi)$. Then, the function ψ is assumed to be extended by zero beyond $[0, 1]$ and a function ξ to be introduced:

$$\xi(\tilde{x}) = \begin{cases} 1\,, & \text{if} \quad \tilde{x} = \bar{x} + \lambda\nu(\bar{x}), \quad \bar{x} \in \Gamma_\psi, \quad 0 \le \lambda \le \frac{\lambda_0}{2}, \\ 0\,, & \text{if} \quad \lambda > \lambda_0. \end{cases}$$

The parameter λ_0 is supposed to be rather small, so that ξ is well defined. In so doing there is no problem to construct a function Ξ in the domain $y > \psi(x)$ provided that φ is also extended by zero along the new graph of the function $y = \psi(x)$. Namely, we put for $y \ge \psi(x)$:

$$\Xi(\tilde{x}) = \varphi(\bar{x})\xi(\tilde{x})\nu(\bar{x}), \quad \tilde{x} = \bar{x} + \lambda\nu(\bar{x}), \quad \bar{x} \in \Gamma_\psi, \quad \lambda \ge 0.$$

In this case one clearly has

$$\Xi(\bar{x})\nu(\bar{x}) = \varphi(\bar{x}), \quad \bar{x} \in \Gamma_\psi.$$

It is seen that the graph of the extended function $y = \psi(x)$ has joint points with Γ and consequently the points like $\tilde{x} = \bar{x} + \lambda\nu(\bar{x})$ may not belong to Ω_ψ. It should present no problems since φ is equal to zero near Γ. As for the domain $y < \psi(x)$ one can easily put

$$\Xi(\tilde{x}) \equiv 0, \quad \tilde{x} \equiv (x, y), \quad y < \psi(x).$$

Hence, the function Ξ is defined in Ω_ψ and the condition

$$[\Xi]\nu = \varphi \quad \text{on} \quad \Gamma_\psi \setminus \partial\Gamma_\psi$$

holds. Moreover, $\Xi \equiv 0$ outside $\mathcal{W}$ for λ_0 small enough. Thus, the inclusion $C_0^1(\Gamma_\psi) \subset U$ is proved. The functional M can be extended on $C_0(\Gamma_\psi)$ by its positiveness (Landkof, 1966). This implies that there exists a nonnegative measure $\gamma \in \sigma(\Gamma_\psi \setminus \partial\Gamma_\psi)$ such that forall $\varphi \in C_0(\Gamma_\psi)$

$$M(\varphi) = \int\limits_{\Gamma_\psi \setminus \partial\Gamma_\psi} \varphi \, d\gamma.$$

Hence, the representation (2.253) follows provided that $\varphi = \bar{W}^*$, and $\bar{W} \in (H^{1,0}(\Omega_\psi) \times H^{1,0}(\Omega_\psi)) \cap C_0(\Gamma_\psi)$, $\bar{W} \equiv 0$ outside $\mathcal{W}$. Lemma 2.6 is proved.

Making use of the above lemmas the following statement can be established.

Theorem 2.26. *Let $\nabla\Phi = 0$ in $\mathcal{W}$. Then, there exist nonnegative measures $\mu \in \sigma(\Omega_\psi)$ and $\gamma \in \sigma(\Gamma_\psi \setminus \partial\Gamma_\psi)$ such that for all $\bar{\chi} = (\bar{W}, \bar{w}) \in H(\Omega_\psi) \cap C_0(\Gamma_\psi)$, $\bar{w} \in C_0(\Omega_\psi)$, a representation*

$$\langle \Pi'(\chi), \bar{\chi} \rangle = \int\limits_{\Omega_\psi} \bar{\chi} N \, d\mu + \int\limits_{\Gamma_\psi \setminus \partial\Gamma_\psi} [\bar{W}]\nu \, d\gamma \tag{2.255}$$

holds.

PROOF. Let us first state that any fixed function $\bar{\chi} = (\bar{W}, \bar{w}) \in H(\Omega_\psi) \cap C_0(\Gamma_\psi)$, $\bar{w} \in C_0(\Omega_\psi)$, can be represented as

$$\bar{\chi} = (\tilde{W}, 0) + (\bar{W} - \tilde{W}, \bar{w}), \tag{2.256}$$

where $\tilde{W} \equiv 0$ outside $\mathcal{W}$ and $\tilde{W} = \bar{W}$ near Γ_ψ. To this end the closed curve

$$l_\psi = \{\bar{x} \mid \ \text{dist}\,(\bar{x}, \Gamma_\psi) = \lambda\}$$

is considered for λ small enough. Let $\bar{W}\ |_{l_\psi}$ be the trace of the function $\bar{W}$ on l_ψ. The function $\bar{W}\ |_{l_\psi}$ can be easily extended in a small outer neighbourhood of l_ψ, being equal to zero beyond this neighbourhood. The extended function is denoted by $\bar{W}_\psi$. Then, we put

$$\tilde{W}(\bar{x}) = \begin{cases} \bar{W}(\bar{x}), & \text{if } \bar{x} \ \text{ is inside } l_\psi, \\ \bar{W}_\psi(\bar{x}), & \text{if } \bar{x} \ \text{ is outside } l_\psi. \end{cases}$$

The function $\tilde{W}$ satisfies all the desired conditions provided that λ is small enough. Hence, the representation (2.256) takes place. Since the function $\bar{W} - \tilde{W}$ has a compact support in Ω_ψ, one can use Lemma 2.5 and Lemma 2.6. So, the following relations hold for all $\bar{\chi}$ satisfying the conditions of Theorem 2.26:

$$\langle \Pi'(\chi), \bar{\chi} \rangle = \langle \Pi'(\chi), (\bar{W} - \tilde{W}, \bar{w}) \rangle + \langle \Pi'(\chi), (\tilde{W}, 0) \rangle$$

$$= \int_{\Omega_\psi} (\bar{W} - \tilde{W}, \bar{w}) N\, d\mu + \int_{\Gamma_\psi \setminus \partial\Gamma_\psi} [\tilde{W}] \nu\, d\gamma = \int_{\Omega_\psi} \bar{\chi} N\, d\mu + \int_{\Gamma_\psi \setminus \partial\Gamma_\psi} [\bar{W}] \nu\, d\gamma.$$

Theorem 2.26 is proved.

2.9.3 Solution regularity

The properties of the constructed measures μ and γ depend largely on the solution regularity. We thus prove some statements.

Theorem 2.27. *Let $x^0 \in \Omega_\psi$ and $|\nabla\Phi(x^0)| > 0$. Then, there exists a neighbourhood $\mathcal{O}(x^0)$ of the point x^0 such that the solution $\chi = (W, w)$ of (2.248) satisfies the inclusion*

$$(W, w) \in H^2(\mathcal{O}(x^0)) \times H^4(\mathcal{O}(x^0)).$$

PROOF. Introduce the following notation:

$$d_{\pm\tau k} p(\bar{x}) = \tau^{-1}(p(\bar{x} \pm \tau j_k) - p(\bar{x})), \quad \Delta_{\tau k} = -d_{-\tau k} d_{\tau k}, \quad k = 1, 2.$$

Here $|\tau| > 0$ and j_k is a unit vector of the axis x_k. We next notice the following. Let $\varphi \in C_0^\infty(\Omega_\psi)$, and p, q be any functions, $0 \le \varphi \le 1$. Then the inequality

$$p + \frac{\tau^2}{2} \varphi \Delta_{\tau k}(\varphi p - \varphi q) \ge q \quad \text{in} \quad \Omega_\psi \tag{2.257}$$

holds for small τ provided that the inequality $p \geq q$ in Ω_ψ takes place. It is evident that the left-hand side of (2.257) makes sense if the distance between the support of φ and $\partial\Omega_\psi$ is more than $|\tau|$. To prove (2.257) we write down the left side of (2.257) taken at the point $\bar{x}$ as follows:

$$(1-\varphi^2(\bar{x}))p(\bar{x}) + \frac{\varphi(\bar{x})}{2}((\varphi p)(\bar{x}+\tau j_k) + (\varphi p)(\bar{x}-\tau j_k))$$

$$-\frac{\tau^2}{2}\varphi(\bar{x})\Delta_{\tau k}(\varphi q)(\bar{x}).$$

It is seen that this quantity is more than or equal to

$$(1-\varphi^2(\bar{x}))q(\bar{x}) + \frac{\varphi(\bar{x})}{2}((\varphi q)(\bar{x}+\tau j_k)$$

$$+(\varphi q)(\bar{x}-\tau j_k)) - \frac{\tau^2}{2}\varphi(\bar{x})\Delta_{\tau k}(\varphi q)(\bar{x}) = q(\bar{x}),$$

i.e. the proof of (2.257) is obtained.

Without decreasing a generality one can assume $\Phi_{x_k}(x^0) \neq 0$, $k=1,2$. Let $B_{2r}(x^0) \subset \Omega_\psi$ be a ball of radius $2r$ centred at the point x^0 such that $|\Phi_{x_k}(\bar{x})| > 0$, $\bar{x} \in B_{2r}(x^0)$, $k=1,2$. We choose a smooth function φ with the properties: $\varphi \equiv 1$ on $B_r(x^0)$, $\varphi \equiv 0$ outside $B_{3r/2}(x^0)$, and $0 \leq \varphi \leq 1$ everywhere. Construct next a vector $\chi_\tau = (w_\tau^1, w_\tau^2, w_\tau)$ with the following components (there is no a summation over i):

$$w_\tau^i = w^i + \frac{\tau^2}{2}\Psi_i, \quad \Psi_i = \varphi\Phi_{x_i}^{-1}\Delta_{\tau k}(w^i\varphi\Phi_{x_i}), \quad i=1,2,$$

$$w_\tau = w + \frac{\tau^2}{2}\varphi\Delta_{\tau k}(\varphi w - \varphi\Phi).$$

It is clear that $(w_\tau^1, w_\tau^2, w_\tau) \in K$ for $|\tau| < r/2$. To verify this, one can apply the arguments used to prove (2.257). This yields the inequality

$$w_\tau - W_\tau \nabla\Phi \geq \Phi \quad \text{in } \Omega_\psi, \quad W_\tau = (w_\tau^1, w_\tau^2).$$

In so doing we take into account the inequality (2.246) for the solution (W, w). Moreover, the relation $\chi_\tau = 0$ on Γ_ψ holds. This implies $\chi_\tau \in K$. Substituting χ_τ in (2.248) as a test function gives the relation

$$B(w, \varphi\Delta_{\tau k}(\varphi w - \varphi\Phi)) + \langle\sigma_{ij}, \varepsilon_{ij}(\Psi)\rangle_\psi \geq 2\tau^{-2}\langle f, \chi_\tau - \chi\rangle_\psi. \qquad (2.258)$$

Herein $\sigma_{ij} = \sigma_{ij}(W)$, $\Psi = (\Psi_1, \Psi_2)$. We have to notice that the difference between the terms (the summation over k is absent)

$$B(w, \varphi\Delta_{\tau k}(\varphi w)) \quad \text{and} \quad -B(d_{\tau k}(\varphi w), d_{\tau k}(\varphi w))$$

can be estimated from the above by the value being on the right-hand side of (2.259) (see below). The difference between the terms

$$\langle \sigma_{ij}, \varepsilon_{ij}(\Psi)\rangle_\psi \quad \text{and} \quad \langle \sigma_{ij}(d_{\tau k}(\varphi W)), \varepsilon_{ij}(d_{\tau k}(\varphi W))\rangle_\psi$$

can also be estimated from the above by the same value. The right-hand side of (2.258) does not depend on τ and can be estimated easier. Thus, it follows from (2.258) that

$$B(d_{\tau k}(\varphi w), d_{\tau k}(\varphi w)) + \langle \sigma_{ij}(d_{\tau k}(\varphi W)), \varepsilon_{ij}(d_{\tau k}(\varphi W))\rangle_\psi \tag{2.259}$$

$$\leq c\left(\|\chi\|^2_{H(\Omega_\psi)} + \|d_{\tau k}(\varphi\chi)\|_{H(\Omega_\psi)}(\|\chi\|_{H(\Omega_\psi)} + \|f\|_{0,\Omega_\psi})\right)$$

with a constant c independent of τ. In view of (2.249), (2.250) the estimate (2.259) results in the relation

$$\|d_{\tau k}(\varphi\chi)\|_{H(\Omega_\psi)} \leq c$$

uniformly in τ. Whence

$$\frac{\partial}{\partial x_k}(\varphi\chi) \in H(\Omega_\psi), \quad k = 1, 2,$$

that is,

$$(W, w) \in H^2(B_r(x^0)) \times H^3(B_r(x^0)).$$

Actually, the function w has four square integrable derivatives in a neighbourhood of x^0. For this assertion to be proved one can choose the ball $B_r(x^0)$, as before. Let $\varphi \in C_0^\infty(B_r(x^0))$ be an arbitrary function, and $\lambda > 0$. It is clear that

$$\chi_\lambda \equiv (w^1 + \lambda\varphi\Phi^{-1}_{x_1}, w^2, w + \lambda\varphi) \in K.$$

Since the functional Π reaches a minimum over the set K at the point (w^1, w^2, w) the inequality

$$\Pi(\chi_\lambda) - \Pi(\chi) \geq 0$$

takes place. Dividing this relation by λ and passing to the limit as $\lambda \to 0$ the following equation is obtained in $B_r(x^0)$:

$$\Delta^2 w - f_3 = \left(\frac{\partial\sigma_{1j}}{\partial x_j} + f_1\right)\Phi^{-1}_{x_1},$$

being fulfilled in the sense of distributions. By deriving this equation we take into account the arbitrariness of φ. According to the foregoing the right side of the above equation belongs to $L^2(B_r(x^0))$; hence

$$w \in H^4(B_{r/2}(x^0)).$$

Theorem 2.27 is proved.

Let us state one more result relating to the regularity of the solution in a neighbourhood of $\Gamma_\psi \setminus \partial\Gamma_\psi$. Let $x^0 \in \Gamma_\psi \setminus \partial\Gamma_\psi$. Assume that the graph Γ_ψ has a rectilinear section near x^0 which is parallel to the axis x.

Theorem 2.28. *Let $\nabla\Phi = 0$ in some neighbourhood of the point x^0. Then there exists a neighbourhood $\mathcal{O}(x^0)$ of the point x^0 such that*

$$W \in H^2(\mathcal{O}(x^0) \cap \Omega_\psi), \quad \frac{\partial w}{\partial x} \in H^2(\mathcal{O}(x^0) \cap \Omega_\psi).$$

PROOF. Let us choose a smooth function φ such that $\varphi \equiv 1$ on $B_r(x^0)$, $\varphi \equiv 0$ outside $B_{3r/2}(x^0)$, $0 \le \varphi \le 1$ everywhere, assuming that r is chosen to be small enough, so that $\nabla\Phi = 0$ in $B_{2r}(x^0)$. Define the functions

$$w_\tau^i = w^i + \frac{\tau^2}{2}\varphi\Delta_\tau(\varphi w^i), \quad i = 1, 2, \quad w_\tau = w,$$

where $d_\tau \equiv d_{\tau 1}$, $\Delta_\tau \equiv \Delta_{\tau 1}$, $|\tau| < \frac{r}{2}$. It is easily checked that $(w_\tau^1, w_\tau^2, w_\tau) \in K$. Indeed, for this inclusion to be valid one has to verify the inequalities (2.246), (2.247). The relation (2.246) obviously takes place since $\nabla\Phi \equiv 0$ in some neighbourhood of x^0. So, it suffices to examine (2.247). To do this, we take into account that the graph Γ_ψ has a rectilinear section near x^0. The last condition implies $[w^2] \ge 0$ on $\Gamma_\psi \cap B_{2r}(x^0)$. Hence, it suffices to prove

$$[w_\tau^2] \ge 0 \quad \text{on} \quad \Gamma_\psi \cap B_{3r/2}(x^0).$$

This inequality holds since

$$[w_\tau^2] = [w^2] + \frac{\tau^2}{2}\varphi[\Delta_\tau(\varphi w^2)] = (1 - \varphi^2)[w^2]$$

$$+ \frac{\varphi}{2}\left[(\varphi w^2)(\bar{x} + \tau j_1) + (\varphi w^2)(\bar{x} - \tau j_1)\right] \ge 0$$

on $\Gamma_\psi \cap B_{3r/2}(x^0)$. Substituting $(w_\tau^1, w_\tau^2, w_\tau)$ in (2.248) as a test function yields the relation

$$\langle \sigma_{ij}, \varepsilon_{ij}(\varphi\Delta_\tau(\varphi W))\rangle_\psi \ge \langle F, \varphi\Delta_\tau(\varphi W)\rangle_\psi. \tag{2.260}$$

Meantime, it is easily seen that the difference between the terms

$$\langle \sigma_{ij}, \varepsilon_{ij}(\varphi\Delta_\tau(\varphi W))\rangle_\psi \quad \text{and} \quad -\langle \sigma_{ij}(d_\tau(\varphi W)), \varepsilon_{ij}(d_\tau(\varphi W))\rangle_\psi$$

can be estimated from the above by the value being on the right-hand side of the inequality (2.261). It enables us to obtain from (2.260) the following relation

$$\langle n_{ij}(d_\tau(\varphi W)), \varepsilon_{ij}(d_\tau(\varphi W))\rangle_\psi \tag{2.261}$$

$$\le c\left(\|W\|_{1,\Omega_\psi}^2 + \|d_\tau(\varphi W)\|_{1,\Omega_\psi}(\|W\|_{1,\Omega_\psi} + \|F\|_{0,\Omega_\psi})\right).$$

Again, by the first Korn inequality, the relation (2.261) results in the estimate

$$\|d_\tau(\varphi W)\|_{1,\Omega_\psi} \leq c,$$

which is uniform in τ. This implies that all the second derivatives of φW with the exception of $(\varphi W)_{yy}$ belong to $L^2(\Omega_\psi)$. At the same time the relation (2.248) provides the validity of the equation

$$W_{yy} = G$$

holding in $B_{2r}(x^0)\cap\Omega_\psi$ in the sense of distributions. To verify this equation, the vector $\bar{\chi} = (\tilde{W}, w)$ is to be substituted in (2.248) as a test function, where w is the third component of the solution (W, w) and $\tilde{W} = (\tilde{w}^1, \tilde{w}^2)$ is an arbitrary smooth function with a compact support in $B_{2r}(x^0) \cap \Omega_\psi$. In view of the proved smoothness of the solution, we have $G \in L^2(B_r(x^0)\cap\Omega_\psi)$. Hence $W_{yy} \in L^2(B_r(x^0) \cap \Omega_\psi)$, that is

$$W \in H^2(B_r(x^0) \cap \Omega_\psi).$$

Let us now prove the second part of Theorem 2.28 concerning the smoothness of w. The function φ is assumed to be chosen as before, $d_\tau = d_{\tau 1}$, $\Delta_\tau = \Delta_{\tau 1}$. We put

$$W_\tau = W, \quad w_\tau = w + \frac{\tau^2}{2}\varphi\Delta_\tau(\varphi w - \varphi\Phi).$$

Taking into account the identity $\nabla\Phi \equiv 0$ fulfilled in $B_{2r}(x^0)$ the inequality

$$w_\tau \geq \Phi \quad \text{in } B_{3r/2}(x^0) \cap \Omega_\psi$$

follows provided that $|\tau| < r/2$. Consequently, the vector (W_τ, w_τ) belongs to the set K. So, one can consider this vector as a test function in (2.248). This implies

$$B(w, \varphi\Delta_\tau(\varphi w - \varphi\Phi)) \geq \langle f_3, \varphi\Delta_\tau(\varphi w - \varphi\Phi)\rangle_\psi.$$

The arguments similar to those used to prove Theorem 2.27 allow us to obtain from here the inequality

$$B(d_\tau(\varphi w), d_\tau(\varphi w)) \leq c$$

with a constant c uniform in τ, whence $\|d_\tau(\varphi w)\|_{2,\Omega_\psi} \leq c$. Thus, one has

$$\frac{\partial w}{\partial x} \in H^2(B_r(x^0) \cap \Omega_\psi).$$

Theorem 2.28 is proved.

Under the conditions of Theorem 2.26, i.e. assuming that $\nabla\Phi = 0$ in W, one can find a density of the measure γ. In this case the measure γ is absolutely continuous with respect to the Lebesgue measure on Γ_ψ. Indeed, let us choose $\bar{\chi}$ in (2.255) as $(\bar{W},0)$, $(\bar{W},0) \in H(\Omega_\psi) \cap C_0(\Gamma_\psi)$, $\bar{W} \equiv 0$ outside $\mathcal{W}$. We arrive at the relation

$$\langle \sigma_{ij}, \varepsilon_{ij}(\bar{W})\rangle_\psi - \langle F, \bar{W}\rangle_\psi = \int\limits_{\Gamma_\psi \setminus \partial\Gamma_\psi} [\bar{W}]\nu \, d\gamma. \tag{2.262}$$

In view of the obtained regularity of W up to $\Gamma_\psi \setminus \partial\Gamma_\psi$ the integration by parts can be fulfilled in (2.262). This yields

$$-\int\limits_{\Gamma_\psi^+} n_{ij}^+ \nu^j \bar{w}^i \, d\Gamma_\psi + \int\limits_{\Gamma_\psi^-} n_{ij}^- \nu^j \bar{w}^i \, d\Gamma_\psi = \int\limits_{\Gamma_\psi \setminus \partial\Gamma_\psi} [\bar{W}]\nu \, d\gamma. \tag{2.263}$$

Of course, in so doing, we take into account the equations

$$\frac{\partial \sigma_{ij}}{\partial x_j} + f_i = 0, \quad i = 1, 2,$$

holding in $\mathcal{W}$. The signs $+, -$ correspond to the positive and negative directions with respect to the normal ν. Since the tangent components $\tau_\nu^\pm$ of the vectors $\{n_{ij}^\pm \nu^j\}$ are equal to zero on $\Gamma_\psi^\pm$ and $n_{ij}^+ \nu^j \nu^i = n_{ij}^- \nu^j \nu^i$, the relation (2.263) implies

$$-\int\limits_{\Gamma_\psi} [\bar{W}]\nu(\sigma_{ij}\nu^j\nu^i) \, d\Gamma_\psi = \int\limits_{\Gamma_\psi \setminus \partial\Gamma_\psi} [\bar{W}]\nu \, d\gamma.$$

Hence, we conclude herefrom that the density p_γ of the measure γ is equal to $-\sigma_{ij}\nu^j\nu^i$ and, moreover, $p_\gamma \in L^2_{loc}(\Gamma_\psi \setminus \partial\Gamma_\psi)$. We should remark at this point that the assumption $\psi \in H_0^3(0,1)$ was used to prove the regularity of W. Meanwhile, Theorem 2.26 still remains valid for the case $\psi \in C_0^1(0,1)$. But in this last case the assertion $p_\gamma \in L^2_{loc}(\Gamma_\psi \setminus \partial\Gamma_\psi)$ is not valid, in general.

The density of the measure μ can be also found in neighbourhoods of points $x^0 \in \Omega_\psi$ satisfying the inequality $| \nabla\Phi(x^0) | > 0$. As we know, the function w has four derivatives near such points. Let a function $\bar{\chi} = (0, 0, \bar{w})$ be substituted in (2.255) as a test element, where $\bar{w} \in H^{2,0}(\Omega_\psi)$ and a support of $\bar{w}$ is situated in a small neighbourhood of the fixed point x^0. This results in the equation

$$B(w, \bar{w}) - \langle f_3, \bar{w}\rangle_\psi = \int\limits_{\Omega_\psi} \bar{w}/\sqrt{1 + |\nabla\Phi|^2} \, d\mu.$$

Integrating by parts on the left-hand side implies

$$\int\limits_{\Omega_\psi} (\Delta^2 w - f_3)\bar{w} \, dx = \int\limits_{\Omega_\psi} \bar{w}/\sqrt{1 + |\nabla\Phi|^2} \, d\mu.$$

Consequently, the density p_μ of the measure μ is equal to

$$(\Delta^2 w - f_3)\sqrt{1 + |\nabla\Phi|^2}.$$

As it is seen from Theorem 2.27, $p_\mu \in L^2(\mathcal{O}(x^0))$ for some neighbourhood $\mathcal{O}(x^0)$ of the point x^0.

2.10 Solving methods for the simplified models

In this section we deal with the simplified nonpenetration condition of the crack faces considered in the previous section. We formulate the model of a plate with a crack accounting for only horizontal displacements and construct approximate equations using penalty and iterative methods. The convergence of these solutions is proved and its application to the one-dimensional problem is discussed. Analytical solutions for the model of a bar with a cut are obtained. The results of this section can be found in (Kovtunenko, 1996c, 1996d).

2.10.1 Formulation of the model

A thin isotropic homogeneous plate is assumed to occupy a bounded domain $\Omega \subset R^2$ with the smooth boundary Γ. The crack Γ_c inside Ω is described by a sufficiently smooth function. The chosen direction of the normal $\nu = (\nu^1, \nu^2)$ to Γ_c defines positive Γ_c^+ and negative Γ_c^- crack faces. Let us denote $\Omega_c = \Omega \backslash \Gamma_c$. The function $W = (w^1, w^2)$ of the plate horizontal displacements is sought to satisfy the following boundary conditions. First, the jam condition $W = 0$ must hold on Γ. Second, the simplified nonpenetration condition of the crack faces is imposed at the internal boundary Γ_c,

$$[W]\nu \geq 0,$$

where $[W]$ is the jump of W on Γ_c, i.e. $[W] = W|_{\Gamma_c^+} - W|_{\Gamma_c^-}$.

Let us define the Hilbert space

$$H(\Omega_c) = \{W = (w^1, w^2) \in H^1(\Omega_c) \mid \quad W = 0 \text{ on } \Gamma\}$$

and the closed convex set

$$K = \{W \in H(\Omega_c) \mid \quad [W]\nu \geq 0 \text{ on } \Gamma_c\}.$$

Using the first Korn inequality

$$\langle \sigma_{ij}(W), \varepsilon_{ij}(W) \rangle \geq c_1 \|W\|^2_{H^1(\Omega_c)},$$

valid for $W \in H(\Omega_c)$, we define the scalar product in $H(\Omega_c)$ by the formula

$$(W, \bar{W}) = \langle \sigma_{ij}(W), \varepsilon_{ij}(\bar{W}) \rangle + \int\limits_{\Gamma_c} ([W]\nu)\,([\bar{W}]\nu)\; d\Gamma_c$$

and the norm in $H(\Omega_c)$ by $\|W\|^2 = (W, W)$. Here $\sigma_{ij}(W)$, $\varepsilon_{ij}(W)$ are the usual linear stress and strain tensors; $\langle \cdot, \cdot \rangle$ denotes integration over Ω_c. Therefore, by Korn's inequality and the continuity of the trace operators, the following estimate is valid:

$$\langle \sigma_{ij}(W), \varepsilon_{ij}(W) \rangle \geq c_2 \|W\|^2, \quad 0 < c_2 < 1. \tag{2.264}$$

Let $f = (f^1, f^2) \in L^2(\Omega_c)$ be given external forces. The equilibrium problem for the thin elastic plate with the crack is formulated as follows:

$$\langle \sigma_{ij}(W), \varepsilon_{ij}(\bar{W} - W) \rangle \geq \langle f, \bar{W} - W \rangle \quad \forall \bar{W} \in K. \tag{2.265}$$

There exists a unique solution $W \in K$ of the variational inequality (2.265).

Assume that the solution W of (2.265) is sufficiently smooth. We use Green's formula

$$\langle \sigma_{ij}(W), \varepsilon_{ij}(\bar{W}) \rangle = \langle -\sigma_{ij,j}(W), \bar{w}^i \rangle - \int\limits_{\Gamma_c} \left[\sigma_\nu(W) \bar{W}\nu + \sigma_\tau(W) \bar{W}\tau \right] d\Gamma_c,$$

where $\sigma_\nu(W) = \sigma_{ij}(W)\nu^j\nu^i$, $\sigma_\tau^i(W) = \sigma_{ij}(W)\nu^j - \sigma_\nu(W)\nu^i$, $i = 1, 2$, and rewrite (2.265) as follows:

$$\langle -\sigma_{ij,j}(W) - f^i, \bar{w}^i - w^i \rangle$$
$$- \int\limits_{\Gamma_c} \left[\sigma_\nu(W)(\bar{W} - W)\nu + \sigma_\tau(W)(\bar{W} - W)\tau \right] d\Gamma_c \geq 0.$$

By varying the test function $\bar{W} \in K$, one can deduce that the variational inequality (2.265) is equivalent to the equilibrium equations in Ω_c,

$$-\sigma_{ij,j}(W) = f^i, \quad i = 1, 2,$$

and boundary conditions at Γ_c,

$$[\sigma_\nu(W)] = 0, \quad \sigma_\tau(W) = 0, \quad [W]\nu \geq 0, \quad \sigma_\nu(W) \leq 0, \quad \sigma_\nu(W)[W]\nu = 0.$$

2.10.2 Approximate equations

To construct a penalty problem, we introduce the penalty operator $\beta : H(\Omega_c) \to H(\Omega_c)^*$ by the relation

$$\langle \beta(W), \bar{W} \rangle = - \int\limits_{\Gamma_c} ([W]\nu)^- \,([\bar{W}]\nu)\; d\Gamma_c.$$

Here $\langle \cdot , \cdot \rangle$ means a duality pairing between $H(\Omega_c)$ and its dual space $H(\Omega_c)^\star$. By the superscribed minus we have denoted the negative part of a function, i.e. $s = s^+ - s^-$; $s^+, s^- \geq 0$; $s^+ s^- = 0$. It is seen that β is a monotonous operator.

By $W^\varepsilon \in H(\Omega_c)$, we denote the unique solution of the following penalty equation depending on a small parameter $\varepsilon > 0$,

$$\langle \sigma_{ij}(W^\varepsilon), \varepsilon_{ij}(\bar{W})\rangle + \varepsilon^{-1}\langle \beta(W^\varepsilon), \bar{W}\rangle = \langle f, \bar{W}\rangle \quad \forall \bar{W} \in H(\Omega_c). \tag{2.266}$$

The last equation is interpreted as follows:

$$-\sigma_{ij,j}(W^\varepsilon) = f^i, \quad i = 1, 2, \quad \text{in } \Omega_c,$$

$$[\sigma_\nu(W^\varepsilon)] = 0, \quad \sigma_\tau(W^\varepsilon) = 0, \quad \sigma_\nu(W^\varepsilon) = -\varepsilon^{-1}\left([W^\varepsilon]\nu\right)^- \quad \text{on } \Gamma_c.$$

Let us fix ε. To linearize the left-hand side of (2.266), we construct the following iterations for an arbitrary $W^{\varepsilon,0} \in H(\Omega_c)$, $n = 0, 1, 2, \ldots$,

$$\begin{aligned} \langle \sigma_{ij}(W^{\varepsilon,n+1}), \varepsilon_{ij}(\bar{W})\rangle &+ \varepsilon^{-1}(W^{\varepsilon,n+1} - W^{\varepsilon,n}, \bar{W}) \\ &= \langle f, \bar{W}\rangle - \varepsilon^{-1}\langle \beta(W^{\varepsilon,n}), \bar{W}\rangle. \end{aligned} \tag{2.267}$$

It is obvious that there exists a unique solution $W^{\varepsilon,n+1} \in H(\Omega_c)$. The appropriate boundary problem is of the form

$$-\sigma_{ij,j}\left(W^{\varepsilon,n+1} + \varepsilon^{-1}(W^{\varepsilon,n+1} - W^{\varepsilon,n})\right) = f^i, \quad i = 1, 2, \quad \text{in } \Omega_c,$$

$$\left[\sigma_\nu\left(W^{\varepsilon,n+1} + \varepsilon^{-1}(W^{\varepsilon,n+1} - W^{\varepsilon,n})\right)\right] = 0,$$

$$\sigma_\tau\left(W^{\varepsilon,n+1} + \varepsilon^{-1}(W^{\varepsilon,n+1} - W^{\varepsilon,n})\right) = 0,$$

$$\begin{aligned} \sigma_\nu\left(W^{\varepsilon,n+1} + \varepsilon^{-1}(W^{\varepsilon,n+1} - W^{\varepsilon,n})\right) &- \varepsilon^{-1}\left[W^{\varepsilon,n+1} - W^{\varepsilon,n}\right]\nu \\ &= -\varepsilon^{-1}\left([W^{\varepsilon,n}]\nu\right)^- . \end{aligned}$$

Theorem 2.29. *We have $W^{\varepsilon,n+1} \to W^\varepsilon$ strongly in $H(\Omega_c)$ as $n \to \infty$ and*

$$\|W^{\varepsilon,n+1} - W^\varepsilon\|^2 \leq (1 + 2c_2\varepsilon)^{-(n+1)}\|W^{\varepsilon,0} - W^\varepsilon\|^2, \tag{2.268}$$

$$W^\varepsilon \to W \quad \text{strongly in } H(\Omega_c) \text{ as } \varepsilon \to 0,$$

where $W^{\varepsilon,n+1}, W^\varepsilon, W$ are the solutions of (2.267), (2.266), (2.265), *respectively.*

Proof. By subtracting (2.266) from (2.267) and adding $-\varepsilon^{-1}(W^\varepsilon, \bar{W})$ to both parts, we obtain

$$\begin{aligned} \langle \sigma_{ij}(W^{\varepsilon,n+1} - W^\varepsilon), \varepsilon_{ij}(\bar{W})\rangle &+ \varepsilon^{-1}(W^{\varepsilon,n+1} - W^\varepsilon, \bar{W}) \\ = \varepsilon^{-1}(W^{\varepsilon,n} - W^\varepsilon, \bar{W}) &- \varepsilon^{-1}\langle \beta(W^{\varepsilon,n}) - \beta(W^\varepsilon), \bar{W}\rangle. \end{aligned}$$

Let us consider this equation with the test function $\bar{W} = W^{\varepsilon,n+1} - W^\varepsilon$ and rewrite its right-hand side as follows:

$$\varepsilon^{-1}\langle\sigma_{ij}(W^{\varepsilon,n} - W^\varepsilon), \varepsilon_{ij}(W^{\varepsilon,n+1} - W^\varepsilon)\rangle \tag{2.269}$$

$$+\varepsilon^{-1}\int_{\Gamma_c} \left([W^{\varepsilon,n} - W^\varepsilon]\nu + ([W^{\varepsilon,n}]\nu)^- - ([W^\varepsilon]\nu)^-\right)[W^{\varepsilon,n+1} - W^\varepsilon]\nu \, d\Gamma_c.$$

Since $s_1 - s_2 + s_1^- - s_2^- = s_1^+ - s_2^+ \le |s_1 - s_2|$, the right-hand side of (2.269), by the Holder inequality, is less than or equal to

$$(2\varepsilon)^{-1}\Big(\|W^{\varepsilon,n} - W^\varepsilon\|^2 + \|W^{\varepsilon,n+1} - W^\varepsilon\|^2\Big).$$

On the other hand, by (2.264), the left-hand side of (2.269) is no less than

$$(c_2 + \varepsilon^{-1})\|W^{\varepsilon,n+1} - W^\varepsilon\|^2.$$

Therefore,

$$\|W^{\varepsilon,n+1} - W^\varepsilon\|^2 \le (1 + 2c_2\varepsilon)^{-1}\|W^{\varepsilon,n} - W^\varepsilon\|^2.$$

By repeating the last estimate as n goes to 0, we derive that (2.268) holds and, therefore, the first assertion of Theorem 2.29 on the convergence is true.

The convergence

$$W^\varepsilon \to W \quad \text{weakly in } H(\Omega_c) \quad \text{as } \varepsilon \to 0 \tag{2.270}$$

is proved by the usual method as in Section 1.3. Indeed, equation (2.266) with $\bar{W} = W^\varepsilon - \xi$, $\xi \in K$ (i.e. $\beta(\xi) = 0$), gives

$$\langle\sigma_{ij}(W^\varepsilon), \varepsilon_{ij}(W^\varepsilon - \xi)\rangle \le \langle\sigma_{ij}(W^\varepsilon), \varepsilon_{ij}(W^\varepsilon - \xi)\rangle$$
$$+\varepsilon^{-1}\langle\beta(W^\varepsilon) - \beta(\xi), W^\varepsilon - \xi\rangle = \langle f, W^\varepsilon - \xi\rangle.$$

Hence, $\|W^\varepsilon\|$ is uniformly bounded in ε, and a subsequence exists such that

$$W^\varepsilon \to \tilde{W} \quad \text{weakly in } H(\Omega_c) \quad \text{as } \varepsilon \to 0.$$

Then

$$\langle\sigma_{ij}(W^\varepsilon), \varepsilon_{ij}(\xi)\rangle \to \langle\sigma_{ij}(\tilde{W}), \varepsilon_{ij}(\xi)\rangle,$$
$$\liminf \langle\sigma_{ij}(W^\varepsilon), \varepsilon_{ij}(W^\varepsilon)\rangle \ge \langle\sigma_{ij}(\tilde{W}), \varepsilon_{ij}(\tilde{W})\rangle,$$
$$\langle\beta(W^\varepsilon), \xi\rangle = \varepsilon\Big(\langle f, \xi\rangle - \langle\sigma_{ij}(W^\varepsilon), \varepsilon_{ij}(\xi)\rangle\Big) \to 0 \quad \text{as } \varepsilon \to 0.$$

Therefore, we can obtain that $\beta(\tilde{W}) = 0$, i.e. $\tilde{W} \in K$, and pass to a lower limit in the following inequality:

$$\langle\sigma_{ij}(W^\varepsilon), \varepsilon_{ij}(\bar{W} - W^\varepsilon)\rangle - \langle f, \bar{W} - W^\varepsilon\rangle$$

$$= \varepsilon^{-1}\langle \beta(\bar{W}) - \beta(W^\varepsilon), \bar{W} - W^\varepsilon \rangle \geq 0 \quad \forall \bar{W} \in K.$$

This provides

$$\langle \sigma_{ij}(\tilde{W}), \varepsilon_{ij}(\bar{W} - \tilde{W}) \rangle \geq \langle f, \bar{W} - \tilde{W} \rangle \quad \forall \bar{W} \in K$$

and $\tilde{W} = W$ owing to the uniqueness property of the solution.

Subtracting $\langle \sigma_{ij}(W), \varepsilon_{ij}(\bar{W}) \rangle$ from (2.266) and considering this equation with the test element $\bar{W} = W^\varepsilon - W$, one obtains

$$\langle \sigma_{ij}(W^\varepsilon - W), \varepsilon_{ij}(W^\varepsilon - W) \rangle - \varepsilon^{-1} \int_{\Gamma_c} ([W^\varepsilon]\nu)^- [W^\varepsilon - W]\nu \, d\Gamma_c$$

$$= \langle f, W^\varepsilon - W \rangle - \langle \sigma_{ij}(W), \varepsilon_{ij}(W^\varepsilon - W) \rangle.$$

Owing to relations

$$-([W^\varepsilon]\nu)^- [W^\varepsilon - W]\nu = \left(([W^\varepsilon]\nu)^-\right)^2 + ([W^\varepsilon]\nu)^- [W]\nu, \quad [W]\nu \geq 0,$$

and the inequality (2.264), we have the following estimate:

$$c_2 \|W^\varepsilon - W\|^2 + \varepsilon^{-1} \int_{\Gamma_c} \left(([W^\varepsilon]\nu)^-\right)^2 d\Gamma_c$$

$$\leq \langle f, W^\varepsilon - W \rangle - \langle \sigma_{ij}(W), \varepsilon_{ij}(W^\varepsilon - W) \rangle.$$

Therefore, (2.270) leads to the second strong convergence shown in Theorem 2.29. The proof is complete.

2.10.3 A bar with a cut

We will consider the one-dimensional crack problem describing a thin bar which occupies the set $\Omega_y = (a, y) \cup (y, b)$. The bar has a cut at the point y, $a < y < b$, i.e. $\Gamma_y = \{y\}$, $\Omega = (a, b)$,

$$H(\Omega_y) = \{W \in H^1(\Omega_y) \mid \quad W(a) = W(b) = 0\},$$

$$K = \{W \in H(\Omega_y) \mid \quad [W] \equiv W(y+0) - W(y-0) \geq 0\}.$$

In this case the equilibrium problem (2.265) takes the form

$$W \in K, \quad \langle W_x, \bar{W}_x - W_x \rangle \geq \langle f, \bar{W} - W \rangle \quad \forall \bar{W} \in K,$$

where $f \in L^2(\Omega_y)$. The corresponding boundary problem is as follows:

$$-W_{xx} = f \quad \text{in } \Omega_y, \tag{2.271}$$

$$[W_x] = 0, \quad [W] \geq 0, \quad W_x(y) \leq 0, \quad W_x(y)[W] = 0.$$

The penalty equation (2.266) is transformed to

$$\langle W^\varepsilon_x, \bar W_x\rangle - \varepsilon^{-1}[W^\varepsilon]^-[\bar W] = \langle f, \bar W\rangle \quad \forall \bar W \in H(\Omega_y)$$

or, equivalently,

$$-W^\varepsilon_{xx} = f \quad \text{in } \Omega_y, \tag{2.272}$$
$$[W^\varepsilon_x] = 0, \quad W^\varepsilon_x(y) + \varepsilon^{-1}[W^\varepsilon]^- = 0.$$

The iterations (2.267) are as follows:

$$(1+\varepsilon^{-1})\langle W^{\varepsilon,n+1}_x, \bar W_x\rangle + \varepsilon^{-1}[W^{\varepsilon,n+1}][\bar W]$$
$$= \langle f, \bar W\rangle + \varepsilon^{-1}\langle W^{\varepsilon,n}_x, \bar W_x\rangle + \varepsilon^{-1}[W^{\varepsilon,n}]^+[\bar W].$$

We can also write the iterative boundary problem

$$-(1+\varepsilon^{-1})W^{\varepsilon,n+1}_{xx} = f - \varepsilon^{-1}W^{\varepsilon,n}_{xx} \quad \text{in } \Omega_y, \quad [W^{\varepsilon,n+1}_x] = 0, \tag{2.273}$$
$$(1+\varepsilon^{-1})W^{\varepsilon,n+1}_x(y) - \varepsilon^{-1}[W^{\varepsilon,n+1}] = \varepsilon^{-1}W^{\varepsilon,n}_x(y) - \varepsilon^{-1}[W^{\varepsilon,n}]^+.$$

Lemma 2.7. *The boundary value problem*

$$-s_{xx} = f \quad \textit{in } \Omega_y,$$
$$s(a) = s(b) = 0, \quad [s_x] = 0, \quad c_1 s_x(y) - c_2[s] = g$$

has a unique solution represented by the formula

$$s(x) = W^0(x) + \frac{g + c_2[W^0]}{c_1 + c_2(b-a)}\alpha(x), \tag{2.274}$$

where $W^0 \in H^2(\Omega_y) \cap H(\Omega_y)$ is a unique solution of the following boundary value problem,

$$-W^0_{xx} = f \quad \text{in } \Omega_y,$$
$$W^0(a) = W^0(b) = 0, \quad [W^0_x] = 0, \quad W^0_x(y) = 0,$$

and the function $\alpha \in C^\infty(\Omega_y) \cap H(\Omega_y)$ is as follows:

$$\alpha(x) = \begin{cases} x - a \;, & x \in (a, y), \\ x - b \;, & x \in (y, b). \end{cases}$$

This lemma can be easily proved in view of the following properties of α:

$$[\alpha] = -(b-a), \quad \alpha_x(x) \equiv 1, \quad \alpha_{xx}(x) \equiv 0.$$

Now it seems to be natural to seek a solution of (2.273) as $W^{\varepsilon,n+1} = W^0 + c^{n+1}(\varepsilon)\alpha$, $c^{n+1}(\varepsilon) \in R$, $n = 0, 1, 2, \dots$. Indeed, then equation (2.273) is fulfilled in the domain Ω_y for any $c^{n+1}(\varepsilon)$:

$$-(1+\varepsilon^{-1})W^{\varepsilon,n+1}_{xx} = (1+\varepsilon^{-1})(-W^0_{xx} - c^{n+1}(\varepsilon)\alpha_{xx}) = (1+\varepsilon^{-1})f$$

$$= f - \varepsilon^{-1}(W^0_{xx} + c^n(\varepsilon)\alpha_{xx}) = f - \varepsilon^{-1}W^{\varepsilon,n}_{xx}.$$

Theorem 2.30. *Solutions of the problems* (2.271), (2.272), (2.273) *have the following presentation:*

$$W = W^0 - \frac{[W^0]^-}{b-a}\alpha, \quad W^\varepsilon = W^0 - \frac{[W^0]^-}{\varepsilon + b - a}\alpha,$$

$$W^{\varepsilon,n+1} = W^0 - \frac{(1-\rho^{n+1})[W^0]^-}{\varepsilon + b - a}\alpha, \quad \rho = \frac{1}{1+\varepsilon+b-a}.$$

PROOF. Let us choose $W^{\varepsilon,0} = W^0$ for simplicity. One can substitute $W^{\varepsilon,0}_x(y) - [W^{\varepsilon,0}]^+ = -[W^0]^+$ in (2.273) and obtain, by (2.274),

$$W^{\varepsilon,1} = W^0 + \frac{-[W^0]^+ + [W^0]}{1+\varepsilon+b-a}\alpha = W^0 - \rho[W^0]^-\alpha.$$

We next find

$$W^{\varepsilon,1}_x(y) - [W^{\varepsilon,1}]^+ = -\rho[W^0]^- - ([W^0] + (b-a)\rho[W^0]^-)^+$$

$$= -\rho[W^0]^- - ([W^0]^+ - (1+\varepsilon)\rho[W^0]^-)^+ = -\rho[W^0]^- - [W^0]^+.$$

Consequently, equations (2.273), (2.274) give

$$W^{\varepsilon,2} = W^0 + \frac{-\rho[W^0]^- - [W^0]^+ + [W^0]}{1+\varepsilon+b-a}\alpha = W^0 - (\rho+\rho^2)[W^0]^-\alpha.$$

Iterating, as n increases, we obtain by a similar method that

$$W^{\varepsilon,n} = W^0 - (\rho + \rho^2 + \ldots + \rho^n)[W^0]^-\alpha$$

$$= W^0 - \frac{\rho(1-\rho^n)}{1-\rho}\alpha = W^0 - \frac{1-\rho^n}{\varepsilon+b-a}[W^0]^-\alpha.$$

Then, by Theorem 2.29, we pass to the limit in the last relation as $n \to \infty$, $\varepsilon \to 0$. The proof is completed.

Theorem 2.30 can also be verified by the direct substitution of the obtained solutions in (2.271), (2.272) and (2.273), respectively. We apply this result to the following example. Let

$$f(x) = \begin{cases} c \quad , & x \in (a,y), \\ -c \quad , & x \in (y,b), \end{cases}$$

which corresponds to the uniform compression for $c > 0$, and to the extension for $c < 0$. Then

$$W^0(x) = \frac{c}{2}\begin{cases} -(x-a)^2 + 2(y-a)(x-a) & , x \in (a,y), \\ (x-b)^2 - 2(y-b)(x-b) & , x \in (y,b), \end{cases}$$

and $[W^0] = -c((y-a)^2 + (y-b)^2)/2$.

If $c \le 0$, then $[W^0] \ge 0$, $[W^0]^- = 0$ and $W = W^0$. If $c > 0$, then $W = W^0 - [W^0]^-/(b-a)\alpha$, i.e. $[W] = 0$ and

$$W(x) = \frac{c}{2}\left\{\begin{array}{l} -(x-a)^2 - \left(\dfrac{(y-a)^2 + (y-b)^2}{b-a} - 2(y-a)\right)(x-a), \\ (x-b)^2 - \left(\dfrac{(y-a)^2 + (y-b)^2}{b-a} + 2(y-b)\right)(x-b). \end{array}\right.$$

2.10.4 Another presentation of the solution

We consider the problem for the bar with the cut and give another presentation of the solution as compared with the previous one. For convenience the dependence of the functions obtained on the cut point y is indicated.

For $f \in L^2(a,b)$, we define the unique solution $U \in H^2(a,b) \cap H_0^1(a,b)$ of the following boundary value problem:

$$-U_{xx} = f \quad \text{in } (a,b), \quad U(a) = U(b) = 0. \tag{2.275}$$

The function U is continuous in (a,b). Let $[W]^y = W(y+0) - W(y-0)$. Introduce the closed convex set describing the nonpenetration condition

$$K^y = \{W \in H(\Omega_y) \mid \; [W]^y \ge 0\}$$

and consider the variational inequality

$$W \in K^y, \quad \langle W_x, \bar{W}_x - W_x\rangle \ge \langle f, \bar{W} - W\rangle \quad \forall \bar{W} \in K^y. \tag{2.276}$$

Theorem 2.31. *The function*

$$W(x) = U(x) - U_x^+(y)\alpha^y(x) \tag{2.277}$$

is a unique solution of the problem (2.276), *where*

$$\alpha^y(x) = \left\{\begin{array}{lll} x-a & , & x \in (a,y), \\ x-b & , & x \in (y,b) \end{array}\right.$$

and U is a solution to the problem (2.275).

PROOF. Recalling the following properties of the function α^y from the space $C^\infty(\Omega_y) \cap H(\Omega_y)$:

$$[\alpha^y]^y = -(b-a), \quad \alpha_x^y(x) \equiv 1, \quad \alpha_{xx}^y(x) \equiv 0 \quad (x \ne y),$$

and integrating by parts, for a test function $\xi \in H(\Omega_y)$ we obtain

$$\langle \alpha_x^y, \xi_x\rangle = -\langle \alpha_{xx}^y, \xi\rangle - [\alpha_x^y \xi]^y = -[\xi]^y.$$

Therefore, it follows from (2.275), (2.277) that

$$\langle f, \xi \rangle = \langle -U_{xx}, \xi \rangle = \langle U_x, \xi_x \rangle + U_x(y)[\xi]^y,$$

$$\langle W_x, \xi_x \rangle = \langle U_x, \xi_x \rangle - U_x^+(y)\langle \alpha_x^y, \xi_x \rangle = \langle U_x, \xi_x \rangle + U_x^+(y)[\xi]^y.$$

One can also calculate

$$[W]^y = [U]^y - U_x^+(y)[\alpha^y]^y = (b-a)U_x^+(y).$$

Thus, we have

$$\langle W_x, \bar{W}_x - W_x \rangle - \langle f, \bar{W} - W \rangle = \Big(U_x^+(y) - U_x(y) \Big) [\bar{W} - W]^y$$

$$= U_x^-(y)[\bar{W}]^y - (b-a)U_x^-(y)U_x^+(y) = U_x^-(y)[\bar{W}]^y \geq 0 \quad \forall \bar{W} \in K^y.$$

Since $[W]^y \geq 0$, the function W given by the formula (2.277) belongs to K^y and is a solution of (2.276). The theorem is proved.

It follows from (2.275) that the function U and, hence, the solution W belong to the space $H^2(\Omega_y)$. Then, by virtue of the properties of U and α^y, the function provided by (2.277) is a solution of the following boundary value problem,

$$-W_{xx} = f, \quad \text{in } \Omega_y,$$

$$[W]^y = (b-a)U_x^+(y), \quad [W_x]^y = 0, \quad W_x(y) = -U_x^-(y).$$

The smoothness of the solution is as follows. If $f \in H^n(\Omega_y)$ $(n \geq 0)$, we have $W \in H^{n+2}(\Omega_y)$.

We can also consider the inverse problem to (2.276). Let an arbitrary function W belong to the space $H^2(\Omega_y) \cap H(\Omega_y)$ and satisfy the relations

$$[W_x]^y = 0, \quad W_x(y)[W]^y = 0, \quad [W]^y \geq 0, \quad W_x(y) \leq 0.$$

Then W is a solution of the problem (2.276) for $f = -W_{xx}$. Indeed, $W \in K^y$, and integrating by parts we obtain

$$\langle W_x, \bar{W}_x - W_x \rangle - \langle f, \bar{W} - W \rangle = \langle -W_{xx} - f, \bar{W} - W \rangle - [W_x(\bar{W} - W)]^y$$

$$= -W_x(y)[\bar{W}]^y + W_x(y)[W]^y = -W_x(y)[\bar{W}]^y \geq 0 \quad \forall \bar{W} \in K^y.$$

Having found the displacement function $W(x) = U(x) - U_x^+(y)\alpha^y(x)$, one can find the other values: the strain $\varepsilon(x)$ and the stress $\sigma(x)$,

$$\varepsilon(x) = \sigma(x) = W_x(x) = U_x(x) - U_x^+(y),$$

which are continuous on (a, b); the contact force is nonnegative, namely

$$p = -\sigma(y) = U_x^-(y) \geq 0.$$

The potential energy is as follows:

$$\Pi(W) = \frac{1}{2}\|W\|^2 - \langle f, W\rangle = -\frac{1}{2}\|W\|^2 - [W_x W]^y = -\frac{1}{2}\|U_x - U_x^+(y)\|_0^2.$$

Here $\|\cdot\|_0$ denotes the norm in $L^2(a,b)$.

EXAMPLE 1. Let $f(x) \equiv c$, $c \geq 0$; then

$$U(x) = -\frac{c}{2}(x-a)(x-b), \quad U_x(y) = \frac{c}{2}(a+b-2y).$$

If $a < y \leq (a+b)/2$, we have

$$W(x) = -\frac{c}{2}\begin{cases} (x-a)(x+a-2y) & , \quad x \in (a,y), \\ (x-b)(x+b-2y) & , \quad x \in (y,b) \end{cases}$$

and $[W]^y = c(b-a)(a+b-2y)/2 \geq 0$, $\sigma(x) = c(y-x)$, $p = 0$.

If $(a+b)/2 \leq y < b$, then

$$W(x) = U(x) = -\frac{c}{2}(x-a)(x-b),$$

and $[W]^y = 0$, $\sigma(x) = c(a+b-2x)/2$, $p = c(2y-(a+b))/2$.

EXAMPLE 2. Let $f(x) = \sin k(x-a)$, $k = \pi/(b-a)$; then

$$U(x) = k^{-2}\sin k(x-a), \quad U_x(x) = k^{-1}\cos k(x-a).$$

Therefore, by Theorem 2.31, we have

$$W(x) = \frac{1}{k^2}\begin{cases} \sin k(x-a) & , \quad y \geq (a+b)/2, \\ \sin k(x-a) - k\cos k(y-a)\alpha^y(x) & , \quad y \leq (a+b)/2, \end{cases}$$

$$\sigma(x) = \frac{1}{k}\begin{cases} \cos k(x-a) & , \quad y \geq (a+b)/2, \\ \cos k(x-a) - \cos k(y-a) & , \quad y \leq (a+b)/2, \end{cases}$$

$$p = \frac{1}{k}\begin{cases} -\cos k(y-a) & , \quad y \geq (a+b)/2, \\ 0 & , \quad y \leq (a+b)/2, \end{cases}$$

$$[W]^y = \frac{b-a}{k}\begin{cases} 0 & , \quad y \geq (a+b)/2, \\ \cos k(y-a) & , \quad y \leq (a+b)/2. \end{cases}$$

2.10.5 Optimal control of the cut

Let us consider the problem of minimization of the crack opening

$$\inf_{a<y<b} [W]^y, \tag{2.278}$$

where W is a solution to the problem (2.276). By virtue of (2.277), the problem (2.278) is equivalent to

$$\inf_{a<y<b} U_x^+(y). \tag{2.279}$$

Let us define the following sets:

$$I^+(U) = \{y \in (a,b) \mid U_x(y) > 0\}, \quad I^-(U) = \{y \in (a,b) \mid U_x(y) \le 0\},$$

where $I^+(U) \cup I^-(U) = (a,b)$. By the imbedding theorems, $U \in C^1(a,b)$, i.e. U_x is a continuous function. Since $U(a) = U(b) = 0$, the set $I^-(U)$ is not empty. Thus, by virtue of $U_x^+(y) \ge 0$, we obtain $\inf_{a<y<b} [W]^y = 0$ for any $y \in I^-(U)$.

EXAMPLE 1. For $f(x) \equiv c$, $c \ge 0$, and for any $(a+b)/2 \le y < b$, the function $W(x) = -c(x-a)(x-b)/2$ is a solution to the problem (2.278).

EXAMPLE 2. For $f(x) = \sin k(x-a)$, $k = \pi/(b-a)$ and for any $(a+b)/2 \le y < b$, the function $W(x) = k^{-2}\sin k(x-a)$ is a solution to the problem (2.278).

We consider next the problem of the stress optimization

$$\inf_{a<y<b} J(y), \qquad J(y) = \|\sigma - \sigma_0\|_0^2, \tag{2.280}$$

where $\sigma_0 \in L^2(a,b)$ is a given stress function, $\sigma = W_x$, and W is a solution of the problem (2.276). In view of (2.277), we can rewrite

$$J(y) = \|U_x - U_x^+(y) - \sigma_0\|_0^2$$

$$= \|U_x - \sigma_0\|_0^2 + 2U_x^+(y)\int_a^b \sigma_0\,dx + (b-a)(U_x^+(y))^2.$$

If $\int_a^b \sigma_0\,dx \ge 0$, then $U_x^+(y) = 0$ provides

$$\inf_{a<y<b} J(y) = \|U_x - \sigma_0\|_0^2 \quad \forall y \in I^-(U).$$

If $\int_a^b \sigma_0\,dx < 0$ and $I^+(U)$ is not empty, then the infimum is attained at $y^\star$ such that

$$U_x^+(y^\star) + \frac{1}{b-a}\int_a^b \sigma_0\,dx \to \inf.$$

If there exists $y^\star \in I^+(U)$ such that the equality

$$U_x^+(y^\star) = -(b-a)^{-1}\int_a^b \sigma_0\,dx$$

is valid, we have

$$J(y^\star) = \|U_x - \sigma_0\|_0^2 - \frac{1}{b-a}\Big(\int_a^b \sigma_0\,dx\Big)^2.$$

We take the case, where $\sigma_0(x) \equiv \text{const}$, $\sigma_0 < 0$; then $(b-a)^{-1}\int\limits_a^b \sigma_0\,dx = \sigma_0$.

EXAMPLE 1. Let $f(x) \equiv c$, $c \geq 0$. Then

$$U_x^+(y) = \begin{cases} c(a+b-2y)/2 & , \quad a < y \leq (a+b)/2, \\ 0 & , \quad (a+b)/2 \leq y < b. \end{cases}$$

If $-\sigma_0 < c(b-a)/2$, then at the point $y^\star = (a+b)/2 + \sigma_0/c$, we have $U_x^+(y^\star) = -\sigma_0$, and the infimum in (2.280) is reached,

$$J(y^\star) = \left\| \frac{c}{2}(a+b-2x) \right\|_0^2 = \frac{c^2(b-a)^3}{12}.$$

If $-\sigma_0 \geq c(b-a)/2$, then $y^\star = a$, the infimum of $J(y)$ is not reached, and it is equal to

$$J(a) = \left\| \frac{c}{2}(a+b-2x) - \frac{c}{2}(b-a) - \sigma_0 \right\|_0^2$$

$$= \frac{c^2(b-a)^3}{12}\left(1 + 3\left(1 + \frac{2\sigma_0}{c(b-a)}\right)^2\right).$$

If $\sigma_0 \geq 0$, then for any $y^\star \in I^-(U)$ the infimum of (2.280) is

$$J(y^\star) = \left\| \frac{c}{2}(a+b-2x) - \sigma_0 \right\|_0^2 = \frac{c^2(b-a)^3}{12}\left(1 + 3\left(\frac{2\sigma_0}{c(b-a)}\right)^2\right).$$

EXAMPLE 2. Let $f(x) = \sin k(x-a)$, $k = \pi/(b-a)$. Then

$$U_x^+(y) = \frac{1}{k}\begin{cases} \cos k(y-a) & , \quad a < y \leq (a+b)/2, \\ 0 & , \quad (a+b)/2 \leq y < b. \end{cases}$$

Let $\sigma_0 < 0$. If $-\sigma_0 < k^{-1}$, we have $U_x^+(y^\star) = -\sigma_0$ at the point $y^\star = a + k^{-1}\arccos(-k\sigma_0)$, and the infimum is reached,

$$J(y^\star) = \|U_x(x)\|_0^2 = \frac{b-a}{2k^2}.$$

If $-\sigma_0 \geq k^{-1}$, the infimum of $J(y)$ is not reached, and it is equal to

$$J(a) = \left\| \frac{1}{k}\cos k(x-a) - \frac{1}{k} - \sigma_0 \right\|_0^2 = (b-a)\left(\frac{1}{2k^2} + \left(\frac{1}{k} + \sigma_0\right)^2\right).$$

If $\sigma_0 \geq 0$, then for any $y^\star \geq (a+b)/2$, the infimum of $J(y)$ is

$$J(y^\star) = \left\| \frac{1}{k}\cos k(x-a) - \sigma_0 \right\|_0^2 = (b-a)\left(\frac{1}{2k^2} + \sigma_0^2\right).$$

Chapter 3

Cracks in complicated plates

3.1 Plate with a crack under the creep condition

We consider a boundary value problem for equations describing an equilibrium of a plate being under the creep law (1.31)–(1.32). The plate is assumed to have a vertical crack. As before, the main peculiarity of the problem is determined by the presence of an inequality imposed on a solution which represents a mutual nonpenetration condition of the crack faces

$$[W]\nu \geq \left| \left[\frac{\partial w}{\partial \nu} \right] \right|,$$

where $W = (w^1, w^2), w$ are horizontal and vertical displacements of mid-surface points of the plate, ν is the normal to the crack shape, and $[\cdot]$ is the jump of a function at crack faces. The presence of a crack alone implies a domain wherein the solution is determined to have a nonsmooth boundary, and boundary conditions given at crack faces are of the inequality type.

An existence theorem to the equilibrium problem of the plate is proved. A complete system of equations and inequalities fulfilled at the crack faces is found. The solvability of the optimal control problem with a cost functional characterizing an opening of the crack is established. The solution is shown to belong to the space C^∞ near crack points provided the crack opening is equal to zero. The results of this section are published in (Khludnev, 1996c).

3.1.1 Equilibrium problem

Let $\Omega \subset R^2$ be a bounded domain with a smooth boundary Γ, and $y = \psi(x)$ describe a crack shape on the (x, y)-plane, $x \in [0, 1]$, $(x, y) \in \Omega$. By Γ_ψ we denote the graph of the function $y = \psi(x)$, $\psi \in H_0^3(0, 1)$. A mid-surface of the plate occupies the domain $\Omega_\psi = \Omega \setminus \Gamma_\psi$. The crack shape as a surface in R^3 can be presented in the form $y = \psi(x)$, $-h \le z \le h$, where z is the distance from the mid-surface, $2h$ is the plate thickness.

Denote by $W = (w^1, w^2)$, w horizontal and vertical displacements of the mid-surface points, respectively, and write down the formulae for strain and integrated stress tensor components $\varepsilon_{ij}(W)$, $\sigma_{ij}(W)$:

$$\varepsilon_{ij}(W) = \frac{1}{2}\left(\frac{\partial w^i}{\partial x_j} + \frac{\partial w^j}{\partial x_i}\right), \quad x_1 = x,\ x_2 = y,$$

$$\sigma_{11}(W) = \varepsilon_{11}(W) + \kappa\varepsilon_{22}(W), \quad \sigma_{22}(W) = \varepsilon_{22}(W) + \kappa\varepsilon_{11}(W),$$

$$\sigma_{12} = (1 - \kappa)\varepsilon_{12}(W), \quad \kappa = \text{ const }, \quad 0 < \kappa < 1/2.$$

Here and everywhere below $i, j = 1, 2$. Let $\chi = (W, w)$ and

$$\chi^\tau(t, x, y) = \chi(t, x, y) + \int_0^t \chi(\tau, x, y)\, d\tau, \tag{3.1}$$

$$B(w, \bar{w}) = \int_{\Omega_\psi} (w_{xx}\bar{w}_{xx} + w_{yy}\bar{w}_{yy} + \kappa w_{xx}\bar{w}_{yy} + \kappa w_{yy}\bar{w}_{xx}$$

$$+ 2(1 - \kappa)w_{xy}\bar{w}_{xy}).$$

We shall consider an equilibrium problem with a constitutive law corresponding to a creep, in particular, the strain and integrated stress tensor components $\varepsilon_{ij}(W^\tau)$, $\sigma_{ij}(W^\tau)$ will depend on $\chi^\tau = (W^\tau, w^\tau)$, where (W^τ, w^τ) are connected with (W, w) by (3.1). In this case, the equilibrium equations will be nonlocal with respect to t.

At the external boundary the following boundary conditions are assumed to be satisfied:

$$w = \frac{\partial w}{\partial n} = W = 0 \quad \text{on } \Gamma \times (0, T). \tag{3.2}$$

These conditions correspond to the clamping of the plate at the boundary.

Let Sobolev space $H^{1,0}(\Omega_\psi)$ consist of functions having the first generalized derivatives square integrable in Ω_ψ and which are equal to zero on Γ; the space $H^{2,0}(\Omega_\psi)$ is introduced analogously and consists of functions equal to zero on Γ with the first derivatives, $H(\Omega_\psi) = H^{1,0}(\Omega_\psi) \times H^{1,0}(\Omega_\psi) \times H^{2,0}(\Omega_\psi)$.

In the domain $Q_\psi = \Omega_\psi \times (0,T)$ (see Fig.3.1) we want to find a function (W, w) satisfying the equilibrium equations

$$-\frac{\partial \sigma_{ij}(W^\tau)}{\partial x_j} = u_i, \tag{3.3}$$

$$\Delta^2 w^\tau = u_3, \tag{3.4}$$

and boundary condition (3.2).

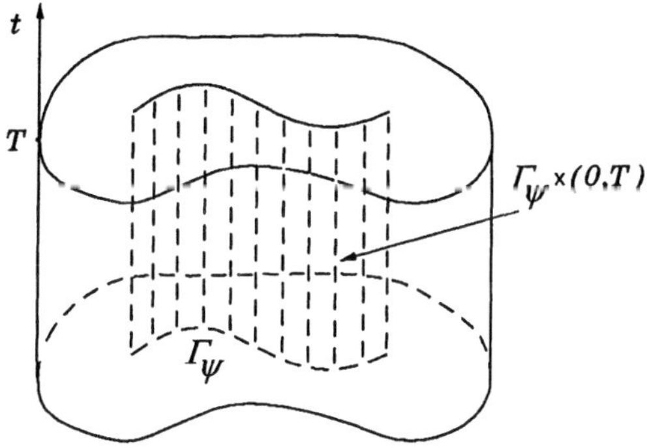

Fig.3.1. Cylinder Q_ψ

At the boundary $\Gamma_\psi \times (0,T)$ a system of equations and inequalities is satisfied whose precise form is found in Section 3.1.3. The function $u = (u_1, u_2, u_3)$ in (3.3), (3.4) is given. The nonpenetration condition of the crack faces can be written as follows:

$$[W]\nu \geq \left| \left[\frac{\partial w}{\partial \nu} \right] \right| \quad \text{on } \Gamma_\psi \times (0,T). \tag{3.5}$$

The structure of the section is as follows. In Section 3.1.2 we prove a solvability of the equilibrium problem. This problem is formulated as a variational inequality holding in Q_ψ. The equations (3.3), (3.4) are fulfilled in the sense of distributions. On the other hand, if the solution is smooth and satisfies (3.3), (3.4) and all the boundary conditions then the above variational inequality holds.

In Section 3.1.3 a complete system of equations and inequalities holding on $\Gamma_\psi \times (0,T)$ is found (i.e. boundary conditions on $\Gamma_\psi \times (0,T)$ are found). Simultaneously, a relationship between two formulations of the problem is established, that is an equivalence of the variational inequality and the equations (3.3), (3.4) with appropriate boundary conditions is proved.

Further, in Section 3.1.4, an optimal control problem is analysed. The external forces u serve as a control. The solution existence of the optimal control problem with a cost functional describing the crack opening is proved. Finally, in Section 3.1.5, we prove C^∞-regularity of the solution near crack points having a zero opening.

3.1.2 Existence of a solution

In this subsection we prove an existence theorem of the equilibrium problem for the plate. The problem is formulated as a variational inequality which together with (3.2), (3.5) contains full information about other boundary conditions holding on $\Gamma_\psi \times (0,T)$. An exact form of these conditions is found in the next subsection.

Let

$$K = \{\chi = (W, w) \in H(\Omega_\psi) \mid [W]\nu \geq \left|\left[\frac{\partial w}{\partial \nu}\right]\right| \quad \text{a.e. on } \Gamma_\psi\}.$$

Introduce the set of admissible displacements of the plate,

$$\mathcal{K} = \{\chi \in L^2(0,T; H(\Omega_\psi)) \mid \chi(t) \in K \quad \text{a.e. on } (0,T)\},$$

and assume that $u = (u_1, u_2, u_3) \in H^1(0,T; L^2(\Omega_\psi))$. Let the brackets $\langle \cdot , \cdot \rangle$ denote the scalar product in $L^2(\Omega_\psi)$. The following statement provides the solution existence of the equilibrium problem.

Theorem 3.1. *There exists a unique function* χ *satisfying the variational inequality*

$$\chi \in \mathcal{K}, \quad \chi_t \in L^2(0,T; H(\Omega_\psi)),$$

$$\int_0^T B(w^\tau, \bar{w} - w)\, dt + \int_0^T \langle \sigma_{ij}(W^\tau), \varepsilon_{ij}(\bar{W} - W)\rangle\, dt \geq \int_0^T \langle u, \bar{\chi} - \chi\rangle\, dt, \quad (3.6)$$

$$\forall \bar{\chi} \in \mathcal{K}.$$

PROOF. Define the linear and continuous operator

$$A : L^2(0,T; H(\Omega_\psi)) \to L^2(0,T; H(\Omega_\psi)^*)$$

by the formula

$$A(\chi)(\bar{\chi}) = \int_0^T (B(w^\tau, \bar{w}) + \langle \sigma_{ij}(W^\tau), \varepsilon_{ij}(\bar{W})\rangle)\, dt,$$

$$\bar{\chi} = (\bar{W}, \bar{w}) \in L^2(0,T; H(\Omega_\psi)),$$

where (W^τ, w^τ) and (W, w) are connected by the formula (3.1), and $H(\Omega_\psi)^*$ is the space dual of $H(\Omega_\psi)$.

Note that the following inequalities hold in Ω_ψ:

$$\langle \sigma_{ij}(W), \varepsilon_{ij}(W)\rangle \geq c\|W\|_1^2, \quad \forall W = (w^1, w^2) \in H^{1,0}(\Omega_\psi), \quad (3.7)$$

$$B(w, w) \geq c\|w\|_2^2, \quad \forall w \in H^{2,0}(\Omega_\psi), \quad (3.8)$$

with the constants c being uniform in W, w, respectively. Hence, owing to the formula

$$A(\chi)(\chi) = \int_0^T \{B(w,w) + \langle \sigma_{ij}(W), \varepsilon_{ij}(W)\rangle\}\, dt$$

$$+\frac{1}{2}B\left(\int_0^T w\, d\tau, \int_0^T w d\tau\right) + \frac{1}{2}\langle \sigma_{ij}\left(\int_0^T W d\tau\right), \varepsilon_{ij}\left(\int_0^T W d\tau\right)\rangle$$

we easily conclude that the operator A is coercive, i.e.

$$\frac{A(\chi)(\chi)}{\|\chi\|_{L^2(0,T;H(\Omega_\psi))}} \to \infty, \quad \|\chi\|_{L^2(0,T;H(\Omega_\psi))} \to \infty.$$

Moreover, the operator A turns out to be monotone. This implies that the problem

$$A(\chi)(\bar{\chi} - \chi) \geq \int_0^T \langle u, \bar{\chi} - \chi\rangle\, dt, \quad \forall \bar{\chi} \in \mathcal{K}, \quad \chi \in \mathcal{K} \tag{3.9}$$

has a solution.

In what follows an additional smoothness of the solution χ of (3.9) with respect to t is proved. To this end the finite differences are used. Let $\varepsilon > 0$ be a parameter and

$$\bar{\chi}_\varepsilon(\xi) = \begin{cases} \bar{\chi} & , \quad \xi \in (t-\varepsilon, t+\varepsilon), \quad \varepsilon > 0, \\ \chi(\xi) & , \quad \text{otherwise} \end{cases}$$

be a test function where $\bar{\chi} \in K$ is a fixed element. We substitute $\bar{\chi}_\varepsilon$ in (3.9) and divide by 2ε the relation obtained. Passing to the limit as $\varepsilon \to 0$ we derive for almost all $t \in (0,T)$

$$B(w^\tau(t), \bar{w} - w(t)) + \langle \sigma_{ij}(W^\tau(t)), \varepsilon_{ij}(\bar{W} - W(t))\rangle \tag{3.10}$$

$$\geq \langle u(t), \bar{\chi} - \chi(t)\rangle, \quad \forall \bar{\chi} = (\bar{W}, \bar{w}) \in K.$$

As seen, the variable t plays the role of a parameter in (3.10). Let us take $\bar{\chi} = \chi(t+h)$ as a test function in (3.10). Then we consider (3.10) at the point $t+h$ and choose $\bar{\chi} = \chi(t)$ as a test function. Summing the obtained inequalities and dividing by h^2 we derive the following relation:

$$B(d_h w(t) + d_h^\tau w(t), d_h w(t)) + \langle \sigma_{ij}(d_h W(t) + d_h^\tau W(t)), \varepsilon_{ij}(d_h W(t))\rangle \tag{3.11}$$

$$\leq \langle d_h u(t), d_h \chi(t)\rangle.$$

Herein the following notation is used:

$$d_h v(t) = \frac{v(t+h) - v(t)}{h}, \quad d_h^\tau v(t) = \frac{1}{h} \int\limits_t^{t+h} v(\tau) d\tau.$$

Let, for instance, $h > 0$. The case $h < 0$ can be considered similarly. In view of (3.7)–(3.8) we have

$$B(w, w) + \langle \sigma_{ij}(W), \varepsilon_{ij}(W) \rangle \geq c \|\chi\|^2_{H(\Omega_\psi)}, \quad \forall\, \chi = (W, w) \in H(\Omega_\psi).$$

Together with (3.11) this entails for almost all $t \in (0, T - h)$

$$\|d_h \chi(t)\|^2_{H(\Omega_\psi)} \leq c \left(\|d_h u(t)\|_0^2 + \|d_h^\tau \chi(t)\|^2_{H(\Omega_\psi)} \right) \tag{3.12}$$

with a constant c uniform in t, h. We next notice that for any smooth function v the following inequalities hold (see Lemma 3.1 below):

$$\int\limits_0^{T-h} \|d_h^\tau v(t)\|_0^2 \, dt \leq \int\limits_0^{T-h} d_h \left(\int\limits_0^t \|v(\tau)\|_0^2 \, d\tau \right) dt \leq \int\limits_0^T \|v(t)\|_0^2 \, dt. \tag{3.13}$$

Hence, the integration of (3.12) with respect to t from 0 to $T - h$ gives the inequality

$$\int\limits_0^{T-h} \|d_h \chi(t)\|^2_{H(\Omega_\psi)} dt \leq c \left(\int\limits_0^{T-h} \|d_h u(t)\|_0^2 dt + \int\limits_0^T \|\chi(t)\|^2_{H(\Omega_\psi)} \, dt \right). \tag{3.14}$$

The constant c in (3.14) is uniform in h. Since $u_t \in L^2(Q_\psi)$ we obtain from (3.14) as $h \to 0$

$$\|\chi_t\|^2_{L^2(0,T;H(\Omega_\psi))} \leq c \left(\|u_t\|^2_{L^2(Q_\psi)} + \|\chi\|^2_{L^2(0,T;H(\Omega_\psi))} \right).$$

Consequently, the existence of the derivative of the solution to (3.9) with respect to t is proved. Moreover, by taking $\bar{\chi} = 0$ in (3.9) we have

$$\|\chi\|^2_{L^2(0,T;H(\Omega_\psi))} \leq c \|u\|^2_{L^2(Q_\psi)}.$$

So, the solution of (3.9) is, in fact, the solution of (3.6).

The uniqueness of the solution to (3.6) can be proved in a standard way. As it follows from (3.6) the difference $\chi = \chi_1 - \chi_2$ of the solutions satisfies the inequality $A(\chi)(\chi) \leq 0$. Hence $\chi \equiv 0$. Theorem 3.1 is proved.

Notice that a substitution in (3.6) of the test functions of the form $\bar{\chi} = \chi + \chi^0$, $\chi^0 \in C_0^\infty(Q_\psi)$, implies that the equations (3.3), (3.4) hold in Q_ψ in the sense of distributions. By virtue of the proved inclusion $\chi_t \in L^2(0, T; H(\Omega_\psi))$ the variational inequality (3.10) is fulfilled for all $t \in (0, T)$.

Now we have to establish the auxiliary statement used to prove Theorem 3.1.

Lemma 3.1. *For any smooth function v the inequalities* (3.13) *hold.*
PROOF. To prove the left inequality of (3.13), we first obtain

$$\|d_h^T v(t)\|_0^2 \leq \frac{1}{h} \int_t^{t+h} \|v(\tau)\|_0^2 \, d\tau.$$

Consequently, we derive the desired inequality since

$$\frac{1}{h} \int_t^{t+h} \|v(\tau)\|_0^2 \, d\tau = d_h \int_0^t \|v(\tau)\|_0^2 \, d\tau.$$

To prove the right inequality of (3.13) it suffices to establish the estimate

$$\int_0^{T-h} d_h \left(\int_0^t f(\tau) \, d\tau \right) dt \leq \int_0^T f(t) \, dt$$

for any smooth nonnegative function $f(t)$. We have

$$\int_0^{T-h} d_h \left(\int_0^t f(\tau) \, d\tau \right) dt = \int_0^{T-h} \frac{1}{h} \left(\int_t^{t+h} f(\tau) \, d\tau \right) dt.$$

Changing the variables $\xi + t = \tau$ in the interior integral implies

$$\int_0^{T-h} \frac{1}{h} \left(\int_0^h f(\xi + t) \, d\xi \right) dt = \int_0^h \frac{1}{h} \left(\int_0^{T-h} f(\xi + t) \, dt \right) d\xi$$

$$\leq \frac{1}{h} \max_{\xi \in (0,h)} \int_0^{T-h} f(\xi + t) \, dt \int_0^h dt = \max_{\xi \in (0,h)} \int_\xi^{\xi + T - h} f(t) \, dt \leq \int_0^T f(t) \, dt,$$

which proves the right inequality of (3.13).

3.1.3 Boundary conditions

This subsection is concerned with searching for boundary conditions holding on $\Gamma_\psi \times (0,T)$ for the solution of (3.10) or, equivalently, of (3.9). Our arguments are formal in that the solution is assumed to be smooth enough.

Let $D \subset R^2$ be a bounded domain, and γ be its smooth boundary with the external normal $n = (n_1, n_2)$. We introduce the operators on the boundary γ,

$$M(w) = \kappa \Delta w + (1 - \kappa) \frac{\partial^2 w}{\partial n^2},$$

$$R(w) = \frac{\partial}{\partial n}\Delta w + (1-\kappa)\frac{\partial^3 w}{\partial n \partial s^2}, \quad s = (-n_2, n_1).$$

For any smooth functions $w, v, W, V = (v_1, v_2)$ the following Green formulae hold:

$$B_D(w, v) = \langle M(w), \frac{\partial v}{\partial n}\rangle_\gamma - \langle R(w), v\rangle_\gamma + \langle \Delta^2 w, v\rangle_D, \tag{3.15}$$

$$\langle \sigma_{ij}(W), \varepsilon_{ij}(V)\rangle_D = \langle \sigma_{ij}(W) n_j, v_i\rangle_\gamma - \langle \frac{\partial \sigma_{ij}(W)}{\partial x_j}, v_i\rangle_D. \tag{3.16}$$

The subscripts D, γ denote the integration over the domain D and the boundary γ, respectively. Note that the boundary $\partial\Omega_\psi$ of Ω_ψ is a combination of the sets $\Gamma, \Gamma_\psi^+, \Gamma_\psi^-$. The formulae (3.15), (3.16) hold true for the domain Ω_ψ despite the absence of regularity of $\partial\Omega_\psi$. To verify this we can extend the graph Γ_ψ so that the domain is divided into two parts. For each part the formulae (3.15), (3.16) are valid, hence the statement follows. We should note at this point that the external normals on Γ_ψ^+, Γ_ψ^- have opposite directions.

To simplify the formulae in this subsection we shall write w^τ, W^τ, w, ... instead of $w^\tau(t)$, $W^\tau(t)$, $w(t)$, ... This means that we fix t and consider the boundary conditions on Γ_ψ for this fixed value t. The same value t is assumed to be chosen in (3.10).

Introduce the notation $U = (u_1, u_2)$ and take the test functions of the form $(\bar{W}, w)$ in (3.10). This implies the variational inequality

$$\langle \sigma_{ij}(W^\tau), \varepsilon_{ij}(\bar{W} - W)\rangle \geq \langle U, \bar{W} - W\rangle \tag{3.17}$$

holding for all functions $\bar{W}$ such that

$$[\bar{W}]\nu \geq \left|\left[\frac{\partial w}{\partial \nu}\right]\right| \quad \text{on } \Gamma_\psi, \quad \bar{W} \in H^{1,0}(\Omega_\psi).$$

On the other hand, we can substitute the test functions of the form $(W, \bar{w})$ in (3.10), which entails the variational inequality

$$B(w^\tau, \bar{w} - w) \geq \langle u_3, \bar{w} - w\rangle. \tag{3.18}$$

The inequality (3.18) holds for all functions $\bar{w}$ satisfying the relation

$$[W]\nu \geq \left|\left[\frac{\partial \bar{w}}{\partial \nu}\right]\right| \quad \text{on } \Gamma_\psi, \quad \bar{w} \in H^{2,0}(\Omega_\psi).$$

At the boundary Γ_ψ^- we can decompose the vector $\{\sigma_{ij}(W^\tau)\nu_j\}$ into the sum of the normal and tangential components,

$$\{\sigma_{ij}(W^\tau)\nu_j\} = \sigma_\nu(W^\tau)\nu + \sigma_s(W^\tau)s, \quad s = (-\nu_2, \nu_1). \tag{3.19}$$

A similar decomposition takes place on Γ_ψ^+. Let us substitute in (3.17) the test functions of the form $W + \tilde{W}$, where smooth functions $\tilde{W}$ belong to $H^{1,0}(\Omega_\psi)$, $[\tilde{W}]\nu \geq 0$ on Γ_ψ, and make use of (3.16). A simple reasoning results in the relations

$$[\sigma_\nu(W^\tau)] = 0, \quad \sigma_s(W^\tau) = 0 \quad \text{on } \Gamma_\psi. \tag{3.20}$$

To proceed, we choose functions of the form $w+\theta$ as test ones in (3.18), where θ is a smooth function in Ω_ψ having support in a neighbourhood of a fixed point of Γ_ψ and such that $[\partial\theta/\partial\nu] = 0$. Note that $[\theta] \neq 0$. By (3.15), this leads to the relations

$$[M(w^\tau)] = 0, \quad R(w^\tau) = 0 \quad \text{on } \Gamma_\psi. \tag{3.21}$$

We next choose in (3.10) the test functions of the form $(\bar{W}, \bar{w}) = (0,0)$, $(\bar{W}, \bar{w}) = 2(W, w)$. Using (3.3), (3.4) and (3.15) one easily gets

$$\langle M(w^\tau), \left[\frac{\partial w}{\partial \nu}\right]\rangle_{\Gamma_\psi} + \langle \sigma_\nu(W^\tau), [W]\nu\rangle_{\Gamma_\psi} = 0. \tag{3.22}$$

On the other hand, a substitution of the test function $(\bar{W}, \bar{w}) = (W, w) + (\tilde{W}, \tilde{w})$ in (3.10) provides the inequality

$$B(w^\tau, \tilde{w}) + \langle \sigma_{ij}(W^\tau), \varepsilon_{ij}(\tilde{W})\rangle \geq \langle u, \tilde{\chi}\rangle, \tag{3.23}$$

where $(\tilde{W}, \tilde{w})$ are smooth functions belonging to K. We can integrate here by (3.3), (3.4), (3.15), which gives

$$\langle M(w^\tau), \left[\frac{\partial \tilde{w}}{\partial \nu}\right]\rangle_{\Gamma_\psi} + \langle \sigma_\nu(W^\tau), [\tilde{W}]\nu\rangle_{\Gamma_\psi} \leq 0. \tag{3.24}$$

Let $(\tilde{W}, \tilde{w})$ be smooth functions having supports in a neighbourhood of a fixed point on Γ_ψ and such that $[\partial\tilde{w}/\partial\nu] = [\tilde{W}]\nu$. We substitute $(\tilde{W}, \tilde{w})$ in (3.24) and derive

$$M(w^\tau) + \sigma_\nu(W^\tau) \leq 0.$$

Analogously, by choosing $[\partial\tilde{w}/\partial\nu] = -[\tilde{W}]\nu$ one easily gets

$$-M(w^\tau) + \sigma_\nu(W^\tau) \leq 0.$$

Thus, in fact, we have the inequality

$$|M(w^\tau)| \leq -\sigma_\nu(W^\tau) \quad \text{on } \Gamma_\psi. \tag{3.25}$$

By virtue of (3.5), (3.22), (3.25) we arrive at the conclusion that

$$M(w^\tau)\left[\frac{\partial w}{\partial \nu}\right] + \sigma_\nu(W^\tau)[W]\nu = 0 \quad \text{on } \Gamma_\psi. \tag{3.26}$$

Hence, the form of the boundary condition on $\Gamma_\psi \times (0,T)$ is completely determined. Together with (3.5), for all $t \in (0,T)$ the conditions (3.20)–(3.21), (3.25)–(3.26) hold on Γ_ψ.

Notice that the variational inequality (3.10) can be derived from (3.3), (3.4) and the above boundary conditions. In fact, let us assume that the solution (W, w) is smooth enough and satisfies (3.3), (3.4) and the boundary conditions obtained. We choose a smooth function $(\bar{W}, \bar{w}) \in K$ and multiply (3.3), (3.4) taken for a fixed $t \in (0,T)$ by $\bar{w}^i - w^i(t), \bar{w} - w(t)$, respectively. We next integrate over Ω_ψ taking into account (3.5), (3.20)–(3.21), (3.25)–(3.26). For the value $t \in (0,T)$ chosen above this implies

$$B(w^\tau, \bar{w} - w) + \langle \sigma_{ij}(W^\tau), \varepsilon_{ij}(\bar{W} - W)\rangle - \langle u, \bar{\chi} - \chi\rangle$$
$$+ \langle M(w^\tau), \left[\frac{\partial \bar{w}}{\partial \nu}\right] - \left[\frac{\partial w}{\partial \nu}\right]\rangle_{\Gamma_\psi} + \langle \sigma_\nu(W^\tau), [\bar{W}]\nu - [W]\nu\rangle_{\Gamma_\psi} = 0.$$

According to the boundary conditions the sum of integrals over Γ_ψ is non-positive here, whence (3.10) follows.

Thus, the boundary problem describing the equilibrium of the plate having the crack can be formulated both in the form (3.10) (or (3.6)) and in the form of equations (3.3), (3.4) with (3.5) and conditions (3.20)–(3.21), (3.25)–(3.26) fulfilled for all $t \in (0,T)$. In this case the latter formulation of the problem is formal in the sense that an additional regularity of the solution is assumed. The solution regularity which follows from (3.6), in general, does not provide the moments $M(w^\tau)$ and transverse forces $R(w^\tau)$ to be clearly identified at the boundary $\Gamma_\psi \times (0,T)$.

3.1.4 Optimal control problem

The goal of this subsection is to prove an existence theorem for the optimal control problem.

Let $\mathcal{W} \subset H^1(0,T; L^2(\Omega_\psi))$ be a convex, bounded and closed set. For any fixed $u \in \mathcal{W}$ we can find the unique solution $\chi = \chi(u)$ of (3.6) and define the cost functional characterizing the opening of the crack,

$$J(u) = \int_0^T \int_{\Gamma_\psi} |[\chi]| d\Gamma_\psi \, dt.$$

Our aim is to minimize this functional:

$$\inf_{u \in \mathcal{W}} J(u). \tag{3.27}$$

The result given below provides the solvability of the optimal control problem formulated.

Theorem 3.2. *There exists a solution of the optimal control problem* (3.27), (3.6).

PROOF. Let a sequence $u^n \in \mathcal{W}$ be a minimizing one. By its boundedness in $H^1(0,T;L^2(\Omega_\psi))$, one can assume that as $n \to \infty$

$$u^n \to u \quad \text{weakly in } H^1(0,T;L^2(\Omega_\psi)), \quad u \in \mathcal{W}. \tag{3.28}$$

For every n there exists a unique solution of the variational inequality

$$A(\chi^n)(\bar{\chi} - \chi^n) \geq \int_0^T \langle u^n, \bar{\chi} - \chi^n \rangle dt, \quad \forall \bar{\chi} \in \mathcal{K}; \quad \chi^n \in \mathcal{K}. \tag{3.29}$$

As we are well aware, $\chi_t^n \in L^2(0,T;H(\Omega_\psi))$ and, moreover, it follows from the proof of Theorem 3.1 that

$$\|\chi^n\|^2_{H^1(0,T;H(\Omega_\psi))} \leq c\|u^n\|^2_{H^1(0,T;L^2(\Omega_\psi))}$$

with a constant c being uniform in n. Without any loss we assume that as $n \to \infty$

$$\chi^n,\ \chi_t^n,\ \int_0^t \chi^n d\tau \to \chi,\ \chi_t,\ \int_0^t \chi d\tau \quad \text{weakly in } L^2(0,T;H(\Omega_\psi)),$$

$$\chi^n \to \chi \quad \text{strongly in } L^2(Q_\psi),$$

$$[\chi^n] \to [\chi] \quad \text{weakly in } L^1(0,T;L^1(\Gamma_\psi)).$$

The last convergence is due to the imbedding continuity of $L^2(0,T;H(\Omega_\psi))$ in $L^2(0,T;L^2(\Gamma_\psi))$. The above convergence and (3.28) allow us to pass to the limit in (3.29) and to get $\chi_t \in L^2(0,T;H(\Omega_\psi))$,

$$A(\chi)(\bar{\chi} - \chi) \geq \int_0^T \langle u, \bar{\chi} - \chi \rangle dt, \quad \forall \bar{\chi} \in \mathcal{K}; \quad \chi \in \mathcal{K}. \tag{3.30}$$

The variational inequality (3.30) precisely means that $\chi = \chi(u)$. On the other hand

$$\inf_{\bar{u} \in \mathcal{W}} J(\bar{u}) = \liminf_{n \to \infty} J(u^n) \geq J(u) \geq \inf_{\bar{u} \in \mathcal{W}} J(\bar{u});$$

consequently, the constructed function u is a solution of the optimal control problem (3.27), (3.6). This completes the proof.

3.1.5 Solution regularity near crack points

When $J(u) = 0$ the crack is said to have a zero opening. As it turns out the solution is infinitely differentiable provided that the crack has a zero opening. This assertion, in particular, means that if we have a zero crack opening the presence of the crack has no influence on the displacement field. In this case the plate behaviour precisely coincides with that of the plate without a crack. This property confirms the removable singularity property. We shall prove that C^∞-regularity is a local property. If the crack opening is zero in the vicinity of some fixed point at Γ_ψ for all $t \in (0, t^0)$, then the solution is infinitely smooth near this point for all $t \in (0, t^0)$. Of course, the external force u is assumed to be infinitely smooth in this case. Also we should remark that the above regularity property holds provided a zero opening takes place since $t = 0$. In general, if the crack opening is zero for $t \in (t^1, t^0)$, $t^1 > 0$, $t^0 > t^1$, the solution does not have C^∞-regularity.

The arguments given below are concerned with a justification of C^∞-regularity of the solution for the crack of zero opening. We shall prove the solution regularity in the neighbourhood of the line $x^0 \times (0, t^0)$, where $x^0 \equiv (0,0)$, $t^0 > 0$, i.e. in the vicinity of the crack tip. The solution regularity near the line $\bar{x} \times (0, t^0)$, where $\bar{x} \in \Gamma_\psi \setminus \partial\Gamma_\psi$, can be easily proved.

So, let $\mathcal{O}(x^0) \subset R^2$ be a neighbourhood of the point x^0, and let $\mathcal{O} = \mathcal{O}(x^0) \times (0, t^0)$. Extend the function $\psi(x)$ beyond $x = 0$ assuming that the extension is smooth enough. Denote by $\tilde{\Gamma}_\psi$ the graph of the extended function. Also, let $\mathcal{O}^+(x^0) = \mathcal{O}(x^0) \cap \{y > \psi(x)\}$, and $\mathcal{O}^-(x^0)$ be defined analogously, $\mathcal{O}^\pm = \mathcal{O}^\pm(x^0) \times (0, t^0)$. As shown, equations (3.3), (3.4) hold in $\mathcal{O}^\pm$ in the sense of distributions.

By the regularity of (W, w) which follows from Theorem 3.1, we conclude that for all $t \in (0, T)$ in $\mathcal{O}^\pm(x^0)$ the following equations are fulfilled,

$$-\frac{\partial \sigma_{ij}(W^\tau(t))}{\partial x_j} = u_i(t), \tag{3.31}$$

$$\Delta^2 w^\tau(t) = u_3(t), \tag{3.32}$$

in the sense of (two-dimensional) distributions. As in the above case, let $D \subset R^2$ be a fixed bounded domain with smooth boundary γ. As we know the values $M(w)$ and $R(w)$ can be correctly defined on γ, namely, $M(w) \in H^{-\frac{1}{2}}(\gamma)$, $R(w) \in H^{-\frac{3}{2}}(\gamma)$ provided that $w \in H^2(D)$, $\Delta^2 w \in L^2(D)$ and, moreover, the following formula holds:

$$B_D(w, v) = \langle M(w), \frac{\partial v}{\partial n}\rangle_{\frac{1}{2},\gamma} - \langle R(w), v\rangle_{\frac{3}{2},\gamma} + \langle \Delta^2 w, v\rangle_D, \tag{3.33}$$

$$\forall v \in H^2(D).$$

Here $\langle \cdot, \cdot \rangle_{s,\gamma}$ stands for the duality pairing between $H^{-s}(\gamma)$ and $H^s(\gamma)$.

If $\sigma_{ij}(W) \in L^2(D)$, $\partial\sigma_{ij}(W)/\partial x_j \in L^2(D)$, the values $\sigma_{ij}(W)n_j$ can be correctly defined on γ as elements of $H^{-\frac{1}{2}}(\gamma)$,

$$\langle \sigma_{ij}(W), \varepsilon_{ij}(V)\rangle_D = \langle \sigma_{ij}(W)n_j, v_i\rangle_{\frac{1}{2},\gamma} - \langle \frac{\partial \sigma_{ij}(W)}{\partial x_j}, v_i\rangle_D, \tag{3.34}$$

$$\forall V = (v_1, v_2) \in H^1(D).$$

Henceforth the boundaries of $\mathcal{O}^{\pm}(x^0)$ are denoted by $\gamma^{\pm}$, respectively.

Let a function $\varphi \equiv (\varphi_1, \varphi_2)$ belong to the space $C_0^{\infty}(\mathcal{O}(x^0))$ and be equal to zero beyond $\mathcal{O}(x^0)$. Then $(W(t) + \varphi, w(t)) \in K$. We substitute $(W(t) + \varphi, w(t))$ in (3.10) as a test function. This implies for all $t \in (0, T)$

$$\langle \sigma_{ij}(W^{\tau}), \varepsilon_{ij}(\varphi)\rangle_{+} + \langle \sigma_{ij}(W^{\tau}), \varepsilon_{ij}(\varphi)\rangle_{-} \geq \langle u_i, \varphi_i \rangle.$$

To simplify the formulae here and below we do not show the dependence of the functions on t. Subscripts $+, -$ denote the integration over $\mathcal{O}^{\pm}(x^0)$, respectively. Owing to the formula (3.34) the last inequality gives, for all $t \in (0, T)$,

$$-\langle [\sigma_{ij}(W^{\tau})\nu_j], \varphi_i\rangle_{\frac{1}{2},\gamma^-} - \langle \frac{\partial \sigma_{ij}(W^{\tau})}{\partial x_j}, \varphi_i\rangle_{\pm} \geq \langle u_i, \varphi_i\rangle. \tag{3.35}$$

The existence of two angular points on $\gamma^{\pm}$ presents no problems since φ has a compact support. Hence, the inequality (3.35) with the equations (3.31) yield the identity

$$\langle [\sigma_{ij}(W^{\tau})\nu_j], \varphi_i\rangle_{\frac{1}{2},\gamma^-} = 0, \quad \forall \varphi \in C_0^{\infty}(\mathcal{O}(x^0))$$

and consequently

$$[\sigma_{ij}(W^{\tau})\nu_j] = 0 \quad \text{on } \tilde{\Gamma}_{\psi} \cap \mathcal{O}(x^0). \tag{3.36}$$

Let $\theta \in C_0^{\infty}(\mathcal{O}(x^0))$. Beyond $\mathcal{O}(x^0)$ the function θ is assumed to be equal to zero. We substitute $(W(t), \theta + w(t))$ as a test function in (3.10). As a result the following inequality being valid for all $t \in (0, T)$ follows:

$$B_{+}(w^{\tau}, \theta) + B_{-}(w^{\tau}, \theta) \geq \langle u_3, \theta\rangle. \tag{3.37}$$

Since equation (3.32) holds in $\mathcal{O}^{\pm}(x^0)$ we easily deduce from (3.37) for all $t \in (0, T)$ that

$$\langle [M(w^{\tau})], \frac{\partial \theta}{\partial n}\rangle_{\frac{1}{2},\gamma} = 0, \quad \langle [R(w^{\tau})], \theta\rangle_{\frac{3}{2},\gamma} = 0, \quad \forall \theta \in C_0^{\infty}(\mathcal{O}(x^0)).$$

Here γ can coincide with γ^+ or γ^-. By the arbitrariness of θ, the above identities imply for all $t \in (0, T)$

$$[M(w^{\tau})] = 0, \quad [R(w^{\tau})] = 0 \quad \text{on } \tilde{\Gamma}_{\psi} \cap \mathcal{O}(x^0). \tag{3.38}$$

Now we are in a position to prove the result on a regularity of the solution near crack faces.

Theorem 3.3. *Let $u \in C^{\infty}(\mathcal{O})$ and*

$$\int_0^{t^0} \int_{\Gamma_{\psi} \cap \mathcal{O}(x^0)} |[\chi]| \, d\Gamma_{\psi} dt = 0.$$

Then

$$\chi \in C^\infty(\mathcal{O}). \tag{3.39}$$

PROOF. The hypotheses of the theorem provide the condition $[\chi] = 0$ on $(\tilde{\Gamma}_\psi \cap \mathcal{O}(x^0)) \times (0, t^0)$, whence

$$[\chi^\tau] = 0 \quad \text{on } (\tilde{\Gamma}_\psi \cap \mathcal{O}(x^0)) \times (0, t^0).$$

Moreover, using (3.5) we obtain

$$\left[\frac{\partial w^\tau}{\partial \nu}\right] = 0 \quad \text{on } (\tilde{\Gamma}_\psi \cap \mathcal{O}(x^0)) \times (0, t^0).$$

Note that $(W^\tau, w^\tau) \in H^1(0, t^0; H^1(\mathcal{O}^\pm(x^0)) \times H^2(\mathcal{O}^\pm(x^0))$. The above observations concerning the jumps $[\chi^\tau]$, $[\partial w^\tau / \partial \nu]$ imply (see Mikhailov, 1976)

$$(W^\tau, w^\tau) \in H^1(0, t^0; H^1(\mathcal{O}(x^0)) \times H^2(\mathcal{O}(x^0)).$$

Following this inclusion and the conditions (3.36), (3.38) we shall prove that the equations (3.3), (3.4) hold in $\mathcal{O}$ in the sense of distributions.

Denote by $(\cdot, \varphi)$ the value of a distribution at the point φ. For any $\varphi \in C_0^\infty(\mathcal{O})$ we have

$$-\left(\frac{\partial \sigma_{ij}(W^\tau)}{\partial x_j} + u_i, \varphi\right) = \int_0^{t^0} \langle \sigma_{ij}(W^\tau), \frac{\partial \varphi}{\partial x_j} \rangle_\pm dt - (u_i, \varphi) \tag{3.40}$$

$$= -\int_0^{t^0} \langle [\sigma_{ij}(W^\tau)\nu_j], \varphi \rangle_{\frac{1}{2},\gamma^-} dt - \int_0^{t^0} \left\langle \frac{\partial \sigma_{ij}(W^\tau)}{\partial x_j} + u_i, \varphi \right\rangle_\pm dt.$$

Owing to (3.36), (3.31) we readily conclude that the right-hand side of (3.40) is equal to zero, which implies the equations

$$-\frac{\partial \sigma_{ij}(W^\tau)}{\partial x_j} = u_i \quad \text{in } \mathcal{O} \tag{3.41}$$

holding in the sense of distributions.

Analogously, for any $\varphi \in C_0^\infty(\mathcal{O})$ we derive

$$(\Delta^2 w^\tau - u_3, \varphi) = \int_0^{t^0} B(w^\tau, \varphi)\, dt - (u_3, \varphi)$$

$$= \int_0^{t^0} B_\pm(w^\tau, \varphi)\, dt - (u_3, \varphi) = -\int_0^{t^0} \langle [M(w^\tau)], \frac{\partial \varphi}{\partial \nu} \rangle_{\frac{1}{2},\gamma^-} dt \tag{3.42}$$

$$+ \int_0^{t^0} \langle [R(w^\tau)], \varphi \rangle_{\frac{3}{2},\gamma^-} \, dt + \int_0^{t^0} \langle \Delta^2 w^\tau - u_3, \varphi \rangle_\pm \, dt.$$

It is evident from (3.32), (3.38) that the right-hand side of (3.42) is equal to zero. Hence, the equation

$$\Delta^2 w^\tau = u_3 \quad \text{in} \quad \mathcal{O} \tag{3.43}$$

holds in the sense of distributions.

The statement (3.39) of the theorem clearly follows from (3.41), (3.43). In fact, one can locally solve the elliptic equations (3.41), (3.43) for each fixed $t \in (0, t^0)$ and get the infinite differentiability with respect to x, y of the functions $\chi^\tau(t) = \chi(t) + \int_0^t \chi(\tau)\, d\tau$ in any fixed subdomain of $\mathcal{O}(x^0)$ (see Fichera, 1972; Lions, Magenes, 1968). The function $\chi^\tau(t)$ is infinitely differentiable with respect to $t \in (0, t^0)$ and hence $\chi^\tau \in C^\infty(\mathcal{O})$, $\chi \in C^\infty(\mathcal{O})$. The proof is complete.

3.2 Contact of two plates one of which has a crack

A contact between two plates is considered provided that one of the plates has a crack. In a stress free state both plates remain at a given distance from each other. The plate displacements satisfy two restrictions of inequality type. The first restriction describes the nonpenetration between the plates, and it is considered in the exterior of the domain. The second one describes the nonpenetration between crack faces.

We prove the existence of the solution and state additional qualitative properties – in particular, a solution regularity near the crack faces and near the external boundary. The results of this section are obtained in (Khludnev, 1997c).

3.2.1 Existence of solutions

Let $\Omega \subset R^2$ be a bounded domain with C^∞- smooth boundary Γ, and Γ_ψ be a graph of the function $y = \psi(x)$, $x \in [0, 1]$, $(x, y) \in \overline{\Omega}$, $\psi \in H_0^3(0, 1)$ (see Fig.3.2).

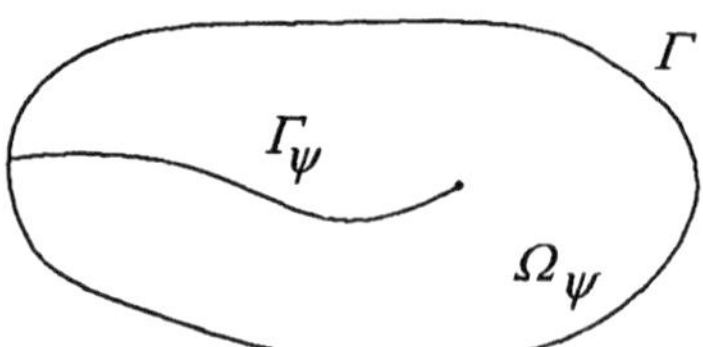

Fig.3.2. Mid-surface of the upper plate

Assume that Γ_ψ and Γ have a joint point– the origin $(0,0)$. The angle between Γ and Γ_ψ is assumed to be positive at the point $(0,0)$. The domain $\Omega_\psi \equiv \Omega \setminus \Gamma_\psi$ corresponds to the mid-surface of the first plate whose mid-surface belongs to the plane $z = 0$; the axis z is orthogonal to the (x,y)-plane.

The second plate (which has no cracks) can be in contact with the first plate (which has the crack). We assume that the plates remain at a distance $\delta \geq 0$ from each other in the stress free state, $\delta = \text{const}$ (see Fig.3.3). They may be in contact due to exterior forces. The mid-surface of the second plate is precisely Ω, which corresponds to the negative value of the coordinate z. By that the first plate is called the upper plate and the second one the lower plate.

Fig.3.3. Plates in a stress-free state

Denote by $\chi = (W, w)$, $\xi = (U, u)$ displacement vectors of mid-surfaces of the upper and the lower plates, respectively, where $W = (w^1, w^2)$, w are horizontal and vertical displacements of the upper plate, and $U = (u^1, u^2)$, u are horizontal and vertical displacements of the lower plate. Let

$$\varepsilon_{ij} = \varepsilon_{ij}(W) = \frac{1}{2}\left(\frac{\partial w^i}{\partial x_j} + \frac{\partial w^j}{\partial x_i}\right), \quad i,j = 1,2, \quad (x_1 = x,\ x_2 = y)$$

and $\sigma_{ij} = \sigma_{ij}(W)$,

$$\sigma_{11} = \varepsilon_{11} + \kappa\varepsilon_{22}, \quad \sigma_{22} = \varepsilon_{22} + \kappa\varepsilon_{11}, \quad \sigma_{12} = (1-\kappa)\varepsilon_{12} \tag{3.44}$$

for $\kappa = \text{const}$, $0 < \kappa < 1/2$.

The energy functional of the upper plate has the form

$$\Pi_f(\chi) = \frac{1}{2}B_\psi(w,w) + \frac{1}{2}\langle\sigma_{ij}(W), \varepsilon_{ij}(W)\rangle_\psi - \langle f, \chi\rangle_\psi,$$

where $\langle\cdot,\cdot\rangle_\psi$ means the integration over Ω_ψ, $f = (f_1, f_2, f_3) \in L^2(\Omega_\psi)$ is the vector of exterior forces and

$$B_\psi(w,\bar{w}) = \int_{\Omega_\psi} (w_{xx}\bar{w}_{xx} + w_{yy}\bar{w}_{yy} + \kappa w_{xx}\bar{w}_{yy} + \kappa w_{yy}\bar{w}_{xx}$$

$$+ 2(1-\kappa)w_{xy}\bar{w}_{xy})\, d\Omega_\psi.$$

Similarly, for the lower plate the energy functional is as follows,

$$\Pi_g(\xi) = \frac{1}{2}B(u,u) + \frac{1}{2}\langle\sigma_{ij}(U), \varepsilon_{ij}(U)\rangle - \langle g, \xi\rangle,$$

where $g = (g_1, g_2, g_3) \in L^2(\Omega)$ and the brackets $\langle\cdot,\cdot\rangle$ denote the integration over Ω,

$$B(u,\bar{u}) = \int_\Omega \Delta u \Delta \bar{u}\, d\Omega.$$

The energy functional for a system consisting of the two plates can be presented as $\Pi_f(\chi) + \Pi_g(\xi)$.

The nonpenetration condition between the crack faces has the form

$$[W]\nu \geq \varepsilon \left|\left[\frac{\partial w}{\partial \nu}\right]\right| \quad \text{on } \Gamma_\psi. \tag{3.45}$$

Here $\nu = (-\psi_x, 1)/\sqrt{1+\psi_x^2}$ is a unit normal vector to the graph Γ_ψ, $\nu = (\nu_1, \nu_2)$. The plates may be in contact such that there is no interpenetration. The nonpenetration condition between the plates can be written as (see Khludnev, Sokolowski, 1997)

$$w \geq u - \delta \quad \text{in } \Omega_\psi. \tag{3.46}$$

We assume that the physical parameters of the lower plate coincide with those of the upper plate; in particular, the stress tensors and strain tensors of the lower plate satisfy (3.44). The thickness of the lower plate is 2ε. The following conditions are considered at the external boundary Γ:

$$w = \partial w/\partial n = W = 0, \quad u = \partial u/\partial n = U = 0 \quad \text{on } \Gamma. \tag{3.47}$$

By n we denote the external normal vector to Γ.

Let us formulate the equilibrium problem of the two plates. We put

$$H(\Omega_\psi) = H^{1,0}(\Omega_\psi) \times H^{1,0}(\Omega_\psi) \times H^{2,0}(\Omega_\psi),$$

$$H_0(\Omega) = H_0^1(\Omega) \times H_0^1(\Omega) \times H_0^2(\Omega), \quad H = H(\Omega_\psi) \times H_0(\Omega).$$

Introducing the admissible set of displacements,

$$K_\varepsilon = \{(\chi, \xi) \in H(\Omega_\psi) \times H_0(\Omega) \mid \quad (\chi, \xi) \quad \text{satisfy } (3.45), (3.46)\},$$

the equilibrium problem admits the variational form

$$\inf_{(\chi,\xi)\in K_\varepsilon} (\Pi_f(\chi) + \Pi_g(\xi)) .$$

By the convexity and differentiability of the functional $\Pi_f(\chi) + \Pi_g(\xi)$ on the space H, the minimization problem is equivalent to the variational inequality

$$\Pi_f'(\chi)(\bar{\chi} - \chi) + \Pi_g'(\xi)(\bar{\xi} - \xi) \geq 0, \quad (\chi, \xi) \in K_\varepsilon, \quad \forall (\bar{\chi}, \bar{\xi}) \in K_\varepsilon, \tag{3.48}$$

where $\Pi_f'(\chi)$, $\Pi_g'(\xi)$ are derivatives of the functionals Π_f, Π_g respectively at the points χ, ξ. We shall use the inequalities

$$B_\psi(w, w) \geq c\|w\|_{2,\Omega_\psi}^2, \quad B(u, u) \geq c\|u\|_{2,\Omega}^2 \tag{3.49}$$

for all $w \in H^{2,0}(\Omega_\psi)$, $u \in H_0^2(\Omega)$,

$$\langle \sigma_{ij}(W), \varepsilon_{ij}(W)\rangle_\psi \geq c\|W\|_{1,\Omega_\psi}^2, \quad \forall W = (w^1, w^2) \in H^{1,0}(\Omega_\psi), \tag{3.50}$$

where $\|\cdot\|_{s,\Omega_\psi}$ is the norm in $H^{s,0}(\Omega_\psi)$, and the inequality

$$\langle \sigma_{ij}(U), \varepsilon_{ij}(U)\rangle \geq c\|U\|_{1,\Omega}^2, \quad \forall U = (u^1, u^2) \in H_0^1(\Omega). \tag{3.51}$$

We introduce one more bilinear form,

$$a(\eta, \bar{\eta}) = B_\psi(w, \bar{w}) + B(u, \bar{u}) + \langle \sigma_{ij}(W), \varepsilon_{ij}(\bar{W})\rangle_\psi + \langle \sigma_{ij}(U), \varepsilon_{ij}(\bar{U})\rangle, \tag{3.52}$$

where $\eta = (\chi, \xi)$, $\chi = (W, w)$, $\xi = (U, u)$ and, respectively, for $\bar{\eta} = (\bar{\chi}, \bar{\xi})$. By (3.49)–(3.51), the following estimate holds:

$$a(\eta, \eta) \geq c\|\eta\|_H^2, \quad \forall \eta \in H. \tag{3.53}$$

The inequality (3.48) can be rewritten as

$$a(\eta, \bar{\eta} - \eta) \geq \langle f, \bar{\chi} - \chi\rangle_\psi + \langle g, \bar{\xi} - \xi\rangle, \quad \forall \bar{\eta} = (\bar{\chi}, \bar{\xi}) \in K_\varepsilon. \tag{3.54}$$

In accordance with (3.53) the functional $\Pi_f(\chi) + \Pi_g(\xi)$ is coercive and weakly lower semicontinuous on the space H, consequently, the problem (3.48) (or the problem (3.54)) has a solution. The solution is unique. Note that the equilibrium equations

$$-\sigma_{ij,j}(W) = f_i, \quad -\sigma_{ij,j}(U) = g_i, \quad i = 1, 2, \tag{3.55}$$

hold in Ω_ψ and Ω, respectively. To verify the first equations it suffices to take the test functions $(\bar{\chi}, \bar{\xi}) = (\chi + \tilde{\chi}, \xi)$ in (3.48), $\tilde{\chi} = (\tilde{W}, 0)$, $\tilde{W} \in (C_0^\infty(\Omega_\psi))^2$. We next take the test functions $(\bar{\chi}, \bar{\xi}) = (\chi, \xi + \tilde{\xi})$ in (3.48), where $\tilde{\xi} = (\tilde{U}, 0)$, $\tilde{U} \in (C_0^\infty(\Omega))^2$, which imply the second equations in (3.55).

In Section 3.2.2 we establish an additional regularity of the solution up to the crack faces. Roughly speaking, we raise the solution smoothness by one as compared to the variational smoothness (i.e. as compared to the inclusion $(\chi, \xi) \in K_\varepsilon$). Section 3.2.3 is devoted to the optimal control problem with the cost functional characterizing the crack opening. In particular, C^∞-regularity is proved provided that the crack opening is zero. The passage to the limit as $\varepsilon \to 0$ is analysed in Section 3.2.4, which corresponds to the transition from the precise nonpenetration condition (3.45) to the approximate nonpenetration condition (when $\varepsilon = 0$ in (3.45)). At this point we have to note that, while passing to the limit, the thickness of the second plate is of no importance.

3.2.2 Solution regularity

Let (χ, ξ) be the solution of the problem (3.48). Additional regularity of the solution in the vicinity of $\Gamma_\psi \setminus \partial\Gamma_\psi$ is proved provided that Γ_ψ is a segment of a straight line. The following statement holds.

Theorem 3.4. *Let $x^0 \in \Gamma_\psi \setminus \partial\Gamma_\psi$, and $D(x^0)$ be a neighbourhood of the point x^0 such that $\Gamma_\psi \cap D(x^0)$ is a segment of a straight line parallel to the axis x. Then there exists $\lambda > 0$ such that*

$$W, w_x \in H^2(R_\lambda(x^0) \cap \Omega_\psi), \quad U, u_x \in H^2(R_\lambda(x^0)).$$

PROOF. We choose a smooth function φ such that $\varphi \equiv 1$ in $R_\lambda(x^0)$, $\varphi \equiv 0$ outside $R_{3\lambda/2}(x^0)$, $0 \le \varphi \le 1$ everywhere, $\partial\varphi/\partial y = 0$ on Γ_ψ. Assume that $R_{2\lambda}(x^0) \subset D(x^0)$.

Introduce the notation

$$d_{\pm\tau} p(\bar{x}) = \tau^{-1}(p(\bar{x} \pm \tau e) - p(\bar{x})), \quad \Delta_\tau = -d_{-\tau} d_\tau, \quad 0 < |\tau| < \lambda/2,$$

where e is a unit vector of the axis x. Consider the vector (χ_τ, ξ_τ) with the components

$$\chi_\tau = \chi + \frac{\tau^2}{2}\varphi^2 \Delta_\tau \chi, \quad \xi_\tau = \xi + \frac{\tau^2}{2}\varphi^2 \Delta_\tau \xi.$$

We have $(\chi_\tau, \xi_\tau) \in K_\varepsilon$. In fact, it suffices to verify (3.45), (3.46). To this end, denote $w - u$ by v. Then $v(\bar{x}) \ge -\delta$ for all $\bar{x} \in \Omega_\psi$. Hence for the function $v_\tau = w_\tau - u_\tau$, the following relation holds:

$$v_\tau(\bar{x}) = (w_\tau - u_\tau)(\bar{x}) = v(\bar{x}) + \frac{\tau^2}{2}\varphi^2(\bar{x})\Delta_\tau v(\bar{x})$$

$$= v(\bar{x})(1 - \varphi^2(\bar{x})) + \frac{\varphi^2(\bar{x})}{2}[v(\bar{x} - \tau e) + v(\bar{x} + \tau e)] \geq -\delta.$$

This means that the inequality like (3.46) holds:

$$w_\tau \geq u_\tau - \delta \quad \text{in } \Omega_\psi. \tag{3.56}$$

Similar to Section 2.4, it can be shown that the function $\chi_\tau = (W_\tau, w_\tau)$ satisfies the following inequality (for $\varepsilon = 1$):

$$[W_\tau]\nu \geq \varepsilon \left| \left[\frac{\partial w_\tau}{\partial \nu} \right] \right| \qquad \text{on } \Gamma_\psi \cap D(x^0). \tag{3.57}$$

Consequently, by the finiteness of the function φ we conclude that (3.57) holds on Γ_ψ. The inequalities (3.56), (3.57), therefore, show that $(\chi_\tau, \xi_\tau) \in K_\varepsilon$. This allows us to substitute (χ_τ, ξ_τ) in (3.54) as a test function which implies the inequality

$$a(\eta, \varphi^2 \Delta_\tau \eta) \geq \langle f, \varphi^2 \Delta_\tau \chi \rangle_\psi + \langle g, \varphi^2 \Delta_\tau \xi \rangle. \tag{3.58}$$

One can verify that the difference between the terms $a(\eta, \varphi^2 \Delta_\tau \eta)$ and $-a(d_\tau(\varphi\eta), d_\tau(\varphi\eta))$ can be estimated from above by the right-hand side of the inequality (3.59) (see below), and hence it follows from (3.58) that

$$a(d_\tau(\varphi\eta), d_\tau(\varphi\eta)) \leq c\Big(\|\eta\|_H^2 \tag{3.59}$$

$$+ \|d_\tau(\varphi\eta)\|_H \left(\|\eta\|_H + \|f\|_{0,\Omega_\psi} + \|g\|_{0,\Omega} \right) \Big)$$

with the constant c independent of τ. By (3.53), the inequality (3.59) implies

$$\|d_\tau(\varphi\chi)\|_{H(\Omega_\psi)} + \|d_\tau(\varphi\xi)\|_{H_0(\Omega)} \leq c, \tag{3.60}$$

where the constant c does not depend on τ. We obtain from (3.60) that

$$\frac{\partial}{\partial x}(\varphi\chi) \in H(\Omega_\psi), \quad \frac{\partial}{\partial x}(\varphi\xi) \in H_0(\Omega)$$

and consequently

$$W_x \in H^1(R_\lambda(x^0) \cap \Omega_\psi), \quad U_x \in H^1(R_\lambda(x^0)), \tag{3.61}$$

$$w_x \in H^2(R_\lambda(x^0) \cap \Omega_\psi), \quad u_x \in H^2(R_\lambda(x^0)).$$

In the domain Ω_ψ the equation (3.55) for W can be written in the form

$$W_{yy} = F.$$

By (3.61), the inclusion $F \in L^2(R_\lambda(x^0) \cap \Omega_\psi)$ holds, and hence

$$W_{yy} \in L^2(R_\lambda(x^0) \cap \Omega_\psi).$$

Besides, by the equation (3.55) for U, in a neighbourhood of the point x^0 the following equation holds

$$U_{yy} = G,$$

and, by (3.61), $G \in L^2(R_\lambda(x^0))$. Theorem 3.4 is proved.

The next theorem provides an additional smoothness of the solution as compared to Theorem 3.4 provided that there is no a contact between two plates in a neighbourhood of a fixed point of the crack.

Theorem 3.5. *Let the conditions of Theorem 3.4 hold and*

$$w^{\pm}(x^0) > u(x^0) - \delta. \tag{3.62}$$

Then

$$\begin{aligned} &W \in H^2(R_\lambda(x^0) \cap \Omega_\psi), \quad U \in H^2(R_\lambda(x^0)), \\ &w \in H^3(R_\lambda(x^0) \cap \Omega_\psi), \quad u \in H^3(R_\lambda(x^0)). \end{aligned} \tag{3.63}$$

PROOF. It follows from (3.62), (3.48) that there exists a neighbourhood $D(x^0)$ of the point x^0 such that the equation

$$\Delta^2 w = f_3 \tag{3.64}$$

holds in $D(x^0) \cap \Omega_\psi$ in the sense of distributions. We shall use the following statement (Duvaut, Lions, 1972). Let $D \subset R^2$ be a bounded domain with a smooth boundary, and v be a distribution on D such that $v, \nabla v \in H^{-1}(D)$. Then $v \in L^2(D)$ and, moreover, there exists a constant c depending on D such that

$$\|v\|_{L^2(D)} \le c\left(\|v\|_{H^{-1}(D)} + \|\nabla v\|_{H^{-1}(D)}\right).$$

It follows from (3.61) that $\partial(\varphi w)/\partial x \in H^{2,0}(\Omega_\psi)$. Hence, the derivatives w_{xxx}, w_{yyx}, w_{xxy} belong to L^2 in a neighbourhood of the point x^0. Equation (3.64) can be written in the form

$$w_{yyyy} = h.$$

By the above arguments, the functions h, w_{yyy}, w_{yyyx} belong to $H^{-1}(\Omega_\psi \cap D)$, where D is a neighbourhood of x^0. Consequently, the function w_{yyy} belongs to $L^2(\Omega_\psi \cap D_1)$ and the following estimate holds:

$$\|w_{yyy}\|^2_{L^2(\Omega_\psi \cap D_1)} \le c\Big(\|w_{yyy}\|^2_{H^{-1}(\Omega_\psi \cap D_1)}$$

$$+\|w_{yyyy}\|^2_{H^{-1}(\Omega_\psi \cap D_1)} + \|w_{yyyx}\|^2_{H^{-1}(\Omega_\psi \cap D_1)}\Big).$$

Here D_1 is a neighbourhood of x^0, $\bar{D}_1 \subset D$. Thus we obtain (3.63) for the function w. Furthermore, the equation

$$\Delta^2 u = g_3 \tag{3.65}$$

holds in $D(x^0)$ in the distribution sense, and by similar arguments the inclusion (3.63) follows for the function u. Theorem 3.5 is proved.

3.2.3 Cracks of minimal opening

In this subsection we analyse an optimal control problem. The exterior forces (f, g) are chosen to minimize the cost functional

$$J_\varepsilon(f, g) = \int_{\Gamma_\psi} |[\chi]| \, d\Gamma_\psi.$$

This functional characterizes an opening of the crack. As before, (χ, ξ) is the solution of (3.48) corresponding to (f, g). At the first step we prove the existence of the optimal control problem. The next step is to prove the C^∞-regularity of the solution provided that the crack opening is zero. We fixed the parameter ε in this subsection; the passage to the limit, as $\varepsilon \to 0$, is analysed in Section 3.2.4.

Let $F \times G \subset L^2(\Omega_\psi) \times L^2(\Omega)$ be a convex closed bounded set, and $(f, g) \in F \times G$. The following statement takes place.

Theorem 3.6. *There exists a solution of the minimization problem*

$$\inf_{F\times G} J_\varepsilon(f, g). \tag{3.66}$$

PROOF. Let $(f_n, g_n) \in F \times G$ be a minimizing sequence. For any n, there exists a unique solution of the problem

$$\Pi'_{f_n}(\chi_n)(\bar{\chi} - \chi_n) + \Pi'_{g_n}(\xi_n)(\bar{\xi} - \xi_n) \geq 0, \quad \forall\, (\bar{\chi}, \bar{\xi}) \in K_\varepsilon. \tag{3.67}$$

By the boundedness of f_n, g_n in $L^2(\Omega)$, it follows from (3.67) that

$$\|\chi_n\|_{H(\Omega_\psi)} + \|\xi_n\|_{H_0(\Omega)} \leq c \tag{3.68}$$

uniformly in n. Choosing a subsequence, if necessary, we can assume that as $n \to \infty$

$$(\chi_n, \xi_n) \to (\chi, \xi) \text{ weakly in } H, \quad \text{strongly in } L^2(\Omega),$$

$$[\chi_n] \to [\chi] \text{ strongly in } L^1(\Gamma_\psi).$$

This convergence allows us to pass to the limit as $n \to \infty$ in (3.67) which implies

$$\Pi'_f(\chi)(\bar{\chi} - \chi) + \Pi'_g(\xi)(\bar{\xi} - \xi) \geq 0, \quad (\chi, \xi) \in K_\varepsilon, \quad \forall\, (\bar{\chi}, \bar{\xi}) \in K_\varepsilon.$$

This variational inequality provides the property $\chi = \chi(f, g)$, $\xi = \xi(f, g)$. Consequently,

$$\inf_{F\times G} J_\varepsilon(\bar{f}, \bar{g}) = \liminf_{n\to\infty} J_\varepsilon(f_n, g_n) \geq J_\varepsilon(f, g) \geq \inf_{F\times G} J_\varepsilon(\bar{f}, \bar{g})$$

which proves that the pair (f, g) solves the optimal control problem (3.66). Theorem 3.6 is proved.

As it turns out, the solution of (3.48) is infinitely differentiable provided that $f, g \in C^\infty$, the crack opening is equal to zero and a contact between plates is absent in the vicinity of the considered point. We prove this assertion in the neighbourhood of a point x^0, $x^0 \in \Gamma \cap \Gamma_\psi$. The case $x^0 \notin \Gamma \cap \Gamma_\psi$ is simpler (see Remark after the proof of Theorem 3.7).

Note that the absence of a contact between the plates at the point $\bar{x} \in \Gamma_\psi \setminus \partial\Gamma_\psi$ means $w^\pm(\bar{x}) > u(\bar{x}) - \delta$. As we know, in this case the following boundary conditions hold in a neighbourhood of the point $\bar{x}$:

$$[\sigma_\nu(W)] = 0, \quad \sigma_s(W) = 0, \quad [m(w)] = 0, \quad t(w) = 0 \quad \text{on } \Gamma_\psi, \tag{3.69}$$

$$|m(w)| \le -\sigma_\nu(W), \quad m(w)\left[\frac{\partial w}{\partial \nu}\right] + \sigma_\nu(W)[W]\nu = 0 \quad \text{on } \Gamma_\psi. \tag{3.70}$$

Here $m(w)$, $t(w)$ are the bending moment and the transverse force,

$$m(w) = \kappa\Delta w + (1-\kappa)\frac{\partial^2 w}{\partial \nu^2}, \quad t(w) = \frac{\partial}{\partial \nu}\Delta w + (1-\kappa)\frac{\partial^3 w}{\partial \nu \partial s^2}, \quad s = (-\nu_2, \nu_1)$$

and $\sigma_\nu(W)$, $\sigma_s(W)$ are the normal and tangential surface forces at Γ_ψ:

$$\{\sigma_{ij}(W)\nu_j\} = \sigma_\nu(W)\nu + \sigma_s(W)s.$$

The above boundary conditions hold provided that the solution $\eta = (\chi, \xi)$ is sufficiently smooth. We shall use only a part of the conditions (3.69), (3.70) to prove the solution regularity.

The main statement related to the cracks of zero opening, i.e. to the cracks with the property $[\chi] = 0$, is as follows.

Theorem 3.7. *Let $\delta > 0$, $x^0 \in \Gamma \cap \Gamma_\psi$. Assume that $[\chi] = 0$ at $\Gamma_\psi \cap D(x^0)$ and $f, g \in C^\infty(D(x^0) \cap \bar{\Omega})$, where $D(x^0)$ is a neighbourhood of the point x^0. Then there exists a neighbourhood $D_1(x^0)$ of the point x^0 such that the solution of the problem* (3.48) *satisfies the inclusion*

$$\chi, \xi \in C^\infty(D_1(x^0) \cap \bar{\Omega}).$$

PROOF. The open set $D(x^0) \cap \Omega_\psi$ can be represented as a union of two domains $D(x^0) \cap \Omega_\psi = D^+ \cup D^-$, where $D^\pm$ correspond to the positive and negative directions of the normal ν, i.e. for $\bar{x} \in D^\pm$ we have $y > \psi(x)$, $y < \psi(x)$, respectively, $\bar{x} = (x, y)$. Since the angle between Γ and Γ_ψ is positive at the point x^0 we can use the imbedding theorem which provides the continuity of w, u in $\bar{\Omega} = \Omega \cup \Gamma$ and $\bar{\Omega}_\psi = \Omega_\psi \cup \Gamma \cup \Gamma_\psi^\pm$, respectively. Hence, the inequality $\delta > 0$ implies the relation $w > u - \delta$ in some neighbourhood $D(x^0)$ of the point x^0. In particular, we have $w^\pm(\bar{x}) > u(\bar{x}) - \delta$, $\bar{x} \in D(x^0) \cap \Gamma_\psi$. Whence, one derives that in D^+, D^- the following equation holds in the distribution sense,

$$\Delta^2 w - f_3 = 0.$$

Let us verify that this equation holds in $D(x^0) \cap \Omega$. Indeed, $[\chi] = 0$ on $\Gamma_\psi \cap D(x^0)$, and by (3.45) we have $[\partial w/\partial \nu] = 0$ on $\Gamma_\psi \cap D(x^0)$. This yields $w \in H^2(D(x^0) \cap \Omega)$ (see Mikhailov, 1976). By the boundary conditions $t(w) = 0$, $[m(w)] = 0$ fulfilled on $\Gamma_\psi \cap D(x^0)$, we obtain (see Section 2.7) that

$$(\Delta^2 w - f_3, \varphi) = 0, \quad \forall \varphi \in C_0^\infty(D(x^0) \cap \Omega) \tag{3.71}$$

which provides the assertion. Here, $(\,\cdot\,, \varphi)$ is a distribution action at the element φ. Analogously, the condition $[\chi] = 0$ on $\Gamma_\psi \cap D(x^0)$ provides the inclusion $W \in H^1(D(x^0) \cap \Omega)$. Hence, by the boundary conditions $[\sigma_{ij}\nu_j] = 0$ holding on $\Gamma_\psi \cap D(x^0)$, $i = 1, 2$, as in Section 2.7, we have

$$\Big(\sigma_{ij,j}(W) + f_i, \varphi\Big) = 0, \quad \forall \varphi \in C_0^\infty(D(x^0) \cap \Omega), \quad i = 1, 2.$$

The inequality $w^\pm(\bar{x}) > u(\bar{x}) - \delta$, $\bar{x} \in D(x^0) \cap \Gamma_\psi$ implies the fulfilment of the equation (3.65) in $D(x^0) \cap \Omega$. All the above arguments show that in $D(x^0) \cap \Omega$, the following equations hold:

$$\Delta^2 w = f_3, \quad \Delta^2 u = g_3, \quad -\sigma_{ij,j}(W) = f_i, \quad -\sigma_{ij,j}(U) = g_i, \quad i = 1, 2.$$

Since the right-hand sides f_i, g_i belong to C^∞ in $D(x^0) \cap \bar{\Omega}$ we obtain the proof of Theorem 3.7.

REMARK. If $x^0 \in \Gamma_\psi$, $x^0 \notin \Gamma \cap \Gamma_\psi$ and $w^\pm(x^0) > u(x^0) - \delta$, the equality $[\chi] = 0$ on $\Gamma_\psi \cap D(x^0)$ also provides C^∞-smoothness of χ, ξ in a neighbourhood $D(x^0)$ under the condition $f, g \in C^\infty(D(x^0))$, i.e.

$$\chi, \xi \in C^\infty(D(x^0)).$$

The proof of this assertion can be fulfilled like that in Theorem 3.7. It suffices to notice that the inequality $w(\bar{x}) > u(\bar{x}) - \delta$ holds for all $\bar{x} \in D_1(x^0) \cap \Omega_\psi$, where $D_1(x^0)$ is a neighbourhood of the point x^0. Moreover, $w^\pm(\bar{x}) > u(\bar{x}) - \delta$, $\bar{x} \in D_1(x^0) \cap \Gamma_\psi$.

3.2.4 The passage to the limit

Consider an approximate description of the nonpenetration condition between the crack faces which can be obtained by putting $\varepsilon = 0$ in (3.45). Similar to the case $\varepsilon > 0$, we can analyse the equilibrium problem of the plates and prove the solution existence of the optimal control problem of the plates with the same cost functional. We aim at the convergence proof of solutions of the optimal control problem as $\varepsilon \to 0$. In this subsection we assume that Γ_ψ is a segment of a straight line parallel to the axis x.

Consider the nonpenetration conditions obtained from (3.45), (3.46) by choosing the parameter $\varepsilon = 0$,

$$[W]\nu \geq 0 \quad \text{on } \Gamma_\psi, \quad w \geq u - \delta \quad \text{in } \Omega_\psi. \tag{3.72}$$

Introduce the set of admissible displacements corresponding to the restriction (3.72):

$$K_0 = \{(\chi, \xi) \in H(\Omega_\psi) \times H_0(\Omega) \mid \ (\chi, \xi) \text{ satisfy (3.72) } \}. \qquad (3.73)$$

Let the set $F \times G$ be chosen as before. For any fixed $(f, g) \in F \times G$ we can find a unique solution of the variational inequality

$$\Pi'_f(\chi)(\bar{\chi} - \chi) + \Pi'_g(\xi)(\bar{\xi} - \xi) \geq 0, \quad (\chi, \xi) \in K_0, \quad \forall (\bar{\chi}, \bar{\xi}) \in K_0. \quad (3.74)$$

The cost functional describing the opening of the crack is as follows:

$$J_0(f, g) = \int_{\Gamma_\psi} |[\chi]| d\Gamma_\psi.$$

In this case the function χ corresponds to (f, g); it is found from (3.74). There exists a unique solution of the optimal control problem,

$$\inf_{F \times G} J_0(f, g). \qquad (3.75)$$

We omit a proof of this statement since it is simpler as compared to the proof of Theorem 3.6.

Let $(\chi_\varepsilon, \xi_\varepsilon, f_\varepsilon, g_\varepsilon)$ correspond to the solution of the optimal control problem (3.66) for fixed ε, i.e. $(f_\varepsilon, g_\varepsilon)$ is the solution of the problem, and $(\chi_\varepsilon, \xi_\varepsilon)$ is defined from (3.48) with $(f, g) = (f_\varepsilon, g_\varepsilon)$. The following result takes place.

Theorem 3.8. *From the sequence $(\chi_\varepsilon, \xi_\varepsilon, f_\varepsilon, g_\varepsilon)$ one can choose a subsequence, with the same notation, such that as $\varepsilon \to 0$*

$$(\chi_\varepsilon, \xi_\varepsilon) \to (\chi, \xi) \quad \textit{weakly in } H(\Omega_\psi) \times H_0(\Omega),$$

$$f_\varepsilon, g_\varepsilon \to f, g \quad \textit{weakly in } L^2(\Omega), \quad m_\varepsilon \to m_0.$$

Here (χ, ξ, f, g) corresponds to the solution of the control problem (3.75) *and*

$$m_\varepsilon = \inf_{F \times G} J_\varepsilon(f, g), \quad m_0 = \inf_{F \times G} J_0(f, g).$$

PROOF. Let $\chi_\varepsilon(f, g)$, $\xi_\varepsilon(f, g)$ be the solutions of the inequality (3.48) with the functions f, g. We take $(\bar{\chi}, \bar{\xi}) \in K_{\varepsilon_0}$. Then $(\bar{\chi}, \bar{\xi}) \in K_\varepsilon$ for all $\varepsilon \leq \varepsilon_0$. Substitution of $(\bar{\chi}, \bar{\xi})$ in (3.48) as a test function implies the estimate

$$\|\chi_\varepsilon(f, g)\|_{H(\Omega_\psi)} + \|\xi_\varepsilon(f, g)\|_{H_0(\Omega)} \leq c \qquad (3.76)$$

uniform in $\varepsilon \leq \varepsilon_0$. Choosing a subsequence, if necessary, one can assume that as $\varepsilon \to 0$

$$\chi_\varepsilon(f, g) \to \tilde{\chi} \text{ weakly in } H(\Omega_\psi), \quad \xi_\varepsilon(f, g) \to \tilde{\xi} \text{ weakly in } H_0(\Omega), \quad (3.77)$$

$$\chi_\varepsilon^\pm(f,g) \to \tilde{\chi}^\pm \quad \text{strongly in } L^1(\Gamma_\psi), \tag{3.78}$$

$$\varepsilon \left\| \left[\frac{\partial w_\varepsilon(f,g)}{\partial \nu} \right] \right\| \to 0 \quad \text{strongly in } L^2(\Gamma_\psi). \tag{3.79}$$

By Lemma 3.2 below, we choose any fixed element $(\bar{\chi}, \bar{\xi}) \in K_0$ and construct a sequence $(\bar{\chi}_\varepsilon, \bar{\xi}_\varepsilon) \in K_\varepsilon$ strongly converging in $H(\Omega_\psi) \times H_0(\Omega)$ to the element $(\bar{\chi}, \bar{\xi})$. We next substitute the elements of this sequence as test functions in the inequality

$$\Pi_f'(\chi_\varepsilon)(\bar{\chi} - \chi_\varepsilon) + \Pi_g'(\xi_\varepsilon)(\bar{\xi} - \xi_\varepsilon) \geq 0, \quad (\chi_\varepsilon, \xi_\varepsilon) \in K_\varepsilon, \quad \forall (\bar{\chi}, \bar{\xi}) \in K_\varepsilon.$$

By (3.77), it is possible to pass to the limit as $\varepsilon \to 0$ in this inequality. Condition (3.79) provides the inclusion $(\tilde{\chi}, \tilde{\xi}) \in K_0$. The limiting variational inequality takes the form

$$\Pi_f'(\tilde{\chi})(\bar{\chi} - \tilde{\chi}) + \Pi_g'(\tilde{\xi})(\bar{\xi} - \tilde{\xi}) \geq 0, \quad (\tilde{\chi}, \tilde{\xi}) \in K_0, \quad \forall (\bar{\chi}, \bar{\xi}) \in K_0$$

which means $\tilde{\chi} = \chi(f,g)$, $\tilde{\xi} = \xi(f,g)$. Consequently, by (3.78), we have

$$J_\varepsilon(f,g) \to J_0(f,g), \quad \varepsilon \to 0. \tag{3.80}$$

Let (f,g) be the solution of the control problem (3.75), (3.74). By (3.80), we obtain $m_\varepsilon \leq J_\varepsilon(f,g) \to J_0(f,g) = m_0$, and hence

$$\limsup m_\varepsilon \leq m_0. \tag{3.81}$$

On the other hand, by the boundedness of the set $F \times G$ in the space $L^2(\Omega_\psi) \times L^2(\Omega)$, the inequality

$$\|(f_\varepsilon, g_\varepsilon)\|_{L^2(\Omega)} \leq c \tag{3.82}$$

is fulfilled being uniform in ε. Consequently, the variational inequalities

$$\Pi_{f_\varepsilon}'(\chi_\varepsilon)(\bar{\chi} - \chi_\varepsilon) + \Pi_{g_\varepsilon}'(\xi_\varepsilon)(\bar{\xi} - \xi_\varepsilon) \geq 0, \tag{3.83}$$

$$(\chi_\varepsilon, \xi_\varepsilon) \in K_\varepsilon, \quad \forall (\bar{\chi}, \bar{\xi}) \in K_\varepsilon$$

yield the uniform in ε estimate

$$\|\chi_\varepsilon\|_{H(\Omega_\psi)} + \|\xi_\varepsilon\|_{H_0(\Omega)} \leq c. \tag{3.84}$$

By (3.82), (3.84), we can assume that as $\varepsilon \to 0$

$$f_\varepsilon, g_\varepsilon \to f, g \quad \text{weakly in } L^2(\Omega), \tag{3.85}$$

$$(\chi_\varepsilon, \xi_\varepsilon) \to (\chi_0, \xi_0) \text{ weakly in } H, \text{ strongly in } L^2(\Omega), \tag{3.86}$$

$$\varepsilon \left\| \left[\frac{\partial w_\varepsilon}{\partial \nu} \right] \right\| \to 0 \quad \text{strongly in } L^2(\Gamma_\psi), \tag{3.87}$$

$$\chi_\varepsilon^{\pm}(f,g) \to \tilde{\chi}_0^{\pm} \quad \text{strongly in } L^1(\Gamma_\psi). \tag{3.88}$$

Again, by Lemma 3.2, we pass to the limit in (3.83) as $\varepsilon \to 0$. As a result, the following variational inequality is obtained,

$$\Pi'_{f_0}(\chi_0)(\bar{\chi} - \chi_0) + \Pi'_{g_0}(\xi_0)(\bar{\xi} - \xi_0) \geq 0, \quad (\chi_0, \xi_0) \in K_0, \quad \forall\, (\bar{\chi}, \bar{\xi}) \in K_0,$$

which means $\chi_0 = \chi(f_0, g_0)$, $\xi_0 = \xi(f_0, \xi_0)$.

As before, we can show that $J_\varepsilon(f_\varepsilon, g_\varepsilon) \to J_0(f_0, g_0)$ as $\varepsilon \to 0$, and consequently

$$\liminf m_\varepsilon \geq J_0(f_0, g_0). \tag{3.89}$$

It follows from (3.81), (3.89) that (f_0, g_0) is a solution of the optimal control problem (3.75), (3.74) and $m_\varepsilon \to m_0$. The proof of Theorem 3.8 is complete.

Now we have to justify an auxiliary statement which has been used in the proof of Theorem 3.8. Let us recall that Γ_ψ is a segment of the axis x.

Lemma 3.2. *For any fixed element* $(\bar{\chi}, \bar{\xi}) \in K_0$ *there exists a sequence* $(\bar{\chi}_\varepsilon, \bar{\xi}_\varepsilon) \in K_\varepsilon$ *such that*

$$(\bar{\chi}_\varepsilon, \bar{\xi}_\varepsilon) \to (\bar{\chi}, \bar{\xi}) \quad \textit{strongly in } H(\Omega_\psi) \times H_0(\Omega).$$

PROOF. Consider a smooth extension of the graph Γ_ψ beyond the point $x = 1$. In so doing we assume that the angle between the boundary Γ and the extended graph is positive. The domain Ω_ψ is divided into two subdomains Ω_1, Ω_2 with Lipschitz boundaries $\partial\Omega_1$, $\partial\Omega_2$. Of course, in the case under consideration the boundaries Γ_ψ^+, Γ_ψ^- are different sets. The inclusion $(\bar{\chi}, \bar{\xi}) \in K_0$ means that the following inequalities are holding true,

$$[\bar{W}]\nu \geq 0 \quad \text{on } \Gamma_\psi, \quad \bar{w} \geq \bar{u} - \delta \quad \text{in } \Omega_\psi,$$

and the inclusion $(\bar{\chi}_\varepsilon, \bar{\xi}_\varepsilon) \in K_\varepsilon$ means that

$$[\bar{W}_\varepsilon]\nu \geq \varepsilon|[\partial\bar{w}_\varepsilon/\partial\nu]| \quad \text{on } \Gamma_\psi, \quad \bar{w}_\varepsilon \geq \bar{u}_\varepsilon - \delta \quad \text{in } \Omega_\psi.$$

To complete the proof of the lemma it suffices to construct a sequence $(\bar{\chi}_\varepsilon, \bar{\xi}_\varepsilon)$ such that $\bar{\xi}_\varepsilon = \bar{\xi}$ and $(\bar{\chi}_\varepsilon, \bar{\xi}) \in K_\varepsilon$,

$$\bar{\chi}_\varepsilon \to \bar{\chi} \quad \text{strongly in } H(\Omega_\psi). \tag{3.90}$$

We have to note at this point that a sequence $\bar{\xi}_\varepsilon, \bar{\chi}$ will be a necessary one provided that we can construct a function $\tilde{W} \in [H^{1,0}(\Omega_\psi)]^2$ such that

$$[\tilde{W}]\nu = |[\partial\bar{w}/\partial\nu]| \quad \text{on } \Gamma_\psi \tag{3.91}$$

and the functions $\bar{\chi}_\varepsilon = (\bar{W}_\varepsilon, \bar{w}_\varepsilon)$ are defined in Ω_ψ by the formula

$$(\bar{W}_\varepsilon, \bar{w}_\varepsilon) = (\bar{W} + \varepsilon\tilde{W}, \bar{w}).$$

Indeed, the convergence (3.90) is obvious and, besides,

$$[\bar{W}_\varepsilon]\nu \geq \varepsilon|[\partial\bar{w}_\varepsilon/\partial\nu]| \quad \text{on } \Gamma_\psi, \quad \bar{w}_\varepsilon \geq \bar{u} - \delta \quad \text{in } \Omega_\psi.$$

So, we have to construct a function $\tilde{W}$ with the above property (3.91). Note that $\nu = (0,1)$ on Γ_ψ. Since $\bar{w} \in H^2(\Omega_\psi)$, we have $\bar{w}_y|_{\Omega_i} \in H^1(\Omega_i)$ $(i = 1,2)$, and consequently $\bar{w}_y|_{\partial\Omega_i} \in H^{1/2}(\partial\Omega_i)$, $i = 1,2$ (see Baiocchi, Capelo, 1984). Consider the following function on the boundary $\partial\Omega_1$,

$$h_\gamma(\bar{x}) = \begin{cases} \min\{-[\bar{w}_y(\bar{x})], [\bar{w}_y(\bar{x})]\} & , \quad \bar{x} \in \Gamma_\psi, \\ 0 & , \quad \bar{x} \notin \Gamma_\psi. \end{cases}$$

Then $h_\gamma \in H^{1/2}(\partial\Omega_1)$. Let $h \in H^1(\Omega_1)$ be an extension of h_γ into the domain Ω_1. Note that zero extension of h into Ω_2 gives the function defined in Ω_ψ, which belongs to $H^1(\Omega_\psi)$. This extension into Ω_ψ is again denoted by h. Now we are in a position to define a vector-function $\tilde{W}$ by the formula $\tilde{W} = (0, h)$ in Ω_ψ. In this case we notice that

$$[\tilde{W}]\nu = \max\{-[\bar{w}_y], [\bar{w}_y]\} = |[\bar{w}_y]| \quad \text{on } \Gamma_\psi$$

and $|[\bar{w}_y]| = |[\partial\bar{w}/\partial\nu]|$ on Γ_ψ.

Consequently, we have built the function $\tilde{W} \in [H^{1,0}(\Omega_\psi)]^2$ with the property (3.91), which completes the proof.

3.3 Thermoelastic plates with cracks

In this section we consider the boundary value problem for model equations of a thermoelastic plate with a vertical crack (see Khludnev, 1996d). The unknown functions in the mathematical model under consideration are such quantities as the temperature θ and the horizontal and vertical displacements $W = (w^1, w^2)$, w of the mid-surface points of the plate. We use the so-called coupled model of thermoelasticity, which implies in particular that we need to solve simultaneously the equations that describe heat conduction and the deformation of the plate. The presence of the crack leads to the fact that the domain of a solution has a nonsmooth boundary. As before, the main feature of the problem as a whole is the existence of a constraint in the form of an inequality imposed on the crack faces. This constraint provides a mutual nonpenetration of the crack faces:

$$[W]\nu \geq \left|\left[\frac{\partial w}{\partial \nu}\right]\right|.$$

Here $[\cdot]$ is the jump of a function across the crack faces and ν is the normal to the surface describing the shape of the crack. Thus, we have to find a solution to the model equations of a thermoelastic plate in a domain with nonsmooth boundary and boundary conditions of the inequality type.

We prove the solvability of the problem. We also find boundary conditions holding on the crack faces and having the form of a system of equations and inequalities and establish some enhanced regularity properties for the solution near the points of the crack. Some other results on thermoelasic problems can be found in (Gilbert et al., 1990; Zuazua, 1995).

3.3.1 Statement of the problem

Let $\Omega \subset R^2$ be a bounded domain with smooth boundary Γ, $\Omega_\psi = \Omega \setminus \Gamma_\psi$, and Γ_ψ be the graph of the function $y = \psi(x)$, $x \in [0,1]$, $(x,y) \in \Omega$. We assume that the mid-surface of the plate coincides with Ω_ψ and that the function ψ describing the shape of the crack on the plane x, y is sufficiently smooth. The crack is assumed vertical. This means that its surface can be given as $y = \psi(x)$, $-h \le z \le h$, where z is the distance to the mid-surface of the plate and $2h$ is the thickness of the plate. We denote by $W = (w^1, w^2)$, w the horizontal and vertical displacements of the mid-surface points of the plate, respectively, and by θ the temperature in the plate. We put $\chi = (W, w)$, $Q_\psi = \Omega_\psi \times (0,T)$, $T > 0$.

In the domain Q_ψ, we consider the following equations describing quasistatic deformation of a plate:

$$\frac{\partial \theta}{\partial t} - \Delta\theta + \delta^2 \frac{\partial}{\partial t}(\operatorname{div} W - \Delta w) = f, \tag{3.92}$$

$$-\sigma_{ij,j} + \delta^2 \theta_{,i} = 0, \quad i = 1,2, \tag{3.93}$$

$$\Delta^2 w + \delta^2 \Delta\theta = 0. \tag{3.94}$$

Here δ is a positive parameter, $\sigma_{ij} = \sigma_{ij}(W)$, $i,j = 1,2$, and

$$\sigma_{11} = \varepsilon_{11} + \kappa\varepsilon_{22}, \quad \sigma_{22} = \varepsilon_{22} + \kappa\varepsilon_{11}, \quad \sigma_{12} = (1-\kappa)\varepsilon_{12},$$

$$\kappa = \text{ const }, \quad 0 < \kappa < 1/2.$$

The symbols $\varepsilon_{ij} = \varepsilon_{ij}(W)$ stand for the components of the strain tensor of the mid-plane of the plate:

$$\varepsilon_{ij}(W) = \frac{1}{2}\left(\frac{\partial w^i}{\partial x_j} + \frac{\partial w^j}{\partial x_i}\right), \quad i,j = 1,2, \quad x_1 = x,\ x_2 = y.$$

The system of equations (3.92)–(3.94) is a model one. More precise (and more bulky) equations for a thermoelastic plate can be found, for instance, in (Nowacki, 1962).

Henceforth we denote by $\nu = (-\psi_x, 1)/\sqrt{1+\psi_x^2}$ the normal to the graph Γ_ψ, $\nu = (\nu_1, \nu_2)$ and $\Gamma_\psi^T = \Gamma_\psi \times (0,T)$. Consequently, the nonpenetration condition on the crack faces can be written as

$$[W]\nu \ge h\left|\left[\frac{\partial w}{\partial \nu}\right]\right| \quad \text{on } \Gamma_\psi^T. \tag{3.95}$$

For simplicity, we further assume $h = 1$. Let some initial temperature distribution be given:

$$\theta = \theta_0 \quad \text{at } t = 0. \tag{3.96}$$

Assume that the temperature and the clamping condition are given on the exterior boundary of the plate:

$$\theta = w = \frac{\partial w}{\partial n} = W = 0 \quad \text{on } \Gamma \times (0, T). \tag{3.97}$$

Let $H^{1,0}(\Omega_\psi)$ stand for the subspace of the Sobolev space $H^1(\Omega_\psi)$ which comprises the functions vanishing on Γ; let $H^{2,0}(\Omega_\psi)$ consist of the functions vanishing on Γ together with their first-order derivatives, $H^{2,0}(\Omega_\psi) \subset H^2(\Omega_\psi)$; and let $H(\Omega_\psi) = H^{1,0}(\Omega_\psi) \times H^{1,0}(\Omega_\psi) \times H^{2,0}(\Omega_\psi)$. We denote the norm in $H^{s,0}(\Omega_\psi)$ by $\|\cdot\|_s$.

Introduce the sets

$$K = \{\chi = (W, w) \in H(\Omega_\psi) \mid \; [W]\nu \geq |[\partial w/\partial \nu]| \quad \text{a.e. on } \Gamma_\psi\},$$

$$\mathcal{K} = \{\chi \in L^2(0, T; H(\Omega_\psi)) \mid \; \chi(t) \in K \quad \text{a.e. on } (0, T)\}$$

of feasible displacements and introduce the bilinear forms

$$B(W, \widetilde{W}) = \langle \sigma_{ij}(W), \varepsilon_{ij}(\widetilde{W}) \rangle,$$

$$b(w, \tilde{w}) = \int\limits_{\Omega_\psi} (w_{xx}\tilde{w}_{xx} + w_{yy}\tilde{w}_{yy} + \kappa w_{xx}\tilde{w}_{yy} + \kappa w_{yy}\tilde{w}_{xx}$$

$$+ 2(1-\kappa) w_{xy}\tilde{w}_{xy}),$$

where $\langle \cdot, \cdot \rangle$ stands for integration over Ω_ψ.

The equilibrium problem for a plate is formulated as some variational inequality. In this case equations (3.92)–(3.94) hold, generally speaking, only in the distribution sense. Alongside (3.95), other boundary conditions hold on the boundary Γ_ψ^T; the form of these conditions is clarified in Section 3.3.3. To derive them, we require the existence of a smooth solution to the variational inequality in question. On the other hand, if we assume that a solution to (3.92)–(3.94) is sufficiently smooth, then the variational inequality is a consequence of equations (3.92)–(3.94) and the initial and boundary conditions. All these questions are discussed in Section 3.3.3. In Section 3.3.2 we prove an existence theorem for a solution to the variational equation and in Section 3.3.4 we establish some enhanced regularity properties for the solution near Γ_ψ^T.

3.3.2 Existence of a solution

We introduce the space $\Xi = \{\theta \in L^2(0, T; H^{1,0}(\Omega_\psi)) \mid \theta_t \in L^2(Q_\psi)\}$ with the norm

$$\|\theta\|_\Xi^2 = \|\theta\|^2_{L^2(0,T;H^{1,0}(\Omega_\psi))} + \|\theta_t\|^2_{L^2(Q_\psi)}.$$

Henceforth it is convenient to use the notations $H = H^1(0,T;H(\Omega_\psi))$ and $U = \Xi\times H$. We shall assume that $\theta_0 \in H^{1,0}(\Omega_\psi)$. Observe that each element $\theta \in \Xi$ has a well-defined trace at $t = 0$; in particular, $\theta(0) \in L^2(\Omega_\psi)$. The operation of taking a trace acts continuously from Ξ into $L^2(\Omega_\psi)$. Consider the closed convex set

$$S = \{(\theta,\chi) \in U \mid \quad \theta(0) = \theta_0 \text{ in } \Omega_\psi, \quad \chi \in \mathcal{K}\}$$

in U. Let $U^\star$ denote the space dual of U. Consider the bounded linear operator $L : U \to U^\star$ acting by the formula

$$\{L(\theta,\chi),(\bar\theta,\bar\chi)\} = \int\limits_{Q_\psi} \left(\theta_t + \delta^2\frac{\partial}{\partial t}(\operatorname{div} W - \Delta w)\right)\bar\theta$$

$$+\int\limits_{Q_\psi} \nabla\theta\nabla\bar\theta + \int\limits_0^T (B(W,\widetilde{W}) + b(w,\widetilde{w}) + \delta^2\langle\theta,\Delta\widetilde{w}\rangle - \delta^2\langle\theta,\operatorname{div}\widetilde{W}\rangle).$$

The braces $\{\cdot,\cdot\}$ stand for the duality pairing between U and $U^\star$.

We can now give an exact statement of the equilibrium problem for a plate. Suppose that $f \in L^2(Q_\psi)$. An element $(\theta,\chi) \in U$ is said to be a solution to the equilibrium problem for a thermoelastic plate with a crack if it satisfies the variational inequality

$$\{L(\theta,\chi),(\bar\theta,\bar\chi) - (\theta,\chi)\} \geq \int\limits_{Q_\psi} f(\bar\theta - \theta), \quad (\theta,\chi) \in S \quad \forall(\bar\theta,\bar\chi) \in S. \tag{3.98}$$

The rationale of this definition of a solution will become clear in the sequel. Observe that the operator L is pseudomonotone (see a definition in Section 1.2) but is not coercive on U. Therefore, solvability of problem (3.98) does not follow from known results.

Theorem 3.9. *For δ small enough, there is a solution to problem* (3.98).

Proof. To prove the existence of a solution, we implement the idea that was earlier used in a simpler case by (Shi, Shillor, 1992). We introduce two closed convex sets

$$S_1 = \{\theta \in \Xi \mid \quad \theta(0) = \theta_0\}, \quad S_2 = \{\chi \in H \mid \quad \chi \in \mathcal{K}\}$$

in the spaces Ξ and H, respectively. Substituting for feasible functions in (3.98) first $(\bar\theta,\chi)$ and next $(\theta,\bar\chi)$, we obtain two variational inequalities

$$\int\limits_{Q_\psi} \left(\frac{\partial\theta}{\partial t} + \delta^2\frac{\partial}{\partial t}(\operatorname{div} W - \Delta w) - f\right)(\bar\theta - \theta) + \int\limits_{Q_\psi} \nabla\theta(\nabla\bar\theta - \nabla\theta) \geq 0, \tag{3.99}$$

$$\theta \in S_1, \quad \forall\bar\theta \in S_1,$$

$$\int_0^T \Big(B(W, \widetilde{W} - W) + b(w, \widetilde{w} - w) + \delta^2 \langle \theta, \Delta \widetilde{w} - \Delta w \rangle \tag{3.100}$$

$$-\delta^2 \langle \theta, \operatorname{div} \widetilde{W} - \operatorname{div} W \rangle \Big) \geq 0, \quad (W, w) \in S_2 \quad \forall (\widetilde{W}, \widetilde{w}) \in S_2.$$

Summing (3.99) and (3.100), we clearly obtain exactly (3.98). Therefore, to prove the solvability of variational inequality (3.98), it suffices to establish solvability for the coupled system of variational inequalities (3.99) and (3.100).

Let $\chi \in S_2$ be an arbitrary fixed element. Using the Galerkin method (see, for instance, Mikhailov, 1976), we can prove that there is a unique function $\theta \in S_1$ satisfying the identity

$$\int_{Q_\psi} \left(\frac{\partial \theta}{\partial t} + \delta^2 \frac{\partial}{\partial t} (\operatorname{div} W - \Delta w) - f \right) \bar{\theta} + \int_{Q_\psi} \nabla \theta \nabla \bar{\theta} = 0 \tag{3.101}$$

$$\forall \bar{\theta} \in L^2(0, T; H^{1,0}(\Omega_\psi)).$$

Moreover, the estimate

$$\|\theta\|_\Xi \leq c_1 \delta \|\chi\|_H + c_2 \tag{3.102}$$

holds, with the constant c_2 depending on the norm of f in $L^2(Q_\psi)$ and the $H^1(\Omega_\psi)$-norm of θ_0, and c_1, c_2 independent of δ, $\delta \leq \delta_0$. From (3.101) we easily see that the function θ satisfies equation (3.92) in Q_ψ in the distribution sense; in particular, $\Delta \theta \in L^2(Q_\psi)$.

We can take a function of the form $\bar{\theta} - \theta$ as $\bar{\theta}$ in (3.101), where $\bar{\theta} \in S_1$. This yields the unique solvability in θ of variational inequality (3.99) for every fixed $\chi \in S_2$. On the other hand, for every fixed $\theta \in S_1$ the problem of minimizing the functional

$$\int_0^T \left(B(W, W) + b(w, w) + 2\delta^2 \langle \theta, \Delta w \rangle - 2\delta^2 \langle \theta, \operatorname{div} W \rangle \right) \tag{3.103}$$

over the set $\mathcal{K}$ has a unique solution. To verify this, it suffices to observe that the inequalities

$$b(w, w) \geq c\|w\|_2^2 \quad \forall w \in H^{2,0}(\Omega_\psi), \tag{3.104}$$

$$B(W, W) \geq c\|W\|_1^2 \quad \forall\ W = (w^1, w^2) \in H^{1,0}(\Omega_\psi) \tag{3.105}$$

hold in Ω_ψ with constants independent of the functions W and w, respectively. In particular, for a fixed $\theta \in S_1$, functional (3.103) is coercive (and weakly lower semicontinuous) on the space $L^2(0, T; H(\Omega_\psi))$. The element

$\chi = (W, w) \in \mathcal{K}$ at which functional (3.103) attains a minimum satisfies the variational inequality

$$\int_0^T \Big(B(W, \widetilde{W} - W) + b(w, \widetilde{w} - w) + \delta^2 \langle \theta, \Delta \widetilde{w} - \Delta w \rangle \tag{3.106}$$

$$- \delta^2 \langle \theta, \operatorname{div} \widetilde{W} - \operatorname{div} W \rangle \Big) \geq 0 \quad \forall (\widetilde{W}, \widetilde{w}) \in \mathcal{K}.$$

Given $\theta \in S_1$, a solution χ to problem (3.106) is unique. Choosing $(\widetilde{W}, \widetilde{w}) = 0$ in (3.106), we easily derive the estimate

$$||\chi||^2_{L^2(0,T;H(\Omega_\psi))} \leq c\delta^2 ||\theta||^2_{L^2(Q_\psi)} \tag{3.107}$$

with some constant uniform in δ, $\delta \leq \delta_0$. From (3.106) we also conclude that equations (3.93) and (3.94) hold in Q_ψ in the distribution sense.

Involving difference relations, we can establish additional smoothness in t for the solution $\chi = (W, w)$ to problem (3.106). We introduce the notations

$$\chi^t = \chi(t), \quad d_\tau \chi^t = \chi^{t+\tau} - \chi^t \tau.$$

It follows from (3.106) that the inequality

$$B(W^t, \widetilde{W} - W^t) + b(w^t, \widetilde{w} - w^t) + \delta^2 \langle \theta^t, \Delta \widetilde{w} - \Delta w^t \rangle \tag{3.108}$$

$$- \delta^2 \langle \theta^t, \operatorname{div} \widetilde{W} - \operatorname{div} W^t \rangle \geq 0, \quad (W^t, w^t) \in K, \quad \forall (\widetilde{W}, \widetilde{w}) \in K$$

holds for almost every $t \in (0, T)$. For definiteness, assume that $\tau > 0$. In inequality (3.108), take $\bar{\chi} = \chi^{t+\tau}$, $0 < t < T - \tau$; then write down (3.108) at the point $t+\tau$; and next take the function χ^t as $\bar{\chi}$. Summing the so-obtained relations and dividing the result by τ^2, we find

$$B(d_\tau W^t, d_\tau W^t) + b(d_\tau w^t, d_\tau w^t) \tag{3.109}$$

$$\leq -\delta^2 \langle d_\tau \theta^t, \Delta d_\tau w^t \rangle + \delta^2 \langle d_\tau \theta^t, \operatorname{div} d_\tau W^t \rangle.$$

Integrate (3.109) with respect to t from 0 to $T - \tau$. Taking (3.104), (3.105) into account, we can easily derive the inequality

$$\int_0^{T-\tau} ||d_\tau \chi^t||^2_{H(\Omega_\psi)} \leq c\delta^2 \int_0^{T-\tau} ||d_\tau \theta^t||^2_0 \tag{3.110}$$

with some constant c independent of τ and δ, $\delta \leq \delta_0$. Since $\theta_t \in L^2(Q_\psi)$, on letting $\tau \to 0$, from (3.110) we infer that

$$||\chi_t||^2_{L^2(0,T;H(\Omega_\psi))} \leq c\delta^2 ||\theta_t||^2_{L^2(Q_\psi)}. \tag{3.111}$$

From (3.107), (3.111) we conclude that the estimate

$$\|\chi\|_H \le c_3\delta\|\theta\|_\Xi \tag{3.112}$$

holds with some constant c_3 independent of δ, $\delta \le \delta_0$.

Observe that variational inequality (3.106) is valid for every function $\bar{\chi} \in S_2$. It means that a solution χ to problem (3.106) with $\theta \in S_1$ coincides with the unique solution to problem (3.100) with the same θ; i.e. problems (3.100) and (3.106) are equivalent. For small δ, we write down an extra variational inequality for which a solution exists, and demonstrate that the solution coincides with the solution of variational inequality (3.98).

Let $c_* = \max\{c_1, c_2, c_3\}$, where c_i are taken from (3.102) and (3.112). Assume the parameter δ to be so small that the quantities

$$m_1 = \frac{c_*}{1-(\delta c_*)^2}, \quad m_2 = \frac{\delta c_*^2}{1-(\delta c_*)^2}$$

satisfy the conditions

$$m_1 > 0, \quad m_2 > 0. \tag{3.113}$$

Obviously, for (3.113) to hold, it suffices to choose δ from the condition $\delta < c_*^{-1}$.

We define the bounded closed convex set

$$S^0 = \{(\theta,\chi) \in S \mid \ \|\theta\|_\Xi \le m_1, \quad \|\chi\|_H \le m_2\}$$

in the space U. Since S^0 is a bounded set and L is a pseudomonotone operator, there is a solution to the problem due to Theorem 1.16:

$$\{L(\theta,\chi), (\bar\theta,\bar\chi)-(\theta,\chi)\} \ge \int\limits_{Q_\psi} f(\bar\theta-\theta), \quad (\theta,\chi) \in S^0, \ \forall\, (\bar\theta,\bar\chi) \in S^0. \tag{3.114}$$

By analogy, we define the sets

$$S_1^0 = \{\theta \in S_1 \mid \ \|\theta\|_\Xi \le m_1\}, \quad S_2^0 = \{\chi \in S_2 \mid \ \|\chi\|_H \le m_2\}. \tag{3.115}$$

Then problem (3.114) can be written down in equivalent form by means of the following two variational inequalities:

$$\int\limits_{Q_\psi} \left(\frac{\partial\theta}{\partial t} + \delta^2\frac{\partial}{\partial t}(\operatorname{div} W - \Delta w) - f\right)(\bar\theta-\theta) + \int\limits_{Q_\psi} \nabla\theta(\nabla\bar\theta - \nabla\theta) \ge 0, \tag{3.116}$$

$$\theta \in S_1^0, \quad \forall \bar\theta \in S_1^0;$$

$$\int\limits_0^T \Big(B(W, \widetilde{W} - W) + b(w, \widetilde{w} - w) + \delta^2\langle\theta, \Delta\widetilde{w} - \Delta w\rangle \tag{3.117}$$

$$-\delta^2\langle\theta, \operatorname{div}\widetilde{W} - \operatorname{div} W\rangle\Big) \geq 0, \quad (W, w) \in S_2^0, \quad \forall\, (\widetilde{W}, \widetilde{w}) \in S_2^0.$$

Now, we are ready to convince ourselves that the coupled system of variational inequalities (3.99) and (3.100) is solvable under condition (3.113). As mentioned, (3.99) and (3.100) yield (3.98).

Indeed, let (θ, χ) be a solution to problem (3.114). Then $\chi \in S_2^0$. We define $\tilde{\theta}$ as a solution to (3.99) with the chosen χ. In accordance with (3.113) and estimate (3.102), we obtain $\tilde{\theta} \in S_1^0$. By the uniqueness of a solution to problem (3.116) given a fixed $\chi \in S_2^0$, we conclude that $\tilde{\theta}$ solves problem (3.116); i.e. $\tilde{\theta} = \theta$. At the same time, $\theta \in S_1^0$. Therefore, the function $\tilde{\chi}$ found from (3.100) at this θ belongs to S_2^0 by virtue of (3.112) and (3.113); i.e. it is a solution to problem (3.117). This means that $\tilde{\chi} = \chi$. It follows that the solution (θ, χ) to problem (3.114) is a solution to problem (3.99), (3.100) and hence is a solution to (3.98). Theorem 3.9 is completely proved.

Observe that, generally speaking, the smallness of δ does not imply that the solution (θ, χ) is small in the norm of the space U.

3.3.3 Boundary conditions at the crack faces

In this subsection we establish the exact form of boundary conditions holding on Γ_ψ^T for the solution (θ, χ) to problem (3.98). The arguments are formal in the sense that the solution is assumed sufficiently smooth. For brevity, hereafter we denote the quantities W^t, w^t, and θ^t by W, w, and θ, indicating each time the value of the variable t at which the corresponding relations hold. Moreover, we assume that $\delta = 1$.

Let the quantities $M(w)$, $R(w)$, $\sigma_\nu(W)$, and $\sigma_s(W)$ be defined in accordance with formulas (3.123) and (3.127) (see below). Prove that, in addition to (3.95)–(3.97), the solution (θ, χ) of problem (3.98) satisfies the following boundary conditions for $t \in (0, T)$:

$$\frac{\partial \theta}{\partial \nu} = 0 \quad \text{on } \Gamma_\psi, \tag{3.118}$$

$$[\sigma_\nu(W) - \theta] = 0, \quad \sigma_s(W) = 0 \quad \text{on } \Gamma_\psi, \tag{3.119}$$

$$[M(w) + \theta] = 0, \quad R(w) = 0 \quad \text{on } \Gamma_\psi, \tag{3.120}$$

$$|M(w) + \theta| \leq -(\sigma_\nu(W) - \theta) \quad \text{on } \Gamma_\psi, \tag{3.121}$$

$$(M(w) + \theta)\left[\frac{\partial w}{\partial \nu}\right] + (\sigma_\nu(W) - \theta)[W]\nu = 0 \quad \text{on } \Gamma_\psi. \tag{3.122}$$

Equality (3.118) is straightforward from identity (3.101); it means that

$$\frac{\partial \theta}{\partial \nu} = 0 \quad \text{on } \Gamma_\psi^+, \quad \frac{\partial \theta}{\partial \nu} = 0 \quad \text{on } \Gamma_\psi^-.$$

The second equalities in (3.119) and (3.120) are understood likewise.

We recall that equations (3.92)–(3.94) are satisfied only in the distribution sense. However, in this subsection we suppose that the solution is smooth enough, so that boundary conditions (3.118)–(3.122) should be regarded as formal consequences of (3.98). By the way, we point out that the results of the forthcoming subsection allow us to attach an exact meaning to some of the relations (3.118)–(3.122).

To test the validity of boundary conditions (3.119)–(3.122), we need two Green formulas. Let $\mathcal{O} \subset R^2$ be a bounded domain with smooth boundary γ and the outward normal $n = (n_1, n_2)$. Introduce the following operators on γ:

$$M(u) = \kappa\Delta u + (1-\kappa)\frac{\partial^2 u}{\partial n^2}, \quad R(u) = \frac{\partial}{\partial n}\Delta u + (1-\kappa)\frac{\partial^3 u}{\partial n \partial s^2}, \tag{3.123}$$

where $s = (-n_2, n_1)$. We know that the formula

$$b_{\mathcal{O}}(u,v) = \left\langle M(u), \frac{\partial v}{\partial n}\right\rangle_\gamma - \langle R(u), v\rangle_\gamma + \langle \Delta^2 u, v\rangle_{\mathcal{O}} \tag{3.124}$$

is valid for sufficiently smooth functions u and v. The subscripts $\mathcal{O}$ and γ signify that the integration is taken over the domain $\mathcal{O}$ and the boundary γ, respectively. Moreover, if $\varphi = (\varphi_1, \varphi_2)$ then

$$\langle \varphi, \nabla u\rangle_{\mathcal{O}} = \langle \varphi n, u\rangle_\gamma - \langle \operatorname{div}\varphi, u\rangle_{\mathcal{O}}. \tag{3.125}$$

The boundary $\partial\Omega_\psi$ of the domain Ω_ψ can be represented as the union of the components Γ, Γ_ψ^+, and Γ_ψ^-. In this connection, we note that formulas like (3.124) and (3.125) are also valid for the domain Ω_ψ. To check this, it suffices to extend the graph Γ_ψ so that Ω_ψ be divided into two parts. On applying formulas (3.124) and (3.125) to both the parts, we can make sure that the formulas are also valid for Ω_ψ.

Fix an arbitrary value $t \in (0,T)$ in (3.108) and choose test functions of the form $(\widetilde{W}, w)$. We come to the inequality

$$B(W, \widetilde{W} - W) - \langle \theta, \operatorname{div}(\widetilde{W} - W)\rangle \ge 0 \tag{3.126}$$

valid for all functions $\widetilde{W}$ satisfying the condition

$$[\widetilde{W}]\nu \ge \left|\left[\frac{\partial w}{\partial \nu}\right]\right| \quad \text{on } \Gamma_\psi, \quad \widetilde{W} \in H^{1,0}(\Omega_\psi).$$

Expand the vector $\{\sigma_{ij}(W)\nu_j\}$, $i = 1,2$, on the boundary Γ_ψ^- in the sum of the normal component $\sigma_\nu(W)$ and the tangent component $\sigma_s(W)$:

$$\{\sigma_{ij}(W)\nu_j\} = \sigma_\nu(W)\nu + \sigma_s(W)s, \quad s = (-\nu_2, \nu_1). \tag{3.127}$$

An analogous representation is also valid on Γ_ψ^+. Insert test functions of the form $W + \widetilde{W}$ in (3.126), with $\widetilde{W} \in H^{1,0}(\Omega_\psi)$ and $[\widetilde{W}]\nu \ge 0$ on Γ_ψ, and make use of formulas like (3.124) and (3.125). Easy arguments lead to relations (3.119).

We can insert test functions of the form $(W, \tilde{w})$ in inequality (3.108) at a given $t \in (0, T)$. We arrive at the relation

$$b(w, \tilde{w} - w) + \langle \theta, \Delta\tilde{w} - \Delta w \rangle \geq 0 \tag{3.128}$$

valid for all functions $\tilde{w}$ satisfying the condition

$$[W]\nu \geq |[\partial\tilde{w}/\partial\nu]| \quad \text{on } \Gamma_\psi, \quad \tilde{w} \in H^{2,0}(\Omega_\psi).$$

In (3.128), choose test functions of the form $w + \varphi$, where φ is a function smooth in Ω_ψ having support in a neighbourhood of some fixed point on Γ_ψ, and such that $[\partial\varphi/\partial\nu] = 0$. Here $[\varphi] \neq 0$ in general. By virtue of formulas (3.124) and (3.125), we obtain (3.120).

Let $(\widetilde{W}, \tilde{w})$ be smooth functions in the set K. Substituting test functions of the form $(W, w) + (\widetilde{W}, \tilde{w})$ in (3.108) yields the inequality

$$B(W, \widetilde{W}) + b(w, \tilde{w}) + \langle \theta, \Delta\tilde{w} \rangle - \langle \theta, \operatorname{div} \widetilde{W} \rangle \geq 0.$$

Performing transformations by formulas like (3.124) and (3.125) and taking (3.118)–(3.120) into account, we come to

$$\left\langle M(w) + \theta, \left[\frac{\partial\tilde{w}}{\partial\nu}\right] \right\rangle_{\Gamma_\psi} + \langle \sigma_\nu(W) - \theta, [\widetilde{W}]\nu \rangle_{\Gamma_\psi} \leq 0. \tag{3.129}$$

In (3.129), we take $\widetilde{W}$ and $\tilde{w}$ as functions smooth in Ω_ψ having support in a neighbourhood of a fixed point on Γ_ψ, and such that $[\partial\tilde{w}/\partial\nu] = [\widetilde{W}]\nu$. Then in some neighbourhood of this point we shall have

$$M(w) + \theta + \sigma_\nu(W) - \theta \leq 0. \tag{3.130}$$

Analogously, choosing $\widetilde{W}$ and $\tilde{w}$ so as to have $[\partial\tilde{w}/\partial\nu] = -[\widetilde{W}]\nu$, we obtain

$$-(M(w) + \theta) + \sigma_\nu(W) - \theta \leq 0. \tag{3.131}$$

From (3.130) and (3.131) we infer exactly (3.121). From (3.121) we deduce in particular that $\sigma_\nu(W) - \theta \leq 0$.

Next, substitutions of the form $(\widetilde{W}, \tilde{w}) = 0$ and $(\widetilde{W}, \tilde{w}) = 2(W, w)$ into inequality (3.108) lead to the equality

$$\left\langle M(w) + \theta, \left[\frac{\partial w}{\partial\nu}\right] \right\rangle_{\Gamma_\psi} + \langle \sigma_\nu(W) - \theta, [W]\nu \rangle_{\Gamma_\psi} = 0. \tag{3.132}$$

Now, (3.122) ensues from (3.95), (3.121), and (3.132). Thus, the form of the boundary conditions on Γ_ψ^T is completely determined. Along with (3.95), for all $t \in (0, T)$ conditions (3.118)–(3.122) hold on Γ_ψ.

It is noteworthy that the original equilibrium problem for a plate with a crack can be stated twofold. On the one hand, it may be formulated as variational inequality (3.98). In this case all the above-derived boundary conditions are formal consequences of such a statement under the supposition of sufficient smoothness of a solution. On the other hand, the problem may be formulated as equations (3.92)–(3.94) given initial and boundary conditions (3.95)–(3.97) and (3.118)–(3.122). Furthermore, if we assume that a solution is sufficiently smooth then from (3.92)–(3.97) and (3.118)–(3.122) we can derive variational inequality (3.98).

Prove the last assertion. Let $(\widetilde{W}, \tilde{w})$ be a smooth function belonging to the set K. Multiply equations (3.93) and (3.94), taken at a fixed $t \in (0, T)$, by $\tilde{w}^i - w^i(t)$ and $\tilde{w} - w(t)$, respectively. Afterwards, integrate over Ω_ψ and apply formulas (3.124) and (3.125) on taking boundary conditions (3.97) and (3.118)–(3.120) into account. At the fixed t, we obtain (recall that in this subsection $\delta = 1$)

$$B(W, \widetilde{W} - W) + b(w, \tilde{w} - w) + \langle \theta, \Delta\tilde{w} - \Delta w\rangle - \langle \theta, \operatorname{div} \widetilde{W} - \operatorname{div} W\rangle$$

$$+ \left\langle M(w) + \theta, \left[\frac{\partial \tilde{w}}{\partial \nu}\right] - \left[\frac{\partial w}{\partial \nu}\right]\right\rangle_{\Gamma_\psi} + \langle \sigma_\nu(W) - \theta, [\widetilde{W}]\nu - [W]\nu\rangle_{\Gamma_\psi} = 0.$$

In accordance with boundary conditions (3.95), (3.121), and (3.122), the sum of the boundary integrals here is nonpositive, whence (3.108) ensues. From (3.108) we infer (3.100). Equation (3.92), together with initial and boundary conditions (3.96)–(3.97) and (3.118), leads to (3.99). As we noted many times, (3.99) and (3.100) imply (3.98), which proves the assertion that (3.92)–(3.97) and (3.118)–(3.122) yield (3.98).

3.3.4 Smoothness of a solution

In this subsection we demonstrate that smoothness of a solution to problem (3.98) near Γ_ψ^T is higher than that guaranteed by Theorem 3.9. As before, we assume $\delta = 1$ for simplicity.

Suppose that, near some fixed point $x^0 \in \Gamma_\psi \setminus \partial\Gamma_\psi$, the graph Γ_ψ is a straight line segment parallel to the x axis. Let $t^0 \in (0, T)$ be an arbitrary fixed point and let $R_\varepsilon \subset R^3$ denote the ball of a sufficiently small radius ε with centre (x^0, t^0). First, we examine the smoothness of the function $\chi = (W, w)$. Let D stand for a first-order derivative and let φ denote an arbitrary smooth function in $R_{2\varepsilon}$ such that $\varphi \equiv 0$ outside $R_{3\varepsilon/2}$, $0 \le \varphi \le 1$, and $\partial\varphi/\partial y = 0$ on Γ_ψ^T.

Theorem 3.10. *The following inclusion holds:*

$$D(\varphi\chi) \in L^2(0, T; H(\Omega_\psi)).$$

Proof. For $D = \partial/\partial t$, the claim of the theorem is already known. Therefore, it suffices to consider the cases $D = \partial/\partial x$ and $D = \partial/\partial y$.

Let e stand for the unit vector along the x axis and let $0 < |h| < \varepsilon/2$. Introduce the notations $d_{\pm h}u(\bar{x}) = h^{-1}(u(\bar{x}\pm he) - u(\bar{x}))$ and $\Delta_h = -d_{-h}d_h$, and define the functions

$$w_h^i = w^i + \frac{h^2}{2}\varphi^2\Delta_h w^i, \quad i = 1, 2, \quad w_h = w + \frac{h^2}{2}\varphi^2\Delta_h w.$$

The normal ν has coordinates $(0, 1)$ near the point x^0. Therefore, for ε small enough, the nonpenetration condition (3.95) on $\Gamma_\psi^T \cap R_{2\varepsilon}$ takes the form

$$[w^2] \geq \left|\left[\frac{\partial w}{\partial y}\right]\right|. \tag{3.133}$$

It is easy to verify that if a function u meets the inequality $u \geq 0$ on $\Gamma_\psi^T \cap R_{2\varepsilon}$, then $u + (h^2/2)\varphi^2\Delta_h u \geq 0$ on $\Gamma_\psi^T \cap R_{2\varepsilon}$. Indeed, take $(\bar{x}, t) \in \Gamma_\psi^T \cap R_{2\varepsilon}$. Then

$$u(\bar{x}, t) + \frac{h^2}{2}\varphi^2(\bar{x}, t)\Delta_h u(\bar{x}, t) = (1 - \varphi^2(\bar{x}, t))u(\bar{x}, t)$$
$$+ \frac{\varphi^2(\bar{x}, t)}{2}[u(\bar{x} - he, t) + u(\bar{x} + he, t)] \geq 0.$$

With this available, it is easy to ascertain that the vector (w_h^1, w_h^2, w_h) obeys constraint (3.133), i.e.

$$[w_h^2] \geq \left|\left[\frac{\partial w_h}{\partial y}\right]\right| \quad \text{on } \Gamma_\psi^T \cap R_{2\varepsilon}.$$

Consequently, for $W_h = (w_h^1, w_h^2)$ we have

$$[W_h]\nu \geq |[\partial w_h \partial y]| \quad \text{on } \Gamma_\psi^T,$$

i.e. $(W_h, w_h) \in \mathcal{K}$. Inserting $(\widetilde{W}, \widetilde{w}) = (W_h, w_h)$ in (3.106), we obtain

$$\int_0^T \Big(B(W, \varphi^2\Delta_h W) + b(w, \varphi^2\Delta_h w) \tag{3.134}$$
$$+ \langle\theta, \Delta\varphi^2\Delta_h w\rangle - \langle\theta, \operatorname{div}\varphi^2\Delta_h W\rangle\Big) \geq 0.$$

At the same time, it is easy to verify that the difference between the integrals

$$\int_0^T (B(W, \varphi^2\Delta_h W) + b(w, \varphi^2\Delta_h w))$$

and

$$-\int_0^T (B(d_h(\varphi W), d_h(\varphi W)) + B(d_h(\varphi w), d_h(\varphi w)))$$

can be estimated from the above by a summand on the right-hand side of inequality (3.135) to be written down below. The two last summands in (3.134) are minor in the sense that they admit an estimate from the above by the right-hand side of (3.135). Thus, (3.134) implies

$$\int_0^T (B(d_h(\varphi W), d_h(\varphi W)) + B(d_h(\varphi w), d_h(\varphi w))) \tag{3.135}$$

$$\leq c \int_0^T \left(\|\chi\|^2_{H(\Omega_\psi)} + \|\theta\|_1^2 + \|d_h(\varphi\chi)\|_{H(\Omega_\psi)} (\|\chi\|_{H(\Omega_\psi)} + \|\theta\|_1) \right),$$

where the constant c depends on φ and the domain Ω_ψ. Grounding on inequalities (3.104) and (3.105), from (3.135) we deduce that

$$\int_0^T \|d_h(\varphi\chi)\|^2_{H(\Omega_\psi)} \leq c$$

uniformly in h. In consequence,

$$\frac{\partial}{\partial x}(\varphi\chi) \in L^2(0, T; H(\Omega_\psi)). \tag{3.136}$$

Observe that equation (3.93) can be rewritten as $W_{yy} = G$, where $G \in L^2(R_\varepsilon \cap Q_\psi)$; thereby, diminishing ε if necessary, we can assume that

$$\frac{\partial}{\partial y}(\varphi W) \in L^2(0, T; H^{1,0}(\Omega_\psi)).$$

We thus prove for W that $D(\varphi W) \in L^2(0, T; H^{1,0}(\Omega_\psi))$.

To estimate the third-order derivatives of the function w with respect to y, we make use of the following fact (see Duvaut, Lions, 1972). Let $\mathcal{O} \subset R^2$ be a bounded domain with smooth boundary and let u be a distribution on $\mathcal{O}$ such that u, $Du \in H^{-1}(\mathcal{O})$. Then $u \in L^2(\mathcal{O})$ and there is a constant c, dependent on $\mathcal{O}$, such that

$$\|u\|_{L^2(\mathcal{O})} \leq c \left(\|u\|_{H^{-1}(\mathcal{O})} + \|Du\|_{H^{-1}(\mathcal{O})} \right).$$

It follows from (3.136) that $\partial(\varphi w)/\partial x \in L^2(0, T; H^{2,0}(\Omega_\psi))$. Therefore, near the point (x^0, t^0), the derivatives w_{xxx}, w_{yyx}, and w_{xxy} belong to L^2. Write down equation (3.94) as

$$w_{yyyy} = g.$$

By the above argument, for almost every $t \in (t_0 - \varepsilon/2, t_0 + \varepsilon/2)$ the functions $g(t)$, $w_{yyy}(t)$, and $w_{yyyx}(t)$ belong to $H^{-1}(\Omega_\psi \cap \mathcal{O})$, where $\mathcal{O}$ is some neighbourhood of the point x^0. Therefore, for almost every $t \in (t_0 - \varepsilon/4, t_0 + \varepsilon/4)$

and some neighbourhood $\mathcal{O}_1$ of the point x^0, $\tilde{\mathcal{O}}_1 \subset \mathcal{O}$, the functions $w_{yyy}(t)$ belong to $L^2(\Omega_\psi \cap \mathcal{O}_1)$ and the estimate

$$\|w_{yyy}(t)\|^2_{L^2(\Omega_\psi \cap \mathcal{O}_1)} \leq c\Big(\|w_{yyy}(t)\|^2_{H^{-1}(\Omega_\psi \cap \mathcal{O}_1)} \tag{3.137}$$

$$+\|w_{yyyy}(t)\|^2_{H^{-1}(\Omega_\psi \cap \mathcal{O}_1)} + \|w_{yyyx}(t)\|^2_{H^{-1}(\Omega_\psi \cap \mathcal{O}_1)}\Big)$$

holds with some constant independent of t. Integrating (3.137) with respect to t from $t_0 - \varepsilon/4$ to $t_0 + \varepsilon/4$, we reach the proof of the claim concerning w_{yyy}. Theorem 3.10 is completely proved.

To conclude the section, we also observe that, for the function φ above, we have the containment

$$\varphi\theta \in L^2(0, T; H^2(\Omega_\psi)). \tag{3.138}$$

Indeed, it follows from (3.101) that the identity

$$\langle \nabla\theta(t), \nabla\bar{\theta}\rangle = \langle F(t), \bar{\theta}\rangle \quad \forall \bar{\theta} \in H^{1,0}(\Omega_\psi)$$

holds for almost every $t \in (0, T)$, with $F \in L^2(Q_\psi)$. Therefore, for almost every $t \in (0, T)$, the following equation holds in Ω_ψ in the sense of (two-dimensional) distributions:

$$-\Delta\theta(t) = F(t). \tag{3.139}$$

Here t plays the role of a parameter. Thus, we can use the results on smoothness up to the boundary for solutions to elliptic equations of the form (3.139) (see Mikhailov, 1976). This yields (3.138).

3.4 Cracks of minimal opening in thermoelastic plates

In this section cracks of minimal opening are considered for thermoelastic plates. It is proved that the cracks of minimal opening provide an equilibrium state of the plate, which corresponds to the state without the crack. This means that such cracks do not introduce any singularity for the solution, and actually we have to solve a boundary value problem without the crack.

3.4.1 Problem formulation

Consider a bounded domain $\Omega \subset R^2$ with a smooth boundary Γ, $\Omega_\psi = \Omega \setminus \Gamma_\psi$, Γ_ψ is the graph of the function $y = \psi(x)$, $x \in [0, 1]$, $(x, y) \in \Omega$, $x_1 = x$, $x_2 = y$. We assume that the mid-surface of the plate coincides with Ω_ψ, and ψ is the smooth function. The plate is supposed to have a vertical crack. Its shape, as a surface in R^3, is defined by the cylinder surface $y = \psi(x)$, $-h \le z \le h$. Herewith z is the distance from the mid-surface, $2h$ is the thickness of the plate, and $z = 0$ corresponds to the mid-surface.

Denote next by $\chi = (W, w)$ the displacement vector, $Q_\psi = \Omega_\psi \times (0, T)$, $T > 0$. Consider the equilibrium equations

$$\frac{\partial \theta}{\partial t} - \Delta\theta + \delta^2 \frac{\partial}{\partial t}(\operatorname{div} W - \Delta w) = f, \tag{3.140}$$

$$\Delta^2 w + \delta^2 \Delta\theta = 0, \tag{3.141}$$

$$-\sigma_{ij,j} + \delta^2 \theta_{,i} = 0, \quad i = 1, 2, \tag{3.142}$$

and the nonpenetration condition

$$[W]\nu \ge \left| \left[\frac{\partial w}{\partial \nu} \right] \right| \tag{3.143}$$

between crack faces. We assume $h = 1$ for simplicity. Consider the initial condition for the temperature

$$\theta = \theta_0, \quad \text{at } t = 0. \tag{3.144}$$

At the external boundary the following conditions are assumed to be satisfied:

$$\theta = w = \frac{\partial w}{\partial n} = W = 0 \quad \text{on } \Gamma \times (0, T). \tag{3.145}$$

As for the crack surface we suppose a continuity of the temperature

$$[\theta] = 0 \quad \text{on } \Gamma_\psi \times (0, T). \tag{3.146}$$

A formulation of the problem to be analysed in this section is as follows. In the domain Q_ψ, we have to find a solution of (3.140)–(3.142) satisfying (3.144)–(3.146) and the inequality (3.143) fulfilled on $\Gamma_\psi \times (0, T)$. In this case the normal ν to Γ_ψ is defined as $\nu = (-\psi_x, 1)/\sqrt{1 + \psi_x^2}$, $\nu = (\nu_1, \nu_2)$, $[U] = U^+ - U^-$, and $U^\pm$ correspond to the positive and negative directions of the normal ν.

The considered problem is formulated as a variational inequality. In general, the equations (3.140)–(3.142) hold in the sense of distributions. In addition to (3.143), complementary boundary conditions will be fulfilled on $\Gamma_\psi \times (0, T)$. The exact form of these conditions is given at the end of the section. The assumption as to sufficient solution regularity requires the variational inequality to be a corollary of (3.140)–(3.142), the initial and all boundary conditions. The relationship between these two problem formulations is discussed in Section 3.4.4. We prove an existence of the solution in Section 3.4.2. In Section 3.4.3 the main result of the section concerned with the cracks of minimal opening is established.

3.4.2 Existence of solutions

Let the brackets $\langle \cdot, \cdot \rangle$ denote the integration over Ω_ψ. Introduce the two bilinear forms used in previous sections,

$$B(W, \bar{W}) = \langle \sigma_{ij}(W), \varepsilon_{ij}(\bar{W}) \rangle,$$

$$b(w, \bar{w}) = \int\limits_{\Omega_\psi} (w_{xx}\bar{w}_{xx} + w_{yy}\bar{w}_{yy} + \kappa w_{xx}\bar{w}_{yy} + \kappa w_{yy}\bar{w}_{xx} + 2(1-\kappa) w_{xy}\bar{w}_{xy}),$$

and denote by $H^{1,0}(\Omega_\psi)$ the subspace of $H^1(\Omega_\psi)$ which consists of functions equal to zero on Γ. Let the functions from $H^{2,0}(\Omega_\psi)$ equal to zero on Γ together with the first derivatives, $H^{2,0}(\Omega_\psi) \subset H^2(\Omega_\psi)$, $H(\Omega_\psi) = H^{1,0}(\Omega_\psi) \times H^{1,0}(\Omega_\psi) \times H^{2,0}(\Omega_\psi)$. The subspace $H_0^1(\Omega)$ of the space $H^1(\Omega)$ consists of functions equal to zero on Γ.

Introduce next the sets of admissible displacements of the plate

$$K = \{\chi = (W, w) \in H(\Omega_\psi) \mid \quad [W]\nu \geq \left| \left[\frac{\partial w}{\partial \nu} \right] \right| \quad \text{a.e. on} \quad \Gamma_\psi\},$$

$$\mathcal{K} = \{\chi \in L^2(0, T; H(\Omega_\psi)) \mid \quad \chi(t) \in K \quad \text{a.e. on} \quad (0, T)\}.$$

Since we assume the continuity conditions (3.146) the equation (3.140) can be considered in the domain $Q = \Omega \times (0, T)$. Indeed, if $\theta \in H^{1,0}(\Omega_\psi)$, $[\theta] = 0$ on Γ_ψ, then the function θ belongs to the space $H_0^1(\Omega)$. In this case the term $\partial(\operatorname{div} W - \Delta w)/\partial t$ will be defined in $L^2(Q_\psi)$, and consequently, it can be considered as the element of the space $L^2(Q)$. Of course, the derivatives $\partial(\operatorname{div} W - \Delta w)/\partial t$ are defined with respect to the domain Q_ψ.

Let $f \in L^2(Q)$, $\theta_0 \in H_0^1(\Omega)$. Introduce the space of functions

$$\Xi = \{\theta \in L^2(0, T; H_0^1(\Omega)) \mid \quad \theta_t \in L^2(Q)\}$$

equipped with the norm

$$\|\theta\|_\Xi^2 = \|\theta\|_{L^2(0,T;H_0^1(\Omega))}^2 + \|\theta_t\|_{L^2(Q)}^2,$$

and denote $H = H^1(0, T; H(\Omega_\psi))$, $\mathcal{H} = H \cap \mathcal{K}$. Consider that for any $\theta \in \Xi$ there exists a trace on the plane $t = 0$ and, in particular, $\theta(0) \in L^2(\Omega)$. The exact formulation of the analysed problem consists in finding the functions $\theta \in \Xi$, $\chi \in \mathcal{H}$ which satisfy the following identity,

$$\int\limits_{Q_\psi} \left(\frac{\partial \theta}{\partial t} + \delta^2 \frac{\partial}{\partial t} (\operatorname{div} W - \Delta w) - f \right) \bar{\theta} + \int\limits_{Q_\psi} \nabla\theta \nabla\bar{\theta} = 0, \quad \forall \bar{\theta} \in \Xi, \tag{3.147}$$

the variational inequality

$$\int\limits_0^T \Big(B(W, \bar{W} - W) + b(w, \bar{w} - w) + \delta^2 \langle \theta, \Delta\bar{w} - \Delta w \rangle \tag{3.148}$$

$$-\delta^2\langle\theta, \operatorname{div}\bar{W} - \operatorname{div}W\rangle\Big) \geq 0, \quad \forall \bar{\chi} \in \mathcal{K},$$

and the initial condition (3.144).

The boundary value problem (3.144), (3.147), (3.148) is analogous to that considered in the previous section. The only difference between these problems is that instead of (3.146), in the previous section the following condition,

$$\frac{\partial\theta}{\partial\nu} = 0 \quad \text{on } \Gamma_\psi \times (0,T),$$

is considered.

Thus, the following result can be established.

Theorem 3.11. *There exists a solution of the problem* (3.144), (3.147), (3.148) *provided that δ is small enough.*

We omit the proof of the theorem since it is analogous to that of Section 3.3 and restrict ourselves to some remarks. When proving the existence theorem the following estimates are obtained:

$$\|\theta\|_\Xi \leq c_1\delta\|\chi\|_H + c_2, \qquad \|\chi\|_H \leq c_3\delta\|\theta\|_\Xi, \tag{3.149}$$

which are valid for all $\delta \leq \delta_0$. The constant c_2 depends on the $L^2(Q)$-norm of f and the $H^1(\Omega)$-norm of θ_0, and the constants c_1, c_3 are independent of δ. The estimates (3.149) hold true for the problem (3.144), (3.147), (3.148). The sufficient condition providing the existence of the solution can be written as $\delta < c_0^{-1}$, where $c_0 = \max\{c_1, c_2, c_3\}$.

The theorem of existence is proved by finding a fixed point of the following operator (which is not compact, in general). Taking $\chi = \chi^0 \in \mathcal{H}$ in (3.147), we can find $\theta = \theta^0$ as a solution of (3.147) satisfying the initial condition $\theta^0(0) = \theta_0$. The function θ^0 is substituted in (3.148), which provides the existence of $\chi = \chi^1 = (W, w)$ for the given $\theta = \theta^0$. It can be demonstrated that there exists a fixed point of the operator $\chi^0 \to \chi^1$ as $\delta < c_0^{-1}$.

We can derive from (3.148) that for all $t \in (0,T)$ the following variational inequality holds:

$$B(W(t), \bar{W} - W(t)) + b(w(t), \bar{w} - w(t)) + \delta^2\langle\theta(t), \Delta\bar{w} - \Delta w(t)\rangle \tag{3.150}$$

$$-\delta^2\langle\theta(t), \operatorname{div}\bar{W} - \operatorname{div}W(t)\rangle \geq 0, \quad (W(t), w(t)) \in K, \quad \forall\ (\bar{W}, \bar{w}) \in K.$$

Also, the inclusion $\theta \in L^2(0,T; H^2(\Omega) \cap H_0^1(\Omega))$ follows from (3.147); in particular,

$$\left[\frac{\partial\theta}{\partial\nu}\right] = 0 \quad \text{on } \Gamma_\psi \times (0,T).$$

3.4.3 Cracks of zero opening

Assume that the solution of the problem (3.144), (3.147), (3.148) has the property

$$\int_0^T \int_{\Gamma_\psi} |[\chi]| \, d\Gamma_\psi \, dt = 0.$$

The crack is said to have a zero opening in this case. As it turned out there is no singularity of the solution provided the crack has a zero opening. What this means is the solution of (3.144), (3.147), (3.148) coincides with the solution of (3.140)–(3.142) found in the domain Q with the initial and boundary conditions (3.144), (3.145) (and without (3.143)). In the last case the equations (3.141), (3.142) hold in Q. This removable singularity property is of local character. Namely, if $\mathcal{O}(x^0)$ is a neighbourhood of the point $x^0 \in \Gamma_\psi$ and

$$\int_a^b \int_{\Gamma_\psi \cap \mathcal{O}(x^0)} |[\chi]| \, d\Gamma_\psi \, dt = 0$$

then the equations (3.141), (3.142) hold in $\mathcal{O}(x^0) \times (a, b)$. The object of further reasoning is to prove the property mentioned. For simplicity let $\delta = 1$.

First, we recall two Green formulae. Let $D \subset R^2$ be a bounded domain with a smooth boundary γ having the external normal $n = (n_1, n_2)$. Consider the following two operators defined on γ:

$$M(w) = \kappa \Delta w + (1 - \kappa) \frac{\partial^2 w}{\partial n^2}, \quad R(w) = \frac{\partial}{\partial n} \Delta w + (1 - \kappa) \frac{\partial^3 w}{\partial n \partial s^2},$$

$$s = (-n_2, n_1).$$

Let $H^{-s}(\gamma)$ be the space dual of $H^s(\gamma)$. The duality pairing between $H^{-s}(\gamma)$ and $H^s(\gamma)$ is denoted by $\langle \cdot , \cdot \rangle_{s,\gamma}$. We know that for any function $w \in H^2(D)$, $\Delta^2 w \in L^2(D)$, the values $M(w)$, $R(w)$ can be correctly defined on γ and, moreover, the following Green formula holds for $v \in H^2(D)$:

$$b_D(w, v) = \langle M(w), \frac{\partial v}{\partial n} \rangle_{\frac{1}{2},\gamma} - \langle R(w), v \rangle_{\frac{3}{2},\gamma} + \langle \Delta^2 w, v \rangle_D. \qquad (3.151)$$

The subscript D means the integration over the domain D. Let $W = (w^1, w^2) \in H^1(D)$, $\sigma_{ij,j}(W) \in L^2(D)$, $i = 1, 2$, (in particular, $\sigma_{ij}(W) \in L^2(D)$). In this case the values $\sigma_{ij}(W) n_j$ are clearly identified on γ in the sense of $H^{-\frac{1}{2}}(\gamma)$, and the Green formula is

$$\langle \sigma_{ij}(W), \varepsilon_{ij}(V) \rangle_D = \langle \sigma_{ij}(W) n_j, v_i \rangle_{\frac{1}{2},\gamma} - \langle \sigma_{ij,j}(W), v_i \rangle_D, \qquad (3.152)$$

$$\forall V = (v_1, v_2) \in H^1(D).$$

Let $x^0 \equiv (1,0)$ be the tip of the crack. We shall state the removable singularity property in a neighbourhood of the line $x^0 \times (a,b)$, $(a,b) \subset (0,T)$. The corresponding result for a neighbourhood of the line $\bar{x} \times (a,b)$, where $\bar{x} \in \Gamma_\psi \setminus \partial\Gamma_\psi$, can be more readily established.

At first we state the continuity of $M(w)$ and $R(w)$ across the crack shape. To this end, denote by $\mathcal{O}(x^0) \subset R^2$ a neighbourhood of x^0; $\mathcal{O} = \mathcal{O}(x^0) \times (a,b)$. Extend the graph of the function $y = \psi(x)$ beyond $x = 1$, denoting it by $\tilde{\Gamma}_\psi$. We assume that $\tilde{\Gamma}_\psi$ is smooth enough. Let $\mathcal{O}^+(x^0) = \mathcal{O}(x^0) \cap \{y > \psi(x)\}$, and $\mathcal{O}^-(x^0)$ be defined analogously, $\mathcal{O}^\pm = \mathcal{O}^\pm(x^0) \times (a,b)$. It is evident that (3.141), (3.142) hold in $\mathcal{O}^\pm$ in the sense of distributions. In fact, one can choose smooth functions (Λ, φ) having compact support in $\mathcal{O}^\pm$ and substitute $(\bar{W}, \bar{w}) = (W, w) + (\Lambda, \varphi)$ in (3.148) as test functions. Moreover, the regularity of the solution (θ, χ) allows us to verify that for all $t \in (0,T)$ the equations

$$\Delta^2 w(t) + \Delta\theta(t) = 0, \tag{3.153}$$

$$-\sigma_{ij,j}(W(t)) + \theta_{,i}(t) = 0, \quad i = 1,2, \tag{3.154}$$

hold in $\mathcal{O}^\pm(x^0)$ in the sense of two-dimensional distributions. In this context we should remember that $\delta = 1$ in this section. In what follows the boundaries of $\mathcal{O}^\pm(x^0)$ are denoted by $\gamma^\pm$, respectively.

Let us take $\varphi \in C_0^\infty(\mathcal{O}(x^0))$ and assume that $\varphi \equiv 0$ outside $\mathcal{O}(x^0)$. We can substitute $(W(t), \varphi + w(t))$ in (3.150) as a test function. This gives the inequality

$$b_+(w(t), \varphi) + b_-(w(t), \varphi) \geq -\langle \Delta\theta(t), \varphi \rangle \tag{3.155}$$

holding for all $t \in (0,T)$. The signs $+, -$ mean the integration over $\mathcal{O}^\pm(x^0)$, respectively. Since the equation (3.153) holds in $\mathcal{O}^\pm(x^0)$, one can use the Green formula (3.151) in order to transform $b_\pm(w(t), \varphi)$. This leads to the following relations:

$$\langle [M(w(t))], \frac{\partial\varphi}{\partial n} \rangle_{\frac{1}{2},\gamma} = 0, \ \langle [R(w(t))], \varphi \rangle_{\frac{3}{2},\gamma} = 0, \ \forall \varphi \in C_0^\infty(\mathcal{O}(x^0)). \tag{3.156}$$

Here γ can be equal to γ^+ as well as to γ^-. We should remark at this point that, in fact, the integration is fulfilled over $\mathcal{O}(x^0)$ in the right-hand side of (3.155). In other words, we integrate over Ω_ψ and use the condition $[\partial\theta(t)/\partial\nu] = 0$ on Γ_ψ holding true due to the regularity of θ. The existence of two angular points on $\gamma^\pm$ presents no problems since the function φ has a compact support. It follows from (3.156) for almost all $t \in (0,T)$ that

$$[M(w(t))] = 0, \quad [R(w(t))] = 0 \quad \text{on } \tilde{\Gamma}_\psi \cap \mathcal{O}(x^0). \tag{3.157}$$

Similarly, let $\Lambda \equiv (\Lambda_1, \Lambda_2) \in C_0^\infty(\mathcal{O}(x^0))$, $\Lambda \equiv 0$ beyond $\mathcal{O}(x^0)$. Then we have $(W(t) + \Lambda, w(t)) \in K$, and hence, a substitution of $(W(t) + \Lambda, w(t))$ in (3.150) as a test function results in the inequality

$$\langle \sigma_{ij}(W(t)), \varepsilon_{ij}(\Lambda) \rangle_+ + \langle \sigma_{ij}(W(t)), \varepsilon_{ij}(\Lambda) \rangle_- \geq \langle \theta(t), \operatorname{div} \Lambda \rangle.$$

By virtue of (3.152) we easily derive

$$-\langle[\sigma_{ij}(W(t))\nu_j],\Lambda_i\rangle_{\frac{1}{2},\gamma^-}-\langle\sigma_{ij,j}(W(t)),\Lambda_i\rangle_{\pm}\geq-\langle\theta_{,i}(t),\Lambda_i\rangle. \tag{3.158}$$

Together with (3.154) we conclude that

$$\langle[\sigma_{ij}(W(t))\nu_j],\Lambda_i\rangle_{\frac{1}{2},\gamma^-}=0,\quad \forall\Lambda\in C_0^\infty(\mathcal{O}(x^0)),$$

and consequently

$$[\sigma_{ij}(W(t))\nu_j]=0\quad\text{on}\ \ \tilde{\Gamma}_\psi\cap\mathcal{O}(x^0),\quad i=1,2. \tag{3.159}$$

By taking into account (3.157), (3.159) we are in a position to prove the result related to the cracks of minimal opening.

Theorem 3.12. *Let* (θ,χ) *be the solution of* (3.144), (3.147), (3.148) *and*

$$\int\limits_a^b\int\limits_{\Gamma_\psi\cap\mathcal{O}(x^0)}|[\chi]|d\Gamma_\psi dt=0.$$

Then the equations

$$\Delta^2 w+\Delta\theta=0, \tag{3.160}$$

$$-\sigma_{ij,j}(W)+\theta_{,i}=0,\quad i=1,2, \tag{3.161}$$

hold in $\mathcal{O}$.

Proof. It is evident that

$$W=(w^1,w^2)\in H^1(a,b;H^1(\mathcal{O}^\pm(x^0)),\quad w\in H^1(a,b;H^2(\mathcal{O}^\pm(x^0)).$$

By (3.143) and the condition of the theorem, we deduce that

$$\left[\frac{\partial w}{\partial\nu}\right]=0\quad\text{on}\ \ (\tilde{\Gamma}_\psi\cap\mathcal{O}(x^0))\times(a,b)$$

and hence (see Mikhailov, 1976)

$$W=(w^1,w^2)\in H^1(a,b;H^1(\mathcal{O}(x^0)),\quad w\in H^1(a,b;H^2(\mathcal{O}(x^0)). \tag{3.162}$$

Starting from (3.157), (3.159), (3.162) we shall prove that the equations (3.160), (3.161) hold in $\mathcal{O}$ in the sense of distributions. Let $(\,\cdot\,,\varphi)$ denote the value of a distribution at the point φ. We choose any function $\varphi\in C_0^\infty(\mathcal{O})$ and verify that

$$(\Delta^2 w+\Delta\theta,\varphi)=0. \tag{3.163}$$

In fact,

$$(\Delta^2 w+\Delta\theta,\varphi)=\int\limits_a^b b_\pm(w(t),\varphi(t))+\int\limits_a^b\langle\Delta\theta(t),\varphi(t)\rangle$$

$$= -\int_a^b \langle [M(w(t))], \frac{\partial \varphi(t)}{\partial \nu} \rangle_{\frac{1}{2},\gamma^-} + \int_a^b \langle [R(w(t))], \varphi(t) \rangle_{\frac{3}{2},\gamma^-} \tag{3.164}$$

$$+ \int_a^b \langle \Delta^2 w(t) + \Delta\theta(t), \varphi(t) \rangle_{\pm}.$$

The right-hand side of (3.164) is equal to zero because of (3.153), (3.157), which completes the proof of (3.163).

Analogously, let $\varphi \in C_0^\infty(\mathcal{O})$. We have

$$-(\sigma_{ij,j}(W) - \theta_{,i}, \varphi) = \int_a^b \langle \sigma_{ij}(W(t)), \varphi(t)_{,i} \rangle_{\pm} + \int_a^b \langle \theta_{,i}(t), \varphi(t) \rangle \tag{3.165}$$

$$= -\int_a^b \langle [\sigma_{ij}(W(t))\nu_j], \varphi(t) \rangle_{\frac{1}{2},\gamma^-} - \int_a^b \langle \sigma_{ij,j}(W(t)) - \theta_{,i}(t), \varphi(t) \rangle_{\pm}, \ i = 1, 2.$$

By (3.154), (3.159), the right-hand side of (3.165) is equal to zero. This means that the equations (3.161) hold in $\mathcal{O}$ in the sense of distributions. The theorem is proved.

REMARK. The equations (3.160), (3.161) hold in Q provided the condition of the theorem takes place and $(a, b) = (0, T)$, $\Gamma_\psi \subset \mathcal{O}(x^0)$.

3.4.4 Boundary conditions at the crack faces

In conclusion we briefly discuss the boundary conditions holding on $\Gamma_\psi \times (0, T)$. To this end, consider the decomposition of the vector $\{\sigma_{ij}(W)\nu_j\}$, $i = 1, 2$, on the boundary $\Gamma_\psi^- \times (0, T)$:

$$\{\sigma_{ij}(W)\nu_j\} = \sigma_\nu(W)\nu + \sigma_s(W)s, \quad s = (-\nu_2, \nu_1).$$

A similar formula applies to $\Gamma_\psi^+ \times (0, T)$. Assume that the solution of (3.144), (3.147), (3.148) is smooth enough. As in Section 3.3, it can be shown that together with (3.143) the following equations and inequalities hold on $\Gamma_\psi \times (0, T)$:

$$[\sigma_\nu^\theta(W)] = 0, \quad \sigma_s(W) = 0, \tag{3.166}$$

$$[M^\theta(w)] = 0, \quad R^\theta(w) = 0, \tag{3.167}$$

$$|M^\theta(w)| \le -\sigma_\nu^\theta(W), \tag{3.168}$$

$$M^\theta(w)\left[\frac{\partial w}{\partial \nu}\right] + \sigma_\nu^\theta(W)[W]\nu = 0. \tag{3.169}$$

Here, $R^\theta(w) = R(w) + \partial\theta/\partial\nu$, $\sigma_\nu^\theta(W) = \sigma_\nu(W) - \theta$, $M^\theta(w) = M(w) + \theta$. The second conditions of (3.166), (3.167) mean that $\sigma_s(W) = R^\theta(w) = 0$

on $\Gamma_\psi^\pm \times (0,T)$. As stated above, we assume $\delta = 1$. It is significant that the equilibrium problem for the thermoelastic plate can be formulated in two different ways. On the one hand, it can be formulated in the form (3.144), (3.147), (3.148). In this case all the written boundary conditions (i.e. (3.166)–(3.169)) are the corollary of this formulation. Of course, in deriving (3.166)–(3.169) we should assume an additional regularity of the solution. On the other hand, the problem admits the formulation in the form of equations (3.140)–(3.142) with the initial and boundary conditions (3.143)–(3.145), (3.166)–(3.169). If a solution of the last boundary problem is smooth enough the formulation (3.144), (3.147), (3.148) follows from (3.140)–(3.145), (3.166)–(3.169). In fact, let $(\bar{W}, \bar{w})$ be a smooth function belonging to K. We multiply equations (3.141), (3.142) taken for a fixed $t \in (0,T)$ by $\bar{w} - w(t)$, $\bar{w}^i - w^i(t)$, respectively. We next integrate the relations over Ω_ψ. By using the Green formulae like (3.151), (3.152) and by (3.145), (3.166)–(3.167) it is easy to derive for the chosen t that

$$B(W(t), \bar{W} - W(t)) + b(w(t), \bar{w} - w(t))$$

$$+ \langle \theta(t), \Delta\bar{w} - \Delta w(t)\rangle - \langle \theta(t), \operatorname{div}\bar{W} - \operatorname{div} W(t)\rangle$$

$$+ \langle M^\theta(w(t)), \left[\frac{\partial \bar{w}}{\partial \nu}\right] - \left[\frac{\partial w(t)}{\partial \nu}\right]\rangle_{\Gamma_\psi} + \langle \sigma_\nu^\theta(W(t)), [\bar{W}]\nu - [W(t)]\nu\rangle_{\Gamma_\psi} = 0.$$

According to (3.143), (3.168)–(3.169) the sum of the boundary integrals is nonpositive here, whence (3.150) follows. It is evident from (3.150) that we obtain (3.148).

3.5 Inclined cracks in plates

In this section we derive a nonpenetration condition between crack faces for inclined cracks in plates and discuss the equilibrium problem. As it turns out, the nonpenetration condition for inclined cracks is of nonlocal character. This means that by writing the condition at a fixed point we have to take into account the displacement values both at the point and at the other point chosen at the opposite crack face. As a corollary of this fact, the equilibrium equations hold only in a domain located outside the crack surface projection on the mid-surface of the plate. This section follows the papers (Khludnev, 1997b; Kovtunenko et al., 1998).

3.5.1 Derivation of the nonpenetration condition

Let the mid-surface of the Kirchhoff–Love plate occupy a domain $\Omega_c = \Omega \setminus \Gamma_c$, where $\Omega \subset R^2$ is a bounded domain with the smooth boundary Γ, and Γ_c is the smooth curve without self-intersections recumbent in Ω (see Fig.3.4). The mid-surface of the plate is in the plane $z = 0$. Coordinate system (x_1, x_2, z) is assumed to be Descartes' and orthogonal, $x = (x_1, x_2)$.

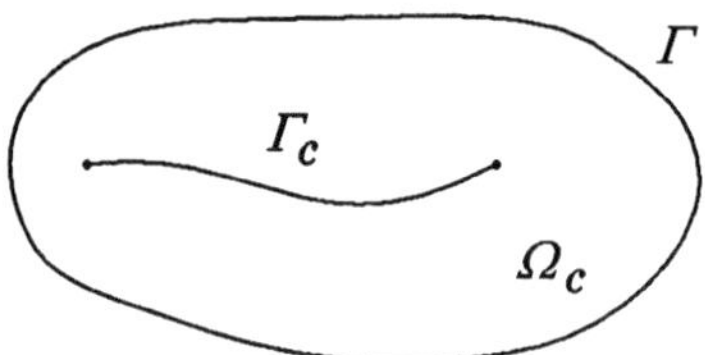

Fig.3.4. Mid-surface of the plate

Let the crack surface Ψ be described by the equation $z = \Phi(x)$, $x \in \Omega_\Psi$ (see Fig.3.5). Here, Ω_Ψ is the orthogonal projection of the crack surface (i.e. of the graph $z = \Phi(x)$) on the plane $z = 0$. The set Ω_Ψ is assumed to be closed in R^2. Denote by

$$n(x) = (-\nabla\Phi(x), 1)/\sqrt{1 + |\nabla\Phi(x)|^2}$$

the unit normal vector to the surface $z = \Phi(x)$, $x \in \Omega_\Psi$. The chosen direction of the normal $n(x)$ defines both positive and negative crack faces to be denoted by $\Psi^\pm$. The curve Γ_ψ is the intersection of the crack surface Ψ with the plane $z = 0$. To simplify the arguments below, we assume $|\nabla\Phi(x)| \neq 0$, $x \in \Omega_\Psi$.

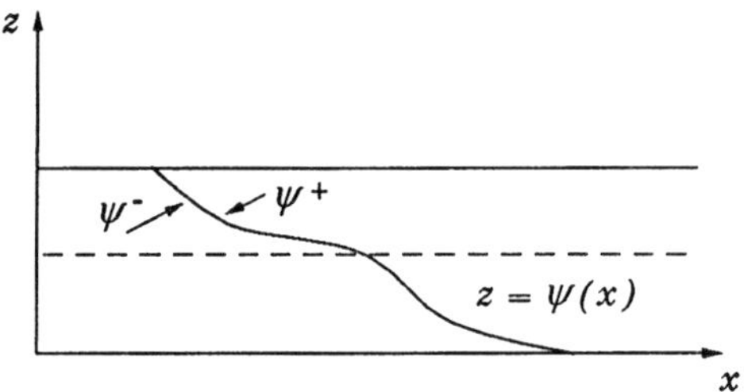

Fig.3.5. Vertical cross-section of the plate

Projection Ω_Ψ of the surface Ψ can be represented as the union of two sets in accordance with the direction of the axis z, namely, $\Omega_\Psi = \Omega_\Psi^+ \cup \Omega_\Psi^-$. We denote by Ω_Ψ^+ the part of the projection of Ψ provided that this part is obtained by moving along the positive direction of the axis z. Respectively, we find Ω_Ψ^-. In particular, the curve Γ_c belongs both to Ω_Ψ^+ and Ω_Ψ^- (see Fig.3.6). We assume the direction of the normal $\nu = (\nu_1, \nu_2)$ to the curve Γ_c in the x-plane is outside Ω_Ψ^- and inside Ω_Ψ^+. Let $x \subset \Omega_\Psi$. Denote by $y = Px$ the orthogonal projection of the point x on the curve Γ_c. The sets $\Omega_\Psi^\pm$ are assumed to be sufficiently small, which means that the value $y = Px$ is uniquely defined for each fixed $x \in \Omega_\Psi^\pm$.

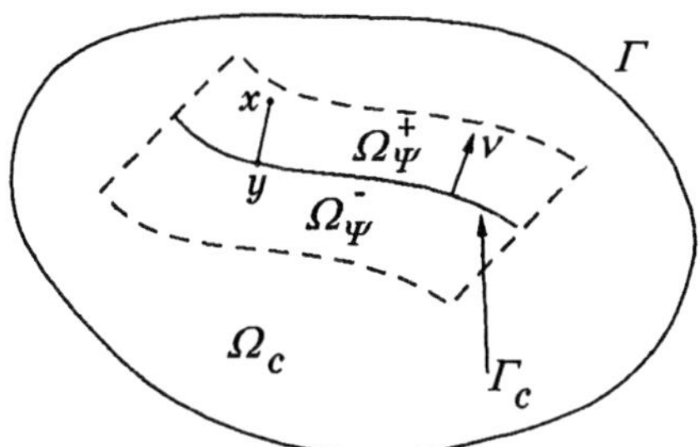

Fig.3.6. Projection of the crack surface

We recall that in the Kirchhoff–Love plate theory the horizontal displacements depend linearly on the coordinate z, i.e.

$$W(z) = W - z\nabla w, \quad |z| \le \varepsilon,$$

where $W = (w^1, w^2)$, w are horizontal and vertical displacements of the mid-surface points, 2ε is the plate thickness. By $\chi = (W, w)$ we denote the full displacement vector, $\chi = \chi(x)$, $x \in \Omega_c$. Within the framework of Kirchhoff–Love's theory, we shall find the plate displacements at the crack faces $\Psi^\pm$ and derive a mutual nonpenetration condition of the crack faces.

Let $(x, z) \in \Psi^+$, $x \in \Omega_\Psi^+$. Then the displacements vector at the point (x, z) is of the form

$$\chi^+(x, z) = (W^+(x) - z\nabla w^+(x), w^+(x)), \quad x \in \Omega_\Psi^+, \quad z = \Phi(x). \tag{3.170}$$

Similarly, the displacement vector has the form

$$\chi^-(x, z) = \left(W^-(y) - z\nabla w^-(y), w^-(y) + |x - y|\frac{\partial w^-(y)}{\partial \nu}\right), \tag{3.171}$$

$$y = Px$$

provided that $(x, z) \in \Psi^-$, $x \in \Omega_\Psi^+$. Formula (3.171) yields that the horizontal displacements at the point $(x, z) \in \Psi^-$, $x \in \Omega_\Psi^+$, coincide with the horizontal displacements at the point (y, z), $y = Px$. The vertical displacements are different, and the difference is equal to $|x - y|\, \partial w^-(y)/\partial \nu$.

The nonpenetration condition of the crack faces at the point $(x, z) \in \Psi$, $x \in \Omega_\Psi^+$, has the following form:

$$(\chi^+(x, z) - \chi^-(x, z))n(x) \ge 0, \quad x \in \Omega_\Psi^+, \quad z = \Phi(x). \tag{3.172}$$

We can substitute the vector $\chi^\pm(x, z)$ in (3.172) in accordance with (3.170), (3.171). This implies

$$(\chi^+(x) - \chi^-(y))n(x) - (\chi_z^+(x) - \chi_z^-(y))n(x) \ge 0, \tag{3.173}$$

$$x \in \Omega_\Psi^+, \quad z = \Phi(x), \quad y = Px,$$

where

$$\chi^{\pm}(s) = (W^{\pm}(s), w^{\pm}(s)), \quad \chi_z^{\pm}(s) = \left(z\nabla w^{\pm}(s), |s - Ps|\frac{\partial w^{\mp}(Ps)}{\partial \nu^{\pm}}\right).$$

We have used the notations $\partial/\partial\nu^+ \equiv \partial/\partial\nu$, $\partial/\partial\nu^- \equiv -\partial/\partial\nu$. It makes sense to notice that $y = Py$ for $y \in \Gamma_c$.

Analogously, we can derive a nonpenetration condition like (3.173) for the points $(x, z) \in \Psi^{\pm}$, $x \in \Omega_{\Psi}^-$. Indeed, let $(x, z) \in \Psi^+$, $x \in \Omega_{\Psi}^-$. Then

$$\chi^+(x, z) = \left(W^+(y) - z\nabla w^+(y), w^+(y) - |x - y|\frac{\partial w^+(y)}{\partial \nu}\right), \tag{3.174}$$

$$x \in \Omega_{\Psi}^-, \quad z = \Phi(x), \quad y = Px.$$

Consider a point $(x, z) \in \Psi^-$, $x \in \Omega_{\Psi}^-$. We have

$$\chi^-(x, z) = (W^-(x) - z\nabla w^-(x), w^-(x)). \tag{3.175}$$

By substituting (3.174), (3.175) in the nonpenetration inequality

$$(\chi^+(x, z) - \chi^-(x, z))n(x) \geq 0, \quad x \in \Omega_{\Psi}^-, \quad z = \Phi(x),$$

it is easy to derive

$$(\chi^+(y) - \chi^-(x))n(x) - (\chi_z^+(y) - \chi_z^-(x))n(x) \geq 0, \tag{3.176}$$

$$x \in \Omega_{\Psi}^-, \quad z = \Phi(x), \quad y = Px.$$

Thus, the mutual nonpenetration condition between the crack faces is described by the inequalities (3.173), (3.176). The inequalities have a nonlocal character; in particular, they contain values of the functions both at the point x and the point $y = Px$; moreover the last values (i.e. at the point $y = Px$) are taken at the opposite crack faces.

It is of importance to note that if the surface $z = \Phi(x)$ transform into the vertical crack corresponding to the cylinder $x \in \Gamma_c$, $-\varepsilon \leq z \leq \varepsilon$, then the conditions (3.173), (3.176) transform into the inequality we have used in the previous sections,

$$[W(x)]\nu(x) \geq \varepsilon\left|\left[\frac{\partial w(x)}{\partial \nu}\right]\right|, \quad x \in \Gamma_c. \tag{3.177}$$

In fact, in this case the normal $n(x)$ is transformed into the vector $(\nu_1, \nu_2, 0)$, and the conditions (3.173), (3.176) imply

$$[W(x)]\nu(x) \geq z[\nabla w(x)]\nu(x), \quad x \in \Gamma_c, \quad -\varepsilon \leq z \leq 0, \tag{3.178}$$

$$[W(x)]\nu(x) \geq z[\nabla w(x)]\nu(x), \quad x \in \Gamma_c, \quad 0 \leq z \leq \varepsilon. \tag{3.179}$$

It is clear that (3.177) is equivalent to (3.178), (3.179). If a crack has both inclined and vertical parts, the nonpenetration condition of the crack faces has the form (3.173), (3.179) or (3.176), (3.178). In a similar way, we can analyse different inclined cracks and derive the formulae as before.

3.5.2 Formulation of the boundary problem

Consider an inclined crack with the nonpenetration condition of the form (3.173), (3.176). Let $\chi = (W, w)$ be the displacement vector of the mid-surface points. Introduce the strain and stress tensor components $\varepsilon_{ij} = \varepsilon_{ij}(W)$, $\sigma_{ij} = \sigma_{ij}(W)$,

$$\varepsilon_{ij}(W) = \frac{1}{2}\left(\frac{\partial w^i}{\partial x_j} + \frac{\partial w^j}{\partial x_i}\right), \quad i,j = 1,2,$$

$$\sigma_{11} = \varepsilon_{11} + \kappa\varepsilon_{22},\ \sigma_{22} = \varepsilon_{22} + \kappa\varepsilon_{11},\ \sigma_{12} = (1-\kappa)\varepsilon_{12},$$

$$\kappa = \text{const}, \quad 0 < \kappa < 1/2.$$

The energy functional of the plate has the form

$$\Pi(\chi) = \frac{1}{2}B(W,W) + \frac{1}{2}b(w,w) - \langle f, \chi\rangle,$$

where

$$B(W,\bar{W}) = \langle \sigma_{ij}(W), \varepsilon_{ij}(\bar{W})\rangle,$$

$$b(w,\bar{w}) = \int\limits_{\Omega_c} (w_{,11}\bar{w}_{,11} + w_{,22}\bar{w}_{,22} + \kappa w_{,11}\bar{w}_{,22} + \kappa w_{,22}\bar{w}_{,11} + 2(1-\kappa)w_{,12}\bar{w}_{,12})$$

and the brackets $\langle\cdot,\cdot\rangle$ denote integration over Ω_c.

Assume that $f = (f_1, f_2, f_3) \in L^2(\Omega_c)$. Let $H(\Omega_c) = H^{1,0}(\Omega_c) \times H^{1,0}(\Omega_c) \times H^{2,0}(\Omega_c)$. The spaces $H^{1,0}(\Omega_c)$, $H^{2,0}(\Omega_c)$ are introduced analogously to the spaces $H^{1,0}(\Omega_\psi)$, $H^{2,0}(\Omega_\psi)$ used in previous sections. Denote by K the set of all functions in $H(\Omega_c)$ satisfying the inequalities (3.173), (3.176). The set is convex and closed in $H(\Omega_c)$. The equilibrium problem for the plate has a variational form

$$\inf_{\chi\in K} \Pi(\chi). \tag{3.180}$$

The functional Π is convex and differentiable on the space $H(\Omega_c)$, and hence the problem (3.180) is equivalent to the variational inequality

$$\chi \in K: \quad \Pi'(\chi)(\bar{\chi} - \chi) \geq 0 \quad \forall \bar{\chi} \in K, \tag{3.181}$$

where $\Pi'(\chi)$ is the derivative of Π at the point χ. By the inequalities

$$b(w,w) \geq c\|w\|^2_{2,\Omega_c} \quad \forall w \in H^{2,0}(\Omega_c),$$

$$B(W,W) \geq c\|W\|^2_{1,\Omega_c} \quad \forall W = (w^1, w^2) \in H^{1,0}(\Omega_c)$$

the functional Π is coercive on the space $H(\Omega_c)$, i.e.

$$\Pi(\chi) \to \infty, \quad \text{as } \|\chi\|_{H(\Omega_c)} \to \infty.$$

In view of the weak lower semicontinuity of Π, we conclude that the problem (3.180) (or the problem (3.181)) has a solution. This solution is unique.

It is seen that in the domain $\Omega_c \setminus \Omega_\Psi$ the following equations,

$$\Delta^2 w = f_3, \quad -\sigma_{ij,j}(W) = f_i, \quad i = 1, 2, \tag{3.182}$$

hold in the distribution sense. To verify this statement we can substitute $\chi + \tilde{\chi}$ in (3.181) as a test function, where $\tilde{\chi} = (\tilde{W}, \tilde{w}) \in C_0^\infty(\Omega_c \setminus \Omega_\Psi)$, and χ is the solution of (3.181). This implies the identity

$$\Pi'(\chi)(\tilde{\chi}) = 0 \quad \forall \tilde{\chi} = (\tilde{W}, \tilde{w}) \in C_0^\infty(\Omega_c \setminus \Omega_\Psi),$$

which means that (3.182) hold in the distribution sense.

Let a point x be interior with respect to Ω_Ψ^+, i.e. there exists a neighbourhood U of the point x such that $U \subset \Omega_\Psi^+$. We choose a smooth function $\tilde{\chi} = (\tilde{W}, \tilde{w})$ in the domain Ω_c such that a support of $\tilde{\chi}$ belongs to U and

$$(\tilde{W}^+(x) - z\nabla\tilde{w}^+(x), \tilde{w}^+(x))n(x) \geq 0, \quad z = \Phi(x), \quad x \in U.$$

In this case $\chi + \tilde{\chi} \in K$, where χ is the solution of (3.181). Substitute $\chi + \tilde{\chi}$ in (3.181) as a test function. This provides the inequality

$$\Pi'(\chi)(\tilde{\chi}) \geq 0,$$

which means that the equilibrium equations (3.182) do not hold in Ω_ψ, in general.

3.5.3 Simplified nonpenetration condition

As before, let the smooth curve Γ_c lie in the plate mid-surface $z = 0$, and Γ_c be intersection of the plane $z = 0$ and the crack surface Ψ described by the equation $z = \Phi(x)$, $x \in \Omega_\Psi$. The unit normal vector to Ψ at a point x is $n(x) = (n_1(x), n_2(x), n_3(x))$. Let y be the projection Px of a point $x \in \Omega_\Psi$ onto Γ_c. The unit normal vector to Γ_c at a point y in the plane $z = 0$ is denoted by $\nu = (\nu_1, \nu_2)$.

Consider a vertical plane Π_y passed over a fixed point $y \in \Gamma_c$ in the direction $n(y)$ (i.e. $n(y)$ lies in Π_y). We assume that the intersection $C_y = \Pi_y \cap \Psi$ is a straight line for every $y \in \Gamma_c$, and denote by $\alpha(y)$ the angle between C_y and $z = 0$. The normal to Ψ in Π_y will not depend on z because of

$$n(x) = n(y) = \Big(\nu(y)\cos\alpha(y), \sin\alpha(y)\Big). \tag{3.183}$$

Next, supposing α to be small enough (leading to small Ω_Ψ), we assume that the displacements in Ω_Ψ do not vary along the normal $\nu(y)$, namely,

$$\chi(x, z) = \chi(y, z), \quad x \in \Omega_\Psi^+, \quad -\varepsilon \leq z \leq 0, \quad y = Px,$$

$$\chi(x,z) = \chi(y,z), \quad x \in \Omega_{\Psi}^{-}, \quad 0 \le z \le \varepsilon, \quad y = Px.$$

This assumption combined with Kirchhoff–Love's formula

$$\chi(y,z) = \Big(W(y) - z\nabla w(y), w(y)\Big), \quad |z| \le \varepsilon,$$

provides the following value of the displacements jump at Ψ,

$$[\chi(x,z)] = [\chi(y,z)] = \Big([W(y) - z\nabla w(y)], [w(y)]\Big), \tag{3.184}$$

$$z = \Phi(x), \quad y = Px, \quad \forall x \in \Omega_{\Psi}.$$

Substituting (3.183), (3.184) into the nonpenetration condition (3.172), we have

$$0 \le \Big([W(y)] - z[\nabla w(y)], [w(y)]\Big) \cdot \Big(\nu(y)\cos\alpha(y), \sin\alpha(y)\Big)$$

$$= [W(y)]\nu(y)\cos\alpha(y) + [w(y)]\sin\alpha(y) - z[\nabla w(y)]\nu(y)\cos\alpha(y), \quad |z| \le \varepsilon.$$

The linearity of this inequality in z means

$$[W(y)]\nu(y)\cos\alpha(y) + [w(y)]\sin\alpha(y) \ge \varepsilon \left| \left[\frac{\partial w(y)}{\partial \nu} \right] \right| \cos\alpha(y).$$

Dividing this inequality by $\cos\alpha(y)$, we finally deduce the required relation

$$[W(y)]\nu(y) + [w(y)]\tan\alpha(y) \ge \varepsilon \left| \left[\frac{\partial w(y)}{\partial \nu} \right] \right|, \quad y \in \Gamma_c. \tag{3.185}$$

The obtained nonpenetration condition (3.185) is local as compared to (3.173), (3.176) since this condition is considered only at the curve Γ_c. Let us recall that we have assumed that the angle between the crack surface Ψ and the axis z is small. By this assumption, the small deflection $|x - Px|$ has been neglected in (3.173), (3.176). It is of importance to deduce (3.177) from (3.185). Indeed, if Ψ is transformed into a vertical crack, then C_y is a straight line, $\alpha(y) \equiv 0$, and from (3.185) we obtain the nonpenetration condition

$$[W(y)]\nu(y) \ge \varepsilon \left| \left[\frac{\partial w(y)}{\partial \nu} \right] \right|, \quad y \in \Gamma_c,$$

used in the previous sections.

3.5.4 Boundary conditions at the crack faces

Let us now define the admissible displacements set

$$K = \{\chi = (W, w) \in H(\Omega_c) \mid \ \chi \text{ satisfies } (3.185)\}$$

which is convex and closed. Consider the equilibrium problem for the plate with the inclined crack in a variational form

$$\inf_{\chi \in K} \Pi(\chi), \tag{3.186}$$

which is equivalent to the variational inequality

$$\chi \in K, \quad \Pi'(\chi)(\bar{\chi} - \chi) \geq 0 \quad \forall \bar{\chi} \in K. \tag{3.187}$$

By arguments of the Section 3.5.2, there exists a unique solution $\chi \in K$ to the problem (3.186). Substituting $\bar{\chi} = \chi \pm \tilde{\chi}$, where $\tilde{\chi} \in C_0^\infty(\Omega_c)$, as a test function in (3.187), we verify that the equilibrium equations

$$\Delta^2 w = f_3, \quad -\sigma_{ij,j}(W) = f_i, \quad i = 1, 2, \tag{3.188}$$

hold in Ω_c in the distribution sense.

Let us now obtain a complete system of boundary conditions fulfilled at Γ_c provided that the simplified nonpenetration condition (3.185) holds. We assume the solution $\chi \in K$ is smooth enough and use Green's formulas for smooth functions (see Section 1.4),

$$B(W, \bar{W}) = -\int\limits_{\Omega_c} \sigma_{ij,j}(W)\bar{u}_i \, d\Omega_c - \int\limits_{\Gamma_c} \Big[\sigma_\nu(W)\bar{W}\nu + \sigma_\tau(W)\bar{W}\tau\Big] \, d\Gamma_c,$$

$$b(w, \bar{w}) = \int\limits_{\Omega_c} \Delta^2 w \, \bar{w} \, d\Omega_c + \int\limits_{\Gamma_c} \left[t(w)\bar{w} - m(w)\frac{\partial \bar{w}}{\partial \nu}\right] d\Gamma_c,$$

$$t(w) = \frac{\partial}{\partial \nu}\left(\Delta w + (1-\kappa)\frac{\partial^2 w}{\partial \tau^2}\right), \quad m(w) = \kappa \Delta w + (1-\kappa)\frac{\partial^2 w}{\partial \nu^2}.$$

Take $\bar{\chi} = \chi \pm \tilde{\chi}$, $\tilde{\chi} \in H_0^1(\Omega) \times H_0^1(\Omega) \times H_0^2(\Omega)$, as a test function, in (3.187), and use (3.188) and the Green formulas. By the independence between $\tilde{W}$, $\tilde{w}$, $\partial \tilde{w}/\partial \nu$ at the boundary, the usual considerations imply the conditions

$$[\sigma_\nu(W)] = [\sigma_\tau(W)] = [t(w)] = [m(w)] = 0 \tag{3.189}$$

holding on Γ_c, and

$$\int\limits_{\Gamma_c} \Bigg(-\sigma_\nu(W)[\bar{W} - W]\nu - \sigma_\tau(W)[\bar{W} - W]\tau + t(w)[\bar{w} - w] \tag{3.190}$$

$$-m(w)\left[\frac{\partial \bar{w}}{\partial \nu} - \frac{\partial w}{\partial \nu}\right]\Bigg) d\Gamma_c \geq 0 \quad \forall \bar{\chi} = (\bar{W}, \bar{w}) \in K.$$

Note that K is a convex cone in $H(\Omega_c)$, i.e. if $\chi \in K$ then $\lambda\chi \in K$, $\lambda \geq 0$ is a constant. Therefore, one can substitute $\bar{\chi} = \lambda\chi$ in (3.190) and deduce

$$\int\limits_{\Gamma_c} \left(-\sigma_\nu(W)[W]\nu - \sigma_\tau(W)[W]\tau + t(w)[w] - m(w)\left[\frac{\partial w}{\partial \nu}\right]\right) d\Gamma_c = 0, \tag{3.191}$$

$$\int_{\Gamma_c} \left(-\sigma_\nu(W)[\bar{W}]\nu - \sigma_\tau(W)[\bar{W}]\tau + t(w)[\bar{w}] - m(w) \left[\frac{\partial \bar{w}}{\partial \nu} \right] \right) d\Gamma_c \geq 0 \quad (3.192)$$

for all $\bar{\chi} \in K$. Let us denote

$$\xi = [\bar{W}], \quad \eta = [\bar{w}], \quad \theta = \left[\frac{\partial \bar{w}}{\partial \nu} \right] \quad \text{on } \Gamma_c. \quad (3.193)$$

By the trace theorems of Section 1.4, smooth enough $\xi = (\xi_1, \xi_2, \xi_3)$, η, θ with compact supports in $\Gamma_c \setminus \partial\Gamma_c$ define a function $\bar{\chi} = (\bar{W}, \bar{w})$ from $H(\Omega_c)$ such that equalities (3.193) take place at the boundary Γ_c. Thus, (3.192) implies

$$\int_{\Gamma_c} \left(-\sigma_\nu(W)\xi\nu - \sigma_\tau(W)\xi\tau + t(w)\eta - m(w)\theta \right) d\Gamma_c \geq 0 \quad (3.194)$$

$$\forall \xi = (\xi_1, \xi_2, \xi_3), \eta, \theta, \quad \xi\nu + \eta \tan\alpha \geq \varepsilon|\theta|.$$

When $\xi\nu = \eta = \theta = 0$, for an arbitrary value of $\xi\tau$, from (3.194) we have

$$\sigma_\tau(W) = 0. \quad (3.195)$$

Consider next the representation

$$-\sigma_\nu(W)\xi\nu + t(w)\eta - m(w)\theta = \frac{1}{2}\left(-\sigma_\nu(W) - \frac{1}{\varepsilon}m(w) \right)(\xi\nu + \eta\tan\alpha + \varepsilon\theta)$$

$$+\frac{1}{2}\left(-\sigma_\nu(W) + \frac{1}{\varepsilon}m(w) \right)(\xi\nu + \eta\tan\alpha - \varepsilon\theta) + \left(\sigma_\nu(W)\tan\alpha + t(w) \right)\eta.$$

Then for arbitrary η, $\theta = 0$, ξ such that $\xi\nu = -\eta\tan\alpha$, we have $\xi\nu + \eta\tan\alpha \pm \varepsilon\theta = 0$, and from (3.194), (3.195) the following inequality is derived,

$$\sigma_\nu(W)\tan\alpha + t(w) = 0. \quad (3.196)$$

After the utilization of (3.195), (3.196), the remaining part of (3.194) provides the inequality

$$\frac{1}{2}\int_{\Gamma_c} \left(\left(-\sigma_\nu(W) - \frac{1}{\varepsilon}m(w) \right)\left(\xi\nu + \eta\tan\alpha + \varepsilon\theta \right) \right.$$

$$\left. + \left(-\sigma_\nu(W) + \frac{1}{\varepsilon}m(w) \right)\left(\xi\nu + \eta\tan\alpha - \varepsilon\theta \right) \right) d\Gamma_c \geq 0$$

$$\forall \xi, \eta, \theta, \quad \xi\nu + \eta\tan\alpha + \varepsilon\theta \geq 0, \quad \xi\nu + \eta\tan\alpha - \varepsilon\theta \geq 0.$$

This means

$$-\sigma_\nu(W) - \frac{1}{\varepsilon}m(w) \geq 0, \quad -\sigma_\nu(W) + \frac{1}{\varepsilon}m(w) \geq 0,$$

i.e.

$$-\sigma_\nu(W) \geq \frac{1}{\varepsilon}|m(w)|. \tag{3.197}$$

We can see that (3.191), by (3.195), (3.196), can be represented in the same way,

$$\int_{\Gamma_c} \left(\left(-\sigma_\nu(W) - \frac{1}{\varepsilon} m(w) \right) \left([W]\nu + [w] \tan\alpha + \varepsilon \left[\frac{\partial w}{\partial \nu} \right] \right) \right.$$

$$\left. + \left(-\sigma_\nu(W) + \frac{1}{\varepsilon} m(w) \right) \left([W]\nu + [w] \tan\alpha - \varepsilon \left[\frac{\partial w}{\partial \nu} \right] \right) \right) d\Gamma_c = 0.$$

Therefore, by (3.185), (3.197), we conclude that

$$\int_{\Gamma_c} \left(-\sigma_\nu(W) - \frac{1}{\varepsilon} m(w) \right) \left([W]\nu + [w] \tan\alpha + \varepsilon \left[\frac{\partial w}{\partial \nu} \right] \right) d\Gamma_c = 0,$$

$$\int_{\Gamma_c} \left(-\sigma_\nu(W) + \frac{1}{\varepsilon} m(w) \right) \left([W]\nu + [w] \tan\alpha - \varepsilon \left[\frac{\partial w}{\partial \nu} \right] \right) d\Gamma_c = 0.$$

Thus, we have proved the following assertion.

Theorem 3.13. *For a smooth solution $\chi = (W, w) \in K$ of the problem (3.187), the following boundary conditions hold at Γ_c,*

$$[\sigma_\nu(W)] = [\sigma_\tau(W)] = [t(w)] = [m(w)] = 0, \quad \sigma_\nu(W) \tan\alpha + t(w) = 0,$$

$$\sigma_\tau(W) = 0, \quad [W]\nu + [w] \tan\alpha \geq \varepsilon \left| \left[\frac{\partial w}{\partial \nu} \right] \right|, \quad -\sigma_\nu(W) \geq \frac{1}{\varepsilon}|m(w)|,$$

$$\left(-\sigma_\nu(W) - \frac{1}{\varepsilon} m(w) \right) \left([W]\nu + [w] \tan\alpha + \varepsilon \left[\frac{\partial w}{\partial \nu} \right] \right) = 0,$$

$$\left(-\sigma_\nu(W) + \frac{1}{\varepsilon} m(w) \right) \left([W]\nu + [w] \tan\alpha - \varepsilon \left[\frac{\partial w}{\partial \nu} \right] \right) = 0.$$

Let us note that, by (3.185), (3.196), the last two equations here are equivalent to the relation

$$\sigma_\nu(W) \Big([W]\nu + [w] \tan\alpha \Big) + m(w) \left[\frac{\partial w}{\partial \nu} \right] = 0,$$

which for $\alpha = 0$ has the same form as for the nonpenetration condition considered in the previous sections.

3.5.5 Inclined cut in a beam

Let us consider a thin homogeneous isotropic beam of thickness 2ε. We assume that the beam mid-line coincides with the segment $(0,1)$ of the axis x. At the point $y = 1/2$, the beam has an inclined cut as a segment having the angle α with the vertical line, $0 \leq \tan\alpha < (2\varepsilon)^{-1}$. We look for the function $\chi = (W, w)$ of horizontal displacements $W(x)$ and vertical displacements $w(x)$ provided that the external forces $g(x), f(x)$ are given.

The condition of clamped edges

$$W = w = w_x = 0 \quad \text{at} \;\; x = 0, 1$$

is assumed to be fulfilled. The nonpenetration condition (3.185) at the cut faces is transformed into

$$[W] + [w]\tan\alpha \geq \varepsilon \Big| [w_x] \Big|, \tag{3.198}$$

where $[s]$ denotes a function s jump at the point y, i.e. $[s] = s(y+0) - s(y-0)$. In the one-dimensional case we have the strain $\varepsilon(W) = W_x$, the stress $\sigma(W) = GW_x$ and the potential energy functional

$$\Pi(\chi) = \int_0^1 \left(\frac{G}{2} W_x^2 + \frac{D}{2} w_{xx}^2 - gW - fw \right) dx,$$

$$G = \frac{E\varepsilon}{1-\kappa^2}, \qquad D = \frac{E\varepsilon^3}{3(1-\kappa^2)}.$$

Denote $\Omega_y = (0, y) \cup (y, 1)$. Introduce the Hilbert space

$$H(\Omega_y) = \left\{ W \in H^1(\Omega_y), w \in H^2(\Omega_y) \mid \;\; W = w = w_x = 0 \quad \text{at} \;\; x = 0, 1 \right\}$$

and the convex closed set of admissible displacements

$$K = \{\chi = (W, w) \in H(\Omega_y) \mid \;\; \chi \;\; \text{satisfies} \;\; (3.198)\}.$$

Let $f, g \in L^2(\Omega_y)$. The equilibrium problem for the beam with the inclined cut is formulated as the following variational inequality:

$$\int_{\Omega_y} \Big(GW_x(\bar{W}_x - W_x) + Dw_{xx}(\bar{w}_{xx} - w_{xx}) \tag{3.199}$$

$$- g(\bar{W} - W) - f(\bar{w} - w) \Big) dx \geq 0 \quad \forall \bar{\chi} = (\bar{W}, \bar{w}) \in K.$$

There exists a unique solution $\chi \in K$ of (3.199). At the cut edges we have

$$\sigma_\nu(W) = G\, W_x(y), \quad m(w) = D\, w_{xx}(y), \quad t(w) = D\, w_{xxx}(y).$$

Therefore, by Theorem 3.13, the problem (3.199) is equivalent to the following boundary value problem:

$$
\begin{gathered}
-GW_{xx} = g, \quad Dw_{xxxx} = f \quad \text{in } \Omega_y, \\
[W_x] = [w_{xx}] = [w_{xxx}] = 0, \quad W_x(y)\tan\alpha + \frac{\varepsilon^2}{3} w_{xxx}(y) = 0, \\
[W] + [w]\tan\alpha \geq \varepsilon |[w_x]|, \quad -W_x(y) \geq \frac{\varepsilon}{3}\left|w_{xx}(y)\right|, \qquad (3.200) \\
\left(W_x(y) + \frac{\varepsilon}{3} w_{xx}(y)\right)\left([W] + [w]\tan\alpha + \varepsilon[w_x]\right) = 0, \\
\left(W_x(y) - \frac{\varepsilon}{3} w_{xx}(y)\right)\left([W] + [w]\tan\alpha - \varepsilon[w_x]\right) = 0, \\
W(0) = w(0) = w_x(0) = W(1) = w(1) = w_x(1) = 0.
\end{gathered}
$$

3.5.6 Construction of analytical solutions

If we construct a solution of the problem (3.200), then it is a solution of (3.199). By the arguments of Section 1.3.5, we seek the solution of (3.200) as a sum $\chi = \chi^0 + \chi^1$, where $\chi^0 = (W^0, w^0)$ is a solution of the boundary value problem

$$
\begin{gathered}
-GW^0_{xx} = g, \quad Dw^0_{xxxx} = f \quad \text{in } \Omega_y, \qquad (3.201) \\
[W^0_x] = [w^0_{xx}] = [w^0_{xxx}] = 0, \quad W^0_x(y) = w^0_{xx}(y) = w^0_{xxx}(y) = 0.
\end{gathered}
$$

Obviously, there exists a solution $\chi^0 = (W^0, w^0)$ from the space $(H^2(\Omega_y) \times H^4(\Omega_y)) \cap H(\Omega_y)$ to (3.201) since it is decomposed into two independent problems in $(0, y)$ and $(y, 1)$ as follows:

$$
\begin{array}{ll}
-GW^0_{xx} = g,\ Dw^0_{xxxx} = f \ \text{in } (0, y), & -GW^0_{xx} = g,\ Dw^0_{xxxx} = f \ \text{in } (y, 1), \\
W^0(0) = w^0(0) = w^0_x(0) = 0, & W^0(1) = w^0(1) = w^0_x(1) = 0, \\
W^0_x(y) = w^0_{xx}(y) = w^0_{xxx}(y) = 0, & W^0_x(y) = w^0_{xx}(y) = w^0_{xxx}(y) = 0.
\end{array}
$$

Note that χ^0 is the solution of the equilibrium problem for the beam as its cut edges are free of the stresses.

For convenience, we introduce the following constant values:

$$\delta = 12\varepsilon^2, \quad \rho = 4\varepsilon^2 + \tan^2\alpha.$$

Having found the solution of (3.201), one can define the values

$$\psi^+ = [W^0] + [w^0]\tan\alpha + \varepsilon[w^0_x], \quad \psi^- = [W^0] + [w^0]\tan\alpha - \varepsilon[w^0_x],$$

$$\psi^+ = [W^0] + [w^0]\tan\alpha + \frac{\varepsilon\rho}{\delta}[w^0_x], \quad \psi^- = [W^0] + [w^0]\tan\alpha - \frac{\varepsilon\rho}{\delta}[w^0_x].$$

Besides, we consider the functions

$$\theta(x) = \begin{cases} x^2 & , x \in (0, y), \\ (x-1)^2 & , x \in (y, 1), \end{cases} \qquad \theta_x(x) = \begin{cases} 2x & , x \in (0, y), \\ 2(x-1) & , x \in (y, 1), \end{cases}$$

$$\beta(x) = \begin{cases} 2x^3 - 3x^2 & , \quad x \in (0, y), \\ 2x^3 - 3x^2 + 1 & , \quad x \in (y, 1). \end{cases}$$

Then the pairs (θ_x, θ), (θ_x, β) belong to the space $(C^\infty(\Omega_y))^2 \cap H(\Omega_y)$, and the following identities hold:

$$\theta_{xx}(x) \equiv 2, \quad \theta_{xxx}(x) \equiv 0, \quad [\theta] = 0, \quad [\theta_x] = -2,$$

$$\beta_x(x) = 6(x^2 - x), \quad \beta_{xx}(x) = 6(2x - 1), \quad \beta_{xxx}(x) \equiv 12,$$

$$\beta_{xxxx}(x) \equiv 0, \quad [\beta] = 1, \quad \beta_{xx}(y) = 0, \quad [\beta_x] = 0.$$

Let us recall that we take $y = 1/2$ in the above formulae.

Theorem 3.14. *The functions*

$$W(x) = W^0(x) + 2\varepsilon^2 A\theta_x(x), \tag{3.202}$$

$$w(x) = w^0(x) + 6\varepsilon B\theta(x) - A\beta(x)\tan\alpha$$

are solutions of the variational inequality (3.199), *where* $\chi^0 = (W^0, w^0)$ *is the solution of* (3.201) *and*

$$(A, B) = \begin{cases} (0,0) & , \text{ if } \phi^+ \geq 0, \quad \phi^- \geq 0, \\ (\delta+\rho)^{-1}(\phi^+, \phi^+) & , \text{ if } \phi^+ < 0, \quad \psi^- \geq 0, \\ (\delta+\rho)^{-1}(\phi^-, -\phi^-) & , \text{ if } \phi^- < 0, \quad \psi^+ \geq 0, \\ ((2\rho)^{-1}(\phi^+ + \phi^-), (2\delta)^{-1}(\phi^+ - \phi^-)) & , \text{ if } \psi^+ < 0, \quad \psi^- < 0. \end{cases}$$

PROOF. To prove the theorem, it suffices to verify conditions (3.200). Indeed, by the properties of the functions θ, β and χ^0, we obtain

$$-GW_{xx} = -GW^0_{xx} - G2\varepsilon^2 A\theta_{xxx} = g \quad \text{in } \Omega_y,$$

$$Dw_{xxxx} = Dw^0_{xxxx} + D6\varepsilon B\theta_{xxxx} - DA\beta_{xxxx}\tan\alpha = f \quad \text{in } \Omega_y,$$

$$[W_x] = [W^0_x] + 2\varepsilon^2 A[\theta_{xx}] = 0, \quad [w_{xx}] = [w^0_{xx}] + 6\varepsilon B[\theta_{xx}] - A[\beta_{xx}]\tan\alpha = 0,$$

$$[w_{xxx}] = [w^0_{xxx}] + 6\varepsilon B[\theta_{xxx}] - A[\beta_{xxx}]\tan\alpha = 0.$$

Now let us calculate the values

$$W_x(y) = W^0_x(y) + 2\varepsilon^2 A\theta_{xx}(y) = 4\varepsilon^2 A,$$

$$w_{xx}(y) = w^0_{xx}(y) + 6\varepsilon B\theta_{xx}(y) - A\beta_{xx}(y)\tan\alpha = 12\varepsilon B,$$

$$w_{xxx}(y) = w^0_{xxx}(y) + 6\varepsilon B\theta_{xxx}(y) - A\beta_{xxx}(y)\tan\alpha = -12A\tan\alpha.$$

Then

$$W_x(y) \pm \frac{\varepsilon}{3} w_{xx}(y) = 4\varepsilon^2(A \pm B), \quad W_x(y)\tan\alpha + \frac{\varepsilon^2}{3} w_{xxx}(y) = 0.$$

Since $[W] = [W^0] - 4\varepsilon^2 A$, $[w] = [w^0] - A\tan\alpha$, $[w_x] = [w_x^0] - 12\varepsilon B$, we have

$$[W] + [w]\tan\alpha \pm \varepsilon[w_x] = \phi^{\pm} - \rho A \mp \delta B.$$

At last, we verify that

$$(A+B)(\phi^+ - \rho A - \delta B) = 0, \quad (A-B)(\phi^- - \rho A + \delta B) = 0,$$

$$\phi^+ \geq \rho A + \delta B, \quad \phi^- \geq \rho A - \delta B, \quad -A \geq |B|.$$

Consideration of the following four admissible cases,

$$A+B=0, \quad A-B=0, \quad \phi^+ - \rho A - \delta B \geq 0, \quad \phi^- - \rho A + \delta B \geq 0,$$

$$A+B<0, \quad A-B=0, \quad \phi^+ - \rho A - \delta B = 0, \quad \phi^- - \rho A + \delta B \geq 0,$$

$$A+B=0, \quad A-B<0, \quad \phi^+ - \rho A - \delta B \geq 0, \quad \phi^- - \rho A + \delta B = 0,$$

$$A+B<0, \quad A-B<0, \quad \phi^+ - \rho A - \delta B = 0, \quad \phi^- - \rho A + \delta B = 0,$$

gives the values of A, B as in Theorem 3.14. The proof is completed.

By the smoothness of χ^0, θ, β, the solution $\chi \in \left(H^2(\Omega_y) \times H^4(\Omega_y)\right) \cap H(\Omega_y)$. The constructed functions θ, β give the correction for the solution χ^0 provided the nonpenetration condition holds and $\chi = \chi^0$ (i.e. $A = B = 0$) only when $\phi^+ \geq 0$, $\phi^- \geq 0$. The presentation (3.202) of χ allows one to find the functions

$$\sigma(x) = G\,W_x(x) = G(W_x^0(x) + 4\varepsilon^2 A),$$

$$m(x) = D\,w_{xx}(x) = D(w_{xx}^0(x) + 12\varepsilon B - 6A(2x-1)\tan\alpha),$$

$$t(x) = D\,w_{xxx}(x) = D(w_{xxx}^0(x) - 12A\tan\alpha),$$

which are continuous in $(0,1)$.

Let us derive corollaries of Theorem 3.14. First, we assume the cut is vertical, i.e. $\alpha = 0$. Then the nonpenetration condition (3.198) is written as

$$[W] \geq \varepsilon\,|[w_x]|,$$

and the boundary problem (3.200) for the variational inequality (3.199) takes the form

$$-GW_{xx} = g, \quad Dw_{xxxx} = f \quad \text{in } \Omega_y,$$

$$[W_x] = [w_{xx}] = [w_{xxx}] = 0, \quad w_{xxx}(y) = 0,$$

$$[W] \geq \varepsilon|[w_x]|, \quad -W_x(y) \geq \frac{\varepsilon}{3}\,|w_{xx}(y)|, \tag{3.203}$$

$$\left(W_x(y) + \frac{\varepsilon}{3} w_{xx}(y)\right)\left([W] + \varepsilon[w_x]\right) = 0,$$

$$\left(W_x(y) - \frac{\varepsilon}{3} w_{xx}(y)\right)\left([W] - \varepsilon[w_x]\right) = 0.$$

For defined values $\phi^{\pm} = [W^0] \pm \varepsilon[w_x^0]$, $\psi^{\pm} = [W^0] \pm \varepsilon/3[w_x^0]$, from (3.202) we deduce the following statement.

Corollary 3.1. *The functions*

$$W(x) = W^0(x) + \frac{A}{2}\theta_x(x), \quad w(x) = w^0(x) + \frac{3B}{2\varepsilon}\theta(x)$$

are solutions to the problem (3.203), *where* $\chi^0 = (W^0, w^0)$ *is the solution of* (3.201) *and*

$$(A,B) = \begin{cases} (0,0) & , \text{ if } \phi^+ \geq 0, \quad \phi^- \geq 0, \\ \frac{1}{4}(\phi^+, \phi^+) & , \text{ if } \phi^+ < 0, \quad \psi^- \geq 0, \\ \frac{1}{4}(\phi^-, -\phi^-) & , \text{ if } \phi^- < 0, \quad \psi^+ \geq 0, \\ \left(\frac{1}{2}(\phi^+ + \phi^-), \frac{1}{6}(\phi^+ - \phi^-)\right) & , \text{ if } \psi^+ < 0, \quad \psi^- < 0. \end{cases}$$

Now let the vertical force be zero, i.e. $f(x) \equiv 0$. Then $w^0(x) \equiv 0$ and, therefore, $\phi^+ = \phi^- = \psi^+ = \psi^- = [W^0]$, $B = 0$. In what follows, we denote $s = s^+ - s^-$, $s^+, s^- \geq 0$, $s^+ s^- = 0$.

Corollary 3.2. *Let* $f = 0$. *Then the functions*

$$W(x) = W^0(x) - \frac{2\varepsilon^2}{\rho}[W^0]^- \theta_x(x), \quad w(x) = \frac{\tan\alpha}{\rho}[W^0]^- \beta(x)$$

are solutions to the problem (3.199).

As we see, the presence of the inclined cut can lead to an appearance of vertical displacements notwithstanding the vertical force is zero. If $\alpha = 0$, then $f = 0$ leads to $w = 0$.

Consider an example. Let $f(x) \equiv 0$, and

$$g(x) = \begin{cases} c & , \quad x \in (0, 1/2), \\ -c & , \quad x \in (1/2, 1). \end{cases}$$

The case $c \geq 0$ corresponds to a compression. One can easily find

$$W^0(x) = \frac{1-\kappa^2}{2E\varepsilon} x(1-x) g(x).$$

Its jump $[W^0] = -c(1-\kappa^2)/(4E\varepsilon) \leq 0$, therefore $[W^0]^- = c(1-\kappa^2)/(4E\varepsilon)$. By Corollary 3.2, the solution χ of (3.199) is as follows,

$$W(x) = \frac{c(1-\kappa^2)}{2E\varepsilon} \begin{cases} -x^2 + \left(1 - \frac{\delta}{6\rho}\right) x & , \quad x \in (0, 1/2), \\ x^2 - \left(1 + \frac{\delta}{6\rho}\right) x + \frac{\delta}{6\rho} & , \quad x \in (1/2, 1), \end{cases}$$

$$w(x) = \frac{c(1-\kappa^2)\tan\alpha}{4E\varepsilon\rho}\begin{cases} 2x^3 - 3x^2 & , \quad x \in (0, 1/2), \\ 2x^3 - 3x^2 + 1 & , \quad x \in (1/2, 1), \end{cases}$$

and $[w] = c(1-\kappa^2)\tan\alpha\big/(4E\varepsilon\rho)^{-1}$. As for the extension, the case $c < 0$ gives $[W^0] > 0$, therefore $[W^0]^- = 0$ and $W(x) = W^0(x)$, $w(x) \equiv 0$.

3.6 Friction problem for plates with cracks

We consider the model of a plate with a crack describing the plate vertical displacements with a given friction between the crack surfaces. The results of this section are published in (Kovtunenko, 1998).

3.6.1 Problem formulation

Let $\Omega \subset R^2$ be a bounded domain with the smooth boundary Γ, which has an inside smooth curve Γ_c without self-intersections. We denote $\Omega_c = \Omega\backslash\Gamma_c$. Let $n = (n_1, n_2)$ be a unit normal vector at Γ, and $\nu = (\nu_1, \nu_2)$ be a unit normal vector at Γ_c, which defines a positive and a negative surface of the crack. We assume that there exists a closed continuation Σ of Γ_c dividing Ω into two domains: the domain Ω^- with the outside normal ν at Σ, and the domain Ω^+ with the outside normal $-\nu$ at Σ (see Section 1.4). By doing so, for a smooth function w in Ω_c, we define the traces of w at boundaries $\partial\Omega^\pm$ and, in particular, the traces $w^\pm$ and the jump $[w] = w^+ - w^-$ at Γ_c.

Let us consider the bilinear form

$$b(w, \bar{w}) = D\int\limits_{\Omega_c}\Big(w_{xx}\bar{w}_{xx} + w_{yy}\bar{w}_{yy} + \kappa(w_{xx}\bar{w}_{yy} + w_{yy}\bar{w}_{xx}) + 2(1-\kappa)w_{xy}\bar{w}_{xy}\Big),$$

where $D = Eh^3/(3(1-\kappa^2))$; $2h$ is the plate thickness; $0 < \kappa < 1/2$. The form $b(w, \bar{w})$ defines a scalar product, and $b(w, w)$ is a square norm $||w||^2$ in the Hilbert space

$$H(\Omega_c) = \{w \in H^2(\Omega_c) \mid \quad w = \partial w/\partial n = 0 \quad \text{on } \Gamma\}.$$

Let $f \in L^2(\Omega_c)$ correspond to external forces in Ω_c, and $g \in L^2(\Gamma_c)$, $g \geq 0$ a.e. on Γ_c, be a known friction force at Γ_c. We introduce the potential energy functional

$$\Pi(w) = \frac{1}{2}b(w, w) + \int\limits_{\Gamma_c} g\big|[w]\big| - \int\limits_{\Omega_c} fw,$$

which is not differentiable. We seek the function $w(x)$ of the plate vertical displacements. The corresponding equilibrium problem implies the minimization of the energy functional

$$\Pi(w) = \inf_{\bar{w} \in H(\Omega_c)} \Pi(\bar{w}), \tag{3.204}$$

or the equivalent variational inequality

$$b(w, \bar{w} - w) + \int_{\Gamma_c} g\left(\left|[\bar{w}]\right| - \left|[w]\right|\right) \geq \int_{\Omega_c} f(\bar{w} - w) \quad \forall \bar{w} \in H(\Omega_c). \tag{3.205}$$

Because g is nonnegative, one can see that Π is a coercive, strongly convex and lower semicontinuous functional. Therefore, there exists a unique solution $w \in H(\Omega_c)$ of the problem (3.204) or (3.205) (see Section 1.2).

Now we are in a position to give a formulation of the boundary value problem. We recall formulae for the transverse force and the bending moment at Γ_c,

$$t(w) = D\frac{\partial}{\partial \nu}\left(\Delta w + (1-\kappa)\frac{\partial^2 w}{\partial \tau^2}\right), \quad m(w) = D\left(\kappa\, \Delta w + (1-\kappa)\frac{\partial^2 w}{\partial \nu^2}\right),$$

with the tangent vector $\tau = (-\nu_2, \nu_1)$ at Γ_c. Let us take $\bar{w} = w \pm \xi$, $\xi \in C_0^\infty(\Omega_c)$, as a test function in (3.205) and deduce the equilibrium equation fulfilled in Ω_c in the distribution sense,

$$D\Delta^2 w = f. \tag{3.206}$$

By $f \in L^2(\Omega_c)$ and (3.206), we obtain for the solution of (3.205) that $\Delta^2 w \in L^2(\Omega_c)$. Thus, in Ω^+ and Ω^-, we have $w \in H^2(\Omega^\pm)$, $\Delta^2 w \in L^2(\Omega^\pm)$ and, therefore, $t^\pm(w) \in H^{-3/2}(\Sigma)$, $m^\pm(w) \in H^{-1/2}(\Sigma)$ are defined, and Green's formula arises:

$$b(w, \bar{w}) = D\int_{\Omega_c} \bar{w}\, \Delta^2 w + \langle t^+(w), \bar{w}^+\rangle_\Sigma - \langle t^-(w), \bar{w}^-\rangle_\Sigma$$

$$-\langle m^+(w), \partial\bar{w}^+/\partial\nu\rangle_\Sigma + \langle m^-(w), \partial\bar{w}^-/\partial\nu\rangle_\Sigma \quad \forall \bar{w} \in H(\Omega_c).$$

Here $\langle\cdot,\cdot\rangle_\Sigma$ denotes the duality pairing between $H^{k+1/2}(\Sigma)$ and its dual space $H^{-k-1/2}(\Sigma)$ for both $k = 0, 1$. Substituting (3.206) in (3.205), by the Green formula, one derives

$$\langle t^+(w), \bar{w}^+ - w^+\rangle_\Sigma - \langle t^-(w), \bar{w}^- - w^-\rangle_\Sigma + \int_{\Gamma_c} g\left(\left|[\bar{w}]\right| - \left|[w]\right|\right) \tag{3.207}$$

$$-\left\langle m^+(w), \frac{\partial(\bar{w}^+ - w^+)}{\partial\nu}\right\rangle_\Sigma + \left\langle m^-(w), \frac{\partial(\bar{w}^- - w^-)}{\partial\nu}\right\rangle_\Sigma \geq 0.$$

Taking here $\bar{w} = w \pm \xi$, $\xi \in H_0^2(\Omega)$, we have $[\xi] = [\partial\xi/\partial\nu] = 0$ and, hence

$$\langle t^+(w) - t^-(w), \xi\rangle_\Sigma - \langle m^+(w) - m^-(w), \partial\xi/\partial\nu\rangle_\Sigma = 0. \qquad (3.208)$$

By the arbitrariness of ξ and $\partial\xi/\partial\nu$ at Σ, the last equality implies

$$t^+(w) = t^-(w), \quad m^+(w) = m^-(w),$$

and we denote $t^\pm(w), m^\pm(w)$ by $t(w), m(w)$, respectively, $t(w) \in H^{-3/2}(\Sigma)$, $m(w) \in H^{-1/2}(\Sigma)$. Thus, (3.207) takes the form

$$\langle t(w), [\bar{w} - w]\rangle_\Sigma - \left\langle m(w), \left[\frac{\partial(\bar{w} - w)}{\partial\nu}\right]\right\rangle_\Sigma + \int_{\Gamma_c} g\left(\left|[\bar{w}]\right| - \left|[w]\right|\right) \geq 0.$$

By $\bar{w} - w \in H(\Omega_c)$, the jumps of $\bar{w} - w$ and $\partial(\bar{w} - w)/\partial\nu$ are zeros at $\Sigma \setminus \Gamma_c$. Consequently, the functionals $t(w)$ and $m(w)$ are defined at Γ_c (see Section 1.4). Hence, the previous inequality can be written in the form

$$\langle t(w), [\bar{w} - w]\rangle_{\Gamma_c} - \left\langle m(w), \left[\frac{\partial(\bar{w} - w)}{\partial\nu}\right]\right\rangle_{\Gamma_c} \qquad (3.209)$$

$$+ \int_{\Gamma_c} g\left(\left|[\bar{w}]\right| - \left|[w]\right|\right) \geq 0.$$

Any given smooth functions $\psi^\pm, \phi^\pm$ with compact supports in $\Gamma_c \setminus \partial\Gamma_c$ define $\xi \in H(\Omega_c)$ such that $\xi^\pm = \psi^\pm$, $\partial\xi^\pm/\partial\nu = \phi^\pm$ at Γ_c. Choosing $\psi^+ = \psi^-$ and substituting $\bar{w} = w \pm \xi$ in (3.209), we conclude that

$$\langle m(w), [\phi]\rangle_{\Gamma_c} = 0 \qquad (3.210)$$

with an arbitrary function $[\phi]$ having compact support in $\Gamma_c \setminus \partial\Gamma_c$. This implies $m(w) = 0$ at Γ_c and, hence, (3.209) gives

$$\langle t(w), [\bar{w} - w]\rangle_{\Gamma_c} + \int_{\Gamma_c} g\left(\left|[\bar{w}]\right| - \left|[w]\right|\right) \geq 0.$$

We replace here $\bar{w}$ by $\pm\lambda\bar{w}$ with the constant $\lambda \geq 0$; then

$$\lambda\left(\pm\langle t(w), [\bar{w}]\rangle_{\Gamma_c} + \int_{\Gamma_c} g\left|[\bar{w}]\right|\right) - \left(\langle t(w), [w]\rangle_{\Gamma_c} + \int_{\Gamma_c} g\left|[w]\right|\right) \geq 0$$

for all $\bar{w} \in H(\Omega_c)$, $\lambda \geq 0$. From this relation one can deduce that

$$\langle t(w), [w]\rangle_{\Gamma_c} + \int_{\Gamma_c} g\left|[w]\right| = 0, \qquad (3.211)$$

$$\pm\langle t(w), [\bar{w}]\rangle_{\Gamma_c} + \int_{\Gamma_c} g\Big|[\bar{w}]\Big| \geq 0 \quad \forall \bar{w} \in H(\Omega_c).$$

By the above arguments, the last inequality means

$$|\langle t(w), \psi\rangle_{\Gamma_c}| \leq \int_{\Gamma_c} g|\psi| \tag{3.212}$$

for all smooth functions ψ having compact supports in $\Gamma_c \setminus \partial\Gamma_c$.

Thus, we obtain the equivalent formulation of (3.205) in the form

$$D\Delta^2 w = f \quad \text{in } \Omega_c,$$

$$m(w) = 0, \quad [t(w)] = 0, \quad |t(w)| \leq g, \tag{3.213}$$

$$t(w)\,[w] + g\Big|[w]\Big| = 0 \text{ on } \Gamma_c,$$

which is fulfilled in the sense (3.208), (3.210)–(3.212). The last two relations correspond to the friction condition

$$\begin{cases} |t(w)| < g & \Longrightarrow & [w] = 0, \\ t(w) = g & \Longrightarrow & [w] \leq 0, \\ t(w) = -g & \Longrightarrow & [w] \geq 0. \end{cases}$$

If $g \equiv 0$ at Γ_c, then (3.213) reduces to $m(w) = t(w) = 0$.

3.6.2 Associated penalty problem

We introduce the function $B \in H^2(R)$ by the formula

$$B(t) = \begin{cases} t & , \quad \text{as } t > 1, \\ -t & , \quad \text{as } t < -1, \\ (1+t^2)/2 & , \quad \text{as } |t| \leq 1. \end{cases}$$

Its derivative $B' \in H^1(R)$ can be found by the formula

$$B'(t) = \begin{cases} 1 & , \quad \text{as } t > 1, \\ -1 & , \quad \text{as } t < -1, \\ t & , \quad \text{as } |t| \leq 1. \end{cases}$$

The functions B, B' are Lipschitz continuous due to

$$|B(t) - B(s)| \leq |t-s|, \quad |B'(t) - B'(s)| \leq |t-s| \quad \forall t, s \in R.$$

Therefore, for any $\phi \in H^1(\Gamma_c \setminus \partial\Gamma_c)$, we have $B(\phi)$, $B'(\phi) \in H^1(\Gamma_c \setminus \partial\Gamma_c)$ (Kinderlehrer, Stampacchia, 1980).

Let us define the penalty functional

$$\Pi_\varepsilon(w) = \frac{1}{2}\|w\|^2_{H(\Omega_c)} + \varepsilon \int_{\Gamma_c} gB\left(\frac{[w]}{\varepsilon}\right) - \int_{\Omega_c} fw, \quad w \in H(\Omega_c)$$

depending on a small parameter $\varepsilon > 0$. Because of $[w] \in H^1(\Gamma_c \setminus \partial\Gamma_c)$ for $w \in H(\Omega_c)$, we have $B([w]/\varepsilon) \in H^1(\Gamma_c \setminus \partial\Gamma_c)$, and the boundary integral in the formula for Π_ε is well defined. The functional Π_ε is coercive due to $g \geq 0$, $B \geq 0$, strictly convex and lower semicontinuous. Hence, there exists a unique solution $w_\varepsilon \in H(\Omega_c)$ of the minimization problem

$$\Pi_\varepsilon(w_\varepsilon) = \inf_{\bar{w} \in H(\Omega_c)} \Pi_\varepsilon(\bar{w}), \tag{3.214}$$

which is equivalent to the variational inequality

$$b(w_\varepsilon, \bar{w} - w_\varepsilon) + \varepsilon \int_{\Gamma_c} g\left(B\left(\frac{[\bar{w}]}{\varepsilon}\right) - B\left(\frac{[w_\varepsilon]}{\varepsilon}\right)\right) \geq \int_{\Omega_c} f(\bar{w} - w_\varepsilon) \tag{3.215}$$

$$\forall\, \bar{w} \in H(\Omega_c).$$

Theorem 3.15. *The strong convergence $w_\varepsilon \to w$ in $H(\Omega_c)$ as $\varepsilon \to 0$ of the solutions w_ε for* (3.215) *yields that w is a solution of* (3.205), *and*

$$\|w_\varepsilon - w\|_{H(\Omega_c)} \leq C\sqrt{\varepsilon}, \qquad C^2 = \frac{1}{2}\|g\|_{L^1(\Gamma_c)}.$$

PROOF. We have the estimate $0 \leq \varepsilon B(t/\varepsilon) - |t| \leq \varepsilon/2$, $t \in R$, which holds also for any continuous function ϕ:

$$0 \leq \varepsilon B(\phi(x)/\varepsilon) - |\phi(x)| \leq \varepsilon/2 \quad \forall\, x. \tag{3.216}$$

Taking $\bar{w} = w_\varepsilon$ in (3.205) and $\bar{w} = w$ in (3.215), we obtain

$$b(w_\varepsilon - w, w_\varepsilon - w) \leq \int_{\Gamma_c} g\left(\left(\varepsilon B\left(\frac{[w]}{\varepsilon}\right) - \big|[w]\big|\right) - \left(\varepsilon B\left(\frac{[w_\varepsilon]}{\varepsilon}\right) - \big|[w_\varepsilon]\big|\right)\right).$$

The imbedding theorems ($[w], [w_\varepsilon] \in C(\Gamma_c)$) and the estimate (3.216) give

$$\|w_\varepsilon - w\|^2_{H(\Omega_c)} \leq \int_{\Gamma_c} g\left(\frac{\varepsilon}{2} - 0\right) \leq \frac{\varepsilon}{2}\|g\|_{L^1(\Gamma_c)}.$$

The theorem is proved.

Now we note that Π_ε is differentiable. Indeed,

$$\lim_{\lambda \to 0} \lambda^{-1}\Big(B(\phi(x) + \lambda h(x)) - B(\phi(x))\Big) = B'(\phi(x))h(x)$$

for any $\phi, h \in C(\Gamma_c)$, and, therefore,

$$b(w_\varepsilon, \bar{w}) + \int_{\Gamma_c} gB'\left(\frac{[w_\varepsilon]}{\varepsilon}\right)[\bar{w}] = \int_{\Omega_c} f\bar{w} \quad \forall\, \bar{w} \in H(\Omega_c). \tag{3.217}$$

By the above Green formula,

$$b(w_\varepsilon, \bar{w}) = D\int_{\Omega_c} \Delta^2 w_\varepsilon \bar{w} + \langle t(w_\varepsilon), [\bar{w}]\rangle_{\Gamma_c} - \langle m(w_\varepsilon), [\partial\bar{w}/\partial\nu]\rangle_{\Gamma_c}, \quad \bar{w} \in H(\Omega_c),$$

we conclude that the problem (3.217) is equivalent to the following boundary value problem,

$$D\Delta^2 w_\varepsilon = f \quad \text{in } \Omega_c, \tag{3.218}$$

$$m(w_\varepsilon) = 0, \quad [t(w_\varepsilon)] = 0, \quad t(w_\varepsilon) + gB'([w_\varepsilon]/\varepsilon) = 0 \quad \text{on } \Gamma_c,$$

in the same sense as before.

3.6.3 Iteration penalty method

Let us fix ε and construct the linear iterations for an arbitrary $u_\varepsilon^0 \in H(\Omega_c)$, $n \geq 0$, as follows:

$$b(w_\varepsilon^{n+1}, \bar{w}) + \int_{\Gamma_c} g\left(\frac{[w_\varepsilon^{n+1} - w_\varepsilon^n]}{\varepsilon}\right)[\bar{w}] \tag{3.219}$$

$$= \int_{\Omega_c} f\bar{w} - \int_{\Gamma_c} gB'\left(\frac{[w_\varepsilon^n]}{\varepsilon}\right)[\bar{w}] \quad \forall\, \bar{w} \in H(\Omega_c).$$

There obviously exists a unique solution $w_\varepsilon^{n+1} \in H(\Omega_c)$ for every n. It satisfies the boundary value problem

$$D\Delta^2 w_\varepsilon^{n+1} = f \quad \text{in } \Omega_c,$$

$$m(w_\varepsilon^{n+1}) = 0, \quad [t(w_\varepsilon^{n+1})] = 0 \quad \text{on } \Gamma_c,$$

$$t(w_\varepsilon^{n+1}) + \frac{1}{\varepsilon} g[w_\varepsilon^{n+1}] = g\left(\frac{1}{\varepsilon}[w_\varepsilon^n] - B'\left(\frac{[w_\varepsilon^n]}{\varepsilon}\right)\right) \quad \text{on } \Gamma_c.$$

Theorem 3.16. *The strong convergence $w_\varepsilon^{n+1} \to w_\varepsilon$ in $H(\Omega_c)$ takes place as $n \to \infty$.*

PROOF. Subtracting (3.217) from (3.219) and taking $w = w_\varepsilon^{n+1} - w_\varepsilon$, we have

$$\|w_\varepsilon^{n+1} - w_\varepsilon\|^2 + \int_{\Gamma_c} \frac{g[w_\varepsilon^{n+1} - w_\varepsilon]^2}{\varepsilon}$$

$$= \int_{\Gamma_c} g\left(\frac{1}{\varepsilon}[w_\varepsilon^n] - \frac{1}{\varepsilon}[w_\varepsilon] - B'\left(\frac{[w_\varepsilon^n]}{\varepsilon}\right) + B'\left(\frac{[w_\varepsilon]}{\varepsilon}\right)\right)[w_\varepsilon^{n+1} - w_\varepsilon].$$

One can verify that $|t - s - B'(t) + B'(s)| \leq |t - s|$, $t, s \in R$. Then, using the Holder inequality, from the previous equality we deduce that

$$2\|w_\varepsilon^{n+1} - w_\varepsilon\|^2 + \frac{1}{\varepsilon}\int_{\Gamma_c} g[w_\varepsilon^{n+1} - w_\varepsilon]^2 \leq \frac{1}{\varepsilon}\int_{\Gamma_c} g[w_\varepsilon^n - w_\varepsilon]^2. \qquad (3.220)$$

By the continuity of the trace operators and the positiveness of g, the estimate

$$c\int_{\Gamma_c} g[w]^2 \leq \|w\|^2 \quad , w \in H(\Omega_c),$$

follows. Therefore, (3.220) implies

$$\int_{\Gamma_c} g[w_\varepsilon^{n+1} - w_\varepsilon]^2 \leq \frac{1}{1+2c\varepsilon}\int_{\Gamma_c} g[w_\varepsilon^n - w_\varepsilon]^2 \leq \frac{1}{(1+2c\varepsilon)^{n+1}}\int_{\Gamma_c} g[w_\varepsilon^0 - w_\varepsilon]^2.$$

By $(1+2c\varepsilon)^{-1} < 1$, the right-hand side of this inequality converges to zero as $n \to \infty$. Consequently,

$$\int_{\Gamma_c} g[w_\varepsilon^{n+1} - w_\varepsilon]^2 \to 0 \quad \text{as } n \to \infty$$

and $\|w_\varepsilon^{n+1} - w_\varepsilon\|^2 \to 0$ as $n \to \infty$ in view of (3.220). The theorem is proved.

Let us note that for $\varepsilon \to 0$ the estimate (3.220) vanishes.

3.6.4 Saddle-point of the problem

By the construction of B, we have the estimate $|B'([w_\varepsilon(x)]/\varepsilon)| \leq 1$ for all $x \in \Gamma_c$ being uniform in ε. Then the sequence $\{B'([w_\varepsilon]/\varepsilon)\}$ is bounded in $L^2(\Gamma_c)$ and there exists a subsequence such that

$$B'([w_\varepsilon]/\varepsilon) \to p \quad \text{weakly in } L^2(\Gamma_c) \quad \text{as } \varepsilon \to 0, \qquad (3.221)$$

$p \in L^2(\Gamma_c)$ and $|p| \leq 1$ almost everywhere in Γ_c. Let us pass to the limit in (3.217) using Theorem 3.15 and (3.221). In doing so we obtain that the solution w of the problem (3.205) satisfies the following identity:

$$b(w, \bar{w}) + \int_{\Gamma_c} gp[\bar{w}] = \int_{\Omega_c} f\bar{w} \quad \forall \bar{w} \in H(\Omega_c). \qquad (3.222)$$

If we take here $\bar{w} = w$, the following equality is obtained:

$$b(w, w) + \int_{\Gamma_c} gp[w] = \int_{\Omega_c} fw.$$

On the other hand, the inequality (3.205) with the test functions $\bar{w} = 0$ and $\bar{w} = 2w$ gives

$$b(w, w) + \int_{\Gamma_c} g|[w]| = \int_{\Omega_c} fw.$$

A comparison of the last two equalities provides

$$\int_{\Gamma_c} g\left(\left|[w]\right| - p[w]\right) = 0. \tag{3.223}$$

In view of $|p(x)| \leq 1$ at almost all $x \in \Gamma_c$, we have

$$p(x)[w(x)] \leq \left|p(x)[w(x)]\right| \leq |p(x)| \left|[w(x)]\right| \leq \left|[w(x)]\right|,$$

i.e. $|[w]| - p[w] \geq 0$ at Γ_c. Consequently, (3.223) yields

$$\left|[w]\right| - p[w] = 0 \quad \text{at } \Gamma_c.$$

So we have obtained $p \in L^2(\Gamma_c)$ such that $|p| \leq 1$, $p[w] = |[w]|$ for the solution w of the problem (3.205). Besides, w is a solution to the problem (3.222), i.e.

$$D\Delta^2 w = f \text{ in } \Omega_c, \quad m(w) = 0, \ [t(w)] = 0, \ t(w) + gp[w] = 0 \text{ on } \Gamma_c.$$

Let us define the convex closed set $K = \{q \in L^2(\Gamma_c) \mid \ |q| \leq 1 \text{ a.e. } \Gamma_c\}$ and the Lagrange function

$$\mathcal{L}(\bar{w}, q) = \frac{1}{2}\|\bar{w}\|^2 + \int_{\Gamma_c} gq[\bar{w}] - \int_{\Omega_c} f\bar{w}, \quad \bar{w} \in H(\Omega_c), \quad q \in K.$$

Theorem 3.17. *The constructed pair* (w, p) *is a unique saddle-point of* $\mathcal{L}(\cdot, \cdot)$ *on the set* $H(\Omega_c) \times K$, *where* w *is a solution of* (3.205) *and* p *is defined by* (3.221).

PROOF. In view of $p[w] = |[w]|$, we have $\mathcal{L}(w, p) = \Pi(w)$. Equation (3.222) coincides with the Euler equation $\mathcal{L}'_w(\bar{w}, p) = 0 \ \forall \bar{w} \in H(\Omega_c)$ for the problem $\mathcal{L}(\bar{w}, p) \overset{\bar{w}}{\to} \text{extr}$. Since $\sup_{\bar{w} \in H(\Omega_c)} \mathcal{L}(\bar{w}, p) = \infty$, then

$$\mathcal{L}(w, p) = \inf_{\bar{w} \in H(\Omega_c)} \mathcal{L}(\bar{w}, p).$$

On the other hand, for an arbitrary $q \in K$, we have $|q(x)| \leq 1$ for almost all $x \in \Gamma_c$, and $\sup_{q \in K}(q(x)[w(x)]) = |[w(x)]| = p(x)[w(x)]$, $x \in \Gamma_c$. By the positiveness of g, this means

$$\mathcal{L}(w, p) = \sup_{q \in K} \mathcal{L}(w, q).$$

Consequently, we have obtained

$$\mathcal{L}(w,q) \le \mathcal{L}(w,p) = \Pi(w) \le \mathcal{L}(\bar{w},p) \quad \forall\, \bar{w} \in H(\Omega_c), \quad q \in K,$$

which proves that (w,p) is the saddle-point of $\mathcal{L}$. The uniqueness of w was discussed before. Let us verify the uniqueness of p. We assume that there exist $p_1, p_2 \in K$ such that $p_1[w] = |[w]| = p_2[w]$ at Γ_c. Then (3.222) for each p_i gives

$$b(w,\bar{w}) + \int\limits_{\Gamma_c} g p_i[\bar{w}] = \int\limits_{\Omega_c} f\bar{w} \quad \forall\, \bar{w} \in H(\Omega_c), \quad i = 1,2.$$

Subtracting these equations as $i = 1,2$, we obtain

$$\int\limits_{\Gamma_c} g(p_1 - p_2)[\bar{w}] = 0 \quad \forall\, \bar{w} \in H(\Omega_c),$$

which implies $p_1 - p_2 = 0$ a.e. at Γ_c. The theorem is proved.

Let us note that $B'(\cdot)$ is the projector of $L^2(\Gamma_c)$ onto the set K:

$$\|\phi - B'(\phi)\|_{L^2(\Gamma_c)} \le \|\phi - \psi\|_{L^2(\Gamma_c)} \quad \forall\, \psi \in K.$$

3.6.5 The friction problem for a bar with a cut

We consider the one-dimensional problem corresponding to $\Omega = (0,1)$; Γ_c is the fixed point y, $0 < y < 1$ and $\Omega_c = (0,y) \cup (y,1)$. The bilinear form b takes the form

$$b(w,\bar{w}) = D\int\limits_{\Omega_c} w_{xx}\bar{w}_{xx}\,dx, \quad w, \bar{w} \in H(\Omega_c),$$

where

$$H(\Omega_c) = \{w \in H^2(\Omega_c) \mid \quad w = w_x = 0 \quad \text{at} \;\; x = 0,1\}.$$

Let $f \in L^2(\Omega_c)$, and the friction coefficient $g \ge 0$ be given. The equilibrium problem for the clamped bar having a vertical cut with the friction between the cut edges is described by the following variational inequality:

$$D\int\limits_{\Omega_c} w_{xx}(\bar{w}_{xx} - w_{xx})\,dx + g\left(\left|[\bar{w}]\right| - \left|[w]\right|\right) \ge \int\limits_{\Omega_c} f(\bar{w} - w)\,dx \qquad (3.224)$$

$$\forall\, \bar{w} \in H(\Omega_c).$$

Here $[\bar{w}] = \bar{w}(y+0) - \bar{w}(y-0)$. Taking $\bar{w} = w \pm \xi$, $\xi \in C_0^\infty(\Omega_c)$, and substituting $\bar{w}$ into (3.224), we have the equilibrium equation fulfilled in Ω_c:

$$Dw_{xxxx} = f. \qquad (3.225)$$

By (3.225) and $f \in L^2(\Omega_c)$, it follows that the problem (3.224) has a solution $w \in H(\Omega_c)$ belonging to the space $H^4(\Omega_c) \subset C^3(\overline{\Omega}_c)$, $\overline{\Omega}_c = [0, y-0] \cup [y+0, 1]$. Therefore, the values of $t^\pm(w) = Dw_{xxx}(y \pm 0)$, $m^\pm(w) = Dw_{xx}(y \pm 0)$ are defined. Similarly to Section 3.6.1, we can obtain $[t(w)] = [m(w)] = 0$ and

$$t(w) = Dw_{xxx}(y), \quad m(w) = Dw_{xx}(y).$$

By the solution smoothness, the Green formula can be used,

$$\int\limits_{\Omega_c} w_{xx}\bar{w}_{xx}\,dx = \int\limits_{\Omega_c} w_{xxxx}\bar{w}\,dx - w_{xx}(y)[\bar{w}_x] + w_{xxx}(y)[\bar{w}] \quad \forall\,\bar{w} \in H(\Omega_c).$$

So, one can obtain the following formulation of the problem considered,

$$Dw_{xxxx} = f \quad \text{in } \Omega_c,$$

$$w_{xx}(y) = 0, \quad D|w_{xxx}(y)| \le g, \quad Dw_{xxx}(y)[w] + g\Big|[w]\Big| = 0.$$

To construct the solution of the problem (3.224), we shall apply the arguments of the previous subsection. Let us define the Lagrange function

$$\mathcal{L}(\bar{w}, q) = \frac{D}{2}\int\limits_{\Omega_c} \bar{w}_{xx}^2\,dx + gq[\bar{w}] - \int\limits_{\Omega_c} f\bar{w}\,dx$$

on the set of functions $\bar{w} \in H(\Omega_c)$, $q \in K = \{t \in R \mid \ |t| \le 1\}$. By Theorem 3.17, the solving of (3.224) is equivalent to the finding of the saddle-point $(w, p) \in H(\Omega_c) \times K$ such that

$$\mathcal{L}(w, p) = \inf_{\bar{w} \in H(\Omega_c)} \sup_{q \in K} \mathcal{L}(\bar{w}, q) = \sup_{q \in K} \inf_{\bar{w} \in H(\Omega_c)} \mathcal{L}(\bar{w}, q). \qquad (3.226)$$

Besides, the property $p[w] = |[w]|$ has already been proved. Let us consider the second (dual) problem in (3.226). We assume that $q \in K$ is a parameter and seek the solution $w^q \in H(\Omega_c)$ of the following problem,

$$\mathcal{L}(w^q, q) = \inf_{\bar{w} \in H(\Omega_c)} \mathcal{L}(\bar{w}, q),$$

which implies

$$D\int\limits_{\Omega_c} w_{xx}^q \bar{w}_{xx}\,dx + gq[\bar{w}] = \int\limits_{\Omega_c} fw\,dx \quad \forall\,\bar{w} \in H(\Omega_c). \qquad (3.227)$$

We consider the presentation $w^q = w^0 + w^{0,q}$ such that

$$D\int\limits_{\Omega_c} w_{xx}^0 \bar{w}_{xx}\,dx = \int\limits_{\Omega_c} f\bar{w}\,dx \quad \forall\,\bar{w} \in H(\Omega_c), \qquad (3.228)$$

$$D \int_{\Omega_c} w_{xx}^{0,q} \bar{w}_{xx}\, dx + gq[\bar{w}] = 0 \quad \forall \bar{w} \in H(\Omega_c). \tag{3.229}$$

By Green's formula, (3.228) gives the boundary value problem

$$Dw_{xxxx}^0 = f \quad \text{in } \Omega_c, \quad w_{xx}^0(y) = w_{xxx}^0(y) = 0.$$

There exists a unique solution $w^0 \in H(\Omega_c) \cap H^4(\Omega_c)$ to this problem since it reduces to the following independent problems:

$$\begin{array}{ll} Dw_{xxxx}^0 = f \quad \text{in } (0, y), & Dw_{xxxx}^0 = f \quad \text{in } (y, 1), \\ w_{xx}^0(y) = w_{xxx}^0(y) = 0, & w_{xx}^0(y) = w_{xxx}^0(y) = 0, \\ w^0(0) = w_x^0(0) = 0, & w^0(1) = w_x^0(1) = 0. \end{array}$$

Now let us consider (3.229), which is equivalent to the following boundary value problem:

$$w_{xxxx}^{0,q} = 0 \quad \text{in } \Omega_c, \quad w_{xx}^{0,q}(y) = 0, \quad Dw_{xxx}^{0,q}(y) + gq = 0.$$

We can construct its solution $w^{0,q} \in H(\Omega_c) \cap H^4(\Omega_c)$ by supposing $w^{0,q}(x) = -gqD^{-1}\alpha(x)$, where $\alpha \in H(\Omega_c)$ must satisfy the relations

$$\alpha_{xxxx} = 0 \quad \text{in } \Omega_c,$$

$$\alpha_{xx}(y) = 0, \quad \alpha_{xxx}(y) = 1, \quad \alpha(0) = \alpha_x(0) = \alpha(1) = \alpha_x(1) = 0.$$

It is easy to find this function

$$\alpha(x) = \frac{1}{6} \begin{cases} x^3 - 3yx^2 & , \quad x \in (0, y), \\ (x-1)^3 - 3(y-1)(x-1)^2 & , \quad x \in (y, 1) \end{cases} \tag{3.230}$$

belonging to the space $H(\Omega_c) \cap C^\infty(\Omega_c)$. Note that $[\alpha] = (3y^2 - 3y + 1)/3 > 0$. Thus, we obtain that the function $w^q(x) = w^0(x) - gqD^{-1}\alpha(x)$ is a unique solution of (3.227).

To find the solution $w = w^p$ of the problem (3.226) (which coincides with the solution w to the problem (3.224)), one needs to find the parameter $q = p$ in (3.227) satisfying

$$p[w^p] = \big|[w^p]\big|, \quad |p| \le 1. \tag{3.231}$$

Substituting here the presentation of w^p, we have

$$p\left([w^0] - gpD^{-1}[\alpha]\right) = \big|[w^0] - gpD^{-1}[\alpha]\big|, \quad |p| \le 1.$$

These conditions imply that

$$p = \begin{cases} 1 & , \quad \text{if } D[w^0](g[\alpha])^{-1} > 1, \\ -1 & , \quad \text{if } D[w^0](g[\alpha])^{-1} < -1, \\ D[w^0](g[\alpha])^{-1} & , \quad \text{if } \big|D[w^0](g[\alpha])^{-1}\big| \le 1. \end{cases}$$

Remembering the function B' of Section 3.6.2, one can see that

$$p = B'\left(\frac{D[w^0]}{g[\alpha]}\right).$$

Thus, we have proved the following statement.

Theorem 3.18. *The solution $w \in H(\Omega_c) \cap H^4(\Omega_c)$ of the problem* (3.224) *is given by*

$$w(x) = w^0(x) - \frac{g}{D}B'\left(\frac{3D[w^0]}{g(3y^2 - 3y + 1)}\right)\alpha(x),$$

where w^0 is a solution of the linear equation (3.228) *and α is a function given by* (3.230).

By the smoothness of α and w^0, we can conclude: if $f \in H^m(\Omega_c)$ then $w \in H^{m+4}(\Omega_c)$. We notice that the solution w is continuous in $(0,1)$, namely, $w \in H^4(0,1)$, provided that

$$\left|[w^0]\right| \leq \frac{g(3y^2 - 3y + 1)}{3D}, \quad [w^0_x] + \frac{3(1-2y)}{2(3y^2 - 3y + 1)}[w^0] = 0.$$

Indeed, one can calculate

$$[w] = \begin{cases} -g[\alpha]/D + [w^0] & , \quad [w^0] > g[\alpha]/D, \\ g[\alpha]/D + [w^0] & , \quad [w^0] < -g[\alpha]/D, \\ 0 & , \quad |[w^0]| \leq g[\alpha]/D, \end{cases}$$

$$[w_x] = [w^0_x] + \begin{cases} -g[\alpha_x]/D & , \quad [w^0] > g[\alpha]/D, \\ g[\alpha_x]/D & , \quad [w^0] < -g[\alpha]/D, \\ -[w^0][\alpha_x]/[\alpha] & , \quad |[w^0]| \leq g[\alpha]/D, \end{cases}$$

and $[w_{xx}] = [w_{xxx}] = 0$.

EXAMPLE. Let $f(x) \equiv kD$, and k be a constant. The solution of (3.228) is

$$w^0(x) = \frac{k}{24}\begin{cases} x^4 - 4yx^3 + 6y^2x^2 & , x \in (0,y), \\ (x-1)^4 - 4(y-1)(x-1)^3 & \\ \quad + 6(y-1)^2(x-1)^2 & , x \in (y,1), \end{cases}$$

with $[w^0] = k((y-1)^4 - y^4)/8$. By finding the value of

$$A \equiv \frac{g}{D}B'\left(\frac{D[w^0]}{g[\alpha]}\right) = \frac{g}{D}B'\left(\frac{3kD((y-1)^4 - y^4)}{8g(3y^2 - 3y + 1)}\right),$$

we find the solution of the problem (3.224) in the form

$$w(x) = \frac{1}{24}\begin{cases} kx^4 - 4(ky + A)x^3 + 6(ky + 2A)yx^2 & , \quad x \in (0,y), \\ k(x-1)^4 - 4\left(k(y-1) + A\right)(x-1)^3 & \\ \quad + 6\left(k(y-1) + 2A\right)(y-1)(x-1)^2 & , \quad x \in (y,1). \end{cases}$$

Chapter 4

Variation of cracks in solids

In this chapter we analyse some problems related to the variations of cracks in solids. The main focus is on the dependence of solutions on the crack length and crack shape. In particular, we find the derivative of the energy functional with respect to the crack length and establish the Griffith formulae widely used in fracture mechanics. The novelty of this result consists in deriving the Griffith formulae subject to nonpenetration conditions holding at the crack faces. We prove convergence of solutions in cases where the crack moves to the boundary or the crack length tends to zero.

4.1 Variation of a crack length

4.1.1 Two-dimensional case

Let $\Omega \subset R^2$ be a bounded domain with a smooth boundary Γ, and the set $\Gamma_\delta = \{(x_1, x_2) \mid 0 < x_1 < \delta,\ x_2 = 0\}$ belongs to Ω (see Fig.4.1).

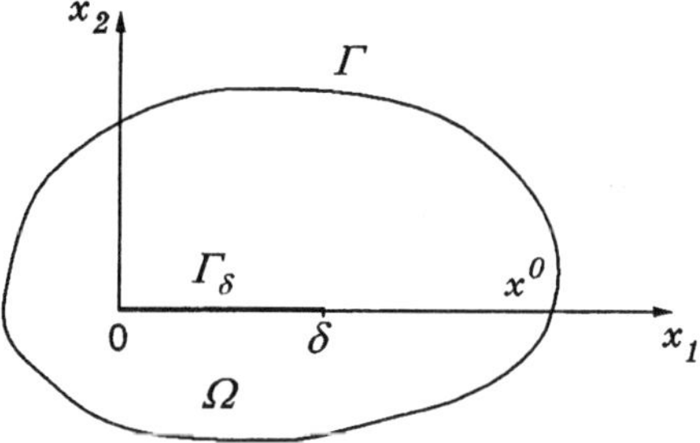

Fig.4.1. Crack length tending to zero

Denote $\Omega_\delta = \Omega \backslash \overline{\Gamma}_\delta$. The problem we analyse in this section is as follows. Let $W = (w^1, w^2)$,

$$\varepsilon_{ij}(W) = \frac{1}{2}(w^i_{,j} + w^j_{,i}),\ i,j = 1,2, \quad \Pi_\delta(W) = \frac{1}{2}\int\limits_{\Omega_\delta} \sigma_{ij}(W)\varepsilon_{ij}(W) - \int\limits_{\Omega_\delta} fW,$$

where $f = (f_1, f_2) \in L^2(\Omega)$ is a given function,

$$\sigma_{ij}(W) = a_{ijkl}\varepsilon_{kl}(W),\ i,j = 1,2, \quad a_{ijkl} = a_{klij} = a_{jikl}, \tag{4.1}$$

$$a_{ijkl}\xi_{kl}\xi_{ij} \geq c|\xi|^2,\ c > 0,$$

and $a_{ijkl} \in L^\infty(\Omega)$, $i, j, k, l = 1, 2$. Introduce the set of admissible displacements

$$K_\delta = \{W = (w^1, w^2) \in H^{1,0}(\Omega_\delta) \mid \ [W]\nu \geq 0 \quad \text{on } \Gamma_\delta\}.$$

Here $H^{1,0}(\Omega_\delta)$ is the Sobolev space of functions having square integrable first derivatives in Ω_δ and equal to zero on Γ, $\nu = (0,1)$ is the normal vector to Γ_δ, and $[W] = W^+ - W^-$, where $W^\pm$ correspond to the positive and negative directions of ν.

Let $B_\delta(W, U) = \int\limits_{\Omega_\delta} \sigma_{ij}(W)\varepsilon_{ij}(U)$. We consider the minimization problem

$$\min_{W \in K_\delta} \Pi_\delta(W).$$

Its solution $W^\delta \in K_\delta$ satisfies the variational inequality

$$B_\delta(W^\delta, V - W^\delta) \geq \langle f, V - W^\delta \rangle_\delta \quad \forall V \in K_\delta. \tag{4.2}$$

Here the brackets $\langle \cdot, \cdot \rangle_\delta$ denote the integration over Ω_δ, in particular, $\langle \cdot, \cdot \rangle_0$ is the integration over Ω. We aim at studying the behaviour of the solution W^δ as $\delta \to 0$. We first extend Γ_δ beyond the point $(\delta, 0)$, so that the extension $\Gamma_\star$ crosses the external boundary Γ at the point x^0. In the sequel δ tends to zero, but we do not assume, in general, that the extension $\Gamma_\star$ is a straight line. We suppose that the angle between the extension $\Gamma_\star$ and Γ at the point x^0 is not equal to zero. Denote by $H_0^1(\Omega)$ the completion of $C_0^\infty(\Omega)$ in the norm of $H^1(\Omega)$.

Now we are in a position to prove the following statement.

Theorem 4.1. *From the sequence W^δ one can choose a subsequence, still denoted by W^δ, such that as $\delta \to 0$*

$$W^\delta \to W \quad \text{weakly in } H^{1,0}(\Omega \backslash \Gamma_\star)$$

and, moreover, $W \in H_0^1(\Omega)$,

$$B_0(W, V) = \langle f, V \rangle_0 \quad \forall V \in H_0^1(\Omega). \tag{4.3}$$

PROOF. It is clear that for all $\delta > 0$, $W^\delta \in H^{1,0}(\Omega\backslash\Gamma_\star)$ and the jump $[W^\delta]$ is equal to zero at $\Gamma_\star\backslash\Gamma_\delta$. We take $V \equiv 0$ in (4.2) as a test function which implies

$$B_\delta(W^\delta, W^\delta) - \langle f, W^\delta\rangle_\delta \leq 0. \tag{4.4}$$

Note that there exists a constant $c > 0$ such that

$$\int\limits_{\Omega\backslash\Gamma_\star} \sigma_{ij}(W)\varepsilon_{ij}(W) \geq c\|W\|^2_{H^{1,0}(\Omega\backslash\Gamma_\star)} \quad \forall W \in H^{1,0}(\Omega\backslash\Gamma_\star). \tag{4.5}$$

From (4.4) it follows that

$$\int\limits_{\Omega\backslash\Gamma_\star} \sigma_{ij}(W^\delta)\varepsilon_{ij}(W^\delta) \leq \int\limits_{\Omega\backslash\Gamma_\star} fW^\delta$$

which, by (4.5), provides the uniform in δ estimate

$$\|W^\delta\|_{H^{1,0}(\Omega\backslash\Gamma_\star)} \leq c. \tag{4.6}$$

We can choose a subsequence, still denoted by W^δ, such that as $\delta \to 0$

$$W^\delta \to W \quad \text{weakly in } H^{1,0}(\Omega\backslash\Gamma_\star). \tag{4.7}$$

From (4.2) we obtain the inequality

$$\int\limits_{\Omega\backslash\Gamma_\star} \sigma_{ij}(W^\delta)\varepsilon_{ij}(V - W^\delta) \geq \int\limits_{\Omega\backslash\Gamma_\star} f(V - W^\delta) \tag{4.8}$$

holding for all $V \in K_\delta$. Let us take $V \in H_0^1(\Omega)$. Then $V \in K_\delta$ for all δ. By (4.7), it is possible to pass the limit in (4.8) as $\delta \to 0$, and we arrive at the relation

$$\int\limits_{\Omega\backslash\Gamma_\star} \sigma_{ij}(W)\varepsilon_{ij}(V - W) \geq \int\limits_{\Omega\backslash\Gamma_\star} f(V - W) \quad \forall V \in H_0^1(\Omega). \tag{4.9}$$

Since $[W] = 0$ on $\Gamma_\star$, the inclusion $W \in H_0^1(\Omega)$ follows. This means that we can integrate over Ω in (4.9), which implies

$$\int\limits_{\Omega} \sigma_{ij}(W)\varepsilon_{ij}(V - W) \geq \int\limits_{\Omega} f(V - W) \quad \forall V \in H_0^1(\Omega).$$

Consequently, the identity (4.3) is obtained. Theorem 4.1 is proved.

4.2 A plate with a crack

We analyse the behaviour of solutions for a plate having a crack provided that the crack length tends to zero. The nonpenetration conditions are assumed to hold at the crack faces.

4.2.1 Convergence of solutions

Let $\Omega \subset R^2$ be a bounded domain with a smooth boundary Γ, and the set $\Gamma_\delta = \{(x_1, x_2) \mid 0 < x_1 < \delta,\ x_2 = 0\}$ belongs to Ω. The domain $\Omega_\delta = \Omega \backslash \Gamma_\delta$ corresponds to the mid-surface of the plate. Consider the space $H(\Omega_\delta) = H^{1,0}(\Omega_\delta) \times H^{1,0}(\Omega_\delta) \times H^{2,0}(\Omega_\delta)$, where $H^{s,0}(\Omega_\delta)$ is the completion in the $H^s(\Omega_\delta)$-norm of smooth functions equal to zero near Γ.

Let $\chi = (W, w)$ be a displacement vector of the mid-surface points of the plate, $W = (w^1, w^2)$. The set of admissible displacements of the plate is as follows

$$K_\delta = \{(W, w) \in H(\Omega_\delta) \mid \ [W]\nu \geq \left|\left[\frac{\partial w}{\partial \nu}\right]\right| \quad \text{a.e. on } \Gamma_\delta\}.$$

Here $\nu = (0, 1)$ is the normal vector to Γ_δ, $[U] = U^+ - U^-$ is the jump of U across Γ_δ, $U^\pm$ fit to the positive and negative directions of ν. Introduce the strain $\varepsilon_{ij} = \varepsilon_{ij}(W)$ and the stress $\sigma_{ij} = \sigma_{ij}(W)$ tensors,

$$\varepsilon_{ij}(W) = \frac{1}{2}(w^i_{,j} + w^j_{,i}), \quad i, j = 1, 2,$$

$$\sigma_{11} = \varepsilon_{11} + \kappa\varepsilon_{22}, \quad \sigma_{22} = \varepsilon_{22} + \kappa\varepsilon_{11}, \quad \sigma_{12} = (1 - \kappa)\varepsilon_{12},$$

$\kappa = \text{const}$, $0 < \kappa < 1/2$, and the bilinear forms

$$B_\delta(W, \bar{W}) = \langle \sigma_{ij}(W), \varepsilon_{ij}(\bar{W}) \rangle_\delta,$$

$$b_\delta(w, \bar{w}) = \int_{\Omega_\delta} (w_{xx}\bar{w}_{xx} + w_{yy}\bar{w}_{yy} + \kappa w_{xx}\bar{w}_{yy} + \kappa w_{yy}\bar{w}_{xx}$$

$$+ 2(1 - \kappa) w_{xy}\bar{w}_{xy}).$$

Here the bracket $\langle \cdot, \cdot \rangle_\delta$ denotes integration over Ω_δ. In the sequel we shall use the notations $B_\Omega(\cdot, \cdot)$, $b_\Omega(\cdot, \cdot)$, $\langle \cdot, \cdot \rangle_\Omega$, which mean integration over Ω.

Introduce the energy functional of the plate

$$\Pi_\delta(\chi) = \frac{1}{2} B_\delta(W, W) + \frac{1}{2} b_\delta(w, w) - \langle f, \chi \rangle_\delta,$$

where $f = (f_1, f_2, f_3) \in L^2(\Omega)$ is a given exterior force, and consider the minimization problem

$$\min_{\chi \in K_\delta} \Pi_\delta(\chi). \tag{4.10}$$

There exists a unique solution $\chi^\delta = (W^\delta, w^\delta) \in K_\delta$ of the problem (4.10) which satisfies the following variational inequality

$$B_\delta(W^\delta, \bar{W} - W^\delta) + b_\delta(w^\delta, \bar{w} - w^\delta) \geq \langle f, \bar{\chi} - \chi^\delta \rangle_\delta \tag{4.11}$$

$$\forall \bar{\chi} = (\bar{W}, \bar{w}) \in K_\delta.$$

We extend Γ_δ beyond the point $(\delta, 0)$ so that the extension $\Gamma_\star$ crosses the external boundary Γ at some point x^0. The angle between $\Gamma_\star$ and Γ at the point x^0 is assumed to be nonzero. Let $H_0(\Omega) = H_0^1(\Omega) \times H_0^1(\Omega) \times H_0^2(\Omega)$. The following assertion holds.

Theorem 4.2. *From the sequence χ^δ one can choose a subsequence, still denoted by χ^δ, such that as $\delta \to 0$*

$$\chi^\delta \to \chi \quad \textit{weakly in } H(\Omega\backslash\Gamma_\star)$$

and, besides, $\chi \in H_0(\Omega)$,

$$B_\Omega(W, \bar{W}) + b_\Omega(w, \bar{w}) = \langle f, \bar{\chi}\rangle_\Omega \quad \forall \bar{\chi} = (\bar{W}, \bar{w}) \in H_0(\Omega). \tag{4.12}$$

PROOF. The restriction of χ^δ to $\Omega\backslash\Gamma_\star$, still denoted by χ^δ, satisfies the inclusion $\chi^\delta = (W^\delta, w^\delta) \in H(\Omega\backslash\Gamma_\star)$, and we have

$$[W^\delta] = 0, \quad [w^\delta] = 0, \quad \left[\frac{\partial w^\delta}{\partial \nu}\right] = 0 \quad \text{on } \Gamma\backslash\Gamma_\delta. \tag{4.13}$$

Substitution of $\bar{\chi} \equiv 0$ in (4.11) as a test function implies the inequality

$$B_\delta(W^\delta, W^\delta) + b_\delta(w^\delta, w^\delta) \leq \langle f, \chi^\delta\rangle_\delta. \tag{4.14}$$

We know that the following inequalities hold,

$$B_{\Omega\backslash\Gamma_\star}(W, W) \geq c\|W\|^2_{H^{1,0}(\Omega\backslash\Gamma_\star)} \quad \forall W \in H^{1,0}(\Omega\backslash\Gamma_\star),$$

$$b_{\Omega\backslash\Gamma_\star}(w, w) \geq c\|w\|^2_{H^{2,0}(\Omega\backslash\Gamma_\star)} \quad \forall w \in H^{2,0}(\Omega\backslash\Gamma_\star),$$

with the constants independent of W, w. Consequently, by (4.14),

$$\|W^\delta\|_{H^{1,0}(\Omega\backslash\Gamma_\star)} + \|w^\delta\|_{H^{2,0}(\Omega\backslash\Gamma_\star)} \leq c \tag{4.15}$$

uniformly in δ. In view of (4.15), one can choose a subsequence (W^δ, w^δ) such that as $\delta \to 0$

$$W^\delta \to W \quad \text{weakly in } H^{1,0}(\Omega\backslash\Gamma_\star), \tag{4.16}$$

$$w^\delta \to w \quad \text{weakly in } H^{2,0}(\Omega\backslash\Gamma_\star). \tag{4.17}$$

We can choose $\bar{\chi} \in H_0(\Omega)$ as a test function in (4.11) and integrate over $\Omega\backslash\Gamma_\star$. This provides

$$B_{\Omega\backslash\Gamma_\star}(W^\delta, \bar{W} - W^\delta) + b_{\Omega\backslash\Gamma_\star}(w^\delta, \bar{w} - w^\delta) \geq \langle f, \bar{w} - w^\delta\rangle_{\Omega\backslash\Gamma_\star}. \tag{4.18}$$

The convergence (4.16), (4.17) allows us to pass to the limit in (4.18) as $\delta \to 0$, which yields

$$B_{\Omega\backslash\Gamma_\star}(W, \bar{W} - W) + b_{\Omega\backslash\Gamma_\star}(w, \bar{w} - w) \geq \langle f, \bar{w} - w\rangle_{\Omega\backslash\Gamma_\star} \tag{4.19}$$

$$\forall \bar{\chi} = (\bar{W}, \bar{w}) \in H_0(\Omega).$$

By (4.13), we have

$$[W] = 0, \quad [w] = 0, \quad \left[\frac{\partial w}{\partial \nu}\right] = 0 \quad \text{on } \Gamma_\star;$$

consequently, $W = (w^1, w^2) \in H_0^1(\Omega)$, $w \in H_0^2(\Omega)$, i.e. $\chi = (W, w) \in H_0(\Omega)$. In this case the inequality (4.19) implies

$$B_\Omega(W, \bar{W} - W) + b_\Omega(w, \bar{w} - w) \geq \langle f, \bar{w} - w \rangle_\Omega \quad \forall \bar{\chi} \in H_0(\Omega),$$

and hence, the identity (4.12) follows. Theorem 4.2 is proved.

4.3 A crack moving to the external boundary

In this section we consider the two-dimensional elastic linear body having a crack which moves to the external boundary. The problem is to analyse the behaviour of the solution – in particular, to prove its convergence.

4.3.1 Equality type boundary conditions

Let $\Omega \subset R^2$ be a bounded domain with a smooth boundary Γ, and $x^0 \in \Gamma$, $x^1 \in \Gamma$, $x^0 \neq x^1$ (see Fig.4.2). We consider the smooth curve $\Gamma_0 \subset \Omega$ such that its ends are the points x^1, x^0. Let $x^\delta \in \Gamma_0$, and the length of the curve with the ends x^δ, x^0 be equal to $\delta > 0$. The curve having the ends x^1, x^δ is denoted by Γ_δ. In our considerations, the ends x^1, x^0 and x^1, x^δ do not belong to the curves Γ_0, Γ_δ, respectively.

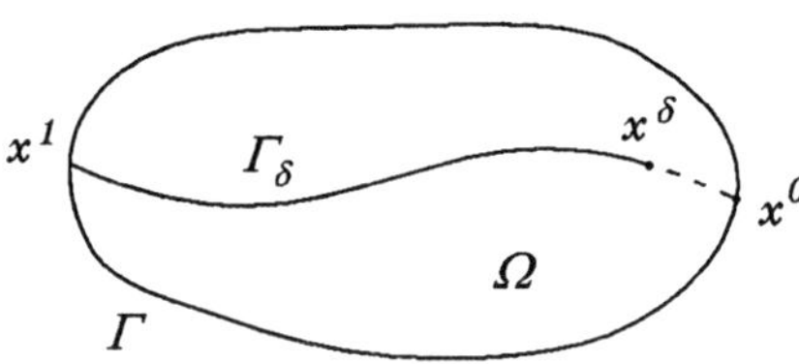

Fig.4.2. Crack moving to the external boundary

We assume that Γ_0 divides the domain Ω into two subdomains Ω_1, Ω_2 with the Lipschitz boundaries $\partial\Omega_1$, $\partial\Omega_2$. Let $\Gamma_i = \Gamma \cap \partial\Omega_i$, $i = 1, 2$, and

$$H^{1,\Gamma_i}(\Omega_i) = \{w \in H^1(\Omega_i) \mid \quad w = 0 \quad \text{on } \Gamma_i\}, \quad i = 1, 2.$$

We consider two boundary value problems

$$W_i = (w_i^1, w_i^2) \in H^{1,\Gamma_i}(\Omega_i), \tag{4.20}$$

$$B_{\Omega_i}(W_i, \bar{W}) = \langle f, \bar{W}\rangle_{\Omega_i} \quad \forall \bar{W} = (\bar{w}^1, \bar{w}^2) \in H^{1,\Gamma_i}(\Omega_i), \quad i = 1, 2. \qquad (4.21)$$

Here $B_{\Omega_i}(W, U) = \langle \sigma_{pq}(W), \varepsilon_{pq}(U)\rangle_{\Omega_i}$, $i = 1, 2$, and $\sigma_{pq}(W)$, $\varepsilon_{pq}(W)$ satisfy the Hooke law (4.1) of Section 4.1, the brackets $\langle \cdot, \cdot \rangle_{\Omega_i}$ mean the integration over Ω_i, and $f = (f_1, f_2) \in L^2(\Omega)$.

For each fixed $\delta > 0$ we consider the boundary value problem in the domain $\Omega_\delta = \Omega\backslash\Gamma_\delta$. Namely, let

$$H^{1,\delta}(\Omega_\delta) = \{w \in H^1(\Omega_\delta) \mid \quad w = 0 \quad \text{on} \quad \Gamma\}.$$

We want to find a function $W^\delta = (w^{1\delta}, w^{2\delta}) \in H^{1,\delta}(\Omega_\delta)$ such that

$$B_\delta(W^\delta, \bar{W}) = \langle f, \bar{W}\rangle_\delta \quad \forall \bar{W} = (\bar{w}^1, \bar{w}^2) \in H^{1,\delta}(\Omega_\delta). \qquad (4.22)$$

Here

$$B_\delta(U, V) = \int\limits_{\Omega_\delta} \sigma_{ij}(U)\varepsilon_{ij}(V), \quad \langle f, \bar{W}\rangle_\delta = \int\limits_{\Omega_\delta} f\bar{W}.$$

It is clear that the problem (4.20), (4.21) and the problem (4.22) admit the variational formulation. Denote by $\nu = (\nu_1, \nu_2)$ the unit normal vector to Γ_0. In this case, it follows from (4.20), (4.21) that

$$\sigma_{ij}(W_k)\nu_j = 0, \quad i, k = 1, 2, \quad \text{on} \ \Gamma_0. \qquad (4.23)$$

Also, it follows from (4.22) that

$$\sigma_{ij}(W^\delta)\nu_j = 0, \quad i = 1, 2, \quad \text{on} \ \Gamma_\delta. \qquad (4.24)$$

Note that conditions (4.23), (4.24) hold in the weak sense. We see that boundary conditions considered at the crack faces have the equality type in this section.

In what follows we prove that the restrictions of W^δ to Ω_i, $i = 1, 2$, denoted by $W^\delta|_{\Omega_i}$, converge to W_1, W_2, respectively, in a proper sense. The following statement holds.

Theorem 4.3. *From the sequence W^δ one can choose a subsequence, still denoted by W^δ, such that as $\delta \to 0$*

$$W^\delta|_{\Omega_i} \to W_i \quad \textit{weakly in} \ H^{1,\Gamma_i}(\Omega_i), \quad i = 1, 2, \qquad (4.25)$$

and W_i satisfy (4.20), (4.21).

Proof. Substitution of $\bar{W} = W^\delta$ in (4.22) as a test function implies

$$B_\delta(W^\delta, W^\delta) = \langle f, W^\delta\rangle_\delta.$$

We can integrate here over $\Omega_1 \cup \Omega_2$. In this case this relation can be rewritten in the form

$$B_{\Omega_1}(W^\delta, W^\delta) + B_{\Omega_2}(W^\delta, W^\delta) = \langle f, W^\delta\rangle_{\Omega_1} + \langle f, W^\delta\rangle_{\Omega_2}. \qquad (4.26)$$

Since

$$B_{\Omega_i}(W,W) \geq c\|W\|^2_{H^{1,\Gamma_i}(\Omega_i)}, \quad \forall W \in H^{1,\Gamma_i}(\Omega_i),$$

from (4.26) we have the estimate

$$\|W^\delta\|_{H^{1,\Gamma_1}(\Omega_1)} + \|W^\delta\|_{H^{1,\Gamma_2}(\Omega_2)} \leq c \tag{4.27}$$

being uniform in δ. The notation W^δ is used in (4.27) instead of $W^\delta|_{\Omega_i}$ to simplify the formula.

Take $\bar{W} \in H^{1,\delta_0}(\Omega_{\delta_0})$ in (4.22) as a test function with a fixed $\delta_0 > 0$. Then $\bar{W} \in H^{1,\delta}(\Omega_\delta)$ for all $\delta < \delta_0$. In this case, (4.22) implies

$$B_{\Omega_1}(W^\delta, \bar{W}) + B_{\Omega_2}(W^\delta, \bar{W}) = \langle f, \bar{W}\rangle_{\Omega_1} + \langle f, \bar{W}\rangle_{\Omega_2}. \tag{4.28}$$

On the other hand, by (4.27), one can choose a subsequence W^δ such that as $\delta \to 0$

$$W^\delta|_{\Omega_1} \to W_1 \quad \text{weakly in } H^{1,\Gamma_1}(\Omega_1), \tag{4.29}$$

$$W^\delta|_{\Omega_2} \to W_2 \quad \text{weakly in } H^{1,\Gamma_2}(\Omega_2). \tag{4.30}$$

Convergences (4.29), (4.30) allow us to pass to the limit in (4.28) as $\delta \to 0$. As a result, the following identity is obtained

$$B_{\Omega_1}(W_1, \bar{W}) + B_{\Omega_2}(W_2, \bar{W}) = \langle f, \bar{W}\rangle_{\Omega_1} + \langle f, \bar{W}\rangle_{\Omega_2} \tag{4.31}$$

as holding for all $\bar{W} \in H^{1,\delta}(\Omega_\delta)$.

By Lemma 4.1, below, for any $\bar{W}_i \in H^{1,\Gamma_i}(\Omega_i)$, $i = 1,2$, we choose a sequence $\bar{W}^\delta \in H^{1,\delta}(\Omega_\delta)$ such that as $\delta \to 0$

$$\bar{W}^\delta|_{\Omega_i} \to \bar{W}_i \quad \text{weakly in } H^{1,\Gamma_i}(\Omega_i), \quad i = 1,2,$$

and substitute $\bar{W} = \bar{W}^\delta$ in (4.31). Next we can pass to the limit in (4.31) as $\delta \to 0$. This provides

$$W_1 \in H^{1,\Gamma_1}(\Omega_1): \quad B_{\Omega_1}(W_1, \bar{W}_1) = \langle f, \bar{W}_1\rangle_{\Omega_1} \quad \forall \bar{W}_1 \in H^{1,\Gamma_1}(\Omega_1),$$

$$W_2 \in H^{1,\Gamma_2}(\Omega_2): \quad B_{\Omega_2}(W_2, \bar{W}_2) = \langle f, \bar{W}_2\rangle_{\Omega_2} \quad \forall \bar{W}_2 \in H^{1,\Gamma_2}(\Omega_2),$$

i.e. we obtain (4.20), (4.21) which proves the theorem.

Now we have to verify the auxiliary statement used to prove Theorem 4.3.

Lemma 4.1. *For any fixed* $\bar{W}_i = (\bar{w}_i^1, \bar{w}_i^2) \in H^{1,\Gamma_i}(\Omega_i)$, $i = 1,2$, *there exists a subsequence* $\bar{W}^\delta \in H^{1,\delta}(\Omega_\delta)$ *such that as* $\delta \to 0$

$$\bar{W}^\delta|_{\Omega_i} \to \bar{W}_i \quad \textit{strongly in } H^{1,\Gamma_i}(\Omega_i), \quad i = 1,2. \tag{4.32}$$

Proof. The condition $\bar{W}_1 \in H^{1,\Gamma_1}(\Omega_1)$ means that there exists a sequence of smooth functions $\tilde{W}_1^\delta = (\tilde{w}_1^{1\delta}, \tilde{w}_1^{2\delta})$ strongly converging to $\bar{W}_1$ in

$H^{1,\Gamma_1}(\Omega_1)$, and each function $\tilde{W}_1^\delta$ is equal to zero in a neighbourhood of Γ_1. A similar statement holds for $\bar{W}_2 \in H^{1,\Gamma_2}(\Omega_2)$: there exists a sequence $\tilde{W}_2^\delta$ strongly converging in $H^{1,\Gamma_2}(\Omega_2)$ to $\bar{W}_2$, and the functions $\tilde{W}_2^\delta$ are equal to zero in a neighbourhood of Γ_2. Define the function

$$\bar{W}^\delta(x) = \begin{cases} \tilde{W}_1^\delta(x)\,, & x \in \Omega_1, \\ \tilde{W}_2^\delta(x)\,, & x \in \Omega_2. \end{cases}$$

Since $\tilde{W}_i^\delta$, $i = 1,2$, are equal to zero on $\Gamma_0 \backslash \Gamma_\delta$, we put $\bar{W}^\delta(x) = 0$ on $\Gamma_0 \setminus \Gamma_\delta$ and this new function, still denoted by $\bar{W}^\delta$, satisfies the condition $\bar{W}^\delta \in H^{1,\delta}(\Omega_\delta)$ which proves the lemma.

4.4 A case of a shallow shell

The problem similar to that considered in the preceding section is analysed here for the linear shallow shell model.

4.4.1 Equality type boundary conditions

Let Ω, Ω_δ, Γ_0, Γ_δ, Γ_i, Ω_i, $i = 1,2$, correspond to those of Section 4.3. We put $H(\Omega_\delta) = H^{1,\delta}(\Omega_\delta) \times H^{1,\delta}(\Omega_\delta) \times H^{2,\delta}(\Omega_\delta)$, where the space $H^{1,\delta}(\Omega_\delta)$ was defined in the previous section, and $H^{2,\delta}(\Omega_\delta)$ is the space of functions from $H^2(\Omega_\delta)$ equal to zero on Γ with the first derivatives. Let $\chi^\delta = (W^\delta, w^\delta)$ be the displacement vector of the mid-surface points of the shell, $W^\delta = (w^{1\delta}, w^{2\delta})$,

$$\varepsilon_{ij}(W) = \frac{1}{2}(w^i_{,j} + w^j_{,i}), \quad e_{ij} = \varepsilon_{ij} + k_{ij}w, \quad i,j = 1,2, \quad W = (w^1, w^2),$$

$$\sigma_{11} = e_{11} + \kappa e_{22}, \quad \sigma_{22} = e_{22} + \kappa e_{11}, \quad \sigma_{12} = (1-\kappa)e_{12},$$

$\kappa = \text{const}$, $0 < \kappa < 1/2$. The functions $e_{ij} = e_{ij}(\chi^\delta)$, $\sigma_{ij} = \sigma_{ij}(\chi^\delta)$ fit to strain and stress components of the shell, $k_{ij} \in C^1(\bar{\Omega})$ are the curvatures of the shell, $f = (f_1, f_2, f_3) \in L^2(\Omega)$ is a given vector of exterior forces.

In the domain Ω_δ, we want to find a solution of the following boundary value problem:

$$\chi^\delta = (W^\delta, w^\delta) \in H(\Omega_\delta), \tag{4.33}$$

$$b_\delta(w^\delta, \bar{w}) + \langle k_{ij}\sigma_{ij}(\chi^\delta), \bar{w}\rangle_\delta = \langle f_3, \bar{w}\rangle_\delta \quad \forall \bar{w} \in H^{2,\delta}(\Omega_\delta), \tag{4.34}$$

$$\langle \sigma_{ij}(\chi^\delta), \varepsilon_{ij}(\bar{W})\rangle_\delta = \langle f_i, \bar{w}^i\rangle_\delta \quad \forall \bar{W} = (\bar{w}^1, \bar{w}^2) \in H^{1,\delta}(\Omega_\delta). \tag{4.35}$$

The bilinear form $b_\delta(w, \bar{w})$ was defined in Section 4.2, and the integration in (4.34), (4.35) is carried out over Ω_δ. Consider the energy functional of the shell

$$\Pi_\delta(\chi) = \frac{1}{2}b_\delta(w,w) + \frac{1}{2}\langle \sigma_{ij}, e_{ij}\rangle_\delta - \langle f, \chi\rangle_\delta.$$

In this case the problem (4.33)–(4.35) fits the variational formulation

$$\min_{\chi \in H(\Omega_\delta)} \Pi_\delta(\chi). \tag{4.36}$$

Problem (4.36) (or the problem (4.33)–(4.35)) has a unique solution $\chi = \chi^\delta$.

We aim at studying the behaviour of the solution χ^δ as $\delta \to 0$. The limiting case $\delta = 0$ corresponds to the cut Γ_0 which divides Ω into two subdomains Ω_1, Ω_2. In domain Ω_1 we can solve the problem

$$\chi_1 = (W_1, w_1) \in H(\Omega_1), \tag{4.37}$$

$$b_{\Omega_1}(w_1, \bar{w}) + \langle k_{ij}\sigma_{ij}(\chi_1), \bar{w}\rangle_{\Omega_1} = \langle f_3, \bar{w}\rangle_{\Omega_1} \quad \forall \bar{w} \in H^{2,\Gamma_1}(\Omega_1), \tag{4.38}$$

$$\langle \sigma_{ij}(\chi_1), \varepsilon_{ij}(\bar{W})\rangle_{\Omega_1} = \langle f_i, \bar{w}^i\rangle_{\Omega_1} \quad \forall \bar{W} = (\bar{w}^1, \bar{w}^2) \in H^{1,\Gamma_1}(\Omega_1). \tag{4.39}$$

Here $H(\Omega_i) = H^{1,\Gamma_i}(\Omega_i) \times H^{1,\Gamma_i}(\Omega_i) \times H^{2,\Gamma_i}(\Omega_i)$, $H^{2,\Gamma_i}(\Omega_i)$ is the completion in the norm of $H^2(\Omega_i)$ of smooth functions equal to zero near Γ_i. The spaces $H^{1,\Gamma_i}(\Omega_i)$ were defined in Section 4.3. Similarly, in the domain Ω_2 the following problem can be solved:

$$\chi_2 = (W_2, w_2) \in H(\Omega_2), \tag{4.40}$$

$$b_{\Omega_2}(w_2, \bar{w}) + \langle k_{ij}\sigma_{ij}(\chi_2), \bar{w}\rangle_{\Omega_2} = \langle f_3, \bar{w}\rangle_{\Omega_2} \quad \forall \bar{w} \in H^{2,\Gamma_2}(\Omega_2), \tag{4.41}$$

$$\langle \sigma_{ij}(\chi_2), \varepsilon_{ij}(\bar{W})\rangle_{\Omega_2} = \langle f_i, \bar{w}^i\rangle_{\Omega_2} \quad \forall \bar{W} = (\bar{w}^1, \bar{w}^2) \in H^{1,\Gamma_2}(\Omega_2). \tag{4.42}$$

The following statement holds.

Theorem 4.4. *From the sequence χ^δ one can choose a subsequence, still denoted by χ^δ, such that as $\delta \to 0$*

$$\chi^\delta|_{\Omega_i} \to \chi_i \quad \textit{weakly in } H(\Omega_i), \quad i = 1, 2, \tag{4.43}$$

and χ_i, $i = 1, 2$, satisfy (4.37)–(4.39) *and* (4.40)–(4.42), *respectively.*

PROOF. Substitute $\chi^\delta = (W^\delta, w^\delta)$ as a test function in (4.34), (4.35). This implies

$$b_\delta(w^\delta, w^\delta) + \langle \sigma_{ij}(\chi^\delta), \bar{\varepsilon}_{ij}(\chi^\delta)\rangle_\delta = \langle f, \chi^\delta\rangle_\delta. \tag{4.44}$$

We can integrate over $\Omega_1 \cup \Omega_2$ in (4.44), which provides

$$b_{\Omega_1}(w^\delta, w^\delta) + b_{\Omega_2}(w^\delta, w^\delta) + \langle \sigma_{ij}(\chi^\delta), e_{ij}(\chi^\delta)\rangle_{\Omega_1} + \langle \sigma_{ij}(\chi^\delta), e_{ij}(\chi^\delta)\rangle_{\Omega_2} \tag{4.45}$$
$$= \langle f, \chi^\delta\rangle_{\Omega_1} + \langle f, \chi^\delta\rangle_{\Omega_2}.$$

It is well known (see Khludnev, Sokolowski, 1997) that as $\|\chi\|_{H(\Omega_p)} \to +\infty$, $\chi = (W, w)$,

$$b_{\Omega_p}(w, w) + \langle \sigma_{ij}(\chi), e_{ij}(\chi)\rangle_{\Omega_p} - \langle f, \chi\rangle_{\Omega_p} \to +\infty, \quad p = 1, 2. \tag{4.46}$$

In this case the relation (4.45) results in the inequality

$$\|\chi^\delta\|_{H(\Omega_1)} + \|\chi^\delta\|_{H(\Omega_2)} \le c \tag{4.47}$$

being uniform in δ. The notation χ^δ is used here instead of $\chi^\delta|_{\Omega_i}$, $i = 1, 2$. Choosing a subsequence, still denoted by χ^δ, we assume that as $\delta \to 0$

$$\chi^\delta \to \chi^1 \quad \text{weakly in } H(\Omega_1), \tag{4.48}$$

$$\chi^\delta \to \chi^2 \quad \text{weakly in } H(\Omega_2). \tag{4.49}$$

We take $\bar\chi \in H(\Omega_{\delta_0})$ as a test function in (4.33)–(4.35) and integrate over $\Omega_1 \cup \Omega_2$. In this case $\bar\chi \in H(\Omega_\delta)$ for all $\delta \le \delta_0$. By (4.48), (4.49), the passage to the limit can be fulfilled in (4.33)–(4.35) as $\delta \to 0$, which implies

$$b_{\Omega_1}(w_1, \bar w) + b_{\Omega_2}(w_2, \bar w) + \langle k_{ij}\sigma_{ij}(\chi_1), \bar w\rangle_{\Omega_1} + \langle k_{ij}\sigma_{ij}(\chi_2), \bar w\rangle_{\Omega_2} \tag{4.50}$$

$$+\langle \sigma_{ij}(\chi_1), \varepsilon(\bar W)\rangle_{\Omega_1} + \langle \sigma_{ij}(\chi_2), \varepsilon(\bar W)\rangle_{\Omega_2} = \langle f, \bar\chi\rangle_{\Omega_1} + \langle f, \bar\chi\rangle_{\Omega_2}.$$

This identity holds for all $\bar\chi \in H(\Omega_{\delta_0})$, and, hence, it holds for all $\bar\chi \in H(\Omega_\delta)$, since δ_0 is arbitrary. By the definition of the space $H(\Omega_i)$, for any fixed $\bar\chi \in H(\Omega_i)$, there exists a sequence $\bar\chi_i^\delta \in H(\Omega_i)$ such that as $\delta \to 0$

$$\bar\chi_i^\delta \to \bar\chi \quad \text{strongly in } H(\Omega_i), \quad i = 1, 2,$$

and $\bar\chi_i^\delta$ are equal to zero in a neighbourhood of Γ_i.

Define the function in Ω_1 and Ω_2

$$\bar\chi^\delta(x) = \begin{cases} \bar\chi_1^\delta(x)\,, & x \in \Omega_1, \\ \bar\chi_2^\delta(x)\,, & x \in \Omega_2. \end{cases}$$

Since $\bar\chi_i^\delta$, $i = 1, 2$, are equal to zero in a neighbourhood of $\Gamma_0\backslash\Gamma_\delta$, we put $\bar\chi^\delta = 0$ on $\Gamma_0\backslash\Gamma_\delta$ and obtain $\bar\chi^\delta \in H(\Omega_\delta)$. Hence, for any fixed $\bar\chi_1 \in H(\Omega_1)$, $\bar\chi_2 \in H(\Omega_2)$ there exists a sequence $\bar\chi^\delta \in H(\Omega_\delta)$ such that as $\delta \to 0$

$$\bar\chi^\delta|_{\Omega_i} \to \bar\chi_i \quad \text{strongly in } H(\Omega_i), \quad i = 1, 2. \tag{4.51}$$

By (4.51), we can substitute the elements $\bar\chi^\delta$ as the test functions $\bar\chi$ in (4.50) and pass to the limit as $\delta \to 0$. In consequence, the relations (4.37)–(4.42) are obtained, which proves the theorem.

4.5 A crack near the boundary

4.5.1 Case of a shell

In this section we consider the model of a shallow shell analysed in the previous section and prove the convergence of solutions provided that the length of the boundary crack tends to zero.

Let $\Omega \subset R^2$ be a bounded domain with a smooth boundary Γ, and $x^0 \in \Gamma$, $x^1 \in \Gamma$, $x^0 \neq x^1$ (see Fig.4.3). Consider the smooth curve $\Gamma_\star$ with the ends x^0, x^1 by assuming $\Gamma_\star \subset \Omega$. Let $x^\delta \in \Gamma_\star$, and the length of the curve Γ_δ with the ends x^0, x^δ be equal to $\delta > 0$. The points x^0, x^1 and x^0, x^δ are assumed not to belong to $\Gamma_\star, \Gamma_\delta$, respectively. The curve $\Gamma_\star$ divides Ω into two subdomain Ω_1, Ω_2. In the domain $\Omega_\delta = \Omega \backslash \Gamma_\delta$ we consider the equilibrium problem for the shallow shell. Our aim is to prove that the solutions of this problem converge to the solution of the equilibrium problem found in the smooth domain Ω as $\delta \to 0$.

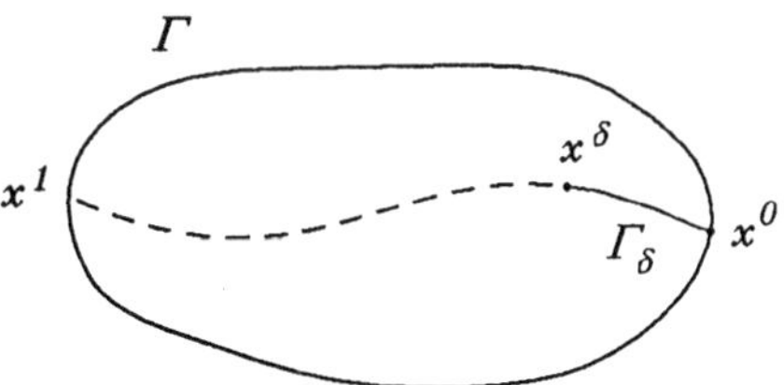

Fig.4.3. Crack near the boundary

Let $\chi = (W, w)$ be the displacement vector of the mid-surface points, and $e_{ij} = e_{ij}(\chi)$, $\sigma_{ij} = \sigma_{ij}(\chi)$ be defined as in the previous section, $H(\Omega_\delta) = H^{1,\delta}(\Omega_\delta) \times H^{1,\delta}(\Omega_\delta) \times H^{2,\delta}(\Omega_\delta)$, where

$$H^{1,\delta}(\Omega_\delta) = \{u \in H^1(\Omega_\delta) \mid \quad u = 0 \quad \text{on } \Gamma\},$$

$$H^{2,\delta}(\Omega_\delta) = \{u \in H^2(\Omega_\delta) \mid \quad u = \frac{\partial u}{\partial n} = 0 \quad \text{on } \Gamma\}.$$

Here n is a unit exterior vector to Γ. As it was indicated in the previous section, in the domain Ω_δ the following problem can be solved,

$$\chi^\delta = (W^\delta, w^\delta) \in H(\Omega_\delta), \tag{4.52}$$

$$b_\delta(w^\delta, \bar{w}) + \langle \sigma_{ij}(\chi_\delta), \varepsilon_{ij}(\bar{W})\rangle_\delta + \langle k_{ij}\sigma_{ij}(\chi^\delta), \bar{w}\rangle_\delta = \langle f, \bar{\chi}\rangle_\delta \tag{4.53}$$
$$\forall \bar{\chi} \in H(\Omega_\delta),$$

where $b_\delta(\cdot, \cdot)$ is the bilinear form used in the previous section with the integration over Ω_δ, $f = (f_1, f_2, f_3) \in L^2(\Omega)$.

Integrating over Ω_1 and Ω_2 in (4.53) we derive

$$b_{\Omega_1}(w^\delta, \bar{w}) + b_{\Omega_2}(w^\delta, \bar{w}) + \langle \sigma_{ij}(\chi^\delta), \varepsilon_{ij}(\bar{W})\rangle_{\Omega_1} + \langle \sigma_{ij}(\chi^\delta), \varepsilon_{ij}(\bar{W})\rangle_{\Omega_2} \tag{4.54}$$
$$+ \langle k_{ij}\sigma_{ij}(\chi^\delta), \bar{w}\rangle_{\Omega_1} + \langle k_{ij}\sigma_{ij}(\chi^\delta), \bar{w}\rangle_{\Omega_2} = \langle f, \bar{\chi}\rangle_{\Omega_1} + \langle f, \bar{\chi}\rangle_{\Omega_2}.$$

Substitution $\bar{\chi} = \chi^\delta$ in (4.54) as a test function implies the equality

$$b_{\Omega_1}(w^\delta, w^\delta) + b_{\Omega_2}(w^\delta, w^\delta) + \langle \sigma_{ij}(\chi^\delta), \varepsilon_{ij}(W^\delta)\rangle_{\Omega_1}$$

$$+\langle\sigma_{ij}(\chi^\delta),\varepsilon_{ij}(W^\delta)\rangle_{\Omega_2}+\langle k_{ij}\sigma_{ij}(\chi^\delta),w^\delta\rangle_{\Omega_1}+\langle k_{ij}\sigma_{ij}(\chi^\delta),w^\delta\rangle_{\Omega_2} \quad (4.55)$$
$$=\langle f,\chi^\delta\rangle_{\Omega_1}+\langle f,\chi^\delta\rangle_{\Omega_2}.$$

By the relations (4.46) of the preceding section, from (4.55) it follows that

$$\|\chi^\delta\|_{H(\Omega_1)}+\|\chi^\delta\|_{H(\Omega_2)}\le c \quad (4.56)$$

with the constant c uniform in δ. Here $H(\Omega_i)=H^{1,\Gamma_i}(\Omega_i)\times H^{1,\Gamma_i}(\Omega_i)\times H^{2,\Gamma_i}(\Omega_i)$, $i=1,2$, $H^{s,\Gamma_i}(\Omega_i)$ is the completion in $H^s(\Omega_i)$ of smooth functions equal to zero near Γ_i, $\Gamma_i=\Gamma\cap\partial\Omega_i$, $i=1,2$.

Choosing a subsequence still denoted by χ^δ we assume that as $\delta\to 0$

$$\chi^\delta|_{\Omega_1}\to\chi_1 \quad \text{weakly in } H(\Omega_1), \quad (4.57)$$

$$\chi^\delta|_{\Omega_2}\to\chi_2 \quad \text{weakly in } H(\Omega_2). \quad (4.58)$$

By our notation, $\Omega_0=\Omega$. We can take $\bar\chi\in H(\Omega)$ as a test function in (4.54). In this case $\bar\chi\in H(\Omega_\delta)$ for all $\delta>0$ (actually, we should consider the restriction of $\bar\chi$ to Ω_δ, but the notation for the functions is not changed). By (4.57), (4.58), the passage to the limit as $\delta\to 0$ in (4.54) implies

$$b_{\Omega_1}(w_1,\bar w)+b_{\Omega_2}(w_2,\bar w)+\langle\sigma_{ij}(\chi_1),\varepsilon_{ij}(\bar W)\rangle_{\Omega_1}+\langle\sigma_{ij}(\chi_2),\varepsilon_{ij}(\bar W)\rangle_{\Omega_2} \quad (4.59)$$
$$+\langle k_{ij}\sigma_{ij}(\chi_1),\bar w\rangle_{\Omega_1}+\langle k_{ij}\sigma_{ij}(\chi_2),\bar w\rangle_{\Omega_2}=\langle f,\bar\chi_1\rangle_{\Omega_1}+\langle f,\bar\chi_2\rangle_{\Omega_2}.$$

Here $\chi_i=(W_i,w_i)$, $i=1,2$. Denote by ν the unit normal vector to $\Gamma_\star$. Since

$$[w^\delta]=\left[\frac{\partial w^\delta}{\partial\nu}\right]=[W^\delta]=0 \quad \text{on } \Gamma\backslash\Gamma_\delta$$

we obtain that the limiting functions χ_1, χ_2 in (4.57), (4.58) satisfy the conditions

$$\chi_1=\chi_2, \quad \frac{\partial w_1}{\partial\nu}=\frac{\partial w_2}{\partial\nu} \quad \text{on } \Gamma_\star.$$

This allows us to define the function $\chi=(W,w)$, such that

$$\chi\in H(\Omega), \quad \chi(x)=\begin{cases}\chi_1(x), & x\in\Omega_1,\\ \chi_2(x), & x\in\Omega_2.\end{cases} \quad (4.60)$$

In this case, from (4.59) it follows that

$$b_\Omega(w,\bar w)+\langle\sigma_{ij}(\chi),\varepsilon_{ij}(\bar W)\rangle_\Omega+\langle k_{ij}\sigma_{ij}(\chi),\bar w\rangle_\Omega=\langle f,\bar\chi\rangle_\Omega \quad (4.61)$$
$$\forall\bar\chi\in H(\Omega).$$

Hence, the following statement has been proved.

Theorem 4.5. *From the solutions χ^δ of the problem* (4.52), (4.53) *one can choose a subsequence still denoted by χ^δ such that as $\delta\to 0$*

$$\chi^\delta|_{\Omega_i}\to\chi_i \quad \textit{weakly in } H(\Omega_i), \quad i=1,2,$$

and the limiting function χ satisfying (4.60) *is the solution of* (4.61).

4.6 Asymptotics of the energy functional for the Poisson equation

It is well known that the derivative of the energy functional is often used to formulate the fracture criterions (see Cherepanov, 1979). As a paradigm for noninterpenetrating crack models, the Poisson equation in a nonsmooth domain in R^2 is considered. The geometrical domain has a cut (a crack) of variable length. At the crack faces, inequality type boundary conditions are prescribed. The behaviour of the energy functional is analysed with respect to the crack length changes. In particular, the derivative of the energy functional with respect to the crack length is obtained. The associated Griffith formula is derived and properties of the solution are investigated. It is shown that the Rice–Cherepanov integral defined for the solutions of the unilateral problem defined in the nonsmooth domain is path-independent. The results of this section can be found in (Khludnev, Sokolowski, 1998b).

4.6.1 Problem formulation

In the present section the differentiability of the energy functional for an elliptic equation with respect to the crack length is shown. The method of proof is different from the proof in the linear case (Destuynder, Jaoua, 1981) since we cannot expect in general that the solution to the variational inequality for the displacement of an elastic membrane with unilateral conditions prescribed on the crack faces is differentiable with respect to the crack length. The method of the proof presented in this section is general and can be applied as well to the energy functionals of the linear elasticity system with the nonpenetration conditions prescribed on the crack faces. In the linear case, i.e. for the homogeneous Neumann boundary conditions prescribed on the crack faces in the scalar case, or for the traction free crack faces in elasticity, the results are known; we refer the reader to (Cherepanov, 1979; Maz'ya, Nazarov, 1987) for the models currently used in the fracture mechanic, and to the paper (Bui, Ehrlacher, 1997) for a review of recent results on the applications to crack propagation. In the linear case, both the first and the second order derivatives of the energy functionals with respect to the crack length are evaluated and used for numerical methods of analysis of crack propagation in solids. However, it seems that we cannot expect in general the second order differentiability of the energy functional with respect to the crack length in the case of the nonlinear problem in which unilateral conditions are prescribed on the crack faces, i.e. only the second order directional differentiability can be obtained. Indeed, from the local point of view, we expect the gradient of the solution to have an inverse square root singularity at the prescribed tips but to be bounded at the edges of the contact set. We refer the reader to (Sokolowski, Zolesio, 1992) for the shape differentiability properties of solutions to variational inequalities in smooth domains. Note that the dependence of the energy functional on the crack length is important in fracture mechanics. The derivative of the functional is often used to formulate fracture criteria.

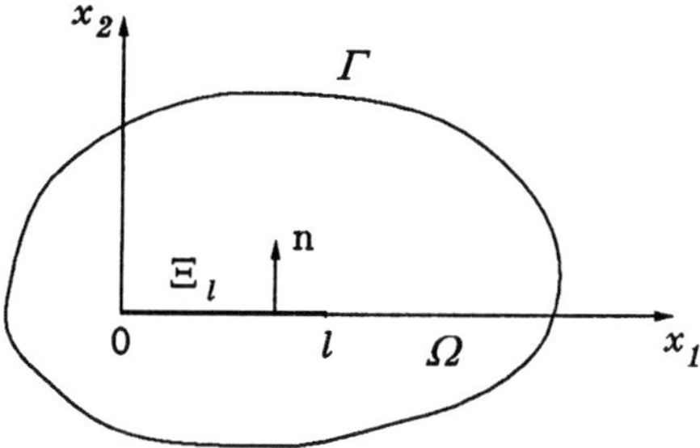

Fig.4.4. Domain with the crack

Let $D \subset R^2$ be a bounded domain with smooth boundary Γ, and $\Xi_{l+\delta}$ be the set $\{(x_1, x_2) \mid 0 < x_1 < l + \delta,\ x_2 = 0\}$. We assume that this set belongs to the domain D for all sufficiently small δ, and $l > 0$. Denote $\Omega_\delta = D \setminus \bar{\Xi}_{l+\delta}$, $\Omega = D \setminus \bar{\Xi}_l$ (see Fig.4.4).

In the domain Ω we consider the problem of finding a function u such that

$$-\Delta u = f, \tag{4.62}$$

$$u = 0 \quad \text{on } \Gamma, \quad [u] \geq 0 \quad \text{on } \Xi_l. \tag{4.63}$$

Here $f \in C^1(\bar{D})$ is the given function, and $[u] = u^+ - u^-$ is the jump of the function u across Ξ_l. The vector $n = (0, 1)$ is orthogonal to Ξ_l, and $u^\pm$ correspond to the positive and negative directions of n. The problem formulation (4.62), (4.63) is not complete to ensure the uniqueness of the solution. In fact, we consider the minimization of the functional

$$\frac{1}{2}\int_\Omega |\nabla\phi|^2 - \int_\Omega f\phi$$

over the set of all functions from $H^1(\Omega)$ satisfying the conditions $u = 0$ on Γ, $[u] \geq 0$ on Ξ_l. The solution of the minimization problem satisfies the variational inequality (4.81) (see below), and, in particular, it satisfies (4.62), (4.63). There are additional relations holding on Ξ_l, and we shall discuss them in the sequel. The energy functional for the problem (4.62)–(4.63) is defined by the formula

$$J(\Omega) = \frac{1}{2}\int_\Omega |\nabla u|^2 - \int_\Omega fu, \tag{4.64}$$

where u is the solution of (4.62)–(4.63).

For a small parameter δ, the family of problems defined in the domain Ω_δ is considered. We want to find a function u^δ such that

$$-\Delta u^\delta = f \quad \text{in } \Omega_\delta, \tag{4.65}$$

$$u^\delta = 0 \quad \text{on } \Gamma, \quad [u^\delta] \geq 0 \quad \text{on } \Xi_{l+\delta}. \tag{4.66}$$

Similar to (4.62), (4.63) the problem formulation (4.65), (4.66) is not complete. Actually, the function u^δ is the solution of variational inequality (4.78) below. The energy functional for the problem (4.65)–(4.66) is equal to

$$J(\Omega_\delta) = \frac{1}{2}\int\limits_{\Omega_\delta} |\nabla u^\delta|^2 - \int\limits_{\Omega_\delta} f u^\delta. \tag{4.67}$$

Our aim is to find the derivative

$$\left.\frac{dJ(\Omega_\delta)}{d\delta}\right|_{\delta=0} = \lim_{\delta\to 0} \frac{J(\Omega_\delta) - J(\Omega)}{\delta} \tag{4.68}$$

which describes the behaviour of the energy functional with respect to variation of the crack length.

4.6.2 Preliminary statement and formulae

In order to find the derivative (4.68) we fulfil a transformation of the domain Ω_δ, so that the domain is mapped onto Ω. Let $\theta \in C_0^\infty(D)$ be any function such that $\theta = 1$ in a neighbourhood of the point $x_l = (l, 0)$. To simplify the arguments the function θ is assumed to be equal to zero in a neighbourhood of the point $(0, 0)$. Consider the transformation of the independent variables

$$y_1 = x_1 - \delta\theta(x_1, x_2), \quad y_2 = x_2, \tag{4.69}$$

where $(x_1, x_2) \in \Omega_\delta$, $(y_1, y_2) \in \Omega$. The Jacobian q_δ of this transformation is equal to

$$\left|\frac{\partial(y_1, y_2)}{\partial(x_1, x_2)}\right| = 1 - \delta\theta_{x_1}.$$

For small δ, the Jacobian q_δ is positive, and hence the transformation (4.69) is one-to-one. Therefore, in view of (4.69) we have $y = y(x, \delta)$, $x = x(y, \delta)$.

Let $u^\delta(x)$ be the solution of (4.65), (4.66), and $u^\delta(x) = u_\delta(y)$, $x = x(y, \delta)$. We have the following formulae:

$$u^\delta_{x_1} = u_{\delta y_1}(1 - \delta\theta_{x_1}), \quad u^\delta_{x_2} = u_{\delta y_1}(-\delta\theta_{x_2}) + u_{\delta y_2}. \tag{4.70}$$

Consequently

$$\int\limits_{\Omega_\delta} |\nabla u^\delta|^2 dx = \int\limits_{\Omega} \langle A_\delta \nabla u_\delta, \nabla u_\delta \rangle dy,$$

where $A_\delta = A_\delta(y)$ is the matrix such that

$$A_\delta(y) = \frac{1}{1-\delta\theta_{x_1}} \begin{pmatrix} (1-\delta\theta_{x_1})^2 + \delta^2\theta_{x_2}^2 & -\delta\theta_{x_2} \\ -\delta\theta_{x_2} & 1 \end{pmatrix}, \quad \theta = \theta(x(y,\delta)).$$

Note that $A_0(y) = E$ is the unit matrix.

It is easy to find the derivative of $A_\delta(y)$ with respect to δ, namely,

$$A'(y) = \frac{dA_\delta(y)}{d\delta}\bigg|_{\delta=0} = \lim_{\delta\to 0} \frac{A_\delta(y) - A_0(y)}{\delta}.$$

We have

$$A'(y) = \begin{pmatrix} -\theta_{y_1}(y) & -\theta_{y_2}(y) \\ -\theta_{y_2}(y) & \theta_{y_1}(y) \end{pmatrix}. \tag{4.71}$$

Consider next the transformation

$$\int\limits_{\Omega_\delta} fu^\delta dx = \int\limits_{\Omega} \frac{f(x(y,\delta))u_\delta(y)}{1-\delta\theta_{x_1}} dy.$$

Denote

$$f^\delta(y) = \frac{f(x(y,\delta))}{1-\delta\theta_{x_1}}$$

and find the derivative

$$f'(y) = \frac{df^\delta(y)}{d\delta}\bigg|_{\delta=0} = \lim_{\delta\to 0} \frac{f^\delta(y) - f^0(y)}{\delta}.$$

Assuming that y, δ are independent variables in (4.69) we have $x = x(y,\delta)$. Differentiation of (4.69) with respect to δ yields

$$0 = \frac{dx_1}{d\delta} - \theta - \delta\theta_{x_1}\frac{dx_1}{d\delta},$$

whence

$$\frac{dx_1}{d\delta} = \frac{\theta}{1-\delta\theta_{x_1}}, \quad \frac{dx_2}{d\delta} = 0. \tag{4.72}$$

Consequently, by (4.72),

$$\frac{\partial f(x(y,\delta))}{\partial\delta}\bigg|_{\delta=0} = f_{x_1}\frac{dx_1}{d\delta}\bigg|_{\delta=0} + f_{x_2}\frac{dx_2}{d\delta}\bigg|_{\delta=0} = f_{y_1}\theta. \tag{4.73}$$

Now we are in a position to find the derivative $f'(y)$. Indeed, by (4.73)

$$f'(y) = \lim_{\delta\to 0}\left[\frac{f(x(y,\delta))}{1-\delta\theta_{x_1}} - f(y)\right]\frac{1}{\delta} = \lim_{\delta\to 0} \frac{f(x(y,\delta)) - f(y)}{\delta} + \theta_{x_1}f(y)\Big|_{\delta=0}$$

$$= f_{y_1}\theta + \theta_{y_1}f = \frac{\partial}{\partial y_1}(\theta f),$$

i.e.

$$f'(y) = (\theta f)_{y_1}(y). \tag{4.74}$$

Since $f \in C^1(\overline{\Omega})$ we can see that as $\delta \to 0$

$$\frac{f^\delta(y) - f^0(y)}{\delta} \to f'(y) \quad \text{in } L^\infty(\Omega). \tag{4.75}$$

Also, notice that, in addition to (4.71), as $\delta \to 0$

$$\frac{A_\delta(y) - A_0(y)}{\delta} \to A'(y) \quad \text{in } L^\infty(\Omega). \tag{4.76}$$

Introduce next the sets of admissible displacements in the problems (4.65)–(4.66), (4.62)–(4.63),

$$K_\delta = \{w \in H^1(\Omega_\delta) \mid \ [w] \geq 0 \quad \text{on } \Xi_{l+\delta}\},$$

$$K_0 = \{w \in H^1(\Omega) \mid \ [w] \geq 0 \quad \text{on } \Xi_l\}.$$

In accordance with (4.69), let $x = x(y, \delta)$. Then $w^\delta(x) = w_\delta(y)$. The inclusion $w^\delta \in K_\delta$ implies $w_\delta \in K_0$, and, conversely, $w_\delta \in K_0$ implies $w^\delta \in K_\delta$. This means that the transformation (4.69) maps K_δ onto K_0 , and it is one-to-one. Now we shall prove an auxiliary statement which is used in the sequel.

Lemma 4.2. *Let u^δ be the solution of* (4.65), (4.66), $u^\delta(x) = u_\delta(y)$, *and u be the solution of* (4.62), (4.63). *Then*

$$\|u_\delta - u\|_{H^1(\Omega)} \to 0, \quad \delta \to 0. \tag{4.77}$$

PROOF. The function $u^\delta \in K_\delta$ is the solution of the variational inequality

$$\int_{\Omega_\delta} \langle \nabla u^\delta, \nabla v - \nabla u^\delta \rangle \geq \int_{\Omega_\delta} f(v - u^\delta) \quad \forall v \in K_\delta. \tag{4.78}$$

We change the variables in (4.78) in accordance with (4.69). To this end we write (4.70) as

$$\nabla_x u^\delta = \nabla_y u_\delta - \delta g D_1 u_\delta,$$

where $D_1 u_\delta = u_{\delta y_1}, g = \nabla_x \theta$, which transforms (4.78) into the inequality

$$\int_\Omega \langle \nabla u_\delta, \nabla \tilde{v} - \nabla u_\delta \rangle \frac{1}{q_\delta} \geq \int_\Omega \langle h_\delta, \nabla \tilde{v} - \nabla u_\delta \rangle + \int_\Omega f^\delta(\tilde{v} - u_\delta) \tag{4.79}$$

$$+\delta \int_\Omega \langle \nabla u_\delta, g D_1 \tilde{v} - g D_1 u_\delta \rangle \frac{1}{q_\delta} + \delta^2 \int_\Omega \langle g D_1 u_\delta, g D_1 \tilde{v} - g D_1 u_\delta \rangle \frac{1}{q_\delta} \quad \forall \tilde{v} \in K_0.$$

Here

$$h_\delta = \frac{\delta g D_1 u_\delta}{q_\delta} \to 0 \quad \text{in } [L^2(\Omega)]^2$$

as $\delta \to 0$. It is of importance that the inequality (4.79) holds for all $\tilde{v} \in K_0$. From (4.78) it follows that

$$\|u^\delta\|_{H^1(\Omega_\delta)} \le c$$

uniformly in δ, consequently,

$$\|u_\delta\|_{H^1(\Omega)} \le c \tag{4.80}$$

uniformly in δ. The solution of the problem (4.62), (4.63) is the solution of the variational inequality

$$u \in K_0 : \quad \int_\Omega \langle \nabla u, \nabla v - \nabla u \rangle \ge \int_\Omega f(v-u) \quad \forall v \in K_0. \tag{4.81}$$

We can substitute $\tilde{v} = u$, $v = u_\delta$ in (4.79), (4.81), respectively, and sum the relations. This implies

$$\int_\Omega \langle \nabla u - \nabla u_\delta, \nabla u - \frac{\nabla u_\delta}{q_\delta} \rangle \le \int_\Omega (f^\delta - f)(u_\delta - u) \tag{4.82}$$

$$+ \int_\Omega \langle h_\delta, \nabla u_\delta - \nabla u \rangle + P(\delta, u, u_\delta, g).$$

By (4.79), (4.80), we have $P(\delta, u, u_\delta, g) \to 0$ as $\delta \to 0$. The inequality (4.82) can be written as

$$\|\nabla u - \nabla u_\delta\|_0^2 + \int_\Omega \langle \nabla u - \nabla u_\delta, \nabla u_\delta - \frac{\nabla u_\delta}{q_\delta} \rangle \le \int_\Omega \langle h_\delta, \nabla u_\delta - \nabla u \rangle$$

$$+ \int_\Omega (f^\delta - f)(u_\delta - u) + P(\delta, u, u_\delta, g),$$

where $\| \cdot \|_0$ is the norm in $L^2(\Omega)$. Hence,

$$\frac{1}{2}\|\nabla u - \nabla u_\delta\|_0^2 \le \|\nabla u_\delta - \frac{\nabla u_\delta}{q_\delta}\|_0^2 + \|h_\delta\|_0 \|\nabla u_\delta - \nabla u\|_0 \tag{4.83}$$

$$+ \|f^\delta - f\|_0 \|u_\delta - u\|_0 + P(\delta, u, u_\delta, g).$$

It is easy to see that

$$\left\| \nabla u_\delta - \frac{\nabla u_\delta}{q_\delta} \right\|_0 \le \delta \frac{\max\limits_\Omega |\theta_x|}{\min\limits_\Omega |q_\delta|} \|\nabla u_\delta\|_0 \to 0, \quad \delta \to 0.$$

In this case the inequality (4.83) implies as $\delta \to 0$

$$||\nabla u_\delta - \nabla u||_0 \to 0$$

which completes the proof of Lemma 4.2.

REMARK. Since f^δ is the smooth function, we have

$$||f^\delta - f||_0 \le c\delta$$

with a constant c being uniform with respect to δ. Taking into account the formulae for $h_\delta, P(\delta, u, u_\delta, g)$, it follows from (4.83) that the result of Lemma 4.2 can be improved, namely, there exists a constant $c > 0$ such that

$$||u_\delta - u||_{H^1(\Omega)} \le c\delta.$$

4.6.3 Griffith formula and Rice–Cherepanov integral

To underline the dependence of the domain Ω on the crack length l we shall write Ω_l instead of Ω in some places of this subsection.

Let $J(\Omega_l)$ be defined by the formula (4.64), and the function θ be chosen as that at the beginning of Section 4.6.2. Our purpose is to prove the following Griffith formula.

Theorem 4.6. *The derivative of $J(\Omega_l)$ with respect to l is given by the formula*

$$\frac{dJ(\Omega_l)}{dl} = -\frac{1}{2}\int\limits_\Omega \left(\theta_{y_1}(u_{y_1}^2 - u_{y_2}^2) + 2\theta_{y_2}u_{y_1}u_{y_2}\right) - \int\limits_\Omega (\theta f)_{y_1} u. \tag{4.84}$$

PROOF. Introduce the notations

$$\Pi(\Omega;\varphi) = \frac{1}{2}\int\limits_\Omega |\nabla\varphi|^2 - \int\limits_\Omega f\varphi, \quad \Pi_\delta(\Omega;\varphi) = \frac{1}{2}\int\limits_\Omega \langle A_\delta\nabla\varphi, \nabla\varphi\rangle - \int\limits_\Omega f^\delta\varphi,$$

$$\Pi(\Omega_\delta;\varphi) = \frac{1}{2}\int\limits_{\Omega_\delta} |\nabla\varphi|^2 - \int\limits_{\Omega_\delta} f\varphi.$$

The solution u of the problem (4.62)–(4.63) satisfies the relation

$$\Pi(\Omega;u) = \min_{\varphi\in K_0} \Pi(\Omega;\varphi),$$

and the solution u^δ of the problem (4.65)–(4.66) satisfies

$$\Pi(\Omega_\delta;u^\delta) = \min_{\varphi\in K_\delta} \Pi(\Omega_\delta;\varphi).$$

We have noted that the transformation (4.69) establishes a one-to-one mapping between K_δ and K_0, hence

$$\min_{\varphi \in K_0} \Pi_\delta(\Omega; \varphi) = \min_{\varphi \in K_\delta} \Pi(\Omega_\delta; \varphi). \tag{4.85}$$

According to our notation,

$$J(\Omega) = \Pi(\Omega; u); \quad J(\Omega_\delta) = \Pi(\Omega_\delta; u^\delta),$$

where u and u^δ are the solutions of (4.62)–(4.63) and (4.65)–(4.66), respectively. Now we can find the limit (4.68). Indeed, by (4.85),

$$\frac{J(\Omega_\delta) - J(\Omega)}{\delta} = \frac{\Pi(\Omega_\delta; u^\delta) - \Pi(\Omega; u)}{\delta}$$

$$= \frac{\Pi_\delta(\Omega; u_\delta) - \Pi(\Omega; u)}{\delta} \leq \frac{\Pi_\delta(\Omega; u) - \Pi(\Omega; u)}{\delta}$$

and consequently,

$$\limsup_{\delta \to 0} \frac{J(\Omega_\delta) - J(\Omega)}{\delta} \leq \limsup_{\delta \to 0} \frac{\Pi_\delta(\Omega; u) - \Pi(\Omega; u)}{\delta} \tag{4.86}$$

$$= \frac{1}{2} \int_\Omega \langle A' \nabla u, \nabla u \rangle - \int_\Omega f' u.$$

On the other hand, by Lemma 4.2 and (4.76), as $\delta \to 0$

$$\frac{\Pi_\delta(\Omega; u_\delta) - \Pi(\Omega; u_\delta)}{\delta} = \frac{1}{2} \int_\Omega \langle \frac{A_\delta - A_0}{\delta} \nabla u_\delta, \nabla u_\delta \rangle \pm \frac{1}{2} \int_\Omega \langle A' \nabla u_\delta, \nabla u_\delta \rangle$$

$$- \frac{1}{\delta} \int_\Omega (f^\delta - f) u_\delta = \frac{1}{2} \int_\Omega \langle (\frac{A_\delta - A_0}{\delta} - A') \nabla u_\delta, \nabla u_\delta \rangle \tag{4.87}$$

$$+ \frac{1}{2} \int_\Omega \langle A' \nabla u_\delta, \nabla u_\delta \rangle - \frac{1}{\delta} \int_\Omega (f^\delta - f) u_\delta \to \frac{1}{2} \int_\Omega \langle A' \nabla u, \nabla u \rangle - \int_\Omega f' u,$$

whence

$$\liminf_{\delta \to 0} \frac{J(\Omega_\delta) - J(\Omega)}{\delta} \geq \liminf_{\delta \to 0} \frac{\Pi_\delta(\Omega; u_\delta) - \Pi(\Omega; u_\delta)}{\delta} \tag{4.88}$$

$$= \frac{1}{2} \int_\Omega \langle A' \nabla u, \nabla u \rangle - \int_\Omega f' u.$$

Comparing (4.86) and (4.88) we find

$$\lim_{\delta \to 0} \frac{J(\Omega_\delta) - J(\Omega)}{\delta} = \frac{1}{2} \int_\Omega \langle A' \nabla u, \nabla u \rangle - \int_\Omega f' u,$$

i.e.

$$\frac{dJ(\Omega_l)}{dl} = \frac{1}{2}\int_\Omega \langle A'\nabla u, \nabla u\rangle - \int_\Omega f'u. \tag{4.89}$$

By (4.71), (4.74), a substitution of A' and f' in (4.89) implies (4.84). The proof of Theorem 4.6 is complete.

We first have to prove that the right-hand side of (4.84) does not depend on θ. As we know (see Yakunina, 1981) the solution of the problem (4.62)–(4.63) has an additional regularity up to the crack faces. For any $x \in \Xi_l$ there exists a neighbourhood V of the point x such that

$$u \in H^2(V \setminus \Xi_l). \tag{4.90}$$

Moreover, the solution u satisfies the following boundary conditions:

$$[u] \geq 0, \quad [u_{y_2}] = 0, \quad u_{y_2} \leq 0, \quad u_{y_2}[u] = 0 \quad \text{on } \Xi_l. \tag{4.91}$$

To prove that the right-hand side of (4.84) is independent of θ, we consider two functions θ_1, θ_2 with the required properties. The difference between right-hand sides of (4.84) corresponding to θ_1, θ_2 is denoted by Λ,

$$\Lambda = -\frac{1}{2}\int_\Omega (\theta_{y_1}(u_{y_1}^2 - u_{y_2}^2) + 2\theta_{y_2}u_{y_1}u_{y_2}) - \int_\Omega (\theta f)_{y_1}u, \tag{4.92}$$

where $\theta = \theta_1 = \theta_2$. Since θ_1, θ_2 are equal to 1 in some neighbourhoods of the point x_l, in (4.92) we have to integrate outside a ball B_{x_l} centred at x_l. Integrating by parts in (4.92) we find

$$\Lambda = \int_{\Omega\setminus B_{x_l}} \theta u_{y_1}(\Delta u + f) + \int_{\Xi_l\setminus B_{x_l}} \theta[u_{y_2}u_{y_1}],$$

and by (4.62), (4.91),

$$\Lambda = \int_{\Xi_l\setminus B_{x_l}} \theta[u_{y_2}u_{y_1}].$$

To prove $\Lambda = 0$ it suffices to establish that

$$u_{y_2}[u_{y_1}] = 0 \quad \text{a.e. on } \Xi_l \cap \{\operatorname{supp}\theta\}. \tag{4.93}$$

Here, by $\{\operatorname{supp}\theta\}$ we denote the support of θ. Introduce the set

$$M = \{x \in \Xi_l \cap \{\operatorname{supp}\theta\} \mid \quad [u(x)] > 0\}.$$

The set M is open, and by (4.90), u is continuous up to Ξ_l. By (4.91), we have

$$u_{y_2} = 0 \quad \text{a.e. on } M. \tag{4.94}$$

The complement of M is characterized by the condition

$$[u] = 0 \quad \text{on} \quad (\Xi_l \cap \{\operatorname{supp}\theta\}) \setminus M.$$

Hence (see Kinderlehrer, Stampacchia, Chapter 2, Theorem A.1, 1980)

$$[u_{y_1}] = 0 \quad \text{a.e. on} \quad (\Xi_l \cap \{\operatorname{supp}\theta\}) \setminus M. \tag{4.95}$$

Consequently, by (4.94), (4.95), we arrive at (4.93), which proves the independence of the right-hand side of (4.84) on θ.

Note that the independence of the right-hand side of (4.84) on θ can be proved more simply. Indeed, since

$$\liminf_{\delta\to 0} \frac{J(\Omega_\delta) - J(\Omega)}{\delta} = \limsup_{\delta\to 0} \frac{J(\Omega_\delta) - J(\Omega)}{\delta}$$

and both sides are independent of θ we conclude that $\left.\frac{dJ(\Omega_\delta)}{d\delta}\right|_{\delta=0}$ exists and does not depend on θ.

The proved assertion means that the right-hand side of (4.84) is, actually, a function of the point x_l and the right-hand side f of (4.62). This allows us to write (4.84) as the Griffith formula

$$\frac{dJ(\Omega_l)}{dl} = k(x_l, f), \tag{4.96}$$

where $\Omega_l = D \setminus \overline{\Xi}_l$, k is a functional depending on x_l, f. In particular, we have

$$J(\Omega_{l+\delta}) = J(\Omega_l) + k(x_l, f)\delta + \alpha(\delta)\delta,$$

where $\Omega_{l+\delta} = D \setminus \overline{\Xi}_{l+\delta}$ and $\alpha(\delta) \to 0$ as $\delta \to 0$.

Note that $k(x_l, f) = 0$ provided that the solution u is sufficiently smooth, i.e.

$$\frac{dJ(\Omega_l)}{dl} = 0. \tag{4.97}$$

In particular, the equality (4.97) holds provided that $u \in H^2(\Omega)$. Indeed, in this case we can extend Ξ_l beyond the points $(0,0)$, $(l,0)$ so that the extension crosses the boundary Γ. As a result, the domain Ω is divided into two subdomains Ω_1, Ω_2. By (4.62), (4.93), (4.84), we have

$$\frac{dJ(\Omega_l)}{dl} = \sum_{i=1}^{2}\Big(-\frac{1}{2}\int\limits_{\Omega_i} \big(\theta_{y_1}(u_{y_1}^2 - u_{y_2}^2) + 2\theta_{y_2}u_{y_1}u_{y_2}\big) \tag{4.98}$$

$$- \int\limits_{\Omega_i} (\theta f)_{y_1} u\Big) = \sum_{i=1}^{2}\int\limits_{\Omega_i} \theta u_{y_1}(\Delta u + f) + \int\limits_{\Xi_*} \theta u_{y_2}[u_{y_1}] = 0,$$

where Ξ_* is the extension of Ξ_l. In fact, to prove (4.98) we need a local regularity of the solution near the point x_l. The inclusion $u \in H^2(\Omega)$ provides sufficient regularity to integrate in (4.98).

An additional regularity of the solution near the point x_l takes place in some particular cases. For example, assume that the solution u satisfies the condition $[u] = 0$ on $B_{x_l} \cap \Xi_l$, where B_{x_l} is a ball centered at x_l. Using the technique of Section 2.5, we can prove that the equation $-\Delta u = f$ holds in B_{x_l}, and consequently, $u \in H^3_{loc}(B_{x_l})$. In this case all the integrals in (4.98) make sense. Consequently, by (4.62), (4.93), the equality (4.97) follows.

We can write (4.84) in the form which does not contain θ. To this end, consider a ball $B_{x_l}(r)$ of radius r with a boundary $\Gamma(r)$ such that $\theta = 1$ on $B_{x_l}(r)$.

Integration by parts in (4.84) implies

$$\frac{dJ(\Omega_l)}{dl} = \int_{\Omega \setminus B_{x_l}(r)} \theta u_{y_1}(\Delta u + f) + \int_{\Xi_l \setminus B_{x_l}(r)} \theta u_{y_2}[u_{y_1}]$$

$$+ \int_{B_{x_l}(r) \setminus \Xi_l} \theta f u_{y_1} + \frac{1}{2} \int_{\Gamma(r)} \theta \left(\nu_1(u^2_{y_1} - u^2_{y_2}) + 2\nu_2 u_{y_1} u_{y_2}\right),$$

where (ν_1, ν_2) is the unit external normal vector to $\Gamma(r)$. Hence, by (4.62), (4.93),

$$\frac{dJ(\Omega_l)}{dl} = \int_{B_{x_l}(r) \setminus \Xi_l} f u_{y_1} + \frac{1}{2} \int_{\Gamma(r)} \left(\nu_1(u^2_{y_1} - u^2_{y_2}) + 2\nu_2 u_{y_1} u_{y_2}\right). \tag{4.99}$$

Now assume that $f = 0$ in some neighbourhood V of the point x_l. For small r, we have $B_{x_l}(r) \subset V$, and the formula (4.99) implies

$$\frac{dJ(\Omega_l)}{dl} = \frac{1}{2} \int_{\Gamma(r)} \left(\nu_1(u^2_{y_1} - u^2_{y_2}) + 2\nu_2 u_{y_1} u_{y_2}\right).$$

The right-hand side of this equality does not depend on r, and consequently, we arrive at the following conclusion. Let u be the solution of the problem (4.62), (4.63), and f be equal to zero in some neighbourhood of the point x_l. Then the integral

$$I = \int_{\Gamma(r)} \left(\nu_1(u^2_{y_1} - u^2_{y_2}) + 2\nu_2 u_{y_1} u_{y_2}\right)$$

is independent of r for all sufficiently small r. Moreover, the above arguments show that the integral

$$I = \int_C \left(\nu_1(u^2_{y_1} - u^2_{y_2}) + 2\nu_2 u_{y_1} u_{y_2}\right) \tag{4.100}$$

does not depend on C for any closed curve C surrounding the point x_l (see Fig.4.5). In this case $\nu = (\nu_1, \nu_2)$ is the normal unit vector to the curve C.

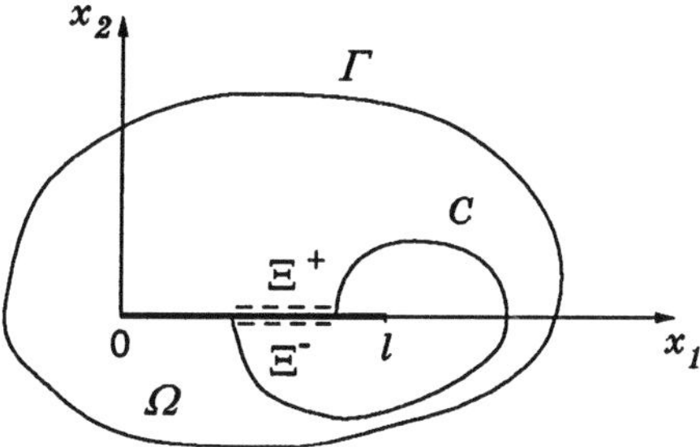

Fig.4.5. Curve C

A part of this curve may belong to Ξ_l. In this last case we can integrate over Ξ^+ or Ξ^- in (4.100), since, in view of (4.91), (4.93), the jump $[u_{y_1} u_{y_2}]$ is equal to zero on Ξ_l. Here $\Xi^\pm = \Xi_l^\pm \cap C$.

Of course, the above independence takes place provided that $f = 0$ in the domain with the boundary C. The integral of the form (4.100) is called the Rice–Cherepanov integral. We have to note that the statement obtained is proved for nonlinear boundary conditions (4.91). This statement is similar to the well-known result in the linear elasticity theory with linear boundary conditions prescribed on Ξ_l (see Bui, Ehrlacher, 1997; Rice, 1968; Rice, Drucker, 1967; Parton, Morozov, 1985; Destuynder, Jaoua, 1981).

4.7 Asymptotics of the energy functional for the Lamé equations

This section is concerned with the two-dimensional elasticity equations. Our aim is to find the derivative of the energy functional with respect to the crack length. The nonpenetration condition is assumed to hold at the crack faces. We derive the Griffith formula and prove the path independence of the Rice–Cherepanov integral. This section follows the publication (Khludnev, Sokolowski, 1998c).

4.7.1 Setting the problem

Let $D \subset R^2$ be a bounded domain with a smooth boundary Γ, and $\Xi_{l+\delta}$ be the set $\{(x_1, x_2) \mid 0 < x_1 < l + \delta,\ x_2 = 0\}$. We assume that this set belongs to the domain D for all sufficiently small δ, and $l > 0$. Introduce the notation $\Omega_\delta = D \backslash \overline{\Xi}_{l+\delta}$, $\Omega = D \backslash \overline{\Xi}_l$. The problem, which we analyse, can be formulated as follows. We want to find a function $W = (u, v)$ such that

$$-\sigma_{ij,j} = f_i, \quad i = 1, 2, \quad \text{in } \Omega, \tag{4.101}$$

$$W = 0 \quad \text{on } \Gamma, \tag{4.102}$$

$$[W]n \geq 0 \quad \text{on } \Xi_l. \tag{4.103}$$

Here $\sigma_{ij} = \sigma_{ij}(W)$ are the stress tensor components, $n = (0,1)$ is a normal vector to Ξ_l, and the Hooke law is assumed to be fulfilled,

$$\sigma_{ij} = 2\mu\varepsilon_{ij} + \lambda \operatorname{div} W\delta_j^i, \quad i,j = 1,2. \tag{4.104}$$

By λ, μ we denote the Lamé parameters, $\varepsilon_{ij} = \varepsilon_{ij}(W)$, δ_j^i is the Kronecker symbol,

$$\varepsilon_{11} = u_{x_1}, \quad \varepsilon_{22} = v_{x_2}, \quad \varepsilon_{12} = 1/2\,(u_{x_2} + v_{x_1}). \tag{4.105}$$

We assume that $f = (f_1, f_2, f_3) \in C^1(\overline{D})$.

The formulation of the problem (4.101)–(4.103) is not complete. Actually, we have to consider the minimization of the functional

$$I(\Omega; U) = \frac{1}{2}\int\limits_{\Omega} \sigma_{ij}(U)\varepsilon_{ij}(U) - \int\limits_{\Omega} fU, \quad U = (u,v), \tag{4.106}$$

over the set

$$K_0 = \{(u,v) \in H^1(\Omega) \mid \quad u = v = 0 \quad \text{on } \Gamma, \quad [v] \geq 0 \quad \text{on } \Xi_l\}. \tag{4.107}$$

In this case the solution W of the minimization problem satisfies (4.101)–(4.103) and some additional boundary conditions holding on Ξ_l. These conditions are analysed below (see (4.128)).

The perturbed problem corresponding to (4.101)–(4.103) is as follows. In the domain Ω_δ, we want to find a function $W^\delta = (u^\delta, v^\delta)$ such that

$$-\sigma_{ij,j} = f_i, \quad i = 1,2, \tag{4.108}$$

$$W^\delta = 0 \quad \text{on } \Gamma, \tag{4.109}$$

$$[W^\delta]n \geq 0 \quad \text{on } \Xi_{l+\delta}. \tag{4.110}$$

Here $\sigma_{ij} = \sigma_{ij}(W^\delta)$, $\varepsilon_{ij} = \varepsilon_{ij}(W^\delta)$, and σ_{ij}, ε_{ij} satisfy the Hooke law (4.104).

Similar to (4.101)–(4.103), the complete formulation of the problem (4.108)–(4.110) is variational. We minimize the functional

$$I(\Omega_\delta; U) = \frac{1}{2}\int\limits_{\Omega_\delta} \sigma_{ij}(U)\varepsilon_{ij}(U) - \int\limits_{\Omega_\delta} fU, \quad U = (u,v), \tag{4.111}$$

over the set

$$K_\delta = \{(u,v) \in H^1(\Omega_\delta) \mid \quad u = v = 0 \quad \text{on } \Gamma, \tag{4.112}$$

$$[v] \geq 0 \quad \text{on } \Xi_{l+\delta}\}.$$

The inequality (4.110) is, in fact, a part of the complete system of boundary conditions holding at $\Xi_{l+\delta}$.

We aim at finding the derivative of the energy functional

$$\lim_{\delta \to 0} \frac{I(\Omega_\delta; W^\delta) - I(\Omega; W)}{\delta}, \tag{4.113}$$

where W^δ, W are the solutions of (4.108)–(4.110) and (4.101)–(4.103), respectively. In this case the limit (4.113) is equal to the derivative

$$\left. \frac{dJ(\Omega_\delta)}{d\delta} \right|_{\delta=0}. \tag{4.114}$$

4.7.2 Auxiliary formulae

Similar to the preceding section we choose a function $\theta \in C_0^\infty(D)$ such that $\theta = 1$ in a neighbourhood of the point $x_l = (l, 0)$. Consider the transformation of independent variables

$$y_1 = x_1 - \delta\theta(x_1, x_2), \quad y_2 = x_2. \tag{4.115}$$

Here $(y_1, y_2) \in \Omega$, $(x_1, x_2) \in \Omega_\delta$. To simplify the arguments the function θ is assumed to be equal to zero in a neighbourhood of the point $(0, 0)$. The transformation (4.115) maps Ω_δ to Ω, and it is one-to-one. The Jacobian q_δ of the transformation is positive for small δ,

$$q_\delta = \left| \frac{\partial(y_1, y_2)}{\partial(x_1, x_2)} \right| = 1 - \delta\theta_{x_1}.$$

Let $x = x(y, \delta)$ correspond to the transformation (4.115), and $W^\delta(x)$ be the solution of (4.108)–(4.110). Then $W^\delta(x) = W_\delta(y)$, $y \in \Omega$. Also, let W be the solution of (4.101)–(4.103). The following statement holds.

Lemma 4.3. *We have, as* $\delta \to 0$,

$$\|W_\delta - W\|_{H^1(\Omega)} \to 0.$$

The proof of this lemma is omitted since it follows the lines of Lemma 4.2. We just indicate that W^δ and W are the solutions of the following variational inequalities,

$$W^\delta \in K_\delta : \int_{\Omega_\delta} \sigma_{ij}(W^\delta)(\varepsilon_{ij}(V) - \varepsilon_{ij}(W^\delta)) \tag{4.116}$$

$$\geq \int_{\Omega_\delta} f(V - W^\delta) \quad \forall V \in K_\delta,$$

$$W \in K_0 : \int_\Omega \sigma_{ij}(W)(\varepsilon_{ij}(V) - \varepsilon_{ij}(W)) \geq \int_\Omega f(V - W) \quad \forall V \in K_0, \tag{4.117}$$

and we can apply the arguments of Lemma 4.2.

Using the transformation (4.115), for $w^\delta(x) = w_\delta(y)$, we obtain

$$\int_{\Omega_\delta} f_i w^\delta \, dx = \int_{\Omega} f_i^\delta w_\delta \, dy, \quad f_i^\delta(y) = \frac{f_i(x(y,\delta))}{1-\delta\theta_{x_1}}, \quad i = 1, 2. \tag{4.118}$$

Similar to Section 4.6, it is possible to find the derivatives

$$f_i'(y) = \lim_{\delta\to 0} \frac{f_i^\delta(y) - f_\delta^0(y)}{\delta} = \left.\frac{df_i^\delta}{d\delta}\right|_{\delta=0}, \quad i = 1, 2. \tag{4.119}$$

We have (see (4.74))

$$f_i'(y) = (\theta f_i)_{y_1}(y), \quad i = 1, 2. \tag{4.120}$$

The formulae (4.120) will be used in getting the derivative of the energy functional.

4.7.3 The Griffith formula

Let $W^\delta = (u^\delta, v^\delta)$ be the solution of the problem (4.108)–(4.110). Denote $u^\delta(x) = \tilde{u}(y)$, $v^\delta(x) = \tilde{v}(y)$, $x \in \Omega_\delta$, $y \in \Omega$, $x = x(y,\delta)$. Here, we use the tilde and omit δ for convenience.

By (4.115), the following formulae arise:

$$\begin{cases} u_{x_1}^\delta = \tilde{u}_{y_1}(1-\delta\theta_{x_1}) \\ u_{x_2}^\delta = \tilde{u}_{y_1}(-\delta\theta_{x_2}) + \tilde{u}_{y_2}, \end{cases} \quad \begin{cases} v_{x_1}^\delta = \tilde{v}_{y_1}(1-\delta\theta_{x_1}) \\ v_{x_2}^\delta = \tilde{v}_{y_1}(-\delta\theta_{x_2}) + \tilde{v}_{y_2}. \end{cases} \tag{4.121}$$

Since

$$\int_{\Omega_\delta} \sigma_{ij}(W^\delta)\varepsilon_{ij}(W^\delta) = \int_{\Omega_\delta} \Big((2\mu+\lambda)\{\varepsilon_{11}^2(W^\delta) + \varepsilon_{22}^2(W^\delta)\}$$

$$+ 2\lambda\varepsilon_{11}(W^\delta)\varepsilon_{22}(W^\delta) + 4\mu\varepsilon_{12}^2(W^\delta)\Big),$$

in view of (4.121), we can change the domain of integration Ω_δ by Ω. This provides

$$\frac{1}{2}\int_{\Omega_\delta} \sigma_{ij}(W^\delta)\varepsilon_{ij}(W^\delta)dx - \int_{\Omega_\delta} fW^\delta \, dx$$

$$= \frac{1}{2}\int_{\Omega} \frac{1}{q_\delta}\Big(\tilde{u}_{y_1}^2\{(2\mu+\lambda)(1-\delta\theta_{x_1})^2 - \mu\delta^2\theta_{x_2}^2\} + \mu\tilde{u}_{y_2}^2$$

$$+ \tilde{v}_{y_1}^2\{(2\mu+\lambda)\delta^2\theta_{x_2}^2 + \mu(1-\delta\theta_{x_1})^2\} + (2\mu+\lambda)\tilde{v}_{y_2}^2 \tag{4.122}$$

$$+ 2\mu\tilde{u}_{y_1}\tilde{u}_{y_2}(-\delta\theta_{x_2}) + 2(\lambda+\mu)\tilde{u}_{y_1}\tilde{v}_{y_1}(-\delta\theta_{x_2})(1-\delta\theta_{x_1})$$

$$+ 2\lambda \tilde{u}_{y_1} \tilde{v}_{y_2}(1 - \delta\theta_{x_1}) + 2\mu \tilde{u}_{y_2} \tilde{v}_{y_1}(1 - \delta\theta_{x_1})$$
$$+ 2(2\mu + \lambda)\tilde{v}_{y_1} \tilde{v}_{y_2}(-\delta\theta_{x_2})\Big) dy - \int\limits_{\Omega} f^\delta \tilde{W}\, dy.$$

Denote by $I_\delta(\Omega; \tilde{W})$ the right-hand side of (4.122). In this case, formula (4.122) provides the transformation of the energy functional

$$I(\Omega_\delta; W^\delta) = I_\delta(\Omega; W_\delta), \tag{4.123}$$

where $W_\delta = \tilde{W}$. Again, let $W^\delta(x) = W_\delta(y)$, $x = x(y, \delta)$. Then $W_\delta \in K$ provided that $W^\delta \in K_\delta$ and conversely. Thus we obtain a one-to-one mapping between K_δ and K_0. In particular, this implies

$$\min_{U \in K_\delta} I(\Omega_\delta; U) = \min_{U \in K_0} I_\delta(\Omega; U). \tag{4.124}$$

By (4.123), (4.124), we have

$$\frac{J(\Omega_\delta) - J(\Omega)}{\delta} = \frac{I(\Omega_\delta; W^\delta) - I(\Omega; W)}{\delta}$$
$$= \frac{I_\delta(\Omega; W_\delta) - I(\Omega; W)}{\delta} \le \frac{I_\delta(\Omega; W) - I(\Omega; W)}{\delta},$$

whence

$$\limsup_{\delta \to 0} \frac{J(\Omega_\delta) - J(\Omega)}{\delta} \le \limsup_{\delta \to 0} \frac{I_\delta(\Omega; W) - I(\Omega; W)}{\delta}. \tag{4.125}$$

On the other hand

$$\liminf_{\delta \to 0} \frac{J(\Omega_\delta) - J(\Omega)}{\delta} \ge \liminf_{\delta \to 0} \frac{I_\delta(\Omega; W_\delta) - I(\Omega; W_\delta)}{\delta}. \tag{4.126}$$

Taking into account (4.122) (or (4.123)) and Lemma 4.3, we can show that the right-hand sides of (4.125) and (4.126) coincide.

Consequently, there exists

$$\lim_{\delta \to 0} \frac{J(\Omega_\delta) - J(\Omega)}{\delta}.$$

We can calculate the right-hand sides of (4.125), (4.126), which give

$$\frac{dJ(\Omega_\delta)}{d\delta}\bigg|_{\delta = 0} = \frac{1}{2} \int\limits_{\Omega} \Big((2\mu + \lambda) u_{y_1}^2 (-\theta_{y_1}) + \mu u_{y_2}^2 \theta_{y_1} + 2\mu u_{y_1} u_{y_2} (-\theta_{y_2})$$
$$+ \mu v_{y_1}^2 (-\theta_{y_1}) + (2\mu + \lambda) v_{y_2}^2 \theta_{y_1} + 2(\lambda + \mu) u_{y_1} v_{y_1} (-\theta_{y_2}) \tag{4.127}$$
$$+ 2(2\mu + \lambda) v_{y_1} v_{y_2} (-\theta_{y_2})\Big) - \int\limits_{\Omega} (\theta f_1)_{y_1} u - \int\limits_{\Omega} (\theta f_2)_{y_1} v.$$

As a result, we have derived the Griffith formula (4.127).

4.7.4 Properties of the derivative of the energy functional

We first analyse boundary conditions holding at Ξ_l in the problem (4.101)–(4.103). According to (Khludnev, Sokolowski, 1997), the following conditions hold at Ξ_l:

$$[v] \geq 0, \quad \sigma_{22} \leq 0, \quad [\sigma_{22}] = 0, \quad \sigma_{12} = 0, \quad [v]\sigma_{22} = 0. \tag{4.128}$$

By (Yakunina, 1981), the solution of the problem (4.101)–(4.103) (i.e. the problem (4.117)) has additional regularity properties up to the crack faces. Namely, for any $x \in \Xi_l$ there exists a neighbourhood $\mathcal{W}$ of the point x such that $W \in H^2(\mathcal{W}\backslash\Xi_l)$. Consequently, the solution W is continuous up to the crack faces, and the conditions (4.128) hold almost everywhere at Ξ_l. Note that

$$\sigma_{22} = (2\mu + \lambda)v_{y_2} + \lambda u_{y_1}, \quad \sigma_{12} = \mu(u_{y_2} + v_{y_1}). \tag{4.129}$$

In addition to (4.128) we prove that

$$\sigma_{22}[v_{y_1}] = 0 \quad \text{a.e. at } \Xi_l. \tag{4.130}$$

Indeed, since v is continuous the set

$$M = \{y \in \Xi_l \mid \quad [v(y)] > 0\}$$

is open. At any point $y \in M$ we have $[v(y)] > 0$, i.e. by (4.128), $\sigma_{22} = 0$, and consequently $\sigma_{22}[v_{y_1}] = 0$ a.e. on M. At the set $\Xi_l\backslash M$ we have $[v] = 0$, whence $[v_{y_1}] = 0$ (see Kinderlehrer, Stampacchia, Chapter 2, Theorem A.1, 1980) which implies $\sigma_{22}[v_{y_1}] = 0$. The equality (4.130) is proved.

Now we prove that the right-hand side of (4.127) is independent of θ. Consider two right-hand sides corresponding to any two functions θ_1, θ_2 with the prescribed properties. The difference between them is denoted by Λ,

$$\begin{aligned}\Lambda = \frac{1}{2}\int\limits_{\Omega}\Big((2\mu + \lambda)u_{y_1}^2(-\theta_{y_1}) + \mu u_{y_2}^2\theta_{y_1} + 2\mu u_{y_1}u_{y_2}(-\theta_{y_2})\\ + \mu v_{y_1}^2(-\theta_{y_1}) + (2\mu + \lambda)v_{y_2}^2\theta_{y_1} + 2(\lambda + \mu)u_{y_1}v_{y_1}(-\theta_{y_2})\\ + 2(2\mu + \lambda)v_{y_1}v_{y_2}(-\theta_{y_2})\Big) - \int\limits_{\Omega}(\theta f_1)_{y_1}u - \int\limits_{\Omega}(\theta f_2)_{y_1})v,\end{aligned} \tag{4.131}$$

where $\theta = \theta_1 - \theta_2$. The functions θ_1, θ_2 are equal 1 in neighbourhoods of the point x_l, consequently, the integration in (4.131) is actually fulfilled over $\Omega\backslash B_{x_l}$. By B_{x_l} we have denoted a ball centered at x_l.

Integrating by parts in (4.131) we derive

$$\Lambda = \int\limits_{\Omega\backslash B_{x_l}} \theta\Big((2\mu + \lambda)u_{y_1}u_{y_1y_1} + \mu u_{y_1}u_{y_2y_2} + \mu v_{y_1}v_{y_1y_1} + (\lambda + \mu)v_{y_1}u_{y_1y_2}$$

$$+(\lambda+\mu)u_{y_1}v_{y_1y_2}+(2\mu+\lambda)v_{y_1}v_{y_2y_2}+f_1u_{y_1}+f_2v_{y_1}\Big) \qquad (4.132)$$

$$+\int_{\Xi_l}\theta\left(\mu[u_{y_1}u_{y_2}]+(\lambda+\mu)[u_{y_1}v_{y_1}]+(2\mu+\lambda)[v_{y_1}v_{y_2}]\right).$$

Note that the equilibrium equations (4.101) can be written as

$$(2\mu+\lambda)u_{y_1y_1}+\mu u_{y_2y_2}+(\lambda+\mu)v_{y_1y_2}=-f_1, \qquad (4.133)$$

$$(2\mu+\lambda)v_{y_2y_2}+(\lambda+\mu)u_{y_1y_2}+\mu v_{y_1y_1}=-f_2.$$

By (4.128)–(4.130), (4.133), the right-hand side of (4.132) is equal to zero, which proves the independence of the right-hand side of (4.127) on θ.

As in the preceding section, the independence follows from simpler arguments. In fact, we have

$$\liminf_{\delta\to 0}\frac{J(\Omega_\delta)-J(\Omega)}{\delta}=\limsup_{\delta\to 0}\frac{J(\Omega_\delta)-J(\Omega)}{\delta} \qquad (4.134)$$

which proves the existence $dJ(\Omega_\delta)/d\delta\big|_{\delta=0}$. Both sides of (4.134) do not depend on θ, hence $\lim_{\delta\to 0}(J(\Omega_\delta)-J(\Omega))\delta^{-1}$ is independent of θ.

Assume that the solution W of the problem (4.101)–(4.103) has the property

$$[W]=0 \quad \text{on } B_{x_l}\cap\Xi_l,$$

where B_{x_l} is any ball. Then the arguments of Section 2.5 used to prove C^∞-regularity of the solution allow us to state that the equations

$$-\sigma_{ij,j}(W)=f_i, \quad i=1,2,$$

hold in B_{x_l} in the distribution sense. Consequently, $W\in H^3_{loc}(B_{x_l})$. In addition to this, by the inclusion $f\in H^1(D)$, we have $W\in H^3_{loc}(\Omega)$. In this case

$$\frac{dJ(\Omega_\delta)}{d\delta}\bigg|_{\delta=0}=0. \qquad (4.135)$$

In fact, we can integrate by parts on the right-hand side of (4.127), which gives

$$\frac{dJ(\Omega_\delta)}{d\delta}\bigg|_{\delta=0}=\int_\Omega\theta\Big((\sigma_{11,1}+\sigma_{12,2})u_{y_1}+(\sigma_{21,1}+\sigma_{22,2})v_{y_1} \qquad (4.136)$$

$$+f_1u_{y_1}+f_2v_{y_1}\Big)+\int_{\Xi_l}\theta\left(\sigma_{22}[v_{y_1}]+[\sigma_{12}u_{y_1}]\right).$$

By (4.101), (4.128), (4.130), the right-hand side of (4.136) is equal to zero, which proves (4.135). To justify (4.135), it suffices, in fact, to have the regularity $W \in H^2(B_{x_l} \backslash \Xi_l)$. In this case we can repeat the above arguments and obtain that the right-hand side of (4.136) is equal to zero.

The formula (4.127) can be written in the form which does not contain the function θ. To show this we choose a ball $B_{x_l}(r)$ of radius r with the boundary $\Gamma(r)$ such that $\theta = 1$ on $B_{x_l}(r)$. In this case the integration by parts in (4.127) yields

$$\left.\frac{dJ(\Omega_\delta)}{d\delta}\right|_{\delta=0} = \int\limits_{B_{x_l}(r)\backslash\Xi_l} (f_1 u_{y_1} + f_2 v_{y_1})$$

$$+\frac{1}{2}\int\limits_{\Gamma(r)} \nu_1 \left((2\mu+\lambda)(u_{y_1}^2 - v_{y_2}^2) + \mu(v_{y_1}^2 - u_{y_2}^2)\right) \tag{4.137}$$

$$+\frac{1}{2}\int\limits_{\Gamma(r)} \nu_2 \Big(2(2\mu+\lambda)v_{y_1}v_{y_2} + 2(\lambda+\mu)u_{y_1}v_{y_1} + 2\mu u_{y_1}u_{y_2}\Big),$$

where (ν_1, ν_2) is the unit normal exterior vector to $\Gamma(r)$.

Now assume that $f = 0$ in some neighbourhood V of the point x_l. For sufficiently small r, $0 < r < r_0$, we have $B_{x_l}(r) \subset V$, and the formula (4.137) implies

$$\left.\frac{dJ(\Omega_\delta)}{d\delta}\right|_{\delta=0} = \frac{1}{2}\int\limits_{\Gamma(r)} \nu_1 \left((2\mu+\lambda)(u_{y_1}^2 - v_{y_2}^2) + \mu(v_{y_1}^2 - u_{y_2}^2)\right) \tag{4.138}$$

$$+\int\limits_{\Gamma(r)} \nu_2 \Big((2\mu+\lambda)v_{y_1}v_{y_2} + (\lambda+\mu)u_{y_1}v_{y_1} + \mu u_{y_1}u_{y_2}\Big).$$

The right-hand side of (4.138) does not depend on r, and consequently, we have the following property. Let W be the solution of the problem (4.101)–(4.103), and f be equal to zero in some neighbourhood of the point x_l. Then the Rice–Cherepanov integral

$$I = \frac{1}{2}\int\limits_{\Gamma(r)} \nu_1 \left((2\mu+\lambda)(u_{y_1}^2 - v_{y_2}^2) + \mu(v_{y_1}^2 - u_{y_2}^2)\right)$$

$$+\int\limits_{\Gamma(r)} \nu_2 \Big((2\mu+\lambda)v_{y_1}v_{y_2} + (\lambda+\mu)u_{y_1}v_{y_1} + \mu u_{y_1}u_{y_2}\Big)$$

is independent of $r > 0$ for all sufficiently small r. Moreover, the above arguments show that the integral

$$I = \frac{1}{2}\int\limits_{C} \nu_1 \left((2\mu+\lambda)(u_{y_1}^2 - v_{y_2}^2) + \mu(v_{y_1}^2 - u_{y_2}^2)\right) \tag{4.139}$$

$$+\int_C \nu_2\Big((2\mu+\lambda)v_{y_1}v_{y_2}+(\lambda+\mu)u_{y_1}v_{y_1}+\mu u_{y_1}u_{y_2}\Big)$$

is path independent for any closed curve C surrounding the point x_l. In this case $\nu = (\nu_1, \nu_2)$ is the normal unit vector to the curve C. Some part of this curve denoted by $\Xi = \Xi_l \cap C$ may belong to Ξ_l. In this case, taking into account (4.130), (4.128), we can integrate over Ξ^+ or Ξ^- in (4.139) and the same value of the integral is obtained. Observe that the above independence takes place provided that $f = 0$ in a domain with the boundary C.

We have to note that the result is obtained for nonlinear boundary conditions (4.128). The well-known path independence of the Rice–Cherepanov integral was previously proved in elasticity theory for linear boundary conditions $\sigma_{22} = 0, \sigma_{12} = 0$ holding on $\Xi_l^\pm$ (see Parton, Morozov, 1985).

4.8 Three-dimensional case

In this section we find the derivative of the energy functional in the three-dimensional linear elasticity model. The derivative characterizes the behaviour of the energy functional provided that the crack length is changed. The crack is modelled by a part of the two-dimensional plane removed from a three-dimensional domain. In particular, we derive the Griffith formula.

4.8.1 Formulation of the problem

Let $D \subset R^3$ be a bounded domain with a smooth boundary Γ (see Fig.4.6). The crack is defined in the form of a two-dimensional surface,

$$\Xi_{l+\delta} = \{(x_1, x_2, x_3) \mid \quad x_3 = 0, \quad -h < x_2 < h, \quad 0 < x_1 < l + \delta\}.$$

Here, $h > 0$, $l > 0$, δ is a small parameter to be tended to zero. Assume that $\Xi_{l+\delta} \subset D$ and denote $\Omega_\delta = D \backslash \overline{\Xi}_{l+\delta}$, $\Omega = D \backslash \overline{\Xi}_l$.

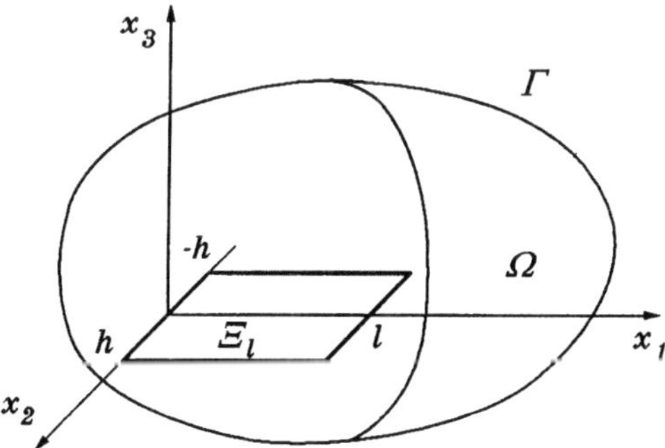

Fig.4.6. Domain Ω and crack Ξ_l in R^3

The equilibrium problem for an elastic body occupying the domain Ω_δ can be formulated as follows. We want to find a function $W = (u, v, w)$

such that

$$-\sigma_{ij,j} = f_i, \quad i = 1, 2, 3, \quad \text{in } \Omega, \tag{4.140}$$

$$W = 0 \quad \text{on } \Gamma, \tag{4.141}$$

$$[W]n \geq 0 \quad \text{on } \Xi_l. \tag{4.142}$$

Here $\sigma_{ij} = \sigma_{ij}(W)$ are stress tensor components, and $n = (0,0,1)$ is a normal vector to the surface $\Xi_{l+\delta}$. We assume that the Hooke law is fulfilled,

$$\sigma_{ij} = 2\mu\varepsilon_{ij} + \lambda \operatorname{div} W \,\delta_j^i, \quad i, j = 1, 2, 3, \tag{4.143}$$

$$\varepsilon_{ij}(W) = \frac{1}{2}(w^i_{,j} + w^j_{,i}), \quad (w^1, w^2, w^3) \equiv (u, v, w),$$

where λ, μ are the Lamé parameters, and δ_j^i is the Kronecker symbol.

Actually, by considering the problem (4.140)–(4.142) we have in mind the minimization of the functional

$$I(\Omega; U) = \frac{1}{2}\int\limits_{\Omega} \sigma_{ij}(U)\varepsilon_{ij}(U) - \int\limits_{\Omega} fU, \quad U = (u, v, w), \tag{4.144}$$

over the set

$$K_0 = \{(u, v, w) \in H^1(\Omega) \mid u = v = w = 0 \text{ on } \Gamma,\ [w] \geq 0 \text{ on } \Xi_l\}, \tag{4.145}$$

where $f = (f_1, f_2, f_3) \in C^1(\overline{D})$.

In this case the conditions (4.140)–(4.142) hold. Moreover, a system of equations and inequalities holds on Ξ_l, and (4.142) is a part of this system (see (4.161) below).

The perturbed problem corresponding to (4.140)–(4.142) is as follows. We want to find a function $W^\delta = (n^\delta, v^\delta, w^\delta)$ such that

$$-\sigma_{ij,j} = f_i, \quad i = 1, 2, 3, \quad \text{in } \Omega_\delta, \tag{4.146}$$

$$W^\delta = 0 \quad \text{on } \Gamma, \tag{4.147}$$

$$[W^\delta]n \geq 0 \quad \text{on } \Xi_{l+\delta}. \tag{4.148}$$

Here $\sigma_{ij} = \sigma_{ij}(W^\delta)$, and $\sigma_{ij}(W^\delta)$, $\varepsilon_{ij}(W^\delta)$ satisfy the Hooke law (4.143). Analogously, by considering the problem (4.146)–(4.148) we, in fact, analyse the minimization of the functional

$$I(\Omega_\delta; U) = \frac{1}{2}\int\limits_{\Omega_\delta} \sigma_{ij}(U)\varepsilon_{ij}(U) - \int\limits_{\Omega_\delta} fU, \quad U = (u, v, w), \tag{4.149}$$

over the set

$$K_\delta = \{(u, v, w) \in H^1(\Omega_\delta) \mid \quad u = v = w = 0 \quad \text{on } \Gamma, \tag{4.150}$$

$$[w] \geq 0 \quad \text{on } \Xi_{l+\delta}\}.$$

Denote $J(\Omega_\delta) = I(\Omega_\delta; W^\delta)$, $J(\Omega) = I(\Omega; W)$, where W^δ, W are the solutions of (4.140)–(4.142) and (4.146)–(4.148), respectively. In this section, we aim at finding the derivative

$$\lim_{\delta \to 0} \frac{J(\Omega_\delta) - J(\Omega)}{\delta}. \tag{4.151}$$

4.8.2 The derivative of the energy functional

Consider a function $\theta \in C_0^\infty(D)$, $\theta = 1$ in a neighbourhood of the set $L = \{(x_1, x_2, x_3) \mid x_1 = l,\ -h < x_2 < h,\ x_3 = 0\}$. The transformation of the variables

$$y_1 = x_1 - \delta\theta(x_1, x_2, x_3), \quad y_2 = x_2, \quad y_3 = x_3 \tag{4.152}$$

maps Ω_δ on Ω. The Jacobian q_δ is positive for all sufficiently small δ,

$$q_\delta = \left| \frac{\partial(y_1, y_2, y_3)}{\partial(x_1, x_2, x_3)} \right| = 1 - \delta\theta_{x_1}, \tag{4.153}$$

hence the transformation is one-to-one. For simplicity, the function θ is assumed to be equal to zero in a neighbourhood of the set $\{(x_1, x_2, x_3) \mid x_1 = 0,\ -h < x_2 < h,\ x_3 = 0\}$. Let $(u^\delta, v^\delta, w^\delta)$ be the solution of the problem (4.146)–(4.148), $(u^\delta(x), v^\delta(x), w^\delta(x)) = (\tilde{u}(y), \tilde{v}(y), \tilde{w}(y))$, and $x = x(y, \delta)$ be the transformation which is inverse to (4.152). We use the tilde instead of δ to simplify the formulae below. By (4.152), the following relations hold:

$$u^\delta_{x_1} = \tilde{u}_{y_1}(1 - \delta\theta_{x_1}),\ u^\delta_{x_2} = \tilde{u}_{y_1}(-\delta\theta_{x_2}) + \tilde{u}_{y_2}, \tag{4.154}$$

$$u^\delta_{x_3} = \tilde{u}_{y_1}(-\delta\theta_{x_3}) + \tilde{u}_{y_3}.$$

Similar formulae are valid for the functions $v^\delta(x)$, $w^\delta(x)$.

Let $\sigma_{ij} = \sigma_{ij}(W^\delta)$, $\varepsilon_{ij} = \varepsilon_{ij}(W^\delta)$. In this case, by (4.143),

$$\sigma_{ij}\varepsilon_{ij} = (2\mu + \lambda)(\varepsilon_{11}^2 + \varepsilon_{22}^2 + \varepsilon_{33}^2) + 4\mu(\varepsilon_{12}^2 + \varepsilon_{13}^2 + \varepsilon_{23}^2)$$
$$+ 2\lambda(\varepsilon_{11}\varepsilon_{22} + \varepsilon_{22}\varepsilon_{33} + \varepsilon_{11}\varepsilon_{33}).$$

This allows us to find the transformation of the energy functional,

$$\frac{1}{2}\int_{\Omega_\delta} \sigma_{ij}\varepsilon_{ij} - \int_{\Omega_\delta} fW^\delta = \frac{1}{2}\int_{\Omega} \frac{1}{q_\delta}\Big((2\mu + \lambda)\{(1 - \delta\theta_{x_1})^2\tilde{u}_{y_1}^2$$
$$+ (\tilde{v}_{y_1}(-\delta\theta_{x_2}) + \tilde{v}_{y_2})^2 + (\tilde{w}_{y_1}(-\delta\theta_{x_3}) + \tilde{w}_{y_3})^2\}$$
$$+ \mu\{(\tilde{u}_{y_1}(-\delta\theta_{x_2}) + \tilde{u}_{y_2} + \tilde{v}_{y_1}(1 - \delta\theta_{x_1}))^2 + (\tilde{u}_{y_1}(-\delta\theta_{x_3}) + \tilde{u}_{y_3}$$
$$+ \tilde{w}_{y_1}(1 - \delta\theta_{x_1}))^2 + (\tilde{v}_{y_1}(-\delta\theta_{x_3}) + \tilde{v}_{y_3} + \tilde{w}_{y_1}(-\delta\theta_{x_2}) + \tilde{w}_{y_2})^2\} \tag{4.155}$$

$$+2\lambda\{\tilde u_{y_1}(1-\delta\theta_{x_1})(\tilde v_{y_1}(-\delta\theta_{x_2})+\tilde v_{y_2})+(\tilde v_{y_1}(-\delta\theta_{x_2})$$

$$+\tilde v_{y_2})(\tilde w_{y_1}(-\delta\theta_{x_3})+\tilde w_{y_3})+\tilde u_{y_1}(1-\delta\theta_{x_1})(\tilde w_{y_1}(-\delta\theta_{x_3})+\tilde w_{y_3})\}\Big)-\int\limits_{\Omega} f^\delta \tilde W,$$

where

$$f^\delta(y)=\frac{f(x(y,\delta))}{1-\delta\theta_{x_1}},\quad \tilde W=(\tilde u,\tilde v,\tilde w)=W_\delta.$$

Consequently, we have obtained the formula

$$I(\Omega_\delta;W^\delta)=I_\delta(\Omega;W_\delta). \tag{4.156}$$

Now we can proceed as in the previous section, which gives

$$\limsup_{\delta\to 0}\frac{J(\Omega_\delta)-J(\Omega)}{\delta}\le\limsup_{\delta\to 0}\frac{I_\delta(\Omega;W)-I(\Omega;W)}{\delta}, \tag{4.157}$$

$$\liminf_{\delta\to 0}\frac{J(\Omega_\delta)-J(\Omega)}{\delta}\ge\liminf_{\delta\to 0}\frac{I_\delta(\Omega;W_\delta)-I(\Omega;W_\delta)}{\delta}. \tag{4.158}$$

Since the right-hand sides of (4.157), (4.158) coincide, we obtain the existence of limit (4.151), and find this limit,

$$\frac{dJ(\Omega_\delta)}{d\delta}\Big|_{\delta=0}=\frac12\int\limits_{\Omega}\Big((2\mu+\lambda)\{u^2_{y_1}(-\theta_{y_1})+v^2_{y_2}\theta_{y_1}+w^2_{y_3}\theta_{y_1}$$

$$\begin{aligned}&+2v_{y_1}v_{y_2}(-\theta_{y_2})+2w_{y_1}w_{y_3}(-\theta_{y_3})\}+\mu\{u^2_{y_2}\theta_{y_1}+v^2_{y_1}(-\theta_{y_1})+u^2_{y_3}\theta_{y_1}\\&+w^2_{y_1}(-\theta_{y_1})+v^2_{y_3}\theta_{y_1}+w^2_{y_2}\theta_{y_1}+2u_{y_1}u_{y_3}(-\theta_{y_3})+2u_{y_1}w_{y_1}(-\theta_{y_3})\\&+2v_{y_1}v_{y_3}(-\theta_{y_3})+2v_{y_1}w_{y_2}(-\theta_{y_3})+2v_{y_3}w_{y_1}(-\theta_{y_2})+2w_{y_1}w_{y_2}(-\theta_{y_2})\\&+2u_{y_1}u_{y_2}(-\theta_{y_2})+2v_{y_1}u_{y_1}(-\theta_{y_2})+2v_{y_3}w_{y_2}\theta_{y_1}\}+2\lambda\{v_{y_2}w_{y_3}\theta_{y_1}\\&+u_{y_1}w_{y_1}(-\theta_{y_3})+v_{y_2}w_{y_1}(-\theta_{y_3})+v_{y_1}w_{y_3}(-\theta_{y_2})+u_{y_1}v_{y_1}(-\theta_{y_2})\}\Big)\\&-\int\limits_{\Omega}(\theta f_1)_{y_1}u-\int\limits_{\Omega}(\theta f_2)_{y_1}v-\int\limits_{\Omega}(\theta f_3)_{y_1}w.\end{aligned} \tag{4.159}$$

We have obtained the Griffith formula (4.159). It is not difficult to show that the right-hand side of (4.159) does not depend on θ. To prove this, consider the difference between right-hand sides of (4.159) corresponding to any two functions θ_1, θ_2. Let $\theta=\theta_1-\theta_2$. We integrate by parts, which implies that the difference Λ between the right-hand sides of (4.159) evaluated for θ_1,θ_2 is equal to

$$\Lambda=\int\limits_{\Omega\setminus S_L}\theta\left((\sigma_{1j,j}+f_1)u_{,1}+(\sigma_{2j,j}+f_2)v_{,1}+(\sigma_{3j,j}+f_3)w_{,1}\right)$$

$$+ \int\limits_{\Xi_l \setminus S_L} \theta \Big((2\mu + \lambda)[w_{,1} w_{,3}] + \mu[u_{,1} u_{,3}] + \mu[u_{,1} w_{,1}] \qquad (4.160)$$

$$+ \mu[v_{,1} v_{,3}] + \mu[v_{,1} w_{,2}] + \lambda[u_{,1} w_{,1}] + \lambda[v_{,2} w_{,1}] \Big).$$

Here S_L is a neighbourhood of the set L. We should recall at this point that $\theta = \theta_1 - \theta_2 = 0$ in some neighbourhood of L. It is known that the solution of the problem (4.140)–(4.142) has an additional regularity up to the crack faces (see Yakunina, 1981). For any point $x \in \Xi_l$ there exists a neighbourhood V of the point x such that

$$W \in H^2(V \setminus \Xi_l).$$

In particular, by the imbedding theorem, W is continuous up to the crack faces. As it was shown in (Khludnev, Sokolowski, 1997) the solution W satisfies the following boundary condition on Ξ_l:

$$[w] \geq 0, \quad [\sigma_{33}] = 0, \quad \sigma_{33} \leq 0, \qquad (4.161)$$

$$[w]\sigma_{33} = 0, \quad \sigma_{13} = 0, \quad \sigma_{23} = 0.$$

It is easy to see that

$$\sigma_{33} = 2\mu w_{,3} + \lambda(u_{,1} + v_{,2} + w_{,3}), \qquad (4.162)$$

$$\sigma_{13} = \mu(u_{,3} + w_{,1}), \quad \sigma_{23} = \mu(v_{,3} + w_{,2}).$$

By (4.140), (4.161), (4.162), from (4.160) it follows that

$$\Lambda = \int\limits_{\Xi_l \setminus S_L} \theta \left(\sigma_{33}[w_{,1}] + [\sigma_{13} u_{,1}] + [\sigma_{23} v_{,1}] \right) \qquad (4.163)$$

and, consequently, by (4.161),

$$\Lambda = \int\limits_{\Xi_l \setminus S_L} \theta \sigma_{33}[w_{,1}].$$

Let us prove that

$$\sigma_{33}[w_{,1}] = 0 \quad \text{a.e. on } \Xi_l \setminus S_L. \qquad (4.164)$$

In accordance with (4.161), on the set

$$M = \{ y \in \Xi_l \setminus S_L \;\big|\; [w(y)] > 0 \},$$

we have $\sigma_{33}(y) = 0$ and, consequently, $\sigma_{33}(y)[w_{,1}(y)] = 0$. The complement $(\Xi_l \setminus S_L) \setminus M$ is characterized by the condition

$$[w(y)] = 0,$$

hence $[w_{,1}(y)] = 0$ a.e. on $(\Xi_l \setminus S_L) \setminus M$ (see Kinderlehrer, Stampacchia, Chapter 2, Theorem A.1, 1980) which provides $\sigma_{33}[w_{,1}] = 0$ a.e. on $\Xi_l \setminus S_L \setminus M$. As a result we obtain (4.164), and hence $\Lambda = 0$ which proves the assertion.

To conclude the section we write the formula (4.159) in the form which does not contain the function θ. To this end, consider a neighbourhood S_L of the set L with a smooth boundary Γ_L assuming that $\theta = 1$ on S_L. Denote by (ν_1, ν_2, ν_3) the unit external normal vector to Γ_L. Integrating by parts in (4.159) we obtain

$$\frac{dJ(\Omega_\delta)}{d\delta}\bigg|_{\delta=0} = \int\limits_{S_L} (f_1 u_{y_1} + f_2 v_{y_1} + f_3 w_{y_1}) + \frac{1}{2}\int\limits_{\Gamma_L} \nu_1 \Big((2\mu+\lambda)(u_{y_1}^2 - v_{y_2}^2 - w_{y_3}^2)$$

$$+ \mu(v_{y_1}^2 - u_{y_2}^2 - u_{y_3}^2 + w_{y_1}^2 - v_{y_3}^2 - w_{y_2}^2 - 2v_{y_3} w_{y_2}) - 2\lambda v_{y_2} w_{y_3} \Big)$$

$$+ \frac{1}{2}\int\limits_{\Gamma_L} \nu_2 \Big(2(2\mu+\lambda) v_{y_1} v_{y_2} + 2\mu(v_{y_3} w_{y_1} + w_{y_1} w_{y_2} + u_{y_1} u_{y_2} + v_{y_1} u_{y_1}) \tag{4.165}$$

$$+ 2\lambda(v_{y_1} w_{y_3} + u_{y_1} v_{y_1}) \Big) + \frac{1}{2}\int\limits_{\Gamma_L} \nu_3 \Big(2(2\mu + \lambda) w_{y_1} w_{y_3}$$

$$+ 2\mu(u_{y_1} u_{y_3} + u_{y_1} w_{y_1} + v_{y_1} v_{y_3} + v_{y_1} w_{y_2}) + 2\lambda(u_{y_1} w_{y_1} + v_{y_2} w_{y_1}) \Big).$$

Denoting the right-hand side of (4.165) by $k(l, h, f)$ we have

$$\frac{dJ(\Omega_\delta)}{d\delta}\bigg|_{\delta=0} = k(l, h, f).$$

Here $k(l, h, f)$ is some functional. Consequently,

$$J(\Omega_\delta) = J(\Omega) + k(l, h, f)\delta + \alpha(\delta)\delta,$$

where $\alpha(\delta) \to 0$ as $\delta \to 0$. Note that $k(l, h, f)$ is independent of S_L.

4.9 Extreme crack shapes in a shallow shell

This section is concerned with an extreme crack shape problem for a shallow shell (see Khludnev, 1997a). The shell is assumed to have a vertical crack the shape of which may change. From all admissible crack shapes with fixed tips we have to find an extreme one. This means that the shell displacements should be as close to the given functions as possible. To be more precise, we consider a functional defined on the set describing crack shapes, which, in particular, depends on the solution of the equilibrium problem for the shell. The purpose is to minimize this functional. We assume that the shell displacements (W, w) found at the opposite crack faces satisfy the nonpenetration condition

$$[W]\nu \geq h \left| \left[\frac{\partial w}{\partial \nu} \right] \right|.$$

Here $[\cdot]$ is a jump of a function at the crack faces, ν is the unit normal vector to the crack shape, and $2h$ is the thickness of the shell. A similar extreme crack shape problem for a plate was considered in Section 2.4.

Note that the problem analysed in this section can be viewed as a shape sensitivity problem (see Sokolowski, Zolesio, 1992; Pironneau, 1984; Ohtsuka, 1986, 1994; Mróz, 1963; Schaeffer, 1975; Kinderlehrer, 1982; Athanasopoulos, 1981).

4.9.1 Convergence of solutions

Let $\Omega \subset R^2$ be a bounded domain with a smooth boundary Γ, $\Omega_\delta = \Omega \setminus \Gamma_\delta$, where Γ_δ is the graph of the function $y = \delta\psi(x)$, $x \in [0, 1]$, $(x, y) \in \Omega$, δ is a small parameter, $\psi \in H_0^3(0, 1)$. We denote by $W = (w^1, w^2)$, w the horizontal and vertical displacements of the mid-surface points, respectively, and $\chi = (W, w)$. Let

$$\varepsilon_{ij}(W) = \frac{1}{2}\left(\frac{\partial w^i}{\partial x_j} + \frac{\partial w^j}{\partial x_i} \right), \quad x_1 = x, \quad x_2 = y,$$

Denote by e_{ij} the strain deformation tensor of the mid-surface,

$$e_{ij} = \varepsilon_{ij}(W) + k_{ij}w, \quad i, j = 1, 2.$$

Here k_{ij} are given smooth functions (the curvatures of the shell). The normal vector to the curve $y = \delta\psi(x)$ is denoted by ν^δ, $\nu^\delta = (-\delta\psi_x, 1)(1 + \delta^2\psi_x^2)^{-1/2}$. We use the Kirchhoff–Love model of the shell. Assume that $z = 0$ corresponds to the mid-surface points, and (x, y, z) are Descartes orthogonal coordinates. The nonpenetration condition can be written as

$$[W - z\nabla w]\,\nu^\delta \geq 0 \quad \text{on } \Gamma_\delta, \quad |z| \leq h. \tag{4.166}$$

Here $[M] = M^+ - M^-$, $M^\pm$ are the quantities of M evaluated at the positive and negative crack faces with respect to ν^δ. We know that (4.166) can be written in the equivalent form

$$[W]\nu^\delta \geq h \left| \left[\frac{\partial w}{\partial \nu^\delta} \right] \right| \quad \text{on } \Gamma_\delta.$$

The energy functional for the shell is of the form

$$\Pi_\delta(\chi) = \frac{1}{2}B_\delta(w, w) + \frac{1}{2}\langle N_{ij}, e_{ij}\rangle_\delta - \langle f, \chi\rangle_\delta.$$

We make use of the following notation:

$$N_{11} = e_{11} + \sigma e_{22}, \quad N_{22} = e_{22} + \sigma e_{11}, \quad N_{12} = (1-\sigma)e_{12},$$

$$0 < \sigma < 1/2, \quad \sigma = \text{const}, \quad f = (f_1, f_2, f_3) \in L^2(\Omega), \quad \langle p, q\rangle_\delta = \int\limits_{\Omega_\delta} pq d\Omega_\delta.$$

The bilinear form $B_\delta(\cdot\,,\,\cdot)$ describing the bending properties of the shell is as follows:

$$B_\delta(w, \bar{w}) = \int\limits_{\Omega_\delta} (w_{xx}\bar{w}_{xx} + w_{yy}\bar{w}_{yy} + \sigma w_{xx}\bar{w}_{yy}$$

$$+ \sigma w_{yy}\bar{w}_{xx} + 2(1-\sigma)w_{xy}\bar{w}_{xy})d\Omega_\delta.$$

Let us fix the conditions on the external boundary corresponding to the clamping of the shell:

$$w = \frac{\partial w}{\partial n} = W = 0 \quad \text{on } \Gamma.$$

We next denote by $H^{1,0}(\Omega_\delta)$ the space of functions from $H^1(\Omega_\delta)$ equal to zero on Γ; the functions from $H^{2,0}(\Omega_\delta)$ are equal to zero on Γ with the first derivatives, $H^{2,0}(\Omega_\delta) \subset H^2(\Omega_\delta)$. Let $H(\Omega_\delta) = H^{1,0}(\Omega_\delta) \times H^{1,0}(\Omega_\delta) \times H^{2,0}(\Omega_\delta)$. Introduce the set of admissible displacements of the shell

$$K_\delta(\Omega_\delta) = \{(W, w) \in H(\Omega_\delta) \mid \ (W, w) \quad \text{satisfy } (4.166)\}.$$

The equilibrium problem for the shallow shell having the vertical crack can be formulated as a variational one,

$$\inf_{\chi \in K_\delta(\Omega_\delta)} \Pi_\delta(\chi). \tag{4.167}$$

Since the functional Π_δ is convex and differentiable on $H(\Omega_\delta)$, the problem (4.167) can be written in the equivalent form: find $\chi = (W, w) \in K_\delta(\Omega_\delta)$ such that

$$B_\delta(w, \bar{w} - w) + \langle k_{ij}N_{ij}, \bar{w} - w\rangle_\delta + \langle N_{ij}, \varepsilon_{ij}(\bar{W} - W)\rangle_\delta \tag{4.168}$$

$$- \langle f, \bar{\chi} - \chi\rangle_\delta \geq 0 \quad \forall \bar{\chi} = (\bar{W}, \bar{w}) \in K_\delta(\Omega_\delta).$$

So, we consider the shallow shell with the distances on the mid-surface coinciding with those on the plane. At the same time the curvatures are not equal to zero, in general. The shells like these are called the weakly curved plates.

At the beginning we study the δ-dependence of the solution and next we consider the problem of finding extreme crack shapes. First, let us note that the problem (4.168) has a solution owing to the coercivity and the weak lower semicontinuity of Π_δ on the space $H(\Omega_\delta)$. The solution is unique for every fixed δ. In order to study the behaviour of the solution as $\delta \to 0$, we accomplish a change of the variables in such a way that the domain Ω_δ maps onto Ω_0. Let Ω_1, Ω_2 be domains such that $\overline{\Omega}_1 \subset \Omega_2$, $\overline{\Omega}_2 \subset \Omega$, $\Gamma_\delta \subset \Omega_1$ for all small δ, and $\xi \in C_0^\infty(\Omega_2)$ be a function identically equal to unity on Ω_1. Extend the function ψ by zero beyond the interval $(0,1)$ and consider the change of variables

$$\tilde{x} = x, \quad \tilde{y} = y - \delta\xi\psi. \tag{4.169}$$

The Jacobian q_δ of this transformation is equal to $1 - \delta\psi\xi_y$, being positive for small δ. Denote $u(\tilde{x}, \tilde{y}) = w(x,y)$, $U(\tilde{x}, \tilde{y}) = W(x,y)$, $\omega = (U,u)$. By substituting $\bar{\chi} = 0$ in (4.168) we find

$$2\Pi_\delta(\chi) + \langle f, \chi\rangle_\delta \le 0. \tag{4.170}$$

This inequality can be written in the new variables. In order to clarify the structure of the relation obtained in this way, we write down one of the second derivatives of w:

$$w_{xx} = u_{\tilde{x}\tilde{x}} - 2\delta u_{\tilde{x}\tilde{y}}(\psi\xi)_x + \delta^2 u_{\tilde{y}\tilde{y}}(\psi\xi)_x^2 - \delta u_{\tilde{y}}(\psi\xi)_{xx}.$$

Hence the inequality (4.170) takes the form

$$\int_{\Omega_0} \left(u_{\tilde{x}\tilde{x}}^2 + u_{\tilde{y}\tilde{y}}^2 + 2\sigma u_{\tilde{x}\tilde{x}} u_{\tilde{y}\tilde{y}} + 2(1-\sigma)u_{\tilde{x}\tilde{y}}^2\right) q_\delta^{-1} d\Omega_0 \tag{4.171}$$

$$+ \langle N_{ij}^\delta, e_{ij}^\delta q_\delta^{-1}\rangle_0 - \langle f^\delta, \omega q_\delta^{-1}\rangle_0 + \delta \int_{\Omega_0} G(\tilde{x}, \tilde{y}, \delta, D^\alpha u, D^\beta U) d\Omega_0 \le 0,$$

where the following notation is used:

$$e_{ij}^\delta = \varepsilon_{ij}(U) + k_{ij}^\delta u, \quad k_{ij}^\delta(\tilde{x}, \tilde{y}) = k_{ij}(x,y), \quad f^\delta(\tilde{x}, \tilde{y}) = f(x,y).$$

In doing so we have $|\alpha| \le 2$, $|\beta| \le 1$. The character of the dependence of G on its arguments is completely determined by the transformation (4.169). It is of importance that the higher order terms have square growth in $D^\alpha u$, $D^\beta U$. Introduce the notation

$$\Pi_0^\delta(\omega) = \frac{1}{2} B_0(u,u) + \frac{1}{2}\langle N_{ij}^\delta, e_{ij}^\delta\rangle_0 - \langle f^\delta, \omega q_\delta^{-1}\rangle_0.$$

Since $q_\delta^{-1} > 1/2$ for small δ, the inequality (4.171) yields

$$\Pi_0^\delta(\omega) + \delta \int_{\Omega_0} G(\tilde{x}, \tilde{y}, \delta, D^\alpha u, D^\beta U) d\Omega_0 \le 0. \tag{4.172}$$

It can be proved that Π_0^δ is coercive functional on the space $H(\Omega_0)$ uniformly in $|\delta| \le \delta_0$, that is

$$\Pi_0^\delta(\omega) \to \infty, \quad \|\omega\|_{H(\Omega_0)} \to \infty, \ \delta \le \delta_0.$$

Indeed, let $S \subset H(\Omega_0)$ be the unit sphere. The following transformation can be defined on S,

$$\bar{U} \to \gamma R\bar{U}, \quad \bar{u} \to R\bar{u}, \quad \bar{U} = (\bar{u}^1, \bar{u}^2), \ (\bar{U}, \bar{u}) \in S,$$

where R is a positive parameter and γ is a positive constant to be chosen below. Assume that

$$||u||^2_{2,\Omega_0} = B_0(u,u), \quad ||U||^2_{1,\Omega_0} = \langle \sigma_{ij}(U), \varepsilon_{ij}(U) \rangle_0.$$

Here the functions $\sigma_{ij}(U)$ coincide with $N_{ij}(U)$ if we put $k_{ij} \equiv 0$. Thus

$$||\omega||^2_{H(\Omega_0)} = ||U||^2_{1,\Omega_0} + ||u||^2_{2,\Omega_0}.$$

Consider the quadratic part of $\Pi_0^\delta(\omega)$:

$$\pi^\delta(\omega) = \frac{1}{2} B_0(u,u) + \frac{1}{2}\langle N_{ij}^\delta, e_{ij}^\delta \rangle_0, \quad \omega = (U, u).$$

Let $(\bar{U}, \bar{u}) \in S$ and $||\bar{u}||^2_{2,\Omega_0} \geq 1/4$. The image of $(\bar{U}, \bar{u})$ for the above transformation is denoted by ω. Then one has $\pi^\delta(\omega) \geq R^2/8$ uniformly in δ. If $||\bar{U}||^2_{1,\Omega_0} \geq 3/4$, then $\pi^\delta(\omega) \geq 3\gamma^2 R^2/8 - c_0\gamma R^2/2$ uniformly in δ. The constant c_0 puts bound to the integral,

$$2\langle u, k_{11}^\delta u_{\bar{x}}^1 + k_{22}^\delta u_{\bar{y}}^2 + \sigma(k_{11}^\delta u_{\bar{y}}^2 + k_{22}^\delta u_{\bar{x}}^1) \rangle_0,$$

on S uniformly in $|\delta| \leq \delta_0$. The above arguments show that the inequality

$$\pi^\delta(\omega) \geq \frac{R^2}{8}$$

holds for any image of S provided the constant γ is chosen from the equation

$$\frac{3}{8}\gamma^2 - \frac{c_0}{2}\gamma = 1.$$

The linear in ω part of Π_0^δ has the linear in R estimate, whence we always have

$$\Pi_0^\delta(\omega) \geq cR^2$$

with the constant c uniform in $|\delta| \leq \delta_0$. Hence the statement on the coercivity of Π_0^δ is proved. As a result we conclude from (4.172) that (denoting ω by ω^δ)

$$||\omega^\delta||_{H(\Omega_0)} \leq c \tag{4.173}$$

uniformly in δ. Now one can write the inequality (4.168) in the new variables. In this case the test functions $\bar{\omega}$ should satisfy the restriction

$$[\bar{U} - z(\bar{u}_{\bar{x}} - \delta\psi_x \bar{u}_{\bar{y}}, \bar{u}_{\bar{y}})] \cdot (-\delta\psi_x, 1) \geq 0 \quad \text{on } \Gamma_0, \quad |z| \leq h, \tag{4.174}$$

obtained from (4.166). A set of functions $(\bar{U}, \bar{u})$ from $H(\Omega_0)$ satisfying (4.174) is denoted by $K_\delta(\Omega_0)$.

By (4.173), we choose a subsequence, still denoted by ω^δ, such that as $\delta \to 0$

$$\omega^\delta \to \omega \quad \text{weakly in } H(\Omega_0). \tag{4.175}$$

In view of Lemma 4.4 below and the convergence (4.175), a passage to the limit can be carried out in the variational inequality (4.168) which is written in the new variables $\tilde{x}, \tilde{y}$. As a result we arrive at the relation

$$\omega = (U, u) \in K_0(\Omega_0):$$

$$B_0(u, \bar{u} - u) + \langle k_{ij}N_{ij}, \bar{u} - u\rangle_0 + \langle N_{ij}, \varepsilon_{ij}(\bar{W} - W)\rangle_0 \tag{4.176}$$

$$-\langle f, \bar{\omega} - \omega\rangle_0 \geq 0 \quad \forall \bar{\omega} = (\bar{U}, \bar{u}) \in K_0(\Omega_0).$$

This obviously means that the limiting function ω is a solution of the equilibrium problem for the shell having the crack shape $y = \psi(x) \equiv 0$, $x \in [0, 1]$. Thus the following statement has been proved.

Theorem 4.7. *From the sequence of solutions $\chi^\delta = \omega^\delta$ of the problem* (4.168) *one can choose a subsequence weakly converging in $H(\Omega_0)$ to ω. The limiting function ω is a solution of the problem* (4.176).

4.9.2 Existence of extreme crack shapes

Consider the problem of finding the extreme crack shapes. The setting of this problem is as follows. Let $\Psi \subset H_0^3(0, 1)$ be a convex, closed and bounded set such that for any $\psi \in \Psi$ the graph of the function $y = \psi(x)$ does not leave Ω. For every fixed $\psi \in \Psi$ a solution of the problem

$$\chi_\psi \in K_\psi(\Omega_\psi): \quad \langle \Pi'_\psi(\chi_\psi), \bar{\chi} - \chi_\psi\rangle \geq 0 \quad \forall \bar{\chi} \in K_\psi(\Omega_\psi)$$

can be found. The functional Π_ψ and the set $K_\psi(\Omega_\psi)$ are introduced similarly to Π_δ, $K_\delta(\Omega_\delta)$, respectively. Let $\chi_0 \in L^2(\Omega)$ be a given function. Define the cost functional

$$J(\psi) = \|\chi_\psi - \chi_0\|_{0,\Omega_\psi}.$$

The problem of finding an extreme crack shape is formulated as follows:

$$\inf_{\psi \in \Psi} J(\psi). \tag{4.177}$$

In what follows we prove the existence of the extreme crack shape.

Theorem 4.8. *Let the above hypotheses be fulfilled. Then there exists a solution of the problem* (4.177).

PROOF. Denote by $\psi^n \in \Psi$ the elements of a minimizing sequence. Without decreasing a generality we assume that as $n \to \infty$

$$\psi^n \to \psi \text{ weakly in } H_0^3(0,1), \quad \psi \in \Psi,$$

$$|\psi^n_{xx}(x) - \psi_{xx}(x)| < \frac{1}{n}, \quad x \in (0,1).$$

For every fixed n the solution of the problem

$$\chi^n \in K_{\psi^n}(\Omega_{\psi^n}) : \quad \langle \Pi'_{\psi^n}(\chi^n), \bar{\chi} - \chi^n \rangle \geq 0 \quad \forall \bar{\chi} \in K_{\psi^n}(\Omega_{\psi^n}) \tag{4.178}$$

obviously exists. Let us choose domains Ω_1, Ω_2 and a function ξ as before, assuming that $\Gamma_{\psi^n} \subset \Omega_1$ for all n. Then we can extend the functions ψ and ψ^n by zero beyond $(0,1)$ and consider the transformation of the independent variables

$$\tilde{x} = x, \quad \tilde{y} = y + (\psi - \psi^n)\xi.$$

This transformation can be written as

$$\tilde{x} = x, \quad \tilde{y} = y - \frac{1}{n}\varphi_n \xi$$

with the functions $\varphi_n = n(\psi^n - \psi)$ bounded in $C^2[0,1]$. Thus, we obtain the one-to-one mapping between Ω_{ψ^n} and Ω_ψ with the Jacobian $q_n = 1 - n^{-1}\varphi_n \xi_y$, which is positive for sufficiently large n. We can argue as in Theorem 4.7. Namely, the problem (4.178) is to be rewritten in the new variables. The appropriate estimates of the solutions are as follows,

$$\|u^n\|_{2,\Omega_\psi} \leq c, \quad \|U^n\|_{1,\Omega_\psi} \leq c,$$

with the constants uniform in n. Suppose that a subsequence still denoted by $\omega^n = (U^n, u^n)$ possesses the property

$$\omega^n \to \omega \quad \text{weakly in } H(\Omega_\psi).$$

The statement analogous to Lemma 4.4 holds true in this case, that is for any fixed $\bar{\omega} \in K_\psi(\Omega_\psi)$ there exists a sequence $\bar{\omega}^n \in K_{\psi^n}(\Omega_\psi)$ such that as $n \to \infty$

$$\bar{\omega}^n \to \bar{\omega} \quad \text{strongly in } H(\Omega_\psi).$$

This allows us to fulfil the passage to the limit in the relations obtained from (4.178) by changing the independent variables and to obtain

$$\omega = (U, u) \in K_\psi(\Omega_\psi) : \quad \langle \Pi'_\psi(\omega), \bar{\omega} - \omega \rangle \geq 0 \quad \forall \bar{\omega} \in K_\psi(\Omega_\psi).$$

This means that the function ω is the solution of the equilibrium problem for the shell with crack shape $y = \psi(x)$, $x \in [0,1]$. At last, denote $\omega_{0n}(\tilde{x}, \tilde{y}) = \chi_0(x, y)$. Then

$$\inf_{\bar{\psi} \in \Psi} J(\bar{\psi}) = \liminf \|\chi^n - \chi_0\|_{0,\Omega_{\psi^n}} = \liminf \|q_n^{-1/2}(\omega^n - \omega_{0n})\|_{0,\Omega_\psi}$$

$$\geq \|\omega - \chi_0\|_{0,\Omega_\psi} = J(\psi).$$

Consequently, the limiting function ψ is a solution of the extreme crack shape problem (4.177). Theorem 4.8 is proved.

In conclusion we establish an auxiliary result used to prove Theorems 4.7, 4.8. It allows us to approximate the elements from $K_0(\Omega_0)$ by elements from $K_\delta(\Omega_0)$. Recall that functions from $K_\delta(\Omega_0)$ should satisfy the inequality (4.174).

Lemma 4.4. *For every element $(u_1, u_2, u) \in K_0(\Omega_0)$ there exists a sequence $(u_1^\delta, u_2^\delta, u^\delta)$ in $K_\delta(\Omega_0)$ such that as $\delta \to 0$*

$$(u_1^\delta, u_2^\delta, u^\delta) \to (u_1, u_2, u) \quad \textit{strongly in } H(\Omega_0).$$

PROOF. Since $(u_1, u_2, u) \in K_0(\Omega_0)$ we have

$$[u_2] - z[u_{\tilde{y}}] \geq 0 \quad \text{on } \Gamma_0, \quad |z| \leq h.$$

The function $u_{\tilde{x}} - \delta\psi_x u_{\tilde{y}}$ belongs to the space $H^{1,0}(\Omega_0)$. Hence, its traces on the lines $\tilde{y} = 0+$, $\tilde{y} = 0-$ are elements of $H^{1/2}(\tilde{y} = 0\pm)$. The difference between these traces belongs to $H^{1/2}(\tilde{y} = 0)$ and coincides with $[u_{\tilde{x}} - \delta\psi_x u_{\tilde{y}}]$ on Γ_0.

Choose an extension of this difference from the space $H^1(R^2)$ and denote it by Q. Consequently, the restriction of the function $\xi h|\delta\psi_x Q|$ to Ω is an element of $H_0^1(\Omega)$. In Ω_0 we define

$$(u_1^\delta, u_2^\delta, u^\delta) = (u_1, u_2, u) + (0,\ \delta\psi_x u_1 + \xi h|\delta\psi_x Q|,\ 0).$$

We first show that $(u_1^\delta, u_2^\delta, u^\delta) \in K_\delta(\Omega_0)$. To this end we notice that the boundary conditions on Γ are fulfilled for $(u_1^\delta, u_2^\delta, u^\delta)$. Hence, it suffices to prove (4.174). From the above considerations it follows that the inequality

$$h\ |\delta\psi_x Q| \geq -\delta z\psi_x[u_{\tilde{x}} - \delta\psi_x u_{\tilde{y}}] \quad \forall z,\ |z| \leq h,$$

holds on Γ_0. Hence, on Γ_0 we have

$$[u_1](-\delta\psi_x) + [u_2] + \delta\psi_x[u_1] + h\ |\delta\psi_x Q| + \delta z\psi_x[u_{\tilde{x}} - \delta\psi_x u_{\tilde{y}}] - z[u_{\tilde{y}}]$$

$$\geq [u_2] - z[u_{\tilde{y}}] \geq 0, \quad |z| \leq h,$$

which means that inequality (4.174) holds for $(u_1^\delta, u_2^\delta, u^\delta)$. The strong convergence of the sequence $(u_1^\delta, u_2^\delta, u^\delta)$ to (u_1, u_2, u) in $H(\Omega_0)$ is obvious. This proves the lemma.

Chapter 5

Cracks in elastoplastic bodies

Concerning the solvability of static and quasistatic elastoplastic boundary problems, a lot of results have been obtained by now (Anzellotti, 1983; Anzellotti, Giaquinta, 1982; Temam, 1983, 1986; Temam, Strang, 1980; Suquet, 1981; Demengel, 1983; Johnson, 1976; Carstensen, 1994); as for numerical results, see (Kovtunenko, 1993, 1996a). For the variational inequality formulations, in particular, there are existence and regularity results, both for the Prandtl–Reuss model in the quasistatic case and for the Hencky model in the static case. It is, however, a drawback of that approach that some of the boundary conditions, which are prescribed originally as part of the boundary value problem, are not easily recovered from the variational inequality, even if the solution is assumed to be regular. The difficulty arises because the set of admissible stresses is not a linear space. The problem has been solved in the one-dimensional case for beams and curvilinear bars; it has been shown that the variational inequality solution satisfies all boundary conditions one expects to hold (Khludnev, Hoffmann, 1992; Khludnev, 1993a, 1993b). In the case of two and three space dimensions, however, no such results are available so far, and the question of whether the solution satisfies all required boundary conditions appears to be open. In the present chapter we show that for the Hencky and Prandtl–Reuss models in three dimensions and those models for plates in two dimensions with Neumann boundary conditions, indeed all boundary conditions, hold true. Our proof combines elliptic or parabolic regularizations with the penalty method in a particular way. It can be used in a variety of elastoplastic problems. We show in particular that, besides the standard situation of three- and two-dimensional domains with a smooth boundary, which we discuss in Sections 5.1–5.4, it also applies both to the case of static and quasistatic interior two-dimensional cracks, modelled by removing a two-dimensional surface from the interior of the domain and to the case of cracks in elastoplastic plates.

5.1 Elastoplastic problems for the Hencky model

We prove the existence of solutions for the three-dimensional elastoplastic problem with Hencky's law and Neumann boundary conditions by elliptic regularization and the penalty method, both for the case of a smooth boundary and of an interior two-dimensional crack (see Brokate, Khludnev, 1998). It is shown in particular that the variational solution satisfies all boundary conditions.

5.1.1 Notation and simple properties

Let $\Omega \subset R^3$ be an open, bounded and connected set with smooth boundary Γ. We define the Banach space

$$LD(\Omega) = \{u = (u_1, u_2, u_3) \mid \quad u_i \in L^1(\Omega),\ i = 1, 2, 3,$$
$$\varepsilon_{ij}(u) \in L^1(\Omega),\ i, j = 1, 2, 3\}$$

equipped with the norm

$$\|u\|_{LD(\Omega)} = \|u\|_{L^1(\Omega)} + \sum_{i,j=1}^{3} \|\varepsilon_{ij}(u)\|_{L^1(\Omega)}. \tag{5.1}$$

Here $\varepsilon_{ij}(u) = (u_{i,j} + u_{j,i})/2$ are the components of the strain tensor. We consider function spaces whose elements are characterized by the conditions

$$\int_\Omega u = 0, \quad \int_\Omega (u_i x_j - u_j x_i) = 0, \quad i, j = 1, 2, 3, \quad u = (u_1, u_2, u_3). \tag{5.2}$$

In particular, we define

$$LD_N(\Omega) = \{u \in LD(\Omega) \mid \quad u \text{ satisfies } (5.2)\}.$$

Note that the linear space $R(\Omega)$ of functions ρ satisfying the conditions $\varepsilon_{ij}(\rho) = 0$ in Ω, $i, j = 1, 2, 3$, can be described as $\rho(x) = c + Bx$, $x \in \Omega$, where $c = (c_1, c_2, c_3)$ is a constant vector, $B = (b_{ij})$ is a constant matrix with $b_{ij} = -b_{ji}$, for all i, j. In componentwise notation, $\rho_i(x) = c_i + b_{ij}x_j$. One sees that the orthogonal complement of the subspace $R(\Omega)$ in $L^2(\Omega)$ coincides with the subspace of all functions from $L^2(\Omega)$ satisfying (5.2). Therefore we see that if $\rho \in R(\Omega)$ satisfies (5.2), then $\rho \equiv 0$.

Since

$$\left|\int_\Omega u\right| + \sum_{i,j=1}^{3}\left|\int_\Omega (u_i x_j - u_j x_i)\right|$$

is a seminorm on the space $LD(\Omega)$ and a norm on $R(\Omega)$, it follows that the

$$|u|_{LD(\Omega)} = \left|\int_\Omega u\right| + \sum_{i,j=1}^{3}\left|\int_\Omega (u_i x_j - u_j x_i)\right| + \sum_{i,j=1}^{3} \|\varepsilon_{ij}(u)\|_{L^1(\Omega)}$$

defines a norm on $LD(\Omega)$ which is equivalent to the original norm (5.1) (see Temam, 1983).

Consider next the space of bounded measures $M^1(\Omega)$. We know that $M^1(\Omega)$ is the space dual of the normed space $C_0(\Omega)$ of continuous functions with compact support, endowed with the uniform convergence topology (see Giusti, 1984; Strang, Temam, 1980). Any ball from $M^1(\Omega)$ is a compact in the weak star topology, and every bounded sequence in $M^1(\Omega)$ has a subsequence which is weakly star convergent. We recall that by definition a sequence $g_m \in M^1(\Omega)$ is weakly star convergent to an element $g \in M^1(\Omega)$ if

$$g_m(\phi) \to g(\phi), \quad m \to \infty$$

for any fixed $\phi \in C_0(\Omega)$. Now we can introduce the Banach space of bounded deformations

$$BD(\Omega) = \{u = (u_1, u_2, u_3) \mid \; u_i \in L^1(\Omega),\ i = 1,2,3,$$
$$\varepsilon_{ij}(u) \in M^1(\Omega),\ i,j = 1,2,3\}$$

equipped with the norm

$$\|u\|_{BD(\Omega)} = \|u\|_{L^1(\Omega)} + \sum_{i,j=1}^{3} \|\varepsilon_{ij}(u)\|_{M^1(\Omega)}.$$

Also, denote by $BD_N(\Omega)$ the subspace of $BD(\Omega)$ which consists of all elements of $BD(\Omega)$ satisfying (5.2). Consider also the space

$$H^1(\Omega) = \{u = (u_1, u_2, u_3) \mid \; u_i \in L^2(\Omega),\ i = 1,2,3;$$
$$u_{i,j} \in L^2(\Omega),\ i,j = 1,2,3\}$$

with the norm

$$\|u\|_{H^1(\Omega)} = \|u\|_0 + \sum_{i,j=1}^{3} \|u_{i,j}\|_0 \tag{5.3}$$

where $\|\cdot\|_0$ is the norm of $L^2(\Omega)$. As usual, to simplify the notations, we write $H^1(\Omega)$ instead of $[H^1(\Omega)]^3$. Let

$$H^1_N(\Omega) = \{u \in H^1(\Omega) \mid \; u \text{ satisfies } (5.2)\}.$$

In the sequel we shall use the following norm in $H^1(\Omega)$,

$$|u|_{H^1(\Omega)} = \left|\int_\Omega u\right| + \sum_{i,j=1}^{3} \left|\int_\Omega (u_i x_j - u_j x_i)\right| + \sum_{i,j=1}^{3} \|\varepsilon_{ij}(u)\|_0,$$

which is equivalent to the norm (5.3). It is easy to see that by

$$|u|_{H^1_N(\Omega)} = \sum_{i,j=1}^{3} \|\varepsilon_{ij}(u)\|_0, \tag{5.4}$$

we obtain a norm on the subspace $H^1_N(\Omega)$.

Let us recall the well-known Green formula. Namely, if $\sigma_{ij} \in L^2(\Omega)$, $\sigma_{ij,j} \in L^2(\Omega)$, $i, j = 1, 2, 3$, then the values $\sigma_{ij} n_j$ can be correctly defined on Γ, and moreover, $\sigma_{ij} n_j \in H^{-1/2}(\Gamma)$,

$$-\langle \sigma_{ij,j}, \theta \rangle = \langle \sigma_{ij}, \theta_{,j} \rangle - \langle \sigma_{ij} n_j, \theta \rangle_{1/2,\Gamma}, \quad \forall \theta \in H^1(\Omega), \quad i = 1, 2, 3. \tag{5.5}$$

Here $n = (n_1, n_2, n_3)$ is the unit outer unit normal to the boundary Γ, the brackets $\langle \cdot, \cdot \rangle$, $\langle \cdot, \cdot \rangle_{1/2,\Gamma}$ denote the integration over Ω and a duality pairing between the spaces $H^{-1/2}(\Gamma)$ and $H^{1/2}(\Gamma)$, respectively.

All functions which carry two lower indices are assumed to be symmetric with respect to those indices, i.e. $\sigma_{ij} = \sigma_{ji}$, etc.

5.1.2 The case of a domain with a smooth boundary

Again, let $\Omega \subset R^3$ be an open, bounded and connected set having a smooth boundary Γ. The formulation of the elastoplastic problem is as follows. In the domain Ω we want to find functions $u = (u_1, u_2, u_3)$, $\sigma = \{\sigma_{ij}\}$, ξ_{ij}, $i, j = 1, 2, 3$, satisfying the following equations and inequalities:

$$-\sigma_{ij,j} = f_i, \quad i = 1, 2, 3, \tag{5.6}$$

$$\varepsilon_{ij}(u) = a_{ijkl}\sigma_{kl} + \xi_{ij}, \quad i, j = 1, 2, 3, \tag{5.7}$$

$$\Phi(\sigma) \le 0, \quad \xi_{ij}(\bar\sigma_{ij} - \sigma_{ij}) \le 0 \quad \forall \bar\sigma, \ \Phi(\bar\sigma) \le 0, \tag{5.8}$$

$$\sigma_{ij} n_j = 0, \quad i = 1, 2, 3, \quad \text{on } \Gamma. \tag{5.9}$$

Here $\Phi : R^6 \to R$ is a continuous convex function describing the plastic yield condition. The equations (5.7) provide a decomposition of the strain tensor $\varepsilon_{ij}(u)$ into a sum of an elastic part $a_{ijkl}\sigma_{kl}$ and a plastic part ξ_{ij}, and (5.6) are the equilibrium equations.

We assume that the functions $a_{ijkl}(x)$ possess the property $a_{ijkl} = a_{jikl} = a_{klij}$, and that there exist constants $c_1, c_2 > 0$ such that

$$c_1|\sigma|^2 \le a_{ijkl}\sigma_{kl}\sigma_{ij} \le c_2|\sigma|^2, \quad \forall \sigma = \{\sigma_{ij}\}. \tag{5.10}$$

The condition (5.10) allows us to solve the equations

$$\varepsilon_{ij}(u) = a_{ijkl}\sigma_{kl}, \quad i,j = 1,2,3$$

with respect to σ_{ij}, and to obtain $\sigma_{ij} = b_{ijkl}\varepsilon_{kl}(u), \quad i,j = 1,2,3$. The functions b_{ijkl} have the same properties as the functions a_{ijkl}. In particular, the inequalities corresponding to (5.10) hold true.

The basic assumption related to the function Φ is that the subset

$$\{\sigma = \{\sigma_{ij}\} \mid \quad \Phi(\sigma) \le 0\}$$

of R^6 contains zero as its interior point.

The functions ξ_{ij} can be eliminated from (5.7), (5.8). Indeed, multiply (5.7) by $\bar{\sigma}_{ij} - \sigma_{ij}$, where $\Phi(\bar{\sigma}) \le 0$ and $\bar{\sigma}_{ij}n_j = 0, \ i = 1,2,3$, on Γ, sum the relations thus obtained over i,j and integrate over Ω. By the second inequality (5.8) this yields the relation

$$\int\limits_\Omega a_{ijkl}\sigma_{kl}(\bar{\sigma}_{ij} - \sigma_{ij}) + \int\limits_\Omega u_i(\bar{\sigma}_{ij,j} - \sigma_{ij,j}) \ge 0,$$

which we will use to define a solution of the problem (5.6)–(5.9).

Consider the space

$$V_0(\Omega) = \{\sigma = \{\sigma_{ij}\} \mid \quad \sigma_{ij} \in L^2(\Omega), \ i,j = 1,2,3,$$
$$\sigma_{ij,j} \in L^3(\Omega), \ i = 1,2,3; \quad \sigma_{ij}n_j = 0, \ i = 1,2,3, \quad \text{on } \Gamma\}.$$

According to the Green formula (5.5), the functions $\sigma_{ij}n_j, i = 1,2,3$, are correctly defined on Γ as elements of $H^{-1/2}(\Gamma)$ provided that $\sigma_{ij} \in L^2(\Omega)$ and $\sigma_{ij,j} \in L^2(\Omega)$, $i,j = 1,2,3$.

We introduce the convex closed subset

$$K = \{\sigma = \{\sigma_{ij}\} \mid \quad \sigma_{ij} \in L^2(\Omega), \ i,j = 1,2,3, \quad \Phi(\sigma(x)) \le 0 \quad \text{a.e. in } \Omega\}$$

of the space $[L^2(\Omega)]^6$, and define a penalty operator p related to the set K by the formula $p(\sigma) = \sigma - \pi\sigma$, where π is the orthogonal projection operator of the space $[L^2(\Omega)]^6$ onto the set K. As it is well known, the operator p is continuous, bounded and monotone.

Finally, we assume that there exists a function $\sigma^0 \in (1+\kappa)^{-1}K$, where $\kappa > 0$ is a constant, such that

$$\langle \sigma^0_{ij}, \varepsilon_{ij}(\bar{u}) \rangle = \langle f, \bar{u} \rangle \quad \forall \bar{u} \in H^1_N(\Omega). \tag{5.11}$$

Now we are in a position to state and prove an existence theorem of the problem (5.6)–(5.9).

Theorem 5.1. *Let $f \in [L^3(\Omega)]^3$ be given such that $\langle f, \rho \rangle = 0$ holds for all $\rho \in R(\Omega)$. Assume that $\sigma^0 \in (1+\kappa)^{-1}K$ satisfies (5.11). Then there exist functions $\sigma = \{\sigma_{ij}\} \in K \cap V_0(\Omega)$, $u \in BD_N(\Omega)$ such that*

$$\langle \sigma_{ij}, \varepsilon_{ij}(\bar{u}) \rangle = \langle f, \bar{u} \rangle \quad \forall \bar{u} \in H^1_N(\Omega), \tag{5.12}$$

$$\langle a_{ijkl}\sigma_{kl}, \bar{\sigma}_{ij} - \sigma_{ij}\rangle + \langle u_i, \bar{\sigma}_{ij,j} - \sigma_{ij,j}\rangle \geq 0 \quad \forall \bar{\sigma} \in K \cap V_0(\Omega). \tag{5.13}$$

PROOF. Using elliptic regularization and the penalty approach, we construct an auxiliary problem which approximates (5.6)–(5.9). Its solution will depend on two positive parameters α and δ which are related to the elliptic regularization and to the penalty approach, respectively. We will obtain a solution σ, u by passing to the limit as $\alpha, \delta \to 0$. So, consider the following boundary value problem in Ω

$$-\alpha(b_{ijkl}\varepsilon_{kl}(u))_{,j} - \sigma_{ij,j} = f_i, \quad i = 1,2,3, \tag{5.14}$$

$$a_{ijkl}\sigma_{kl} - \varepsilon_{ij}(u) + \frac{1}{\delta}p(\sigma)_{ij} = 0, \quad i,j = 1,2,3, \tag{5.15}$$

$$\sigma_{ij}n_j + \alpha b_{ijkl}\varepsilon_{kl}(u)n_j = 0, \quad i = 1,2,3, \quad \text{on } \Gamma. \tag{5.16}$$

Our purpose is to prove the existence of a solution of (5.14)–(5.16) for any $\alpha, \delta > 0$ and to obtain appropriate a priori estimates with respect to α, δ. To simplify the notation, during the first step we do not indicate the dependence of solutions on α, δ. The second step of reasoning is concerned with passage to the limit as $\alpha, \delta \to 0$.

We shall prove that problem (5.14)–(5.16) is solvable in the following sense:

$$u \in H^1_N(\Omega), \quad \sigma_{ij} \in L^2(\Omega), \quad i,j = 1,2,3, \tag{5.17}$$

$$\alpha\langle b_{ijkl}\varepsilon_{kl}(u), \varepsilon_{ij}(\bar{u})\rangle + \langle \sigma_{ij}, \varepsilon_{ij}(\bar{u})\rangle = \langle f, \bar{u}\rangle \quad \forall \bar{u} \in H^1_N(\Omega), \tag{5.18}$$

$$a_{ijkl}\sigma_{kl} - \varepsilon_{ij}(u) + \frac{1}{\delta}p(\sigma)_{ij} = 0, \quad i,j = 1,2,3. \tag{5.19}$$

First of all, let us establish an a priori estimate for solutions of (5.14)–(5.16) in a purely formal manner. To this end, we substitute u as a test function in (5.18) and multiply (5.19) by $\sigma_{ij} - \sigma^0_{ij}$. Summing in i, j and integrating over Ω we obtain the inequality

$$\alpha\langle b_{ijkl}\varepsilon_{kl}(u), \varepsilon_{ij}(u)\rangle + \langle \sigma_{ij}, \varepsilon_{ij}(u)\rangle - \langle f, u\rangle \tag{5.20}$$

$$+ \langle a_{ijkl}\sigma_{kl}, \sigma_{ij} - \sigma^0_{ij}\rangle - \langle \varepsilon_{ij}(u), \sigma_{ij} - \sigma^0_{ij}\rangle \leq 0.$$

In so doing we have omitted the nonnegative term $\delta^{-1}\langle p(\sigma)_{ij}, \sigma_{ij} - \sigma^0_{ij}\rangle$. Since σ^0 satisfies (5.11), and because the coefficients a_{ijkl}, b_{ijkl} have the positive definiteness property (5.10), the inequality (5.20) results in the estimate

$$\alpha \sum_{i,j=1}^{3} \|\varepsilon_{ij}(u)\|_0^2 + \|\sigma\|_0^2 \leq c,$$

with a constant c uniform in α, δ. Hence

$$\alpha|u|^2_{H^1_N(\Omega)} + \|\sigma\|_0^2 \leq c. \tag{5.21}$$

To prove the solvability of the boundary value problem (5.17)–(5.19), we write it in the form

$$A(w) = F, \tag{5.22}$$

with an operator A which maps a Hilbert space V to its dual space V', and where F is a given element of V'. Here we choose $V = H^1_N(\Omega) \times [L^2(\Omega)]^6$ and define A by

$$A(w)(\bar{w}) = \langle \alpha b_{ijkl}\varepsilon_{kl}(u) + \sigma_{ij}, \varepsilon_{ij}(\bar{u})\rangle + \langle a_{ijkl}\sigma_{kl} - \varepsilon_{ij}(u) + \frac{1}{\delta}p(\sigma)_{ij}, \bar{\sigma}_{ij}\rangle,$$

where $w = (u,\sigma), \bar{w} = (\bar{u},\bar{\sigma})$, and we set $F(\bar{w}) = \langle f, \bar{u}\rangle$. The operator A is bounded, monotone and semicontinuous; actually, the computations leading to the estimate (5.21) also provide the coercivity of A in the sense

$$\frac{A(w)(w)}{\|w\|_V} \to \infty, \quad \|w\|_V \to \infty.$$

Thus, the solvability of the equation (5.22), or, equivalently, of the problem (5.17)–(5.19) follows from Theorem 1.14. The boundary conditions (5.16) are preserved in the identity (5.18); see also the comments at the end of this section.

In addition to the estimate (5.21) one can prove (see Khludnev, Sokolowski, 1997) that the estimate

$$\frac{1}{\delta}\|p(\sigma)\|_{L^1(\Omega)} \le c$$

holds uniformly in α and δ. It then follows from (5.19) that the inequality

$$\sum_{i,j=1}^{3} \|\varepsilon_{ij}(u)\|_{L^1(\Omega)} \le c \tag{5.23}$$

is uniform in α, δ. Since $u \in LD_N(\Omega)$, inequality (5.23) yields

$$|u|_{LD_N(\Omega)} \le c \tag{5.24}$$

with the constant c being uniform with respect to α and δ. Taking into account the continuity of the imbedding $LD(\Omega) \subset L^{3/2}(\Omega)$, which holds in the three-dimensional case (see Temam, 1983), we find from (5.24) that

$$\|u\|_{L^{3/2}(\Omega)} \le c. \tag{5.25}$$

Now we can justify a passage to the limit as $\alpha, \delta \to 0$ in (5.17)–(5.19). Denote by $u^{\alpha\delta}, \sigma^{\alpha\delta}$ a solution of (5.17)–(5.19) corresponding to given values of α and δ. From (5.19) it follows that for any fixed δ there exists a constant $c(\delta)$ depending, in general, on δ such that

$$\sum_{i,j=1}^{3} \|\varepsilon_{ij}(u^{\alpha\delta})\|_0 \le c(\delta)$$

and hence

$$|u^{\alpha\delta}|_{H^1_N(\Omega)} \le c(\delta).$$

Consequently, for any fixed δ, from the sequence $u^{\alpha\delta}, \sigma^{\alpha\delta}$ one can choose a subsequence still denoted by $u^{\alpha\delta}, \sigma^{\alpha\delta}$ such that as $\alpha \to 0$

$$u^{\alpha\delta} \to u^{\delta} \quad \text{weakly in } H^1_N(\Omega), \tag{5.26}$$

$$\sigma^{\alpha\delta}_{ij} \to \sigma^{\delta}_{ij} \quad \text{weakly in } L^2(\Omega), \quad i,j = 1,2,3. \tag{5.27}$$

Note that the subsequence depends on δ and that for any fixed δ the α-subsequences are different, in general.

In view of (5.26)–(5.27) we can turn to the limit as $\alpha \to 0$ in (5.18)–(5.19) and obtain

$$u^{\delta} \in H^1_N(\Omega), \quad \sigma^{\delta}_{ij} \in L^2(\Omega), \quad i,j = 1,2,3, \tag{5.28}$$

$$\langle \sigma^{\delta}_{ij}, \varepsilon_{ij}(\bar{u}) \rangle = \langle f, \bar{u} \rangle \quad \forall \bar{u} \in H^1_N(\Omega), \tag{5.29}$$

$$a_{ijkl}\sigma^{\delta}_{kl} - \varepsilon_{ij}(u^{\delta}) + \frac{1}{\delta} p(\sigma^{\delta})_{ij} = 0, \quad i,j = 1,2,3. \tag{5.30}$$

The passage from $p(\sigma^{\alpha\delta})$ to $p(\sigma^{\delta})$ is justified by a standard monotonicity argument, the details of which we omit here.

Let us consider the passage to the limit in (5.28)–(5.30) as $\delta \to 0$. Because the imbedding $L^1(\Omega) \subset M^1(\Omega)$ is continuous, due to (5.21), (5.23) and (5.25) we can find a subsequence $u^{\delta}, \sigma^{\delta}$ converging to a limit u, σ in the sense

$$u^{\delta} \to u \quad \text{weakly in } L^{3/2}(\Omega),$$

$$\varepsilon_{ij}(u^{\delta}) \to \varepsilon_{ij}(u) \quad \star\text{–weakly in } M^1(\Omega), \quad i,j = 1,2,3,$$

$$\sigma^{\delta}_{ij} \to \sigma_{ij} \quad \text{weakly in } L^2(\Omega), \quad i,j = 1,2,3.$$

Again, we just write $u^{\delta}, \sigma^{\delta}$ for the subsequence. As $\delta \to 0$, the identity (5.12) results from (5.29).

Notice that the space $H^1(\Omega)$ equals the direct sum

$$H^1(\Omega) = R(\Omega) \oplus H^1_N(\Omega)$$

of the subspace $R(\Omega)$ and $H^1_N(\Omega)$ which are orthogonal with respect to the scalar product

$$(u, v) = \langle u, v \rangle + \langle \varepsilon_{ij}(u), \varepsilon_{ij}(v) \rangle, \quad u, v \in H^1(\Omega).$$

In fact, by the second Korn inequality this scalar product induces a norm which is equivalent to the norm given by (5.3). Hence, because $\langle f, \rho \rangle = 0$ for all $\rho \in R(\Omega)$, the identity (5.29) actually holds for every $\bar{u} \in H^1(\Omega)$). Therefore, the equilibrium equations

$$-\sigma^{\delta}_{ij,j} = f_i, \quad i = 1,2,3, \tag{5.31}$$

hold in Ω in the sense of distributions. Consequently, $\sigma^{\delta}_{ij,j} \in L^2(\Omega)$, $i = 1,2,3$, and the use of Green's formula (5.5) gives $\sigma^{\delta}_{ij}n_j = 0$, $i = 1,2,3$, on Γ. Taking into account the boundary conditions obtained for σ^{δ} it follows from (5.30) that for any $\bar{\sigma} \in K \cap V_0(\Omega)$ the inequality

$$\langle a_{ijkl}\sigma^{\delta}_{kl}, \bar{\sigma}_{ij} - \sigma^{\delta}_{ij}\rangle + \langle u^{\delta}_i, \bar{\sigma}_{ij,j} - \sigma^{\delta}_{ij,j}\rangle \geq 0 \tag{5.32}$$

holds. Owing to (5.31) the values $\sigma^{\delta}_{ij,j}$ can be replaced by $-f_i$ in (5.32), and we can easily pass to the limit in (5.32) as $\delta \to 0$. This provides (5.13).

The inclusion $\sigma \in K$ can be obtained in the usual way. The property $\sigma \in V_0(\Omega)$ is contained in (5.12). Indeed, as above, we conclude that the equations

$$-\sigma_{ij,j} = f_i, \quad i = 1,2,3,$$

hold in the sense of distributions, whence $\sigma_{ij,j} \in L^2(\Omega), i = 1,2,3$. The boundary conditions (5.9) we readily obtain from (5.12), using the Green formula (5.5). Theorem 5.1 is completely proved.

REMARK. The specific choice of b_{ijkl} as the inverse of the a_{ijkl} for the elliptic regularization appears to be natural, since in the case of pure elastic (with $K = [L^2(\Omega)]^6$, respectively $p(\sigma) \equiv 0$), the boundary condition (5.16) reduces to (5.9). However, the proof of Theorem 5.1 works with any other choice of b_{ijkl} as long as requirements of symmetry, boundedness and coercivity are met.

5.1.3 The case of two-dimensional cracks

In this section we shall prove the existence of a solution of the elastoplastic boundary value problem for the particular case of a nonsmooth boundary which arises if we remove a two-dimensional surface from the interior of the body.

Let $\Omega \subset R^3$ be an open, bounded and connected set with a smooth boundary Γ, and $\Gamma_c \subset \Omega$ be a smooth orientable two-dimensional surface. We assume that this surface can be extended up to the outer boundary Γ in such a way that Ω is divided into two subdomains Ω_1, Ω_2 with Lipschitz boundaries. We assume that this inner surface Γ_c is described parametrically by the equations

$$x_i = x_i(y_1, y_2), \quad i = 1,2,3, \tag{5.33}$$

where (y_1, y_2) belong to the closure of an open bounded connected set $\omega \subset R^2$ having a smooth boundary γ. We suppose that the rank of the Jacobi matrix $\partial x_i/\partial y_j$ equals 2 at every point $(y_1, y_2) \in \omega \cup \gamma$, and that the map (5.33) is one-to-one. Let $\nu = (\nu_1, \nu_2, \nu_3)$ be a unit normal vector to Γ_c, for example

$$\nu = \frac{\frac{\partial x}{\partial y_1} \times \frac{\partial x}{\partial y_2}}{\left|\frac{\partial x}{\partial y_1} \times \frac{\partial x}{\partial y_2}\right|}.$$

Denote $\Omega_c = \Omega \setminus \Gamma_c$. The formulation of the elastoplastic problem for a body occupying the domain Ω_c in its undeformed state is as follows. In the domain Ω_c we have to find functions $u = (u_1, u_2, u_3)$, $\sigma = \{\sigma_{ij}\}$, ξ_{ij}, $i, j = 1, 2, 3$, which satisfy the following equations and inequalities:

$$-\sigma_{ij,j} = f_i, \quad i = 1, 2, 3, \tag{5.34}$$

$$\varepsilon_{ij}(u) = a_{ijkl}\sigma_{kl} + \xi_{ij}, \quad i, j = 1, 2, 3, \tag{5.35}$$

$$\Phi(\sigma) \leq 0, \quad \xi_{ij}(\bar{\sigma}_{ij} - \sigma_{ij}) \leq 0 \quad \forall \bar{\sigma},\ \Phi(\bar{\sigma}) \leq 0, \tag{5.36}$$

$$\sigma_{ij}n_j = 0, \quad i = 1, 2, 3, \quad \text{on } \Gamma, \tag{5.37}$$

$$\sigma_{ij}\nu_j = 0, \quad i = 1, 2, 3, \quad \text{on } \Gamma_c^{\pm}. \tag{5.38}$$

We use the same notation as in the previous subsection. The boundary of Ω_c consists of three components $\Gamma, \Gamma_c^+, \Gamma_c^-$, where $\Gamma_c^{\pm}$ correspond to the positive and negative directions of the normal ν, respectively. We introduce the space

$$H^1(\Omega_c) = \{u = (u_1, u_2, u_3) \mid \ u_i \in L^2(\Omega_c),\ i = 1, 2, 3;$$
$$u_{i,j} \in L^2(\Omega_c),\ i, j = 1, 2, 3\}.$$

Notice that boundary values on Γ_c^+ and Γ_c^- of any element $u \in H^1(\Omega_c)$ (which we may think of as one-sided limits) are different, in general. Accordingly, for all functions on Ω_c to be discussed below, their traces, if they exist, will in general differ on Γ_c^+ and Γ_c^-.

As before, the Neumann boundary conditions (5.37) and (5.38) enforce a function space decomposition based on the conditions

$$\int_{\Omega_c} u = 0, \quad \int_{\Omega_c} (u_i x_j - u_j x_i) = 0, \quad i, j = 1, 2, 3, \quad u = (u_1, u_2, u_3). \tag{5.39}$$

In particular, we define

$$H_N^1(\Omega_c) = \{u \in H^1(\Omega_c) \mid \ u \text{ satisfies } (5.39)\}.$$

We also introduce the spaces

$$LD(\Omega_c) = \{u = (u_1, u_2, u_3) \mid \ u_i \in L^1(\Omega_c),\ i = 1, 2, 3;$$
$$\varepsilon_{ij}(u) \in L^1(\Omega_c),\ i, j = 1, 2, 3\},$$
$$LD_N(\Omega_c) = \{u \in LD(\Omega_c) \mid \ u \text{ satisfies } (5.39)\},$$
$$BD_N(\Omega_c) = \{u \in BD(\Omega_c) \mid \ u \text{ satisfies } (5.39)\}.$$

In the spaces $H^1(\Omega_c)$ and $LD(\Omega_c)$ the following norms will be considered:

$$|u|_{H^1(\Omega_c)} = \left|\int_{\Omega_c} u\right| + \sum_{i,j=1}^{3} \left|\int_{\Omega_c} (u_i x_j - u_j x_i)\right| + \sum_{i,j=1}^{3} \|\varepsilon_{ij}(u)\|_{0,c}, \tag{5.40}$$

$$|u|_{LD(\Omega_c)} = \left|\int_{\Omega_c} u\right| + \sum_{i,j=1}^{3} \left|\int_{\Omega_c} (u_i x_j - u_j x_i)\right| + \sum_{i,j=1}^{3} \|\varepsilon_{ij}(u)\|_{L^1(\Omega_c)}. \quad (5.41)$$

Here $\|\cdot\|_{0,c}$ is the norm in $L^2(\Omega_c)$. Accordingly, on the subspaces $H^1_N(\Omega_c)$, $LD_N(\Omega_c)$ we consider the following norms which are equivalent to (5.40), (5.41), respectively,

$$|u|_{H^1_N(\Omega_c)} = \sum_{i,j=1}^{3} \|\varepsilon_{ij}(u)\|_{0,c}, \qquad |u|_{LD_N(\Omega_c)} = \sum_{i,j=1}^{3} \|\varepsilon_{ij}(u)\|_{L^1(\Omega_c)}.$$

In order to prove that the norms (5.40)–(5.41) are equivalent to the standard ones given by (5.3), (5.1) in $H^1(\Omega_c)$ and $LD(\Omega_c)$, respectively, we extend Γ_c in such a way that it divides Ω_c into two subdomains Ω_1, Ω_2 with Lipschitz boundaries (see the assumption at the beginning of this subsection), and argue in the usual manner.

Let us introduce one more notation, namely

$$V_0(\Omega) = \{\sigma = \{\sigma_{ij}\} \mid \ \sigma_{ij} \in L^2(\Omega),\ i,j = 1,2,3;$$
$$\sigma_{ij,j} \in L^3(\Omega),\ i = 1,2,3;$$
$$\sigma_{ij} n_j = 0,\ i = 1,2,3, \quad \text{on } \Gamma; \quad \sigma_{ij}\nu_j = 0,\ i = 1,2,3, \quad \text{on } \Gamma_c^{\pm}\}.$$

The set K is defined just as before,

$$K = \{\sigma = \{\sigma_{ij}\} \mid \ \sigma_{ij} \in L^2(\Omega_c),\ i,j = 1,2,3,$$
$$\Phi(\sigma(x)) \le 0 \quad \text{a.e. in } \Omega_c\}.$$

We moreover assume that there exists a function $\sigma^0 = \{\sigma^0_{ij}\}$ such that $\sigma^0 \in (1+\kappa)^{-1}K$, where $\kappa > 0$ is a constant, and

$$\langle \sigma^0_{ij}, \varepsilon_{ij}(\bar u)\rangle_c = \langle f, \bar u\rangle_c \quad \forall \bar u \in H^1_N(\Omega_c). \quad (5.42)$$

The brackets $\langle\cdot,\cdot\rangle_c$ represent integration over Ω_c. Also, we recall that the set $\{\sigma = \{\sigma_{ij}\} \mid \ \Phi(\sigma) \le 0\}$ in R^6 has to include zero as an interior point.

Denote by $R(\Omega_c)$ the set of all functions $\rho = (\rho_1, \rho_2, \rho_3)$ which represent rigid motions, that is $\rho_i(x) = c_i + b_{ij}x_j$, where $c_i = \text{const}$ and b_{ij} are constant with $b_{ij} = -b_{ji}$. As before we see that if $\rho \in R(\Omega_c)$ and if ρ satisfies (5.39), then $\rho \equiv 0$.

Now we can present our main existence theorem for problem (5.34)–(5.38).

Theorem 5.2. *Let $f \in [L^3(\Omega_c)]^3$ such that $\langle f, \rho\rangle_c = 0$ for all $\rho \in R(\Omega_c)$, and σ^0 have the properties stated above. Then there exist functions $\sigma \in K$, $u \in BD_N(\Omega_c)$, such that*

$$\langle \sigma_{ij}, \varepsilon_{ij}(\bar u)\rangle_c = \langle f, \bar u\rangle_c \quad \forall \bar u \in H^1_N(\Omega_c), \quad (5.43)$$

$$\langle a_{ijkl}\sigma_{kl}, \bar{\sigma}_{ij} - \sigma_{ij}\rangle_c + \langle u_i, \bar{\sigma}_{ij,j} - \sigma_{ij,j}\rangle_c \geq 0 \quad \forall \bar{\sigma} \in K \cap V_0(\Omega_c). \tag{5.44}$$

PROOF. The general scheme of reasoning is the same as that in the previous subsection; we will pay attention to those details related to the interior nonsmooth boundary Γ_c.

Let $p(\sigma) = \sigma - \pi\sigma$ be the penalty operator, where $\pi : [L^2(\Omega_c)]^6 \to K$ is the operator of orthogonal projection. Consider the following auxiliary boundary value problem which includes two positive parameters α and δ. In the domain Ω_c we want to find functions $u = (u_1, u_2, u_3)$, $\sigma = \{\sigma_{ij}\}$, $i, j = 1, 2, 3$, such that

$$-\alpha(b_{ijkl}\varepsilon_{kl}(u))_{,j} - \sigma_{ij,j} = f_i, \quad i = 1, 2, 3, \tag{5.45}$$

$$a_{ijkl}\sigma_{kl} - \varepsilon_{ij}(u) + \frac{1}{\delta}p(\sigma)_{ij} = 0, \quad i, j = 1, 2, 3, \tag{5.46}$$

$$\sigma_{ij}n_j + \alpha b_{ijkl}\varepsilon_{kl}(u)n_j = 0, \quad i = 1, 2, 3, \quad \text{on } \Gamma, \tag{5.47}$$

$$\sigma_{ij}\nu_j + \alpha b_{ijkl}\varepsilon_{kl}(u)\nu_j = 0, \quad i = 1, 2, 3, \quad \text{on } \Gamma_c^{\pm}. \tag{5.48}$$

The solvability of the problem (5.45)–(5.48) for fixed parameters α, δ will be proved in the following sense:

$$u \in H_N^1(\Omega_c), \quad \sigma_{ij} \in L^2(\Omega_c), \quad i, j = 1, 2, 3, \tag{5.49}$$

$$\alpha\langle b_{ijkl}\varepsilon_{kl}(u), \varepsilon_{ij}(\bar{u})\rangle_c + \langle \sigma_{ij}, \varepsilon_{ij}(\bar{u})\rangle_c = \langle f, \bar{u}\rangle_c \quad \forall \bar{u} \in H_N^1(\Omega_c), \tag{5.50}$$

$$a_{ijkl}\sigma_{kl} - \varepsilon_{ij}(u) + \frac{1}{\delta}p(\sigma)_{ij} = 0, \quad i, j = 1, 2, 3. \tag{5.51}$$

To obtain an a priori estimate of the solution to (5.49)–(5.51) we can argue as in the previous subsection. Namely, we substitute $\bar{u} = u$ in (5.50) and multiply (5.51) by $\sigma_{ij} - \sigma_{ij}^0$. This gives the estimate

$$\alpha \sum_{i,j=1}^{3} \|\varepsilon_{ij}(u)\|_{0,c}^2 + \|\sigma\|_0^2 \leq c. \tag{5.52}$$

Since $u \in H_N^1(\Omega_c)$ it follows from (5.52) that

$$\alpha|u|_{H_N^1(\Omega_c)}^2 + \|\sigma\|_{0,c}^2 \leq c. \tag{5.53}$$

The constant c does not depend on α and δ.

The estimate (5.53) allows us to prove the solvability of (5.49)–(5.51) for any fixed parameters α, δ. The boundary conditions (5.47)–(5.48) are a consequence of the identity (5.50).

From (5.51) we can derive an additional estimate. Indeed, for any fixed $\delta > 0$ there exists a constant $c(\delta)$ depending on δ such that

$$\sum_{i,j=1}^{3} \|\varepsilon_{ij}(u)\|_{0,c} \leq c(\delta)$$

and hence

$$|u|_{H^1_N(\Omega_c)} \le c(\delta). \tag{5.54}$$

Now we can pass to the limit in (5.49)–(5.51) as $\alpha, \delta \to 0$. Denote by $u^{\alpha\delta}, \sigma^{\alpha\delta}$ the solution of (5.49)–(5.51) corresponding to given parameters α, δ. Due to the estimates (5.53) and (5.54), we can choose a subsequence, still denoted by $u^{\alpha\delta}, \sigma^{\alpha\delta}$, such that for $\alpha \to 0$ and any fixed δ

$$u^{\alpha\delta} \to u^\delta \quad \text{weakly in } H^1_N(\Omega_c),$$
$$\sigma^{\alpha\delta}_{ij} \to \sigma^\delta_{ij} \quad \text{weakly in } L^2(\Omega_c), \quad i,j = 1,2,3.$$

On passing to the limit as $\alpha \to 0$, the equations (5.49)–(5.51) become

$$u^\delta \in H^1_N(\Omega_c), \quad \sigma^\delta_{ij} \in L^2(\Omega_c), \quad i,j = 1,2,3, \tag{5.55}$$

$$\langle \sigma^\delta_{ij}, \varepsilon_{ij}(\bar{u})\rangle_c = \langle f, \bar{u}\rangle_c \quad \forall \bar{u} \in H^1_N(\Omega_c), \tag{5.56}$$

$$a_{ijkl}\sigma^\delta_{kl} - \varepsilon_{ij}(u^\delta) + \frac{1}{\delta} p(\sigma^\delta)_{ij} = 0, \quad i,j = 1,2,3. \tag{5.57}$$

As before, we have

$$\frac{1}{\delta}\|p(\sigma^\delta)_{ij}\|_{L^1(\Omega_c)} \le c, \quad i,j = 1,2,3,$$

uniformly in δ. Consequently, the equations (5.57) imply that

$$\sum_{i,j=1}^{3} \|\varepsilon_{ij}(u^\delta)\|_{L^1(\Omega_c)} \le c$$

and since $u^\delta \in H^1_N(\Omega_c)$, this inequality gives

$$|u^\delta|_{LD_N(\Omega_c)} \le c.$$

The imbeddings $LD(\Omega_c) \subset L^{3/2}(\Omega_c)$, $L^1(\Omega_c) \subset M^1(\Omega_c)$ are continuous, hence

$$\|u^\delta\|_{L^{3/2}(\Omega_c)} \le c, \quad \sum_{i,j=1}^{3} \|\varepsilon_{ij}(u^\delta)\|_{M^1(\Omega_c)} \le c. \tag{5.58}$$

Due to the estimates (5.53) and (5.58), we can assume that a subsequence u^δ, σ^δ possesses the properties

$$u^\delta \to u \quad \text{weakly in } L^{3/2}(\Omega_c),$$

$$\varepsilon_{ij}(u^\delta) \to \varepsilon_{ij}(u) \quad \star\text{-weakly in } M^1(\Omega_c), \quad i,j = 1,2,3,$$

$$\sigma^\delta_{ij} \to \sigma_{ij} \quad \text{weakly in } L^2(\Omega_c), \quad i,j = 1,2,3.$$

The identity (5.56) easily yields (5.43). We can next derive from (5.56) that the equations

$$-\sigma^{\delta}_{ij,j} = f_i, \quad i = 1, 2, 3,$$

hold in the sense of distributions in the domain Ω_c, whence

$$\langle \sigma^{\delta}_{ij}, \varepsilon_{ij}(u^{\delta}) \rangle_c = \langle f, u^{\delta} \rangle_c = -\langle \sigma^{\delta}_{ij,j}, u^{\delta}_i \rangle_c. \tag{5.59}$$

Note that

$$\langle \varepsilon_{ij}(u^{\delta}), \bar{\sigma}_{ij} \rangle_c = -\langle u^{\delta}_i, \bar{\sigma}_{ij,j} \rangle_c \quad \forall \bar{\sigma} \in V_0(\Omega_c). \tag{5.60}$$

Let us multiply (5.57) by $\bar{\sigma}_{ij} - \sigma^{\delta}_{ij}$, where $\bar{\sigma} \in K \cap V_0(\Omega_c)$. Taking into account (5.59), (5.60) we have

$$\langle a_{ijkl}\sigma^{\delta}_{kl}, \bar{\sigma}_{ij} - \sigma^{\delta}_{ij} \rangle_c + \langle u^{\delta}_i, \bar{\sigma}_{ij,j} - \sigma^{\delta}_{ij,j} \rangle_c \geq 0, \quad \forall \bar{\sigma} \in K \cap V_0(\Omega_c). \tag{5.61}$$

The values $\sigma^{\delta}_{ij,j}$ can be replaced by $-f_i$ in (5.61). This allows us to pass to the limit as $\delta \to 0$, and we arrive at (5.44).

The inclusion $\sigma \in K$ is proved by a standard method. The boundary conditions (5.37)–(5.38) are a consequence of (5.43). Indeed, consider the scalar product in $H^1(\Omega_c)$

$$(u, v)_c = \langle u, v \rangle_c + \langle \varepsilon_{ij}(u), \varepsilon_{ij}(v) \rangle_c, \quad u, v \in H^1(\Omega_c).$$

Then the space $H^1(\Omega_c)$ can be written as a sum

$$H^1(\Omega_c) = R(\Omega_c) \oplus H^1_N(\Omega_c)$$

of orthogonal subspaces, and $\langle f, \rho \rangle_c = 0 \quad \forall \rho \in R(\Omega_c)$. Hence, the equality in (5.43) actually holds for all test functions from $H^1(\Omega_c)$, and in particular, the equilibrium equations (5.34) hold in Ω_c in the sense of distributions. Consequently, the boundary conditions (5.37) are satisfied in the sense of $H^{-1/2}(\Gamma)$. As for the conditions (5.38), they are valid in the weak sense, namely they hold at all points of $\Gamma^{\pm}_c$ where the solution of (5.43)–(5.44) is smooth enough.

5.2 Elastoplastic problems for the Prandtl–Reuss model

In this section the existence of a solution to the three-dimensional elastoplastic problem with the Prandtl–Reuss constitutive law and the Neumann boundary conditions is obtained. The proof is based on a suitable combination of the parabolic regularization of equations and the penalty method for the elastoplastic yield condition. The method is applied in the case of the domain with a smooth boundary as well as in the case of an interior two-dimensional crack. It is shown that the weak solutions to the elastoplastic problem satisfying the variational inequality meet all boundary conditions. The results of this section can be found in (Khludnev, Sokolowski, 1998a).

5.2.1 Domain with a smooth boundary

We start with notations and preliminary remarks. Let $\Omega \subset R^3$ be a bounded domain with a smooth boundary Γ having an exterior unit normal vector $n = (n_1, n_2, n_3)$.

We know that if $v = (v_1, v_2, v_3)$, $\varepsilon_{ij}(v) = 0$ in Ω, $i, j = 1, 2, 3$, then $v_i(x) = c_i + b_{ij}x_j$, $i = 1, 2, 3$, where $c_i, b_{ij} \in R$, $b_{ij} = -b_{ji}$, $i, j = 1, 2, 3$. The linear space of all vectors $v = (v_1, v_2, v_3)$, $v_i(x) = c_i + b_{ij}x_j$, is called the space of rigid displacements. We denote it by $R(\Omega)$.

In the sequel we consider different functional spaces. To simplify the notation we write $L^2(\Omega)$, $H^1(\Omega)$ instead of $[L^2(\Omega)]^3$, $[H^1(\Omega)]^3$ and so on.

Consider all functions from $L^2(\Omega)$ satisfying the conditions

$$\int_\Omega v = 0, \quad \int_\Omega (v_i x_j - v_j x_i) = 0, \quad i, j = 1, 2, 3, \quad v = (v_1, v_2, v_3). \tag{5.62}$$

It is clear that if $\rho \in R(\Omega)$ satisfies (5.62) then $\rho \equiv 0$. In fact, we have $L^2(\Omega) = R(\Omega) \oplus R(\Omega)^\perp$, and $R(\Omega)^\perp$ coincides with all functions from $L^2(\Omega)$ satisfying (5.62). Let

$$H^1(\Omega) = \{v = (v_1, v_2, v_3) \mid \quad v_i \in L^2(\Omega),\ i = 1, 2, 3;$$
$$v_{i,j} \in L^2(\Omega),\ i, j = 1, 2, 3\}.$$

In the space $H^1(\Omega)$ we shall consider different equivalent norms, in particular

$$||v||_1^2 = ||v||_0^2 + \sum_{i,j=1}^{3} ||\varepsilon_{ij}(v)||_0^2, \tag{5.63}$$

$$|v|_1 = \left|\int_\Omega v\right| + \sum_{i,j=1}^{3} \left|\int_\Omega (v_i x_j - v_j x_i)\right| + \sum_{i,j=1}^{3} ||\varepsilon_{ij}(v)||_0. \tag{5.64}$$

Here, $||\cdot||_0$ is the norm in $L^2(\Omega)$. The norm (5.63) is equivalent to the usual norm in $H^1(\Omega)$ due to the second Korn inequality. As for the norm (5.64) it is easy to see that the two first terms in the right-hand side of (5.64) give a seminorm on $H^1(\Omega)$ being a norm on $R(\Omega)$, and the statement follows from (Temam, 1983).

Denote by $\langle \cdot, \cdot \rangle$ the scalar product in $L^2(\Omega)$. We can consider the scalar product in $H^1(\Omega)$ inducing the norm (5.63),

$$(u, v) = \langle u, v\rangle + \langle \varepsilon_{ij}(u), \varepsilon_{ij}(v)\rangle, \quad u, v \in H^1(\Omega). \tag{5.65}$$

In this case $H^1(\Omega) = R(\Omega) \oplus H^1_N(\Omega)$, where

$$H^1_N(\Omega) = \{v = (v_1, v_2, v_3) \in H^1(\Omega) \mid \quad v \text{ satisfies } (5.62)\}.$$

This means that in $H^1_N(\Omega)$ one can consider the equivalent norm

$$|v|_1 = \sum_{i,j=1}^{3} \|\varepsilon_{ij}(v)\|_0 \,.$$

Introduce some additional notations which are useful in the sequel. Consider the space

$$LD(\Omega) = \{v = (v_1, v_2, v_3) \mid \ v_i \in L^1(\Omega),\ i = 1,2,3;$$
$$\varepsilon_{ij}(v) \in L^1(\Omega),\ i,j = 1,2,3\}$$

equipped with the norm

$$\|v\|_{LD(\Omega)} = \|v\|_{L^1(\Omega)} + \sum_{i,j=1}^{3} \|\varepsilon_{ij}(v)\|_{L^1(\Omega)} \,. \tag{5.66}$$

Let

$$LD_N(\Omega) = \{v \in LD(\Omega) \mid \ v \text{ satisfies } (5.62)\} \,.$$

Along with the usual norm (5.66) we shall consider the following norm in $LD(\Omega)$:

$$|v|_{LD(\Omega)} = \left|\int_\Omega v\right| + \sum_{i,j=1}^{3}\left|\int_\Omega (v_i x_j - v_j x_i)\right| + \sum_{i,j=1}^{3} \|\varepsilon_{ij}(v)\|_{L^1(\Omega)} \,.$$

Consequently, the subspace $LD_N(\Omega)$ of the space $LD(\Omega)$ is defined by the norm

$$\|v\|_{LD_N(\Omega)} = \sum_{i,j=1}^{3} \|\varepsilon_{ij}(v)\|_{L^1(\Omega)} \,.$$

Also, we consider the space of bounded measures $M^1(\Omega)$. Introduce the Banach space of bounded deformation

$$BD(\Omega) = \{v = (v_1, v_2, v_3) \mid \ v_i \in L^1(\Omega),\ i = 1,2,3;$$
$$\varepsilon_{ij}(v) \in M^1(\Omega),\ i,j = 1,2,3\}$$

endowed with the norm

$$\|v\|_{BD(\Omega)} = \|v\|_{L^1(\Omega)} + \sum_{i,j=1}^{3} \|\varepsilon_{ij}(v)\|_{M^1(\Omega)} \,.$$

As above,

$$BD_N(\Omega) = \{v \in BD(\Omega) \mid \ v \text{ satisfies } (5.62)\}. \tag{5.67}$$

Let $\Omega \subset R^3$ be a bounded domain with a smooth boundary Γ, $Q = \Omega \times (0, T)$, $x = (x_1, x_2, x_3) \in \Omega$, $t \in (0, T)$. The formulation of the elastoplastic problem for a body occupying the domain Ω in the nondeformed state is as follows. In the domain Q, we want to find the functions $v = (v_1, v_2, v_3)$, $\sigma = \{\sigma_{ij}\}$, η_{ij}, $i, j = 1, 2, 3$, satisfying the following equations and inequalities:

$$-\sigma_{ij,j} = f_i, \quad i = 1, 2, 3, \tag{5.68}$$

$$\varepsilon_{ij}(v) = c_{ijkl}\dot{\sigma}_{kl} + \eta_{ij}, \quad i, j = 1, 2, 3, \tag{5.69}$$

$$\Phi(\sigma) \leq 0, \quad \eta_{ij}(\bar{\sigma}_{ij} - \sigma_{ij}) \leq 0 \quad \forall \bar{\sigma}, \ \Phi(\bar{\sigma}) \leq 0, \tag{5.70}$$

$$\sigma_{ij} n_j = 0, \quad i = 1, 2, 3, \quad \text{on } \Gamma \times (0, T), \tag{5.71}$$

$$\sigma = 0, \quad t = 0. \tag{5.72}$$

The functions $v, \sigma_{ij}, \varepsilon_{ij}(v)$ represent the velocity, components of the stress tensor and components of the rate strain tensor. The dot denotes the derivative with respect to t. The convex and continuous function Φ describes the plasticity yield condition. It is assumed that the set

$$\{\sigma = \{\sigma_{ij}\} \in R^6 \mid \ \Phi(\sigma) \leq 0\} \tag{5.73}$$

contains zero as its interior point. We assume that $c_{ijkl}(x) = c_{jikl}(x) = c_{klij}(x)$ for $i, j, k, l = 1, 2, 3$ and there exist two positive constants c_1, c_2 such that

$$c_1|\sigma|^2 \leq c_{ijkl}\sigma_{kl}\sigma_{ij} \leq c_2|\sigma|^2 \qquad \forall \sigma = \{\sigma_{ij}\}. \tag{5.74}$$

To simplify the formulae below we assume that $c_{ijkl} = \delta^i_k \delta^j_l$, δ^i_j is the Kronecker symbol. Nevertheless, all the results obtained in the section are valid in the general case (5.74).

The values η_{ij} can be eliminated from (5.69), (5.70). In fact, multiply (5.69) by $\bar{\sigma}_{ij} - \sigma_{ij}$ and sum in i, j. This provides

$$\Phi(\sigma) \leq 0, \quad (\dot{\sigma}_{ij} - \varepsilon_{ij}(v))(\bar{\sigma}_{ij} - \sigma_{ij}) \geq 0 \quad \forall \bar{\sigma}, \quad \Phi(\bar{\sigma}) \leq 0. \tag{5.75}$$

Inequality (5.75) will be used in the definition of a solution to the problem (5.68)–(5.72).

Consider the set of admissible stresses

$$K = \{\sigma = \{\sigma_{ij}\} \mid \ \sigma_{ij} \in L^2(\Omega), \ i, j = 1, 2, 3, \quad \Phi(\sigma(x)) \leq 0 \quad \text{a.e. in } \Omega\}$$

and the penalty operator p related to the set K. The operator can be constructed by the formula $p(\sigma) = \sigma - \pi\sigma$, where $\pi : [L^2(\Omega)]^6 \to K$ is the operator of orthogonal projection. Recall that the operator p is bounded, monotone and continuous.

Let the brackets $(\cdot, \cdot)$ denote the scalar product in $L^2(Q)$, and $f = (f_1, f_2, f_3)$,

$$V_0(\Omega) = \{\sigma = \{\sigma_{ij}\} \mid \ \sigma_{ij} \in L^2(\Omega), \ i, j = 1, 2, 3;$$

$\sigma_{ij,j} \in L^3(\Omega)$, $i = 1,2,3$; $\sigma_{ij}n_j = 0$, $i = 1,2,3$, on $\Gamma\}$.

Suppose that there exists a function $\xi = \{\xi_{ij}\}$ satisfying the equation (5.68) such that $\xi \in C^2(\bar{Q})$, $\xi(0) = \dot{\xi}(0) = 0$ and $(1+\kappa)\xi(t) \in K \cap V_0(\Omega)$, where $\kappa = \text{const} > 0$, $t \in [0,T]$.

Now we can prove the following existence theorem for the problem (5.68)–(5.72).

Theorem 5.3. *Let $f \in L^3(Q)$, $\dot{f} \in L^2(Q)$, $f(0) = 0$, $\langle f(t), \rho\rangle = 0$ $\forall \rho \in R(\Omega), t \in [0,T]$, and the above assumption on ξ hold. Then there exist functions $v = (v_1, v_2, v_3)$, $\sigma = \{\sigma_{ij}\}$ such that*

$$v \in L^2(0,T; BD_N(\Omega)), \quad \sigma \in L^2(0,T; V_0(\Omega)),$$

$$\dot{\sigma} \in L^2(Q), \quad \sigma(t) \in K, \quad t \in (0,T),$$

$$(\sigma_{ij}, \varepsilon_{ij}(\bar{v})) = (f, \bar{v}) \quad \forall \bar{v} \in L^2(0,T; H^1(\Omega)), \tag{5.76}$$

$$(\dot{\sigma}_{ij}, \bar{\sigma}_{ij} - \sigma_{ij}) + (v_i, \bar{\sigma}_{ij,j} - \sigma_{ij,j}) \geq 0 \quad \forall \bar{\sigma} \in L^2(0,T; V_0(\Omega)), \tag{5.77}$$

$$\bar{\sigma}(t) \in K \quad \text{a.e. in } (0,T),$$

$$\sigma = 0, \quad t = 0. \tag{5.78}$$

PROOF. We consider a parabolic regularization of the problem approximating (5.68)–(5.72). The auxiliary boundary value problem will contain two positive parameters α, δ. The first parameter is responsible for the parabolic regularization and the second one characterizes the penalty approach. Our aim is first to prove an existence of solutions for the fixed parameters α, δ and second to justify a passage to limits as $\alpha, \delta \to 0$. A priori estimates uniform with respect to α, δ are needed to analyse the passage to the limits, and we shall obtain all necessary estimates while the theorem of existence is proved.

Now we consider in the domain Q an auxiliary boundary value problem: to find functions $v = (v_1, v_2, v_3)$, $\sigma = \{\sigma_{ij}\}$ such that

$$\alpha \dot{v}_i - \alpha \varepsilon_{ij}(v)_{,j} - \sigma_{ij,j} = f_i, \quad i = 1,2,3, \tag{5.79}$$

$$\dot{\sigma}_{ij} + \alpha \sigma_{ij} - \varepsilon_{ij}(v) + \frac{1}{\delta} p(\sigma)_{ij} = 0, \quad i,j = 1,2,3, \tag{5.80}$$

$$\sigma_{ij}n_j + \alpha \varepsilon_{ij}(v) n_j = 0, \quad i = 1,2,3, \quad \text{on } \Gamma \times (0,T), \tag{5.81}$$

$$v = 0, \quad \sigma = 0, \quad t = 0. \tag{5.82}$$

The dependence of solutions to (5.79)–(5.82) on the parameters α, δ is not indicated at this step in order to simplify the formulae. Note that boundary conditions (5.81) do not coincide with (5.71); the conditions (5.81) can be viewed as a regularization of (5.71) connected with the proposed regularization of the equilibrium equations (5.68). Also, the artificial initial condition for v is introduced.

Solvability of the problem (5.79)–(5.82) will be proved in the following sense:

$$v \in L^2(0,T;H^1(\Omega)), \quad \dot{v}, \sigma, \dot{\sigma} \in L^2(Q), \tag{5.83}$$

$$\alpha(\dot{v},\bar{v}) + \alpha(\varepsilon_{ij}(v), \varepsilon_{ij}(\bar{v})) + (\sigma_{ij}, \varepsilon_{ij}(\bar{v})) = (f,\bar{v}) \tag{5.84}$$
$$\forall \bar{v} \in L^2(0,T;H^1(\Omega)),$$

$$\dot{\sigma}_{ij} + \alpha\sigma_{ij} - \varepsilon_{ij}(v) + \frac{1}{\delta}p(\sigma)_{ij} = 0, \quad i,j = 1,2,3, \tag{5.85}$$

$$v = 0, \quad \sigma = 0, \quad t = 0. \tag{5.86}$$

In this case the boundary conditions (5.81) are included in (5.84). At the first step we get a priori estimates. Assume that the solutions of (5.79)–(5.82) are smooth enough. Multiply (5.79), (5.80) by v_i, $\sigma_{ij} - \xi_{ij}$, respectively, and integrate over Ω. Taking into account that the penalty term is nonnegative this provides the inequality

$$\frac{1}{2}\frac{d}{dt}\left(\alpha\|v\|_0^2 + \|\sigma\|_0^2\right) + \alpha\|\sigma\|_0^2 - \langle \alpha\varepsilon_{ij}(v)_{,j} + \sigma_{ij,j}, v_i\rangle \tag{5.87}$$
$$-\langle \varepsilon_{ij}(v), \sigma_{ij} - \xi_{ij}\rangle - \langle f, v\rangle \le \langle \alpha\sigma + \dot{\sigma}, \xi\rangle.$$

We do not show the dependence of v, σ, ξ, f on t in (5.87). The integration by parts in the third term of the left-hand side of (5.87) can be done. Recall that ξ satisfies the equation (5.68). As a result the following inequality is obtained:

$$\frac{1}{2}\left(\alpha\|v(t)\|_0^2 + \|\sigma(t)\|_0^2\right) + \alpha\sum_{i,j=1}^{3}\int_0^t \|\varepsilon_{ij}(v)\|_0^2 d\tau \tag{5.88}$$

$$\le \langle \sigma(t), \xi(t)\rangle - \int_0^t \langle \sigma(\tau), \dot{\xi}(\tau)\rangle d\tau + \frac{\alpha}{2}\int_0^t \|\xi(\tau)\|_0^2 d\tau.$$

Since $\langle \sigma, \xi\rangle \le 1/4\,\|\sigma\|_0^2 + \|\xi\|_0^2$, the integration of (5.88) implies the estimate

$$\sup_{0\le t\le T}\|\sigma(t)\|_0^2 + \alpha\|v\|_{L^2(Q)}^2 + \alpha\sum_{i,j=1}^{3}\int_0^T \|\varepsilon_{ij}(v)\|_0^2 d\tau \le c \tag{5.89}$$

with the constant c being uniform in α, δ, $\alpha \le \alpha_0$. Hence

$$\sup_{0\le t\le T}\|\sigma(t)\|_0^2 + \alpha\|v\|_{L^2(0,T;H^1(\Omega))}^2 \le c. \tag{5.90}$$

A derivation of the next estimate requires the (α,δ)-uniform boundedness of $\delta^{-1}p(\sigma)$ in the space $L^1(Q)$. By (5.90), it is easy to see that uniformly in α, δ

$$\frac{1}{\delta}\int_0^T \langle p(\sigma), \sigma - \xi\rangle dt \le c$$

provided that the penalty term is not neglected in (5.87). Due to the monotonicity of p

$$\frac{1}{\delta}\int_0^T \langle p(\sigma), \bar{\sigma} - \sigma\rangle dt \le 0 \quad \forall \bar{\sigma} \in L^2(Q),\ \bar{\sigma}(t) \in K.$$

Combining the two last inequalities we have

$$\frac{1}{\delta}\int_0^T \langle p(\sigma), \bar{\sigma} - \xi\rangle dt \le c.$$

We can take here $\bar{\sigma} = \xi + \bar{\xi}$, $\|\bar{\xi}\|_{L^\infty(Q)} \le \mu$. By the hypothesis imposed on ξ the inclusions $\bar{\sigma}(t) \in K$, $t \in (0, T)$, hold provided that μ is small enough, hence

$$\frac{1}{\delta}\int_0^T \langle p(\sigma), \bar{\xi}\rangle dt \le c \quad \forall \bar{\xi},\ \|\bar{\xi}\|_{L^\infty(Q)} \le \mu$$

and, consequently,

$$\frac{1}{\delta}\|p(\sigma)\|_{L^1(Q)} \le c. \tag{5.91}$$

In the sequel this estimate will be improved, namely, we state that $\delta^{-1}p(\sigma)$ is, in fact, bounded in $L^2(0, T; L^1(\Omega))$.

Next, it follows from (5.79), (5.80), (5.82) that

$$\dot{v}_i(0) = 0, \quad i = 1, 2, 3; \quad \dot{\sigma}_{ij}(0) = 0, \quad i, j = 1, 2, 3.$$

Differentiate with respect to t the equations (5.79), (5.80) and multiply by $\dot{v}_i, \dot{\sigma}_{ij} - \dot{\xi}_{ij}$, respectively. Since the term

$$\frac{1}{\delta}\langle \frac{d}{dt} p(\sigma(t)), \dot{\sigma}(t)\rangle$$

is nonnegative for almost all $t \in (0, T)$ (see Lions, 1969) the above multiplication and integration over Ω result in the inequality

$$\frac{1}{2}\frac{d}{dt}\left(\alpha\|\dot{v}\|_0^2 + \|\dot{\sigma}\|_0^2\right) + \alpha\|\dot{\sigma}\|_0^2 - \langle \alpha\varepsilon_{ij}(\dot{v})_{,j} + \dot{\sigma}_{ij,j}, \dot{v}_i\rangle$$

$$-\langle \varepsilon_{ij}(\dot{v}), \dot{\sigma}_{ij} - \dot{\xi}_{ij}\rangle - \langle \dot{f}, \dot{v}\rangle \le \langle \alpha\dot{\sigma} + \ddot{\sigma}, \dot{\xi}\rangle + \frac{1}{\delta}\langle \frac{d}{dt} p(\sigma), \dot{\xi}\rangle.$$

Boundary conditions (5.81) can be taken into account here in order to integrate by parts in the left-hand side. Next we can integrate the inequality obtained in t from 0 to t. This implies

$$\frac{1}{2}\left(\alpha\|\dot{v}(t)\|_0^2 + \|\dot{\sigma}(t)\|_0^2\right) + \alpha\sum_{i,j=1}^{3}\int_0^t \|\varepsilon_{ij}(\dot{v})\|_0^2 d\tau \le \frac{1}{\delta}\langle p(\sigma), \dot{\xi}\rangle|_0^t$$

$$-\frac{1}{\delta}\int_0^t \langle p(\sigma), \ddot{\xi}\rangle d\tau + \langle \dot{\sigma}, \dot{\xi}\rangle|_0^t - \int_0^t \langle \ddot{\sigma}, \ddot{\xi}\rangle d\tau + \frac{\alpha}{2}\int_0^t \|\dot{\xi}(\tau)\|_0^2 d\tau\,.$$

By the estimates (5.90), (5.91) and the condition $\dot{\xi}(0) = 0$ the (α, δ)-uniform estimate follows:

$$\|\dot{\sigma}\|^2_{L^2(Q)} + \alpha\|\dot{v}\|^2_{L^2(Q)} \le c. \tag{5.92}$$

Let $v = v_N + \rho_N$ be the decomposition of the function v into the sum of two orthogonal elements, $v_N \in L^2(0,T; H^1_N(\Omega))$, $\rho_N \in L^2(0,T; R(\Omega))$. We should note at this point that $L^2(0,T;H^1(\Omega)) = L^2(0,T;R(\Omega)) \oplus L^2(0,T;H^1_N(\Omega))$ provided that the scalar product (5.65) is considered in $H^1(\Omega)$. For almost all $t \in (0,T)$

$$\int_\Omega v_N(t) = 0, \quad \int_\Omega (v_{Ni}(t)x_j - v_{Nj}(t)x_i) = 0, \quad i,j = 1,2,3.$$

Hence, by (5.90), (5.92), it follows from (5.80) that $\varepsilon_{ij}(v_N)$ are bounded in $L^2(Q)$ uniformly in α for any fixed δ. This implies the estimate

$$\|v_N\|_{L^2(0,T;H^1_N(\Omega))} \le c(\delta) \tag{5.93}$$

with the constant $c(\delta)$ depending on δ, in general.

Now, observe that in view of the estimates (5.90), (5.92) we can use the Galerkin approach for parabolic problems with monotone operators (Lions, 1969) and obtain that for any fixed α, δ a solution to (5.79)–(5.82) exists in the sense of (5.83)–(5.86). The estimates obtained allow us to pass to the limit as $\alpha \to 0$. Indeed, denote the solution of (5.83)–(5.86) by v^α, σ^α and consider the decomposition $v^\alpha = v^\alpha_N + \rho^\alpha_N$, $v^\alpha_N \in L^2(0,T;H^1_N(\Omega))$, $\rho^\alpha_N \in L^2(0,T;R(\Omega))$. Note that the solution $(v^\alpha, \sigma^\alpha)$ satisfies the estimates (5.90), (5.92), (5.93). Hence, from the sequence v^α, σ^α one can choose a subsequence (with the previous notation for the subsequence) such that for any fixed $\delta > 0$ and $\alpha \to 0$

$$\alpha v^\alpha \to 0 \quad \text{weakly in } L^2(0,T;H^1(\Omega)),$$

$$v^\alpha_N \to v^\delta \quad \text{weakly in } L^2(0,T;H^1_N(\Omega)),$$

$$\alpha\dot{v}^\alpha \to 0 \quad \text{weakly in } L^2(Q), \quad \sigma^\alpha, \dot{\sigma}^\alpha \to \sigma^\delta, \dot{\sigma}^\delta \quad \text{weakly in } L^2(Q).$$

Passing to the limit in (5.79), (5.80) as $\alpha \to 0$ one derives

$$(\sigma^\delta_{ij}, \varepsilon_{ij}(\bar{v})) = (f, \bar{v}) \quad \forall \bar{v} \in L^2(0,T;H^1(\Omega)), \tag{5.94}$$

$$\dot{\sigma}^\delta_{ij} - \varepsilon_{ij}(v^\delta) + \frac{1}{\delta}p(\sigma^\delta)_{ij} = 0, \quad i,j = 1,2,3. \tag{5.95}$$

A justification of the convergence $p(\sigma^\alpha) \to p(\sigma^\delta)$ can be done by the monotonicity arguments. We omit the details.

Now let us prove that $\delta^{-1}p(\sigma^\delta)$ are bounded in $L^2(0,T,L^1(\Omega))$ uniformly in δ. It follows from (5.94) that

$$-\langle\sigma^\delta_{ij}(t),\varepsilon_{ij}(\bar v)\rangle = \langle f(t),\bar v\rangle \quad \forall \bar v\in H^1(\Omega).$$

Hence, for almost all $t\in(0,T)$

$$-\langle\sigma^\delta_{ij}(t),\varepsilon_{ij}(v^\delta(t))\rangle = \langle f(t),v^\delta(t)\rangle. \tag{5.96}$$

Multiply (5.95) by $\sigma^\delta_{ij}-\xi_{ij}$ and integrate over Ω. This provides

$$\langle\varepsilon_{ij}(v^\delta(t)),\sigma^\delta_{ij}(t)-\xi_{ij}(t)\rangle + \frac{1}{\delta}\langle p(\sigma^\delta(t)),\xi(t)-\sigma^\delta(t)\rangle \tag{5.97}$$

$$= \langle\dot\sigma^\delta(t),\sigma^\delta(t)-\xi(t)\rangle.$$

By (5.90), (5.92), the right-hand side of (5.97) is bounded in $L^2(0,T)$ uniformly in δ. Combining (5.96) and (5.97) we obtain that

$$\frac{1}{\delta}\langle p(\sigma^\delta(t)),\sigma^\delta(t)-\xi(t)\rangle \quad \text{are bounded in} \quad L^2(0,T) \tag{5.98}$$

uniformly in δ. Introduce next the convex functional on the space $[L^2(\Omega)]^6$,

$$F(\sigma)=\|\sigma-\pi\sigma\|_0^2,\quad \sigma=\{\sigma_{ij}\},\quad i,j=1,2,3.$$

The derivative of the functional F can be found by the formula $F'(\sigma)=2p(\sigma)$. Let us take a function $\tilde\sigma=\{\tilde\sigma_{ij}\}\in L^\infty(Q)$. Then it follows from the conditions imposed on ξ that $\xi(t)+\tilde\sigma(t)$ belongs to the set K, $t\in(0,T)$, provided that the norm $\|\tilde\sigma\|_{L^\infty(Q)}$ is small enough. By the convexity of F we have

$$\frac{1}{\delta}\langle p(\sigma^\delta(t)),\tilde\sigma(t)\rangle \le \frac{1}{\delta}\langle p(\sigma^\delta(t)),\sigma^\delta(t)-\xi(t)\rangle \tag{5.99}$$

$$+\frac{1}{2\delta}F(\xi(t)+\tilde\sigma(t))-\frac{1}{2\delta}F(\sigma^\delta(t)).$$

The second term of the right-hand side of (5.99) equals zero by the inclusion $\xi(t)+\tilde\sigma(t)\in K$ and consequently, by (5.98),

$$\frac{1}{\delta}\langle p(\sigma^\delta(t)),\tilde\sigma(t)\rangle \quad \text{are bounded in} \quad L^2(0,T).$$

Since $\tilde\sigma$ is an arbitrary element of the space $L^\infty(Q)$ with a small norm we infer that the desired estimate

$$\frac{1}{\delta}p(\sigma^\delta(t)) \quad \text{is bounded in} \quad L^2(0,T;L^1(\Omega)).$$

Hence, it follows from (5.95) that

$$\|\varepsilon_{ij}(v^\delta)\|_{L^2(0,T;L^1(\Omega))}\le c,\quad i,j=1,2,3. \tag{5.100}$$

By (5.100) the inclusion $v^\delta \in L^2(0,T;H^1_N(\Omega))$ yields the estimate

$$\|v^\delta\|_{L^2(0,T;LD_N(\Omega))} \le c\,,$$

being uniform in δ. Moreover, the space $L^1(\Omega)$ is continuously imbedded in $M^1(\Omega)$, and consequently

$$\|v^\delta\|_{L^2(0,T;BD_N(\Omega))} \le c. \tag{5.101}$$

It is useful to bear in mind that the inequality

$$\|\sigma^\delta\|_{L^2(Q)} + \|\dot\sigma^\delta\|_{L^2(Q)} \le c \tag{5.102}$$

holds true uniformly in δ. Recall that $BD(\Omega) \subset L^{3/2}(\Omega)$ in the three-dimensional case. Moreover, the estimate

$$\|\sigma^\delta(T)\|_0 \le c\left(\|\sigma^\delta\|_{L^2(Q)} + \|\dot\sigma^\delta\|_{L^2(Q)}\right) \tag{5.103}$$

holds with the constant c independent of functions.

By (5.101), (5.102), (5.103), we can choose a subsequence with the previous notation for the subsequence such that, as $\delta \to 0$,

$$\sigma^\delta, \dot\sigma^\delta \to \sigma, \dot\sigma \text{ weakly in } L^2(Q), \quad v^\delta \to v \text{ weakly in } L^2(0,T;L^{3/2}(\Omega)),$$

$$\varepsilon_{ij}(v^\delta) \to \varepsilon_{ij}(v) \quad \star\text{–weakly in } L^2(0,T;M^1(\Omega)), \quad i,j = 1,2,3,$$

$$\sigma^\delta(T) \to \sigma(T) \quad \text{weakly in } L^2(\Omega).$$

As a result, passing to the limit as $\delta \to 0$ in (5.94), we obtain

$$(\sigma_{ij}, \varepsilon_{ij}(\bar v)) = (f, \bar v) \quad \forall \bar v \in L^2(0,T;H^1(\Omega)). \tag{5.104}$$

The equations (5.95) imply

$$(\dot\sigma^\delta_{ij}, \bar\sigma_{ij} - \sigma^\delta_{ij}) - (\varepsilon_{ij}(v^\delta), \bar\sigma_{ij} - \sigma^\delta_{ij}) \ge 0 \quad \forall \bar\sigma \in L^2(0,T;V_0(\Omega)), \tag{5.105}$$

$$\bar\sigma(t) \in K, \quad t \in (0,T).$$

The identity (5.104) provides a fulfilment of the equilibrium equations

$$-\sigma^\delta_{ij,j} = f_i\,, \quad i = 1,2,3, \tag{5.106}$$

in the distribution sense.

Consider next that by (5.94)

$$(\varepsilon_{ij}(v^\delta), \sigma^\delta_{ij}) = -(\sigma^\delta_{ij,j}, v^\delta_i)$$

and, moreover,

$$(\varepsilon_{ij}(v^\delta), \bar\sigma_{ij}) = -(v^\delta_i, \bar\sigma_{ij,j}) \quad \forall \bar\sigma \in L^2(0,T;V_0(\Omega)).$$

Hence, the inequality (5.105) can be rewritten in the form

$$(\dot{\sigma}^\delta_{ij}, \bar{\sigma}_{ij} - \sigma^\delta_{ij}) + (v^\delta_i, \bar{\sigma}_{ij,j} - \sigma^\delta_{ij,j}) \geq 0 \quad \forall \bar{\sigma} \in L^2(0,T; V_0(\Omega)), \tag{5.107}$$

$$\bar{\sigma}(t) \in K, \quad t \in (0,T).$$

By (5.106), the values $\sigma^\delta_{ij,j}$ can be replaced by $-f_i$ and, as a result, from (5.107) it follows that

$$(\dot{\sigma}^\delta_{ij}, \bar{\sigma}_{ij}) + (v^\delta_i, \bar{\sigma}_{ij,j} + f_i) \geq \frac{1}{2}\|\sigma^\delta(T)\|^2_0 . \tag{5.108}$$

Passing on to the lower limit on both sides of (5.108) and next changing f_i by $-\sigma_{ij,j}$ we easily arrive at (5.77). The inclusion $\sigma(t) \in K$, $t \in (0,T)$, can be verified by the standard arguments. Since $v^\delta \in L^2(0,T; H^1_N(\Omega))$, the convergence of v^δ to v provides the inclusion $v \in L^2(0,T; BD_N(\Omega))$. The property $\sigma(t) \in V_0(\Omega)$, $t \in (0,T)$, actually follows from (5.104) provided that we take into account the equations

$$-\sigma_{ij,j} = f_i, \quad i = 1,2,3,$$

and the Green formula (5.5). Theorem 5.3 is proved.

5.2.2 Domain with a crack

In this subsection we prove an existence theorem for the elastoplastic problem in the case where the domain has a nonsmooth boundary.

Again, let $\Omega \subset R^3$ be a bounded domain with a smooth boundary Γ and $\Gamma_c \subset \Omega$ be a smooth orientable two-dimensional surface with a regular boundary. We assume that Γ_c can be extended in such a way that the domain Ω is divided into two parts with Lipschitz boundaries. The surface Γ_c can be described parametrically

$$x_i = x_i(y_1, y_2), \qquad i = 1,2,3, \tag{5.109}$$

where $(y_1, y_2) \in \bar{\omega}, \omega \subset R^2$ is a bounded domain with smooth boundary γ, $\bar{\omega} = \omega \cup \gamma$. Assume that for any point $(y_1, y_2) \in \bar{\omega}$ the rank of the Jacobi matrix $\partial x_i / \partial y_j$ equals 2 and the map (5.109) is one-to-one. In this case one can choose a unit normal vector to the surface Γ_c,

$$\nu = \frac{\frac{\partial x}{\partial y_1} \times \frac{\partial x}{\partial y_2}}{|\frac{\partial x}{\partial y_1} \times \frac{\partial x}{\partial y_2}|}.$$

Denote $\Omega_c = \Omega \setminus \Gamma_c$, $Q_c = \Omega_c \times (0,T)$, $T > 0$. Formulation of the equilibrium problem for an elastoplastic body occupying the domain Ω_c is as follows. In the domain Q_c we want to find functions $v = (v_1, v_2, v_3)$, $\sigma = \{\sigma_{ij}\}$, η_{ij}, $i,j = 1,2,3$, satisfying the following equations and inequalities:

$$-\sigma_{ij,j} = f_i, \quad i = 1,2,3, \tag{5.110}$$

$$\varepsilon_{ij}(v) = c_{ijkl}\dot{\sigma}_{kl} + \eta_{ij}, \quad i,j = 1,2,3, \tag{5.111}$$

$$\Phi(\sigma) \le 0, \quad \eta_{ij}(\bar{\sigma}_{ij} - \sigma_{ij}) \le 0 \quad \forall \bar{\sigma},\ \Phi(\bar{\sigma}) \le 0, \tag{5.112}$$

$$\sigma_{ij} n_j = 0, \quad i = 1,2,3, \quad \text{on } \Gamma \times (0,T), \tag{5.113}$$

$$\sigma_{ij} \nu_j = 0, \quad i = 1,2,3, \quad \text{on } \Gamma_c^{\pm} \times (0,T), \tag{5.114}$$

$$\sigma = 0, \quad t = 0. \tag{5.115}$$

All notations fit those used in the preceding subsection. As we see, in this case the boundary of the domain Ω_c consists of the parts Γ, Γ_c^+, Γ_c^-, where $\Gamma_c^{\pm}$ correspond to the positive and negative directions of the normal ν, respectively. Introduce the space

$$H^1(\Omega_c) = \{v = (v_1, v_2, v_3) \mid \ v_i \in L^2(\Omega_c),\ i = 1,2,3;$$
$$v_{i,j} \in L^2(\Omega_c),\ i,j = 1,2,3\}.$$

In this subsection we shall consider functions satisfying the relations

$$\int\limits_{\Omega_c} v = 0, \quad \int\limits_{\Omega_c} (v_i x_j - v_j x_i) = 0, \quad i,j = 1,2,3, \quad v = (v_1, v_2, v_3). \tag{5.116}$$

Let

$$LD(\Omega_c) = \{v = (v_1, v_2, v_3) \mid \ v_i \in L^1(\Omega_c),\ i = 1,2,3;$$
$$\varepsilon_{ij}(v) \in L^1(\Omega_c),\ i,j = 1,2,3\}.$$

The subspaces $H_N^1(\Omega_c)$, $LD_N(\Omega_c)$ consist of all functions from $H^1(\Omega_c)$ and $LD(\Omega_c)$, respectively, satisfying (5.116). In the subspaces $H_N^1(\Omega_c)$, $LD_N(\Omega_c)$ we can consider the norms

$$|v|_{H_N^1(\Omega_c)} = \sum_{i,j=1}^{3} \|\varepsilon_{ij}(v)\|_{0,c}\,, \quad |v|_{LD_N(\Omega_c)} = \sum_{i,j=1}^{3} \|\varepsilon_{ij}(v)\|_{L^1(\Omega_c)}\,.$$

which are equivalent to the standard ones. Here $\|\cdot\|_{0,c}$ stands for the norm in $L^2(\Omega_c)$. The proof of the equivalency is based on the compactness of imbeddings $H^1(\Omega_c) \subset L^2(\Omega_c)$, $LD(\Omega_c) \subset L^1(\Omega_c)$ which take place under the conditions imposed on Γ_c and Γ. Consider two more spaces,

$$BD_N(\Omega_c) = \{v \in BD(\Omega_c) \mid \ v \ \text{ satisfies } \ (5.116)\}\,,$$

$$U_0(\Omega_c) = \{\sigma = \{\sigma_{ij}\} \mid$$
$$\sigma_{ij} \in H^1(\Omega_c),\ i,j = 1,2,3; \quad \sigma_{ij,j} \in L^3(\Omega_c),\ i = 1,2,3;$$
$$\sigma_{ij} n_j = 0,\ i = 1,2,3, \quad \text{on } \Gamma; \quad \sigma_{ij}\nu_j = 0,\ i = 1,2,3, \quad \text{on } \Gamma_c^{\pm}\}\,.$$

Again, to simplify the formulae we assume $c_{ijkl} = \delta_k^i \delta_l^j$. Recall that the set (5.73) contains zero as its interior point. The set K is introduced as before,

$$K = \{\sigma = \{\sigma_{ij}\} \mid \ \sigma_{ij} \in L^2(\Omega_c),\ i,j = 1,2,3;$$

$$\Phi(\sigma(x)) \leq 0 \quad \text{a.e. in } \Omega_c\}.$$

The scalar products in $L^2(\Omega_c)$ and $L^2(Q_c)$ are denoted by $\langle \cdot, \cdot \rangle_c$, $(\cdot, \cdot)_c$, respectively. The space of all rigid displacements on Ω_c is denoted by $R(\Omega_c)$.

We assume that there exists a function $\xi = \{\xi_{ij}\}$, $\xi \in C^2(\bar{Q}_c)$, such that $\xi(0) = \dot{\xi}(0) = 0$, satisfying the equation

$$(\xi_{ij}, \varepsilon_{ij}(\bar{v}))_c = (f, \bar{v})_c \quad \forall \bar{v} \in L^2(0, T; H^1(\Omega_c)) \tag{5.117}$$

and $(1 + \kappa)\xi(t) \in K \cap U_0(\Omega_c)$, $\kappa = \text{const} > 0$, $t \in [0, T]$. Now we are in a position to prove the theorem of existence of the problem (5.110)–(5.115).

Theorem 5.4. Let $f \in L^3(Q_c)$, $\dot{f} \in L^2(Q_c)$, $f(0) = 0$, $\langle f(t), \rho \rangle_c = 0$ for all $\rho \in R(\Omega_c)$, $t \in [0, T]$, and the above assumption on ξ hold. Then there exist functions $v = (v_1, v_2, v_3)$, $\sigma = \{\sigma_{ij}\}$ such that

$$v \in L^2(0, T; BD_N(\Omega_c)), \quad \sigma, \dot{\sigma} \in L^2(Q_c), \quad \sigma(t) \in K, \quad t \in (0, T),$$

$$(\sigma_{ij}, \varepsilon_{ij}(\bar{v}))_c = (f, \bar{v})_c \quad \forall \bar{v} \in L^2(0, T; H^1(\Omega_c)), \tag{5.118}$$

$$(\dot{\sigma}_{ij}, \bar{\sigma}_{ij} - \sigma_{ij})_c + (v_i, \bar{\sigma}_{ij,j} - \sigma_{ij,j})_c \geq 0 \quad \forall \bar{\sigma} \in L^2(0, T; U_0(\Omega_c)), \tag{5.119}$$

$$\bar{\sigma}(t) \in K \quad \text{a.e. on } (0, T),$$

$$\sigma = 0, \quad t = 0. \tag{5.120}$$

PROOF. The general scheme of reasoning coincides with that used in the proof of Theorem 5.3 and our attention now focuses on details related to the nonsmoothness of the boundary.

Let p be the penalty operator related to the set K, $p(\sigma) = \sigma - \pi\sigma$; π is the orthogonal projection operator of the space $[L^2(\Omega_c)]^6$ onto the set K. Consider two positive parameters α, δ and the auxiliary boundary value problem in Q_c for finding $v = (v_1, v_2, v_3)$ and $\sigma = \{\sigma_{ij}\}$,

$$\alpha \dot{v}_i - \alpha \varepsilon_{ij}(v)_{,j} - \sigma_{ij,j} = f_i, \quad i = 1, 2, 3, \tag{5.121}$$

$$\dot{\sigma}_{ij} + \alpha \sigma_{ij} - \varepsilon_{ij}(v) + \frac{1}{\delta} p(\sigma)_{ij} = 0, \quad i, j = 1, 2, 3, \tag{5.122}$$

$$\sigma_{ij} n_j + \alpha \varepsilon_{ij}(v) n_j = 0, \quad i = 1, 2, 3, \quad \text{on } \Gamma \times (0, T), \tag{5.123}$$

$$\sigma_{ij} n_j + \alpha \varepsilon_{ij}(v) \nu_j = 0, \quad i = 1, 2, 3, \quad \text{on } \Gamma_c^{\pm} \times (0, T), \tag{5.124}$$

$$v = 0, \quad \sigma = 0, \quad t = 0. \tag{5.125}$$

We first obtain a priori estimates of solutions to the problem (5.121)–(5.125). Multiply (5.121), (5.122) by v_i, $\sigma_{ij} - \xi_{ij}$ and integrate over Ω. As for obtaining (5.90) we derive

$$\sup_{0 \leq t \leq T} \|\sigma(t)\|_{0,c}^2 + \alpha \|v\|_{L^2(Q_c)}^2 + \alpha \sum_{i,j=1}^{3} \int_0^T \|\varepsilon_{ij}(v)\|_{0,c}^2 d\tau \leq c \tag{5.126}$$

with a constant c uniform in $\alpha, \delta, \alpha \leq \alpha_0$.

From (5.121), (5.122), (5.125) we have

$$\dot{v}_i(0) = 0, \quad i = 1,2,3, \quad \dot{\sigma}_{ij}(0) = 0, \quad i,j = 1,2,3.$$

Hence, a differentiation of (5.121), (5.122) with respect to t and multiplication by $\dot{v}_i$, $\dot{\sigma}_{ij} - \dot{\xi}_{ij}$ result in the estimate

$$\|\dot{\sigma}\|^2_{L^2(Q_c)} + \alpha\|\dot{v}\|^2_{L^2(Q_c)} \leq c. \tag{5.127}$$

Moreover, if $v = v_N + \rho_N$, $v_N \in L^2(0,T; H^1_N(\Omega_c))$, $\rho_N \in L^2(0,T; R(\Omega_c))$, then (5.122) provides

$$\|v_N\|_{L^2(0,T;H^1_N(\Omega_c))} \leq c(\delta) \tag{5.128}$$

where the constant $c(\delta)$ depends, in general, on δ.

The estimates (5.126)–(5.127) allow us to prove the solvability of the system (5.121)–(5.125) for the fixed parameters α, δ in the following sense:

$$v^\alpha \in L^2(0,T; H^1(\Omega_c)), \qquad \dot{v}^\alpha, \sigma^\alpha, \dot{\sigma}^\alpha \in L^2(Q_c),$$

$$\alpha(\dot{v}^\alpha, \bar{v})_c + \alpha(\varepsilon_{ij}(v^\alpha), \varepsilon_{ij}(\bar{v}))_c + (\sigma^\alpha_{ij}, \varepsilon_{ij}(\bar{v}))_c = (f, \bar{v})_c \tag{5.129}$$
$$\forall \bar{v} \in L^2(0,T; H^1(\Omega_c)),$$

$$\dot{\sigma}^\alpha_{ij} + \alpha\sigma^\alpha_{ij} - \varepsilon_{ij}(v^\alpha) + \frac{1}{\delta}p(\sigma^\alpha)_{ij} = 0, \quad i,j = 1,2,3, \tag{5.130}$$

$$v^\alpha = 0, \quad \sigma^\alpha = 0, \quad t = 0. \tag{5.131}$$

The solution of the above problem is denoted by v^α, σ^α since the following step of our reasoning is a passage to the limit as $\alpha \to 0$. Note that boundary conditions (5.123), (5.124) are included in the identity (5.129).

In accord with the estimates (5.126)–(5.128) for any fixed $\delta > 0$ one can choose a subsequence v^α, σ^α such that as $\alpha \to 0$

$$\alpha v^\alpha \to 0 \quad \text{weakly in } L^2(0,T; H^1(\Omega_c)),$$

$$v^\alpha_N \to v^\delta \quad \text{weakly in } L^2(0,T; H^1_N(\Omega_c)),$$

$$\alpha\dot{v}^\alpha \to 0 \quad \text{weakly in } L^2(Q_c), \quad \sigma^\alpha, \dot{\sigma}^\alpha \to \sigma^\delta, \dot{\sigma}^\delta \quad \text{weakly in } L^2(Q_c).$$

Having fulfilled the passage to the limit as $\alpha \to 0$ we obtain

$$(\sigma^\delta_{ij}, \varepsilon_{ij}(\bar{v}))_c = (f, \bar{v})_c \quad \forall \bar{v} \in L^2(0,T; H^1(\Omega_c)), \tag{5.132}$$

$$\dot{\sigma}^\delta_{ij} - \varepsilon_{ij}(v^\delta) + \frac{1}{\delta}p(\sigma^\delta)_{ij} = 0, \quad i,j = 1,2,3. \tag{5.133}$$

Analogously to (5.101) the following estimate holds

$$\|v^\delta\|_{L^2(0,T;BD_N(\Omega_c))} \leq c \tag{5.134}$$

being uniform in δ. Consequently, without any loss we can assume that there exists a subsequence still denoted by v^δ, σ^δ such that as $\delta \to 0$

$$\sigma^\delta, \dot{\sigma}^\delta \to \sigma, \dot{\sigma} \quad \text{weakly in } L^2(Q_c),$$

$$v^\delta \to v \quad \text{weakly in } L^2(0,T; L^{3/2}(\Omega_c)),$$

$$\varepsilon_{ij}(v^\delta) \to \varepsilon_{ij}(v) \quad *\text{-weakly in } L^2(0,T; M^1(\Omega_c)),$$

$$\sigma^\delta(T) \to \sigma(T) \quad \text{weakly in } L^2(\Omega_c).$$

From (5.132) we obtain

$$(\sigma_{ij}, \varepsilon_{ij}(\bar{v}))_c = (f, \bar{v})_c \quad \forall \bar{v} \in L^2(0,T; H^1(\Omega_c)), \tag{5.135}$$

and, hence, the equations

$$-\sigma_{ij,j} = f_i, \quad i = 1,2,3,$$

hold in Q_c in the distribution sense. Also, (5.132) implies

$$-\sigma^\delta_{ij,j} = f_i, \quad i = 1,2,3. \tag{5.136}$$

Hence, by (5.132), (5.136) we have

$$(\varepsilon_{ij}(v^\delta), \sigma^\delta_{ij})_c = -(\sigma^\delta_{ij,j}, v^\delta_i)_c = (f_i, v^\delta_i)_c.$$

Moreover, for $\bar{\sigma} \in L^2(0,T; U_0(\Omega_c))$ we have $(\varepsilon_{ij}(v^\delta), \bar{\sigma}_{ij})_c = -(v^\delta_i, \bar{\sigma}_{ij,j})_c$. As a result, it follows from (5.133) for any $\bar{\sigma} \in L^2(0,T; U_0(\Omega_c))$, $\bar{\sigma}(t) \in K$ a.e. on $(0,T)$, that

$$(\dot{\sigma}^\delta_{ij}, \bar{\sigma}_{ij} - \sigma^\delta_{ij})_c + (v^\delta_i, \bar{\sigma}_{ij,j} - \sigma^\delta_{ij,j})_c \geq 0. \tag{5.137}$$

A passage to the limit as $\delta \to 0$ can be fulfilled in (5.137) as that in (5.107). Therefore, we arrive at (5.119). The property $\sigma(t) \in K$, $t \in (0,T)$, is obtained in a standard way. Boundary conditions (5.113), (5.114) are included in the identity (5.118). Theorem 5.4 is proved.

5.3 Elastoplastic problems for plates with cracks

We prove an existence theorem for elastoplastic plates having cracks. The presence of the cracks entails the domain to have a nonsmooth boundary. The proof of the theorem combines an elliptic regularization and the penalty method. We show that the solution satisfies all boundary conditions imposed at the external boundary and at the crack faces. The results of this section follow the paper (Khludnev, 1998).

5.3.1 The Hencky model. Problem formulation

Let $\Omega \subset R^2$ be a bounded domain with a smooth boundary Γ, and Γ_c be a smooth curve without selfintersections, $\Gamma_c \subset \Omega$. We assume that Γ_c contains its end points. Denote by Ω_c the mid-surface of the plate, $\Omega_c = \Omega \setminus \Gamma_c$. We choose a unit normal vector $\nu = (\nu_1, \nu_2)$ to the curve Γ_c. The curve Γ_c corresponds to the crack in the plate. The crack shape as a surface in R^3 can be described as $x \in \Gamma_c$, $-h \leq z \leq h$, where $x = (x_1, x_2) \in \Omega$, $2h$ is the thickness of the plate, z is a distance to Ω. The domain Ω_c contains, therefore, three components of the boundary: Γ, Γ_c^+, Γ_c^-. Here $\Gamma_c^\pm$ fit to the positive and negative directions of the normal ν, respectively. Let $n = (n_1, n_2)$ be the external unit normal vector to Γ.

Denote by $H^{1,0}(\Omega_c)$ the Sobolev space of functions having the first square integrable derivatives in Ω_c and equal to zero on the external boundary Γ. The space $H^2(\Omega_c)$ contains all functions having derivatives up to the second order square integrable in Ω_c.

Hereafter the known Green formula will be used, namely, for all smooth functions w, $\{m_{ij}\}$, $i, j = 1, 2$, we have

$$\int\limits_\Omega w m_{ij,ij} = \int\limits_\Omega w_{,ij} m_{ij} + \int\limits_\Gamma R_n(m_{ij}) w - \int\limits_\Gamma m_{ij} n_j n_i \frac{\partial w}{\partial n}, \tag{5.138}$$

where $R_n(m_{ij})$ is the transverse force on the boundary Γ defined by the formula

$$R_n(m_{ij}) = m_{ij,j} n_i - \frac{\partial}{\partial \tau}[(m_{11} - m_{22}) n_1 n_2 + m_{12}(n_2^2 - n_1^2)], \quad \tau = (-n_2, n_1).$$

The same formula is valid for the domain Ω_c. In this case the additional integrals over Γ_c^+, Γ_c^- will appear. By $M^1(\Omega_c)$ we denote the space of bounded measures on Ω_c.

Formulation of the elastoplastic problem for the plate having the crack is as follows. In the domain Ω_c we want to find functions w, $m = \{m_{ij}\}$, ξ_{ij}, $i, j = 1, 2$, satisfying the following equations and inequalities:

$$-m_{ij,ij} = f, \tag{5.139}$$

$$-w_{,ij} = a_{ijkl} m_{kl} + \xi_{ij}, \quad i, j = 1, 2, \tag{5.140}$$

$$\Phi(m_{ij}) \leq 0, \quad \xi_{ij}(\bar{m}_{ij} - m_{ij}) \leq 0 \quad \forall \bar{m}, \ \Phi(\bar{m}_{ij}) \leq 0, \tag{5.141}$$

$$w = 0, \quad m_{ij} n_j n_i = 0 \quad \text{on } \Gamma, \tag{5.142}$$

$$m_{ij} \nu_j \nu_i = 0, \quad R_\nu(m_{ij}) = 0 \quad \text{on } \Gamma_c^\pm. \tag{5.143}$$

Here $\Phi : R^3 \to R$ is the convex and continuous function describing a plasticity yield condition. The function w describes vertical displacements of the plate, m_{ij} are bending moments, (5.139) is the equilibrium equation, and equations (5.140) give a decomposition of the curvatures $-w_{,ij}$ as a sum of elastic and plastic parts $a_{ijkl}m_{kl}$, ξ_{ij}, respectively. Let $a_{ijkl}(x) = a_{jikl}(x) = a_{klij}(x)$, $i,j,k,l = 1,2$, and there exist two positive constants c_1, c_2 such that

$$c_2|m|^2 \le a_{ijkl}m_{kl}m_{ij} \le c_1|m|^2, \quad \forall m = \{m_{ij}\}.$$

As for the function Φ, the main assumption is that the following set in R^3

$$\{m \mid \quad \Phi(m_{ij}) \le 0\}$$

contains zero as its interior point.

The functions ξ_{ij} can be eliminated from (5.140), (5.141), which gives

$$\Phi(m_{ij}) \le 0, \quad (a_{ijkl}m_{kl} + w_{,ij})(\bar{m}_{ij} - m_{ij}) \ge 0 \quad \forall \bar{m}, \ \Phi(\bar{m}_{ij}) \le 0.$$

These inequalities will be used in definition of solutions to the problem (5.139)–(5.143).

5.3.2 Solution existence

Introduce the notation

$$U(\Omega_c) = \{m = \{m_{ij}\} \in H^2(\Omega_c) \mid \quad m_{ij}n_jn_i = 0 \quad \text{on} \quad \Gamma;$$

$$m_{ij}\nu_j\nu_i = R_\nu(m_{ij}) = 0 \quad \text{on} \quad \Gamma_c^\pm\},$$

$$K = \{m = \{m_{ij}\} \in L^2(\Omega_c) \mid \quad \Phi(m_{ij}(x)) \le 0 \quad \text{a.e. in} \ \ \Omega_c\}.$$

Assume that there exists a function $m^0 = \{m^0_{ij}\}$, $(1+\kappa)m^0 \in K$, $\kappa = \text{const} > 0$, such that the equation (5.139) is fulfilled in the following sense:

$$-\langle m^0_{ij}, \bar{w}_{,ij}\rangle = \langle f, \bar{w}\rangle \quad \forall \bar{w} \in H^2(\Omega_c) \cap H^{1,0}(\Omega_c). \tag{5.144}$$

The brackets $\langle \cdot, \cdot \rangle$ denote the integration over Ω_c.

The main result of this section can be formulated as follows.

Theorem 5.5. *Assume that $f \in L^2(\Omega_c)$ and the above assumption on m^0 holds. Then there exist functions $w, m = \{m_{ij}\}$ such that*

$$w \in H^{1,0}(\Omega_c), \quad w_{,ij} \in M^1(\Omega_c), \ i,j = 1,2, \quad m \in K,$$

$$-\langle m_{ij}, \bar{w}_{,ij}\rangle = \langle f, \bar{w}\rangle \quad \forall \bar{w} \in H^2(\Omega_c) \cap H^{1,0}(\Omega_c), \tag{5.145}$$

$$\langle a_{ijkl}m_{kl}, \bar{m}_{ij} - m_{ij}\rangle + \langle w, \bar{m}_{ij,ij} - m_{ij,ij}\rangle \ge 0 \quad \forall \bar{m} \in U(\Omega_c) \cap K. \tag{5.146}$$

PROOF. The idea of the proof is to use an elliptic regularization for the penalty equations approximating (5.139)–(5.143). Solutions of the auxiliary problem will depend on two positive parameters ε, δ. The first parameter is responsible for the elliptic regularization and the second one characterizes the penalty approach. More precisely, in the domain Ω_c we want to find the functions $w, m = \{m_{ij}\}$ such that

$$\varepsilon w_{,ijij} - m_{ij,ij} = f, \tag{5.147}$$

$$a_{ijkl}m_{kl} + w_{,ij} + \frac{1}{\delta}p(m)_{ij} = 0, \quad i,j = 1,2, \tag{5.148}$$

$$w = 0, \quad (m_{ij} - \varepsilon w_{,ij})n_j n_i = 0 \quad \text{on } \Gamma, \tag{5.149}$$

$$(m_{ij} - \varepsilon w_{,ij})\nu_j \nu_i = 0 \quad \text{on } \Gamma_c^{\pm}, \tag{5.150}$$

$$R_\nu(m_{ij}) - \varepsilon R_\nu(w_{,ij}) = 0 \quad \text{on } \Gamma_c^{\pm}. \tag{5.151}$$

Here $p(m) = m - \pi(m)$ is the penalty operator, where $\pi : [L^2(\Omega_c)]^3 \to K$ is the orthogonal projection operator. Note that p is monotone, continuous and bounded.

We do not show the dependence of the solution to (5.147)–(5.151) on the parameters to simplify the notation. Our aim is first to prove the solution existence of the problem (5.147)–(5.151) and second to pass to the limit as $\varepsilon \to 0, \delta \to 0$.

Let us derive a priori estimates for solutions of (5.147)–(5.151) assuming that solutions are sufficiently smooth. Multiply (5.147), (5.148) by $w, m_{ij} - m_{ij}^0$, sum and integrate over Ω_c. This gives

$$\varepsilon\langle w_{,ij}, w_{,ij}\rangle + \langle a_{ijkl}m_{kl}, m_{ij}\rangle + \langle w_{,ij}, m_{ij}\rangle + \frac{1}{\delta}\langle p(m)_{ij}, m_{ij} - m_{ij}^0\rangle \tag{5.152}$$

$$-\langle m_{ij,ij}, w\rangle - \langle w_{,ij}, m_{ij}^0\rangle - \langle f, w\rangle = \langle a_{ijkl}m_{kl}, m_{ij}^0\rangle.$$

Integrate by parts in the fifth and sixth terms of the left-hand side of (5.152) taking into account the boundary conditions (5.149)–(5.151) and the Green formula like (5.138) for the domain Ω_c. The penalty term is nonnegative and m_{ij}^0 satisfy the equation (5.144). Hence the uniform in the ε, δ estimate follows,

$$\varepsilon\langle w_{,ij}, w_{,ij}\rangle + \langle a_{ijkl}m_{kl}, m_{ij}\rangle \le c,$$

and consequently

$$\varepsilon\|w\|_2^2 + \|m\|_0^2 \le c. \tag{5.153}$$

Here $\|\cdot\|_s$ stands for the norm in $H^s(\Omega_c)$. The estimate (5.153) allows us to prove the solvability of the problem (5.147)–(5.151) for the fixed parameters ε, δ in the following sense:

$$w \in H^2(\Omega_c) \cap H^{1,0}(\Omega_c), \quad m_{ij} \in L^2(\Omega_c), \quad i,j = 1,2, \tag{5.154}$$

$$\varepsilon\langle w_{,ij}, \bar{w}_{,ij}\rangle - \langle m_{ij}, \bar{w}_{,ij}\rangle = \langle f, \bar{w}\rangle \quad \forall \bar{w} \in H^2(\Omega_c) \cap H^{1,0}(\Omega_c), \tag{5.155}$$

$$\langle a_{ijkl}m_{kl} + w_{,ij} + \frac{1}{\delta}p(m)_{ij}, \bar{m}_{ij}\rangle = 0 \quad \forall \bar{m}_{ij} \in L^2(\Omega_c). \tag{5.156}$$

Indeed, introduce the space $V(\Omega_c) = (H^2(\Omega_c) \cap H_0^1(\Omega_c)) \times [L^2(\Omega_c)]^3$. The elements of this space are denoted by u, where $u = (w, m_{ij})$. Consider the operator $B : V(\Omega_c) \to (V(\Omega_c))'$ defined by the formula

$$B(u)(\bar{u}) = \langle \varepsilon w_{,ij} - m_{ij}, \bar{w}_{,ij} \rangle + \langle a_{ijkl} m_{kl} + w_{,ij} + \frac{1}{\delta} p(m)_{ij}, \bar{m}_{ij} \rangle,$$

where $\bar{u} = (\bar{w}, \bar{m}_{ij})$. Define the linear and continuous functional on $V(\Omega_c)$ by the formula $F(\bar{u}) = \langle f, \bar{w} \rangle$. In this case the identities (5.154)–(5.156) can be written in the form

$$B(u)(\bar{u}) = F(\bar{u}) \quad \forall \bar{u} \in V(\Omega_c),$$

which means

$$B(u) = F. \tag{5.157}$$

Note that the derivation of the estimate (5.153), actually provides the coercivity of the operator B in the following sense:

$$\frac{B(u)(u)}{\|u\|_{V(\Omega_c)}} \to \infty, \quad \|u\|_{V(\Omega_c)} \to \infty.$$

Moreover, B is monotone, bounded and semicontinuous. Hence, the solvability of the equation (5.157) or, equivalently, of the problem (5.154)–(5.156) follows from Theorem 1.14. We should recall that the parameters ε, δ are fixed at this point.

Now we can pass to the limit as $\varepsilon \to 0, \delta \to 0$. Denote the solution of (5.154)–(5.156) by $w^{\varepsilon\delta}, m^{\varepsilon\delta}$. The estimate (5.153) provides the inequality

$$\|m^{\varepsilon\delta}\|_0 \le c. \tag{5.158}$$

From (5.156) the following equations are obtained:

$$-w^{\varepsilon\delta}_{,ij} = a_{ijkl} m^{\varepsilon\delta}_{kl} + \frac{1}{\delta} p(m^{\varepsilon\delta})_{ij}, \quad i, j = 1, 2.$$

Hence, in view of zero boundary conditions for $w^{\varepsilon\delta}$, these equations imply

$$\|w^{\varepsilon\delta}\|_2 \le c(\delta), \tag{5.159}$$

where the constant $c(\delta)$ depends on δ, in general.

By (5.158), (5.159), we choose a subsequence, still denoted by $w^{\varepsilon\delta}, m^{\varepsilon\delta}$, such that for any fixed δ as $\varepsilon \to 0$

$$m^{\varepsilon\delta}_{ij} \to m^{\delta}_{ij} \quad \text{weakly in } L^2(\Omega_c), \quad i, j = 1, 2,$$

$$w^{\varepsilon\delta} \to w^{\delta} \quad \text{weakly in } H^2(\Omega_c) \cap H^{1,0}(\Omega_c).$$

Passing to the limit as $\varepsilon \to 0$ in (5.154)–(5.156) we have

$$-\langle m_{ij}^{\delta}, \bar{w}_{,ij}\rangle = \langle f, \bar{w}\rangle \quad \forall \bar{w} \in H^2(\Omega_c) \cap H^{1,0}(\Omega_c), \tag{5.160}$$

$$\langle a_{ijkl} m_{kl}^{\delta} + w_{,ij}^{\delta} + \frac{1}{\delta} p(m^{\delta})_{ij}, \bar{m}_{ij}\rangle = 0 \quad \forall \bar{m}_{ij} \in L^2(\Omega_c). \tag{5.161}$$

Convergence $p(m^{\varepsilon\delta}) \to p(m^{\delta})$ can be justified by the monotonicity arguments.

Let us prove that uniformly in δ

$$\sum_{i,j=1}^{2} \|w_{,ij}^{\delta}\|_{L^1(\Omega_c)} \le c. \tag{5.162}$$

First we notice from (5.160), (5.161) that uniformly in δ

$$\frac{1}{\delta}\langle p(m^{\delta})_{ij}, m_{ij}^{\delta} - m_{ij}^{0}\rangle \le c. \tag{5.163}$$

Consider the convex functional P on $[L^2(\Omega_c)]^3$,

$$P(m) = \frac{1}{2\delta}\|p(m)\|_0^2.$$

The Gateaux derivative P' of the functional P can be found by the formula $P'(m) = \delta^{-1} p(m)$. Hence, by the convexity of P, we have

$$P(m^0 + q) - P(m^{\delta}) \ge P'(m^{\delta})(m^0 + q - m^{\delta}), \tag{5.164}$$

$$q = \{q_{ij}\} \in [L^2(\Omega_c)]^3.$$

Let $\|q\|_{L^\infty(\Omega_c)} \le \alpha$, where α is chosen to be small enough so that $m^0 + q \in K$. Here we use the conditions imposed on m^0 and the set $\{m = \{m_{ij}\} \mid \Phi(m_{ij}) \le 0\}$. Since $P(m^0 + q) = 0$, it follows from (5.164) that

$$\frac{1}{\delta}\langle p(m^{\delta}), q\rangle \le \frac{1}{\delta}\langle p(m^{\delta}), m^{\delta} - m^0\rangle.$$

In view of the inequality (5.163) we have

$$\frac{1}{\delta}\langle p(m^{\delta}), q\rangle \le c \quad \forall q, \quad \|q\|_{L^\infty(\Omega_c)} \le \alpha,$$

which completes the proof of (5.162).

Taking into account Lemma 5.2 (see below) we conclude from (5.162) that

$$\|w^{\delta}\|_{W_1^2(\Omega_c)} \le c. \tag{5.165}$$

Here, $W_1^2(\Omega_c)$ is the Sobolev space of functions having derivatives up to the second order belonging to $L^1(\Omega_c)$.

Extend the curve Γ_c outside both ends so that each extension cuts the boundary Γ; therefore, the domain Ω_c is divided into two subdomains Ω_1

and Ω_2 with Lipschitz boundaries Γ_1, Γ_2. We assume that meas $\Gamma_i \cap \Gamma \neq \emptyset$, $i = 1, 2$, and each set $\Gamma_i \cap \Gamma$ is not a subset of a straight line. It is known that the imbedding $W_1^1(D) \subset L^2(D)$ is continuous provided that $D \subset R^2$ is a bounded domain with a Lipschitz boundary. Hence

$$\sum_{i=1}^{2} \|w^\delta_{,i}\|_{W_1^1(\Omega_c)} = \sum_{i=1}^{2} \|w^\delta_{,i}\|_{W_1^1(\Omega_1)} + \sum_{i=1}^{2} \|w^\delta_{,i}\|_{W_1^1(\Omega_2)} \tag{5.166}$$

$$\geq c\left(\sum_{i=1}^{2} \|w^\delta_{,i}\|_{L^2(\Omega_1)} + \sum_{i=1}^{2} \|w^\delta_{,i}\|_{L^2(\Omega_2)}\right) \geq c\sum_{i=1}^{2} \|w^\delta_{,i}\|_{L^2(\Omega_c)}.$$

Analogously, we have

$$\|w^\delta\|_{W_1^1(\Omega_c)} \geq c\|w^\delta\|_{L^2(\Omega_c)}. \tag{5.167}$$

Hence, from (5.165), (5.166), (5.167) the boundedness of w^δ follows, i.e.

$$\|w^\delta\|_1 \leq c.$$

The imbedding $L^1(\Omega_c) \subset M^1(\Omega_c)$ is continuous, and consequently, by (5.162), from equations (5.161) it is clear that

$$\sum_{i,j=1}^{2} \|w^\delta_{,ij}\|_{M^1(\Omega_c)} \leq c.$$

As a result we derive the following uniform in δ estimate for the solution w^δ, m^δ of the problem (5.160), (5.161),

$$\|m^\delta\|_0 + \|m^\delta_{ij,ij}\|_0 + \|w^\delta\|_1 + \sum_{i,j=1}^{2} \|w^\delta_{,ij}\|_{M^1(\Omega_c)} \leq c.$$

Now we can pass to the limit as $\delta \to 0$. Choosing a subsequence w^δ, m^δ we can assume that as $\delta \to 0$

$$m^\delta_{ij} \to m_{ij} \quad \text{weakly in } L^2(\Omega_c),\ i,j = 1,2,$$

$$m^\delta_{ij,ij} \to m_{ij,ij} \quad \text{weakly in } L^2(\Omega_c),$$

$$w^\delta \to w \quad \text{weakly in } H^{1,0}(\Omega_c),$$

$$w^\delta_{,ij} \to w_{,ij} \quad \star\text{-weakly in } M^1(\Omega_c),\ i,j = 1,2.$$

It follows from (5.160) that $-m^\delta_{ij,ij} = f$ in the sense of distributions, whence

$$-m_{ij,ij} = f. \tag{5.168}$$

Moreover, from (5.160) we obtain the identity (5.145). Next, by the monotonicity of p, from (5.161) the following inequality is derived:

$$\langle a_{ijkl}m^\delta_{kl}, \bar{m}_{ij} - m^\delta_{ij}\rangle + \langle w^\delta_{,ij}, \bar{m}_{ij} - m^\delta_{ij}\rangle \geq 0 \quad \forall \bar{m} \in U(\Omega_c) \cap K. \quad (5.169)$$

We see that for $\bar{m} \in U(\Omega_c)$ the relation

$$\langle w^\delta_{,ij}, \bar{m}_{ij}\rangle = \langle w^\delta, \bar{m}_{ij,ij}\rangle$$

holds. Furthermore, by (5.160), (5.168) the equalities

$$-\langle m^\delta_{ij}, w^\delta_{,ij}\rangle = \langle f, w^\delta\rangle = -\langle m_{ij,ij}, w^\delta\rangle$$

take place. Consequently, (5.169) implies

$$\langle a_{ijkl}m^\delta_{kl}, \bar{m}_{ij} - m^\delta_{ij}\rangle + \langle w^\delta, \bar{m}_{ij,ij} - m_{ij,ij}\rangle \geq 0 \quad \forall \bar{m} \in U(\Omega_c) \cap K. \quad (5.170)$$

Passing to the limit as $\delta \to 0$ in (5.170) we arrive at (5.146).

The inclusion $m \in K$ can be proved by standard arguments. Note that the second boundary condition (5.142) and the conditions (5.143) are included in the identity (5.145). This means that it is possible to obtain these conditions by integrating by parts provided that the solution is sufficiently smooth. Actually, we can prove that the second condition (5.142) holds in the sense $H^{-1/2}(\Gamma)$, but the arguments are omitted here. The theorem is proved.

Now we have to prove an auxiliary statement which was used in proving the theorem.

Assume that $D \subset R^2$ is a bounded domain with a Lipschitz boundary γ, and γ_0 is a curve being a part of γ such that the length of γ_0 is positive. Denote by $W^{2,\gamma_0}_1(D)$ the subspace of the space $W^2_1(D)$ consisting of all functions equal to zero on γ_0. Furthermore, we assume that γ_0 is not a segment of a straight line. The following statement holds.

Lemma 5.1. *There exists a constant* $c > 0$ *such that*

$$\sum_{i,j=1}^{2} \|w_{,ij}\|_{L^1(D)} \geq c\|w\|_{W^2_1(D)} \quad \forall w \in W^{2,\gamma_0}_1(D). \quad (5.171)$$

PROOF. Assume that the inequality (5.171) is not valid. Then there exists a sequence $w^k \in W^{2,\gamma_0}_1(D)$ such that

$$\|w^k\|_{W^2_1(D)} = 1, \quad (5.172)$$

$$\sum_{i,j=1}^{2} \|w^k_{,ij}\|_{L^1(D)} < \frac{1}{k}. \quad (5.173)$$

Since the imbedding $W_1^2(D) \subset W_1^1(D)$ is compact, by (5.172) we conclude that

$$w^k \to w \quad \text{strongly in } W_1^1(D).$$

Consequently, in view of (5.173)

$$w^k \to w \quad \text{strongly in } W_1^2(D), \tag{5.174}$$

and, besides, $w_{,ij}(x) \equiv 0$ in D. Whence, $w(x) = c_0 + c_1x_1 + c_2x_2$, where c_0, c_1, c_2 are constants. Since $w \in W_1^{2,\gamma_0}(D)$, we derive $w(x) \equiv 0$ – a contradiction to the equality $\|w\|_{W_1^2(D)} = 1$ which follows from (5.172), (5.174). Lemma 5.1 is proved.

Consider the subspace $W_1^{2,\Gamma}(\Omega_c)$ of the space $W_1^2(\Omega_c)$ which contains all functions from $W_1^2(\Omega_c)$ equal to zero on Γ.

Lemma 5.2. *There exists a constant $c > 0$ such that*

$$\sum_{i,j=1}^{2} \|w_{,ij}\|_{L^1(\Omega_c)} \geq c\|w\|_{W_1^2(\Omega_c)} \quad \forall w \in W_1^{2,\Gamma}(\Omega_c).$$

PROOF. We divide the domain Ω_c into two subdomains Ω_1, Ω_2 with Lipschitz boundaries as in the proof of Theorem 5.5. Then, by Lemma 5.1

$$\sum_{i,j=1}^{2} \|w_{,ij}\|_{L^1(\Omega_c)} = \sum_{i,j=1}^{2} \|w_{,ij}\|_{L^1(\Omega_1)} + \sum_{i,j=1}^{2} \|w_{,ij}\|_{L^1(\Omega_2)}$$

$$\geq c\Big(\|w\|_{W_1^2(\Omega_1)} + \|w\|_{W_1^2(\Omega_2)}\Big) = c\|w\|_{W_1^2(\Omega_c)} \quad \forall w \in W_1^{2,\Gamma}(\Omega_c),$$

which completes the proof of Lemma 5.2.

5.4 The Prandtl–Reuss elastoplastic plates

We prove an existence of solutions for the Prandtl–Reuss model of elastoplastic plates with cracks. The proof is based on a special combination of a parabolic regularization and the penalty method. With the appropriate a priori estimates, uniform with respect to the regularization and penalty parameters, a passage to the limit along the parameters is fulfilled. Both the smooth and nonsmooth domains are considered in the present section. The results obtained provide a fulfilment of all original boundary conditions.

5.4.1 Domain with a smooth boundary

Let $\Omega \subset R^2$ be a bounded domain with a smooth boundary Γ, $Q = \Omega \times (0, T)$, $x = (x_1, x_2) \in \Omega$, $t \in (0, T)$. Formulation of the elastoplastic problem for a plate is as follows. In the domain Q we have to find functions $v, m = \{m_{ij}\}, \eta_{ij}$, $i, j = 1, 2$, satisfying the following equations and inequalities:

$$-m_{ij,ij} = f, \tag{5.175}$$

$$-v_{,ij} = a_{ijkl}\dot{m}_{kl} + \eta_{ij}, \quad i, j = 1, 2, \tag{5.176}$$

$$\Phi(m_{ij}) \leq 0, \quad \eta_{ij}(\bar{m}_{ij} - m_{ij}) \leq 0 \quad \forall \bar{m},\ \Phi(\bar{m}_{ij}) \leq 0, \tag{5.177}$$

$$v = 0, \quad m_{ij}n_jn_i = 0 \quad \text{on } \Gamma \times (0, T), \tag{5.178}$$

$$m = 0, \quad t = 0. \tag{5.179}$$

Here $\Phi : R^3 \to R$ is the convex and continuous function describing a plasticity yield condition, the dot denotes a derivative with respect to t, $n = (n_1, n_2)$ is the unit normal vector to the boundary Γ. The function v describes a vertical velocity of the plate, m_{ij} are bending moments, (5.175) is the equilibrium equation, and equations (5.176) give a decomposition of the curvature velocities $-v_{ij}$ as a sum of elastic and plastic parts $a_{ijkl}\dot{m}_{kl}$, η_{ij}, respectively. Let $a_{ijkl}(x) = a_{jikl}(x) = a_{klij}(x)$, $i, j, k, l = 1, 2$, and there exist two positive constants c_1, c_2 such that for all $m = \{m_{ij}\}$

$$c_2|m|^2 \leq a_{ijkl}m_{kl}m_{ij} \leq c_1|m|^2. \tag{5.180}$$

As for the function Φ, the main assumption is that the following set in R^3

$$\{m \mid \quad \Phi(m_{ij}) \leq 0\}$$

contains zero as its interior point.

In the sequel the known Green formula is used, namely, for all smooth functions v, $\{m_{ij}\}$, $i, j = 1, 2$, we have

$$\int_\Omega v m_{ij,ij} = \int_\Omega v_{,ij}m_{ij} + \int_\Gamma R_n(m_{ij})v - \int_\Gamma m_{ij}n_jn_i\frac{\partial v}{\partial n}, \tag{5.181}$$

where $R_n(m_{ij})$ is the transverse force on the boundary Γ defined by the formula

$$R_n(m_{ij}) = m_{ij,j}n_i - \frac{\partial}{\partial \tau}[(m_{11} - m_{22})n_1n_2 + m_{12}(n_2^2 - n_1^2)], \quad \tau = (-n_2, n_1).$$

The Green formula (5.181) can be specified for the case where the functions v, m_{ij} are not smooth enough. To this end, introduce the Hilbert space

$$V(\Omega) = \{m = \{m_{ij}\} \mid \quad m_{ij} \in L^2(\Omega),\ i, j = 1, 2; \quad m_{ij,ij} \in L^2(\Omega)\}$$

equipped with the norm

$$\|m\|^2_{V(\Omega)} = \|m\|^2_0 + \|m_{ij,ij}\|^2_0$$

where $\|\cdot\|_s$ is the norm in the Sobolev space $H^s(\Omega)$. It can be proved (see Section 1.4) that for $m \in V(\Omega)$ the values $m_{ij}n_jn_i$ and $R_n(m_{ij})$ are correctly defined on Γ as elements of $H^{-1/2}(\Gamma), H^{-3/2}(\Gamma)$, respectively. Moreover, for $v \in H^2(\Omega)$ the following formula arises:

$$\langle v, m_{ij,ij}\rangle = \langle m_{ij}, v_{,ij}\rangle + \langle R_n(m_{ij}), v\rangle_{3/2,\Gamma} - \langle m_{ij}n_jn_i, \frac{\partial v}{\partial n}\rangle_{1/2,\Gamma}. \quad (5.182)$$

Here, $H^{-s}(\Gamma)$ is the space dual of the space $H^s(\Gamma)$, and $\langle \cdot, \cdot \rangle$, $\langle \cdot, \cdot \rangle_{s,\Gamma}$ denote the scalar product in $L^2(\Omega)$ and the duality pairing between $H^{-s}(\Gamma)$, $H^s(\Gamma)$, respectively.

Introduce the closed convex set in $[L^2(\Omega)]^3$

$$K = \{m = \{m_{ij}\} \mid \quad m_{ij} \in L^2(\Omega),\ i,j = 1,2,$$
$$\Phi(m_{ij}(x)) \le 0 \quad \text{a.e. in } \Omega\}.$$

Let $\pi : [L^2(\Omega)]^3 \to K$ be the operator of orthogonal projection, and $p(m) = m - \pi m$ be the penalty operator which is monotonous, bounded and continuous from $[L^2(\Omega)]^3$ to $[L^2(\Omega)]^3$.

To simplify the formulae the assumption $a_{ijkl} = \delta^i_k \delta^j_l$ will be used, where δ^i_k is the Kronecker symbol. Nevertheless, we have to note that all the results obtained in the section hold true in the general case (5.180).

Let $H^1_0(\Omega)$ be the subspace of the Sobolev space $H^1(\Omega)$ which consists of all functions equal to zero on Γ; $H^{2,0}(\Omega) = H^2(\Omega) \cap H^1_0(\Omega)$.

Denote by $M^1(\Omega)$ the bounded measures on Ω. Consider next the subspace in $V(\Omega)$:

$$V_0(\Omega) = \{m = \{m_{ij}\} \in V(\Omega) \mid \quad m_{ij}n_jn_i = 0 \quad \text{on } \Gamma\}.$$

In accordance with the above remarks the values $m_{ij}n_jn_i$ are defined on Γ in the sense of $H^{-1/2}(\Gamma)$.

The functions η_{ij} can be eliminated from (5.176), (5.177) which gives

$$\Phi(m_{ij}) \le 0, \quad (a_{ijkl}\dot{m}_{kl} + v_{,ij})(\bar{m}_{ij} - m_{ij}) \ge 0 \quad \forall \bar{m}, \ \Phi(\bar{m}_{ij}) \le 0. \quad (5.183)$$

Inequalities (5.183) will be used in definition of solutions to the problem (5.175)–(5.179).

Let the brackets $(\cdot, \cdot)$ denote the scalar product in $L^2(Q)$. Assume that there exists a function $M = \{M_{ij}\} \in L^\infty(0,T; V_0(\Omega))$, $\dot{M}, \ddot{M} \in L^\infty(Q)$, satisfying equation (5.175) and, besides, $(1+\kappa)M(t) \in K$, $t \in (0,T)$, $\kappa =$ const > 0, $M(0) = \dot{M}(0) = 0$. Now we are in a position to formulate the existence theorem related to the problem (5.175)–(5.179).

Theorem 5.6. *Let $f, \dot{f} \in L^2(Q)$, $f(0) = 0$, and the above assumption on M hold. Then there exist functions $v, m = \{m_{ij}\}$ such that*

$$v \in L^2(0,T; H^1_0(\Omega)), \quad v_{,ij} \in L^2(0,T; M^1(\Omega)),\ i,j = 1,2,$$
$$m \in L^2(0,T; V_0(\Omega)), \quad \dot{m} \in L^2(Q), \quad m(t) \in K, \quad t \in (0,T),$$

$$(\dot{m}_{ij}, \bar{m}_{ij} - m_{ij}) + (v, \bar{m}_{ij,ij} - m_{ij,ij}) \geq 0 \tag{5.184}$$
$$\forall \bar{m} \in L^2(0,T; V_0(\Omega)), \quad \bar{m}(t) \in K, \quad t \in (0,T),$$

and the equation (5.175) *and initial condition* (5.179) *hold.*

PROOF. The idea of the proof is to use a parabolic regularization for the penalty equations approximating (5.175)–(5.179). Solutions of the auxiliary problem will depend on two positive parameters ε, δ. The first parameter is responsible for the parabolic regularization and the second one characterizes the penalty approach. Namely, in the domain Q we want to find functions $v, m = \{m_{ij}\}$ such that

$$\varepsilon\dot{v} + \varepsilon\Delta^2 v - m_{ij,ij} = f, \tag{5.185}$$

$$\dot{m}_{ij} + \varepsilon m_{ij} + v_{,ij} + \frac{1}{\delta} p(m)_{ij} = 0, \quad i,j = 1,2, \tag{5.186}$$

$$v = 0, \quad (m_{ij} - \varepsilon v_{,ij}) n_j n_i = 0 \quad \text{on } \Gamma \times (0,T), \tag{5.187}$$

$$v = 0, \quad m = 0, \quad t = 0. \tag{5.188}$$

To simplify the notations we do not indicate the dependence of the solutions on the parameters ε, δ. Our aim is first to prove the existence of solutions to (5.185)–(5.188) and next to justify the passage to limits as $\varepsilon, \delta \to 0$. A priori estimates uniform with respect to ε, δ are needed to study the passage to the limits, and we shall derive all the necessary inequalities while the existence theorem is proved.

Assume first that the solution of (5.185)–(5.188) is smooth enough. Multiply (5.185), (5.186) by $v, m_{ij} - M_{ij}$, respectively, and integrate over Ω. This gives

$$\frac{1}{2}\frac{d}{dt}\left(\varepsilon\|v(t)\|_0^2 + \|m(t)\|_0^2\right) + \varepsilon\|m(t)\|_0^2 + \langle \varepsilon v_{,ijij}(t) - m_{ij,ij}(t), v(t)\rangle \tag{5.189}$$

$$+ \langle v_{,ij}(t), m_{ij}(t)\rangle - \langle v_{,ij}(t), M_{ij}(t)\rangle - \langle f(t), v(t)\rangle \leq \langle \varepsilon m(t) + \dot{m}(t), M(t)\rangle.$$

In so doing we have omitted the nonnegative term containing the penalty operator. Using the formula (5.181), the integration by parts can be done in the third and the fifth terms of the left-hand side of (5.189). Also, note that M_{ij} satisfy equation (5.175). Integration of (5.189) in t from 0 to t results in the inequality

$$\frac{1}{2}\left(\varepsilon\|v(t)\|_0^2 + \|m(t)\|_0^2\right) + \varepsilon\sum_{i,j=1}^{2}\int_0^t \|v_{,ij}(\tau)\|_0^2 d\tau \leq \langle m(t), M(t)\rangle$$

$$-\int_0^t \langle m(\tau), \dot{M}(\tau)\rangle d\tau + \frac{\varepsilon}{2}\int_0^t \|M(\tau)\|_0^2 d\tau.$$

To evaluate the right-hand side we can use the inequality $\langle \xi, \eta \rangle \le 1/4\,\|\xi\|_0^2 + \|\eta\|_0^2$. Also, we use the well-known estimate

$$\sum_{i,j=1}^{2} \|v_{,ij}\|_0 \ge c\, \|v\|_{H^{2,0}(\Omega)} \quad \forall v \in H^{2,0}(\Omega),$$

where the constant c does not depend on v. Next we can apply the Grownwall lemma which gives

$$\|m\|^2_{L^2(Q)} + \varepsilon \|v\|^2_{L^2(0,T;H^{2,0}(\Omega))} \le c \tag{5.190}$$

with the constant c being uniform in $\varepsilon, \delta, \varepsilon \le \varepsilon_0$.

A derivation of the next estimate requires the boundedness of the penalty term in $L^1(Q)$, i.e. uniformly in ε, δ

$$\frac{1}{\delta}\|p(m)\|_{L^1(Q)} \le c_0. \tag{5.191}$$

Let us establish (5.191). By (5.190), it is clear that uniformly in ε, δ

$$\frac{1}{\delta}\int_0^T \langle p(m), m - M\rangle dt \le c$$

provided that the penalty term is not neglected in (5.189). Due to the monotonicity of p,

$$\frac{1}{\delta}\int_0^T \langle p(m), \bar{m} - m\rangle dt \le 0 \quad \forall \bar{m} \in L^2(Q), \quad \bar{m}(t) \in K.$$

Summing the two last inequalities we have

$$\frac{1}{\delta}\int_0^T \langle p(m), \bar{m} - M\rangle dt \le c.$$

We can take here $\bar{m} = M + \bar{M}$, $\|\bar{M}\|_{L^\infty(Q)} \le \lambda$. By the hypothesis imposed on M the inclusions $\bar{m}(t) \in K$, $t \in (0,T)$, hold provided that λ is small enough, and hence

$$\frac{1}{\delta}\int_0^T \langle p(m), \bar{M}\rangle dt \le c \quad \forall \bar{M}, \ \|\bar{M}\|_{L^\infty(Q)} \le \lambda,$$

which implies (5.191). Notice that in the sequel we shall improve the estimate (5.191), namely, we prove that $\delta^{-1}p(m)$ are bounded in $L^2(0,T;L^1(\Omega))$ uniformly with respect to δ.

It is seen from the equations (5.185), (5.186) considered for $t = 0$ and the initial conditions (5.188) that

$$\dot{v}(0) = 0, \quad \dot{m}_{ij}(0) = 0, \quad i,j = 1,2. \tag{5.192}$$

Differentiate with respect to t the equations (5.185), (5.186) and multiply by $\dot{v}, \dot{m}_{ij} - \dot{M}_{ij}$, respectively. The nonnegative term (see Lions, 1969)

$$\frac{1}{\delta}\langle \frac{d}{dt}p(m(t)), \dot{m}(t)\rangle, \quad t \in (0,T),$$

can be neglected. Hence, the integration over Ω yields

$$\frac{1}{2}\frac{d}{dt}\left(\varepsilon\|\dot{v}(t)\|_0^2 + \|\dot{m}(t)\|_0^2\right) + \varepsilon\|\dot{m}(t)\|_0^2 + \langle \varepsilon\dot{v}_{,ijij}(t) - \dot{m}_{ij,ij}(t), \dot{v}(t)\rangle$$
$$+ \langle \dot{v}_{,ij}(t), \dot{m}_{ij}(t)\rangle - \langle \dot{v}_{,ij}(t), \dot{M}_{ij}(t)\rangle - \langle \dot{f}(t), \dot{v}(t)\rangle$$
$$\leq \langle \varepsilon\dot{m}(t) + \ddot{m}(t), \dot{M}(t)\rangle + \frac{1}{\delta}\langle \frac{d}{dt}p(m(t)), \dot{M}(t)\rangle.$$

Taking into account the conditions (5.187), we first integrate by parts in the left-hand side of the inequality obtained and next we integrate in t. Simultaneously, the integration by parts in t is fulfilled. This gives the inequality

$$\frac{1}{2}\left(\varepsilon\|\dot{v}(t)\|_0^2 + \|\dot{m}(t)\|_0^2\right) + \varepsilon\sum_{i,j=1}^{2}\int_0^t \|\dot{v}_{,ij}(\tau)\|_0^2 d\tau$$
$$\leq \langle \dot{m}(t), \dot{M}(t)\rangle - \int_0^t \langle \dot{m}(\tau), \ddot{M}(\tau)\rangle d\tau + \frac{1}{\delta}\langle p(m(t)), \dot{M}(t)\rangle$$
$$- \int_0^t \frac{1}{\delta}\langle p(m(\tau)), \ddot{M}(\tau)\rangle d\tau + \frac{\varepsilon}{2}\int_0^t \|\dot{M}(\tau)\|_0^2 d\tau.$$

As a result the ε, δ-uniform estimate is derived:

$$\varepsilon\|\dot{v}\|_{L^2(Q)}^2 + \|\dot{m}\|_{L^2(Q)}^2 \leq c\,. \tag{5.193}$$

Moreover, by the estimate (5.190), from (5.186) it follows that $v_{,ij}$ are bounded in $L^2(Q)$ uniformly in ε for any fixed δ. Hence, by the first boundary condition (5.187)

$$\|v\|_{L^2(0,T;H^{2,0}(\Omega))} \leq c(\delta), \tag{5.194}$$

where the constant $c(\delta)$ depends on δ, in general.

So the necessary estimates are obtained, and we can use the Galerkin method to prove the solvability of the parabolic boundary value problem (5.185)–(5.188) (see Lions, 1969). This proves that the solution exists in the following sense,

$$v^{\varepsilon\delta} \in L^2(0,T;H^{2,0}(\Omega)), \quad \dot{v}^{\varepsilon\delta},\, m^{\varepsilon\delta},\, \dot{m}^{\varepsilon\delta} \in L^2(Q), \tag{5.195}$$

$$\varepsilon(\dot{v}^{\varepsilon\delta},\bar{v}) + \varepsilon(v^{\varepsilon\delta}_{,ij},\bar{v}_{,ij}) - (m^{\varepsilon\delta}_{ij},\bar{v}_{,ij}) = (f,\bar{v}) \tag{5.196}$$
$$\forall \bar{v} \in L^2(0,T;H^{2,0}(\Omega)),$$

$$\dot{m}^{\varepsilon\delta}_{ij} + \varepsilon m^{\varepsilon\delta}_{ij} + v^{\varepsilon\delta}_{,ij} + \frac{1}{\delta}p(m^{\varepsilon\delta})_{ij} = 0, \quad i,j=1,2, \tag{5.197}$$

and the initial conditions (5.188) hold.

Let us pass on to the limit as $\varepsilon \to 0$ in (5.195)–(5.197). By the estimates (5.190), (5.193), (5.194), one can choose a subsequence still denoted by $v^{\varepsilon\delta}, m^{\varepsilon\delta}$ such that for any fixed δ and $\varepsilon \to 0$

$$\varepsilon\dot{v}^{\varepsilon\delta} \to 0 \text{ weakly in } L^2(Q), \quad v^{\varepsilon\delta} \to v^\delta \text{ weakly in } L^2(0,T;H^{2,0}(\Omega)),$$

$$m^{\varepsilon\delta}_{ij},\, \dot{m}^{\varepsilon\delta}_{ij} \to m^\delta_{ij},\, \dot{m}^\delta_{ij} \quad \text{weakly in } L^2(Q), \quad i,j=1,2.$$

We have to note that for any fixed δ the ε-subsequences are different, in general. Passing on to the limit in (5.195)–(5.197), the following relations are derived:

$$-(m^\delta_{ij},\bar{v}_{,ij}) = (f,\bar{v}) \quad \forall \bar{v} \in L^2(0,T;H^{2,0}(\Omega)), \tag{5.198}$$

$$\dot{m}^\delta_{ij} + v^\delta_{,ij} + \frac{1}{\delta}p(m^\delta)_{ij} = 0, \quad i,j=1,2. \tag{5.199}$$

A convergence justification of the nonlinear terms $\delta^{-1}p(m^{\varepsilon\delta})_{ij}$ to the term $\delta^{-1}p(m^\delta)_{ij}$ can be done by monotonicity arguments. The details are omitted here.

Now we shall derive an additional a priori estimate which improves (5.191). It follows from (5.198) that for almost all $t \in (0,T)$

$$-\langle m^\delta_{ij}(t),\bar{v}_{,ij}\rangle = \langle f(t),\bar{v}\rangle \quad \forall \bar{v} \in H^{2,0}(\Omega).$$

Consequently, for almost all $t \in (0,T)$

$$-\langle m^\delta_{ij}(t), v^\delta_{,ij}(t)\rangle = \langle f(t), v^\delta(t)\rangle. \tag{5.200}$$

Multiply (5.199) by $m^\delta_{ij} - M_{ij}$ and integrate over Ω. This implies

$$\langle v^\delta_{,ij}(t), m^\delta_{ij}(t) - M_{ij}(t)\rangle + \frac{1}{\delta}\langle p(m^\delta(t)), m^\delta(t) - M(t)\rangle \tag{5.201}$$
$$= \langle \dot{m}^\delta(t), M(t) - m^\delta(t)\rangle.$$

Notice that the right-hand side of (5.201) is the function bounded in $L^2(0,T)$ uniformly in δ. Let us integrate by parts in the left-hand side of (5.201) and use the fact that M_{ij} satisfy equation (5.175). Combining (5.201) with (5.200) we infer that

$$\frac{1}{\delta}\langle p(m^\delta(t)), m^\delta(t) - M(t)\rangle \quad \text{are bounded in} \quad L^2(0,T) \tag{5.202}$$

uniformly in δ.

Consider next the convex functional on the space $[L^2(\Omega)]^3$,

$$P(m) = \|m - \pi m\|_0^2, \quad m = \{m_{ij}\}.$$

Its derivative can be found (see Khludnev, Sokolowski, 1997) by the formula $P'(m) = 2(m - \pi m)$, i.e. $P'(m) = 2p(m)$. We take any function $\hat{m} = \{\hat{m}_{ij}\} \in L^\infty(Q)$ such that $M(t) + \hat{m}(t)$ belongs to the set K, $t \in (0,T)$. This can be done provided the norm $\|\hat{m}\|_{L^\infty(Q)}$ is small enough. At this point we need the assumptions imposed on M and the set $\{m \mid \Phi(m_{ij}) \leq 0\}$.

For almost all $t \in (0,T)$, due to the convexity of P, we have

$$\delta^{-1}\langle p(m^\delta(t)), \hat{m}(t)\rangle \leq \delta^{-1}\langle p(m^\delta(t)), m^\delta(t) - M(t)\rangle \tag{5.203}$$

$$+\frac{1}{2}\delta^{-1}P(M(t) + \hat{m}(t)) - \frac{1}{2}\delta^{-1}P(m^\delta(t)).$$

The second term on the right-hand side of (5.203) equals zero by the inclusion $M(t) + \hat{m}(t) \in K$ and hence, owing to (5.202),

$$\frac{1}{\delta}\langle p(m^\delta(t)), \hat{m}(t)\rangle \quad \text{are bounded in} \quad L^2(0,T)\,.$$

Since $\hat{m}$ is an arbitrary element of $L^\infty(Q)$ with a small norm we conclude that

$$\frac{1}{\delta}p(m^\delta(t))_{ij} \quad \text{are bounded in} \quad L^2(0,T;L^1(\Omega)), \quad i,j = 1,2, \tag{5.204}$$

uniformly in δ. Then it follows from (5.199), (5.204) that

$$\|v^\delta_{,ij}\|_{L^2(0,T;L^1(\Omega))} \leq c, \quad i,j = 1,2. \tag{5.205}$$

Let $W_1^2(\Omega)$ be the Sobolev space of functions having derivatives up to the second order in Ω which are integrable with power one. By continuously imbedding $W_1^2(\Omega) \subset H^1(\Omega)$, the estimate (5.205) yields

$$\|v^\delta\|_{L^2(0,T;H_0^1(\Omega))} \leq c. \tag{5.206}$$

Besides, the space $L^1(\Omega)$ is continuously imbedded in $M^1(\Omega)$, whence

$$\|v^\delta_{,ij}\|_{L^2(0,T;M^1(\Omega))} \leq c, \quad i,j = 1,2. \tag{5.207}$$

Note that the estimate

$$\|m^\delta(T)\|_0 \le c\left(\|m^\delta\|_{L^2(Q)} + \|\dot{m}^\delta\|_{L^2(Q)}\right)$$

holds with a constant uniform in δ, and, moreover, by (5.190), (5.193), the right-hand side is bounded in δ.

Now we can choose a subsequence v^δ, m^δ denoted just as the sequence such that as $\delta \to 0$

$$v^\delta \to v \quad \text{weakly in } L^2(0,T;H_0^1(\Omega)),$$

$$m_{ij}^\delta,\ \dot{m}_{ij}^\delta \to m_{ij},\ \dot{m}_{ij} \quad \text{weakly in } L^2(Q), \quad i,j=1,2,$$

$$v_{,ij}^\delta \to v_{,ij} \quad \star\text{-weakly in } L^2(0,T;M^1(\Omega)), \quad i,j=1,2,$$

$$m^\delta(T) \to m(T) \quad \text{weakly in } L^2(\Omega).$$

From (5.198), (5.199) it follows that

$$-(m_{ij}, \bar{v}_{,ij}) = (f, \bar{v}) \quad \forall \bar{v} \in L^2(0,T;H^{2,0}(\Omega)), \tag{5.208}$$

$$(\dot{m}_{ij}^\delta, \bar{m}_{ij} - m_{ij}^\delta) + (v^\delta, \bar{m}_{ij,ij} - m_{ij,ij}^\delta) \ge 0 \tag{5.209}$$

$$\forall \bar{m} \in L^2(0,T;V_0(\Omega)), \quad \bar{m}(t) \in K, \quad t \in (0,T).$$

Hence, the equilibrium equation (5.175) holds in the sense (5.208). The second boundary condition (5.178) is contained in the identity (5.208) and, consequently, $m \in L^2(0,T;V_0(\Omega))$.

By (5.198), the equation

$$-m_{ij,ij}^\delta = f$$

holds in the sense of distributions, and hence (5.209) can be written in the form

$$(\dot{m}_{ij}^\delta, \bar{m}_{ij}) + (v^\delta, \bar{m}_{ij,ij} + f) \ge \frac{1}{2}\|m^\delta(T)\|_0^2. \tag{5.210}$$

By passing on to the lower limit on both sides of (5.210) and next changing f by $-m_{ij,ij}$ we readily arrive at (5.184).

The inclusion $m(t) \in K$, $t \in (0,T)$, can be obtained in the standard way. Theorem 5.6 is proved.

5.4.2 Domain with the crack

In this subsection we prove the solvability of the elastoplastic problem for a plate having a nonsmooth boundary. A solution of the problem will satisfy all boundary conditions both at the exterior boundary and at the crack faces.

Let $\Omega \subset R^2$ be a bounded domain with a smooth boundary Γ, and $\Gamma_c \subset \Omega$ be a smooth curve without selfintersections. Assume that Γ_c contains its end points. We put $\Omega_c = \Omega \setminus \Gamma_c$; the domain Ω_c corresponds to the mid-surface points of the plate. Denote by $\nu = (\nu_1, \nu_2)$ a unit normal vector to Γ_c. Boundary of Ω_c consists of three components $\Gamma, \Gamma_c^+, \Gamma_c^-$, where $\Gamma_c^\pm$ fit the positive and negative directions of ν. Formulation of the elastoplastic problem in this case is as follows. In the domain $Q_c = \Omega_c \times (0,T)$, $T > 0$, we have to find functions $v, m = \{m_{ij}\}, \eta_{ij}$, $i,j = 1,2$, satisfying the following equations and inequalities:

$$-m_{ij,ij} = f, \tag{5.211}$$

$$-v_{,ij} = a_{ijkl}\dot{m}_{kl} + \eta_{ij}, \quad i,j = 1,2, \tag{5.212}$$

$$\Phi(m_{ij}) \le 0, \quad \eta_{ij}(\bar{m}_{ij} - m_{ij}) \le 0 \quad \forall \bar{m}, \ \Phi(\bar{m}_{ij}) \le 0, \tag{5.213}$$

$$v = 0, \quad m_{ij}n_jn_i = 0 \quad \text{on } \Gamma \times (0,T), \tag{5.214}$$

$$m_{ij}\nu_j\nu_i = 0, \quad R_\nu(m_{ij}) = 0 \quad \text{on } \Gamma_c^\pm \times (0,T), \tag{5.215}$$

$$m = 0, \quad t = 0. \tag{5.216}$$

All notations fit those in the previous subsection. Some arguments are required to explain in which sense boundary conditions (5.215) hold. This will be done later on. Note that conditions (5.215) will be contained in an integral identity.

Denote by $H^{1,0}(\Omega_c)$ the subspace of $H^1(\Omega_c)$ consisting of functions equal to zero at the external boundary Γ; $H^{2,0}(\Omega_c) = H^2(\Omega_c) \cap H^{1,0}(\Omega_c)$.

Consider the Sobolev space $W_1^2(\Omega_c)$ of functions whose derivatives up to the second order in Ω_c are integrable with the first power. Introduce the notation

$$U(\Omega_c) = \{m = \{m_{ij}\} \mid \ m_{ij} \in H^2(\Omega_c), \ i,j = 1,2; \quad m_{ij}n_jn_i = 0 \quad \text{on } \Gamma;$$
$$m_{ij}\nu_j\nu_i = R_\nu(m_{ij}) = 0 \quad \text{on } \Gamma_c^\pm\}.$$

Let $(\cdot\,,\,\cdot)_c$ denote the scalar product in $L^2(Q_c)$ and

$$K = \{m = \{m_{ij}\} \mid \ m_{ij} \in L^2(\Omega), \ i,j = 1,2,$$
$$\Phi(m_{ij}(x)) \le 0 \ \text{ a.e. in } \Omega_c\}.$$

Assume that there exists a function $M = \{M_{ij}\} \in L^\infty(Q_c)$, $\dot{M}, \ddot{M} \in L^\infty(Q_c)$, satisfying the equation (5.211) in the following sense,

$$-(M_{ij}, \bar{v}_{,ij})_c = (f, \bar{v})_c \quad \forall \bar{v} \in L^2(0,T; H^{2,0}(\Omega_c)), \tag{5.217}$$

and, moreover, $(1+\kappa)M(t) \in K$, $t \in (0,T)$, $\kappa = \text{const} > 0$, $M(0) = \dot{M}(0) = 0$.

Now we can prove the existence of solutions to (5.211)–(5.216).

Theorem 5.7. *Let $f, \dot{f} \in L^2(Q_c)$, $f(0) = 0$, and the above assumption on M satisfying* (5.217) *hold. Then there exist functions $v, m = \{m_{ij}\}$ such that*

$$v \in L^2(0,T; H^{1,0}(\Omega_c)), \quad v_{,ij} \in L^2(0,T; M^1(\Omega_c)), \quad i,j = 1,2,$$

$$m,\ \dot{m} \in L^2(Q_c), \quad m(t) \in K, \quad t \in (0,T),$$

$$-(m_{ij}, \bar{v}_{,ij})_c = (f, \bar{v})_c \quad \forall \bar{v} \in L^2(0,T; H^{2,0}(\Omega_c)), \tag{5.218}$$

$$(\dot{m}_{ij}, \bar{m}_{ij} - m_{ij})_c + (v, \bar{m}_{ij,ij} - m_{ij,ij})_c \geq 0 \tag{5.219}$$

$$\forall \bar{m} \in L^2(0,T; U(\Omega_c)), \quad \bar{m}(t) \in K,\ t \in (0,T)$$

and the initial condition (5.216) *holds.*

PROOF. A scheme of reasoning is the same as that used in the previous subsection, and our attention now focuses on the details related to the nonsmoothness of the boundary.

Let ε, δ be positive parameters and $p : [L^2(\Omega_c)]^3 \to [L^2(\Omega_c)]^3$ be the penalty operator defined as before. In the domain Q_c, consider the following problem with the parameters ε, δ. We want to find functions $v, m = \{m_{ij}\}$, $i,j = 1,2$, such that

$$\varepsilon \dot{v} + \varepsilon \Delta^2 v - m_{ij,ij} = f, \tag{5.220}$$

$$\dot{m}_{ij} + \varepsilon m_{ij} + v_{,ij} + \frac{1}{\delta} p(m)_{ij} = 0, \quad i,j = 1,2, \tag{5.221}$$

$$v = 0, \quad (m_{ij} - \varepsilon v_{,ij}) n_j n_i = 0 \quad \text{on } \Gamma \times (0,T), \tag{5.222}$$

$$(m_{ij} - \varepsilon v_{,ij}) \nu_j \nu_i = 0, \quad R_\nu (m_{ij} - \varepsilon v_{,ij}) = 0 \quad \text{on } \Gamma_c^\pm \times (0,T), \tag{5.223}$$

$$v = 0, \quad m = 0, \quad t = 0. \tag{5.224}$$

Multiplying (5.220), (5.221) by $v, m_{ij} - M_{ij}$ we can argue as in the previous subsection where the estimate (5.190) is derived. This provides the uniform in the ε, δ inequality

$$\varepsilon \|v\|^2_{L^2(Q_c)} + \|m\|^2_{L^2(Q_c)} + \varepsilon \int_0^T \sum_{i,j=1,2}^{2} \|v_{,ij}\|^2_{0,c} \leq c. \tag{5.225}$$

In this section $\|\cdot\|_{s,c}$ denotes the norm in $H^s(\Omega_c)$.

The arguments used to prove Lemma 5.2 allow us to obtain the estimate

$$\sum_{i,j=1,2}^{2} \|v_{,ij}\|^2_{0,c} \geq c\, \|v\|^2_{2,c} \quad \forall v \in H^{2,0}(\Omega_c).$$

Hence, it follows from (5.225) that

$$\|m\|^2_{L^2(Q_c)} + \varepsilon \|v\|^2_{L^2(0,T; H^{2,0}(\Omega_c))} \leq c \tag{5.226}$$

with the constant c being uniform in $\varepsilon, \delta, \varepsilon \leq \varepsilon_0$.

Again, we have the initial conditions

$$\dot{v}(0) = 0, \quad \dot{m}(0) = 0.$$

This gives from (5.220)–(5.224) two more estimates

$$\varepsilon\|\dot{v}\|^2_{L^2(Q_c)} + \|\dot{m}\|^2_{L^2(Q_c)} \le c\,, \tag{5.227}$$

$$\|v\|_{L^2(0,T;H^{2,0}(\Omega_c))} \le c(\delta), \tag{5.228}$$

where the constant $c(\delta)$ depends on δ, in general.

Now we can use the Galerkin approach to prove the existence of solutions to (5.220)–(5.224). The solutions $v^{\varepsilon\delta}$, $m^{\varepsilon\delta} = \{m^{\varepsilon\delta}_{ij}\}$ exist in the following sense,

$$v^{\varepsilon\delta} \in L^2(0,T;H^{2,0}(\Omega_c)), \quad \dot{v}^{\varepsilon\delta}, m^{\varepsilon\delta}, \dot{m}^{\varepsilon\delta} \in L^2(Q_c), \tag{5.229}$$

$$\varepsilon(\dot{v}^{\varepsilon\delta},\bar{v})_c + \varepsilon(v^{\varepsilon\delta}_{,ij},\bar{v}_{,ij})_c - (m^{\varepsilon\delta}_{ij},\bar{v}_{,ij})_c = (f,\bar{v})_c \tag{5.230}$$

$$\forall \bar{v} \in L^2(0,T;H^{2,0}(\Omega_c)),$$

$$\dot{m}^{\varepsilon\delta}_{ij} + \varepsilon m^{\varepsilon\delta}_{ij} + v^{\varepsilon\delta}_{,ij} + \frac{1}{\delta}p(m^{\varepsilon\delta})_{ij} = 0, \quad i,j=1,2, \tag{5.231}$$

and the initial conditions (5.224) hold.

Let us pass on to the limit as $\varepsilon \to 0$ for any fixed $\delta > 0$. By the estimates (5.226)–(5.228), we can choose a subsequence denoted just as the sequence $v^{\varepsilon\delta}$, $m^{\varepsilon\delta}$ such that for any fixed $\delta > 0$ and $\varepsilon \to 0$

$$\varepsilon\dot{v}^{\varepsilon\delta} \to 0 \text{ weakly in } L^2(Q_c), \quad v^{\varepsilon\delta} \to v^\delta \text{ weakly in } L^2(0,T;H^{2,0}(\Omega_c)),$$

$$m^{\varepsilon\delta}_{ij}, \dot{m}^{\varepsilon\delta}_{ij} \to m^\delta_{ij}, \dot{m}^\delta_{ij} \quad \text{weakly in } L^2(Q_c), \quad i,j=1,2.$$

Then the relations (5.230)–(5.231) result in the following identity and equations:

$$-(m^\delta_{ij},\bar{v}_{,ij})_c = (f,\bar{v})_c \quad \forall \bar{v} \in L^2(0,T;H^{2,0}(\Omega_c)), \tag{5.232}$$

$$\dot{m}^\delta_{ij} + v^\delta_{,ij} + \frac{1}{\delta}p(m^\delta)_{ij} = 0, \quad i,j=1,2. \tag{5.233}$$

Similarly to (5.205) we can derive

$$\|v^\delta_{,ij}\|_{L^2(0,T;L^1(\Omega_c))} \le c, \quad i,j=1,2. \tag{5.234}$$

By Lemma 5.2, from (5.234) it follows that

$$\|v^\delta\|_{L^2(0,T;W^2_1(\Omega_c))} \le c \tag{5.235}$$

with the constant uniform in δ. It is known that for a bounded domain $D \subset R^2$ with the Lipschitz boundary the imbedding $W^1_1(D) \subset L^2(D)$ is continuous. Consequently, if subdomains Ω_1, Ω_2 are chosen as those in Lemma 5.2 we find for each $v \in W^2_1(\Omega_c)$,

$$\sum_{i=1}^2 \|v_{,i}\|_{W^1_1(\Omega_c)} = \sum_{i=1}^2 \|v_{,i}\|_{W^1_1(\Omega_1)} + \sum_{i=1}^2 \|v_{,i}\|_{W^1_1(\Omega_2)} \tag{5.236}$$

$$\geq c\Big(\sum_{i=1}^{2}\|v_{,i}\|_{L^2(\Omega_1)}+\sum_{i=1}^{2}\|v_{,i}\|_{L^2(\Omega_2)}\Big)\geq c\sum_{i=1}^{2}\|v_{,i}\|_{L^2(\Omega_c)}\,.$$

The same reasoning provides the inequality

$$\|v\|_{W_1^1(\Omega_c)}\geq c\,\|v\|_{L^2(\Omega_c)}. \tag{5.237}$$

By (5.236)–(5.237), the estimate (5.235) implies

$$\|v^\delta\|_{L^2(0,T;H^{1,0}(\Omega_c))}\leq c \tag{5.238}$$

and, besides, the inequality (5.234) gives

$$\|v^\delta_{,ij}\|_{L^2(0,T;M^1(\Omega_c))}\leq c,\quad i,j=1,2. \tag{5.239}$$

So, we have the following estimate,

$$\int_0^T\Big(\|m^\delta\|_{0,c}^2+\|\dot m^\delta\|_{0,c}^2+\|m^\delta_{ij,ij}\|_{0,c}^2+\|v^\delta\|_{1,c}^2+\sum_{i,j=1,2}\|v^\delta_{,ij}\|^2_{M^1(\Omega_c)}\Big)\leq c,$$

which allows us to choose a subsequence still denoted by v^δ, m^δ such that as $\delta\to 0$

$$\begin{aligned}
&v^\delta\ \to\ v\quad \text{weakly in}\ \ L^2(0,T;H^{1,0}(\Omega_c)),\\
&m^\delta_{ij},\ \dot m^\delta_{ij}\ \to\ m_{ij},\ \dot m_{ij}\quad \text{weakly in}\ \ L^2(Q_c),\quad i,j=1,2,\\
&v^\delta_{,ij}\ \to\ v_{,ij}\quad \star\text{-weakly in } L^2(0,T;M^1(\Omega_c)),\quad i,j=1,2,\\
&m^\delta(T)\ \to\ m(T)\ \text{ weakly in }\ L^2(\Omega_c),\\
&m^\delta_{ij,ij}\ \to\ m_{ij,ij}\ \text{ weakly in }\ L^2(Q_c).
\end{aligned}$$

Here, the last line is a corollary of the equation

$$-m^\delta_{ij,ij}=f \tag{5.240}$$

holding in Q_c in the sense of distributions.

The required identity (5.218) is readily derived from (5.232). The equations (5.233) give

$$(\dot m^\delta_{ij},\bar m_{ij}-m_{ij})_c+(v^\delta_{,ij},\bar m_{ij}-m^\delta_{ij})_c\geq 0 \tag{5.241}$$

$$\forall \bar m\in L^2(0,T;U(\Omega_c)),\quad \bar m(t)\in K, t\in(0,T).$$

For each $\bar m\in L^2(0,T;U(\Omega_c))$ we have

$$(v^\delta_{,ij},\bar m_{ij})_c=(v^\delta,\bar m_{ij,ij})_c. \tag{5.242}$$

By (5.240), the equation

$$-m_{ij,ij} = f \tag{5.243}$$

holds in Q_c. Consequently, in view of (5.232), (5.233) the following equalities hold:

$$-(m_{ij}^{\delta}, v_{,ij}^{\delta})_c = (f, v^{\delta})_c = -(m_{ij,ij}, v^{\delta})_c\,. \tag{5.244}$$

Relations (5.242), (5.244) allow us to rewrite (5.241) in the form

$$(\dot{m}_{ij}^{\delta}, \bar{m}_{ij} - m_{ij}^{\delta})_c + (v^{\delta}, \bar{m}_{ij,ij} - m_{ij,ij})_c \geq 0 \tag{5.245}$$

$$\forall \bar{m} \in L^2(0,T;U(\Omega_c)), \quad \bar{m}(t) \in K, \quad t \in (0,T).$$

Passage to the limit as $\delta \to 0$ can be fulfilled in (5.245) as in (5.209). As a result we easily arrive at (5.219). In a regular way the inclusion $m(t) \in K$, $t \in (0,T)$, follows.

The second boundary condition (5.214) and the conditions (5.215) are involved in (5.218). This means that those conditions hold at any point of Γ, $\Gamma_c^{\pm}$, respectively, provided the solution v, m_{ij} is smooth enough. The statement can be verified by integrating by parts. Theorem 5.7 is proved.

5.5 Contact elastoplastic problem for the curvilinear Kirchhoff rod

5.5.1 Formulation of the problem

Let the rod axis be described by the functions $x = a(\alpha)$, $y = b(\alpha)$, $\alpha \in I$, $I = (0,1)$, and the rigid punch surface be defined as $\psi(x,y) = 0$ (see Fig.5.1). We assume that a bounded domain $G \subset R^2$ is chosen such that the curve $x = a(\alpha)$, $y = b(\alpha)$, $\alpha \in I$, belongs to G. Moreover, we assume that the surface line $\psi(x,y) = 0$ is smooth and divides G into two subdomains with the properties $\psi(x,y) > 0$ for the first part and $\psi(x,y) < 0$ for the second one. Let $v = v(\alpha)$, $w = w(\alpha)$ be tangential and normal displacements of the rod, respectively, $m = m(\alpha)$ be the bending moment, and $n = n(\alpha)$ be the stress integrated across the rod.

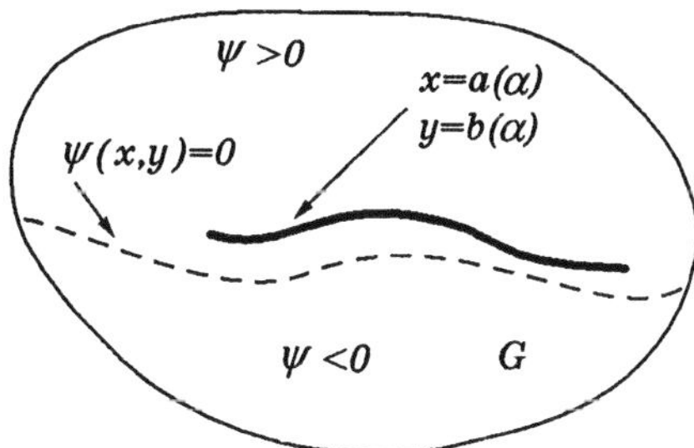

Fig.5.1. Curvilinear rod and the punch shape

A linear approximation for the nonpenetration condition between the rod and rigid punch can be written as follows:

$$\psi(a(\alpha), b(\alpha)) + \nabla\psi(a(\alpha), b(\alpha)) \cdot \Pi \begin{pmatrix} v(\alpha) \\ w(\alpha) \end{pmatrix} \geq 0, \tag{5.246}$$

$$\nabla\psi = (\psi_x, \psi_y), \quad \Pi = \frac{1}{\Lambda}\begin{pmatrix} a_\alpha & -b_\alpha \\ b_\alpha & a_\alpha \end{pmatrix}, \quad \Lambda = \sqrt{a_\alpha^2 + b_\alpha^2}.$$

On the segment I we want to find the functions v, w, m, n, ξ, η, which satisfy the inequality (5.246) and the following relations:

$$(n_\alpha + km_\alpha + g\Lambda)(\bar{v} - v) + \left(\left(\frac{1}{\Lambda}m_\alpha\right)_\alpha - k\Lambda n + f\Lambda\right)(\bar{w} - w) \leq 0, \tag{5.247}$$

$$v_\alpha + k\Lambda w = c_1\Lambda n + \xi, \tag{5.248}$$

$$(kv)_\alpha - \left(\frac{1}{\Lambda}w_\alpha\right)_\alpha = c_2\Lambda m + \eta, \tag{5.249}$$

$$\Phi(n, m) \leq 0, \tag{5.250}$$

$$\xi(\bar{n} - n) + \eta(\bar{m} - m) \leq 0 \quad \forall(\bar{n}, \bar{m}), \quad \Phi(\bar{n}, \bar{m}) \leq 0, \tag{5.251}$$

$$v = w = w_\alpha = 0, \quad \alpha = 0, 1. \tag{5.252}$$

The inequality (5.247) should be fulfilled for all functions $(\bar{v}, \bar{w})$ satisfying the restriction (5.246) and the boundary conditions (5.252). The function $\Phi : R^2 \to R$ describes the plastic yield condition. We assume Φ to be convex and continuous. The relations (5.248), (5.249) give the representations of the mid-axis strains of the rod and the curvature variations as the sum of elastic and plastic parts; inequality (5.251) defines the direction of a plastic strain vector (ξ, η) with respect to the yield surface $\Phi(n, m) = 0$. The functions g, f are the densities of exterior forces; c_1, c_2 are positive constants; k is the curvature of the line $x = a(\alpha)$, $y = b(\alpha)$.

Note that from (5.247) we obtain two equilibrium equations

$$n_\alpha + km_\alpha + g\Lambda = 0, \quad \left(\frac{1}{\Lambda}m_\alpha\right)_\alpha - k\Lambda n + f\Lambda = 0$$

provided that there is no a contact between the punch and the rod (i.e. the solution u, w satisfies a strict inequality (5.246)). Moreover, (5.248), (5.249) imply the elastic constitutive law provided that there are no plastic deformations, i.e. $\xi = \eta = 0$ (see the elastic shell model in Section 1.1.3).

In the sequel the values Λg, Λf, Λk, $c_1\Lambda$, $c_2\Lambda$ are denoted by g_1, f_1, k_1, Λ_1, Λ_2, respectively. Assume that the set

$$\{(n, m) \in R^2 \mid \Phi(n, m) \leq 0\}$$

contains zero as its interior point, and $g_1, f_1 \in L^2(I)$, $k \in H^1(I)$, $\Lambda_i \geq c_0$, $i = 1,2$, c_0 being a positive constant, $\psi \in H^2(G)$, $a, b \in H^2(I)$, $\nabla\psi(a(\alpha), b(\alpha)) \neq 0$. Moreover we assume that there exists a solution n^0, m^0 of the equations

$$n^0_\alpha + km^0_\alpha + g_1 = 0, \quad \left(\frac{1}{\Lambda} m^0_\alpha\right)_\alpha - k_1 n^0 + f_1 = 0, \tag{5.253}$$

satisfying the inclusion $(1+\kappa)(n^0, m^0) \in K$, where

$$K = \{(n,m) \in L^2(I) \mid \quad \Phi(n(x), m(x)) \leq 0 \quad \text{a.e. in } I\},$$

$\kappa > 0$ is a constant. The set of functions $(v, w) \in H^1_0(I) \times H^2_0(I)$ satisfying the inequality (5.246) is denoted by B. Here $H^s_0(I)$ is the completion of the space $C^\infty_0(I)$ in the norm of $H^s(I)$, $s = 1, 2$. The results of Sections 5.5–5.8 can be partly found in (Khludnev, 1992, 1993a, 1993b).

5.5.2 Existence of the solution

Assuming a sufficient regularity of the solution to (5.247)–(5.252), we can deduce relations considered as a corollary from the exact formulation of the problem. In what follows the theorem of existence of these relations is established. Substituting the values ξ, η from (5.248), (5.249) in (5.251) and summing the resulting inequality with (5.247), we obtain, after integration over I,

$$\Phi(n,m) \leq 0, \quad \psi(a(\alpha), b(\alpha)) + \nabla\psi \cdot \Pi \begin{pmatrix} v(\alpha) \\ w(\alpha) \end{pmatrix} \geq 0,$$

$$\begin{aligned} &\langle n, \bar{v}_\alpha \rangle + \langle m, (k\bar{v})_\alpha \rangle - \langle g_1, \bar{v} - v \rangle - \langle m, \left(\frac{1}{\Lambda}\bar{w}_\alpha\right)_\alpha \rangle \\ &+ \langle k_1 n, \bar{w} \rangle - \langle f_1, \bar{w} - w \rangle + \langle \Lambda_1 n, \bar{n} - n \rangle - \langle k_1 w, \bar{n} \rangle \\ &+ \langle v, \bar{n}_\alpha \rangle + \langle \Lambda_2 m, \bar{m} - m \rangle + \langle kv, \bar{m}_\alpha \rangle + \langle w, \left(\frac{1}{\Lambda}\bar{m}_\alpha\right)_\alpha \rangle \geq 0. \end{aligned} \tag{5.254}$$

This inequality should be fulfilled for all sufficiently smooth functions $(\bar{n}, \bar{m})$, $\Phi(\bar{n}, \bar{m}) \leq 0$ and $(\bar{v}, \bar{w}) \in B$. In the sequel the space of bounded measures on I is denoted by $M^1(I)$.

Theorem 5.8. *Let the above assumptions be fulfilled and at least one function* $(v_0, w_0) \in B$ *can be found. Then, there exists a solution of the variational inequality* (5.254) *such that* $n, m, v \in L^2(I)$, $w \in H^1_0(I)$, v_α, $(\Lambda^{-1}w_\alpha)_\alpha \in M^1(I)$.

PROOF. We introduce the penalty operator $q : [L^2(I)]^2 \to [L^2(I)]^2$ connected with the set of functions (v, w) satisfying (5.62). Let π be the operator of the orthogonal projection of the space $\left[L^2(I)\right]^2$ on the set K.

Assume that $p(n,m) = (n,m) - \pi(n,m)$. Consider the auxiliary boundary problem containing three positive parameters $\varepsilon, \delta, \lambda$ without stating the dependence of the solution on these parameters:

$$-\varepsilon v_{\alpha\alpha} - n_\alpha - km_\alpha + \frac{1}{\lambda} q_1(v,w) = g_1, \tag{5.255}$$

$$\varepsilon w_{\alpha\alpha\alpha\alpha} - \left(\frac{1}{\Lambda} m_\alpha\right)_\alpha + k_1 n + \frac{1}{\lambda} q_2(vw) = f_1, \tag{5.256}$$

$$\Lambda_1 n - v_\alpha - k_1 w + \frac{1}{\delta} p_1(n,m) = 0, \tag{5.257}$$

$$\Lambda_2 m + \left(\frac{1}{\Lambda} w_\alpha\right)_\alpha - (kv)_\alpha + \frac{1}{\delta} p_2(n,m) = 0, \tag{5.258}$$

$$v = w = w_\alpha = 0, \quad \alpha = 0, 1. \tag{5.259}$$

Here, the components of the penalty operators are denoted by q_i, p_i, $i = 1, 2$. To obtain a priori estimates, we multiply (5.255)–(5.258) by $v - v^0$, $w - w^0$, $n - n^0$, $m - m^0$, respectively, and integrate over I. Using the monotonicity of the operator q and the equations (5.253), we arrive at the inequality

$$\varepsilon\|v_\alpha\|^2 + \varepsilon\|w_{\alpha\alpha}\|^2 + \int_0^1 (\Lambda_1 n^2 + \Lambda_2 m^2) d\alpha + \frac{1}{\delta}\langle p(n,m), (n,m) - (n^0, m^0)\rangle$$

$$\le -\langle g_1, v^0\rangle - \langle f_1, w^0\rangle + \langle n, v^0_\alpha\rangle - \langle m, \left(\frac{1}{\Lambda} w^0_\alpha\right)_\alpha\rangle + \langle \Lambda_1 n, n^0\rangle \tag{5.260}$$

$$+ \langle \Lambda_2 m, m^0\rangle + \varepsilon\langle v_\alpha, v^0_\alpha\rangle + \varepsilon\langle w_{\alpha\alpha}, w^0_{\alpha\alpha}\rangle + \langle k_1 n, w^0\rangle + \langle m, (kv^0)_\alpha\rangle.$$

It is easy to see that simple arguments allow us to deduce from (5.260) the estimate

$$\varepsilon\|v_\alpha\|^2 + \varepsilon\|w_{\alpha\alpha}\|^2 + \|n\|^2 + \|m\|^2 \le c$$

uniform in $\varepsilon \le \varepsilon_0$, δ, λ. This estimate and the properties of the operators p, q admit the use of Theorem 1.14, so that the solution of (5.255)–(5.259) exists for all fixed ε, δ, λ. In addition, we obtain from (5.260) one more estimate

$$\frac{1}{\delta}\langle p_1, n - n^0\rangle + \frac{1}{\delta}\langle p_2, m - m^0\rangle \le c \tag{5.261}$$

with the constant independent of $\varepsilon \le \varepsilon_0$, δ, λ. From here, it follows that $\delta^{-1} p_1$, $\delta^{-1} p_2$ are bounded in the space $L^1(I)$. Indeed, let us introduce the functional $H(n,m) = (2\delta)^{-1}\|(n,m) - \pi(n,m)\|^2$. Its derivative at the point (n,m) is calculated by the formula $H'(n,m) = \delta^{-1}(p_1(n,m), p_2(n,m))$ (see Khludnev, Sokolowski, 1997). It follows from the assumption on the solution (n^0, m^0) that $(n^0, m^0) + (\bar{n}, \bar{m}) \in K$ provided that $\|(\bar{n}, \bar{m})\|_{L^\infty(I)} \le \gamma$ and

$\gamma > 0$ is sufficiently small. The convexity of the functional H provides the inequality

$$\langle H'(n,m),(\bar{n},\bar{m})\rangle \le \langle H'(n,m),(n,m)-(n^0,m^0)\rangle$$

$$+H((n^0,m^0)+(\bar{n},\bar{m}))-H(n,m).$$

According to the above arguments, we have $H((n^0,m^0)+(\bar{n},\bar{m}))=0$. Besides, by (5.261), the right-hand side of the obtained inequality is bounded from above and nonnegative. Therefore

$$\frac{1}{\delta}\|p_1\|_{L^1(I)}+\frac{1}{\delta}\|p_2\|_{L^1(I)}\le c$$

uniformly in $\varepsilon \le \varepsilon_0$, δ, λ. This estimate is among the major ones, and on its basis it is possible to obtain the estimates for v, w. Namely, from (5.257), (5.258) and the boundedness of n, m in $L^2(I)$, it follows that $v_\alpha + k_1 w$, $(\Lambda^{-1}w_\alpha)_\alpha - (kv)_\alpha$ are bounded in $L^1(I)$ uniformly in $\varepsilon \le \varepsilon_0$, δ, λ. Following these considerations, we can show that v are bounded in $L^1(I)$ and w are bounded in $H_0^1(I)$. Actually, (5.257), (5.258) can be written as follows,

$$v_\alpha + k_1 w = b_1, \quad \left(\frac{1}{\Lambda}w_\alpha - kv\right)_\alpha = b_2, \tag{5.262}$$

where b_1, b_2 are the functions bounded in $L^1(I)$ uniformly in the parameters. From the first equation of (5.262) one has

$$v(\alpha) = -\int_0^\alpha (k_1 w - b_1)d\alpha,$$

and from the second equation we obtain

$$\frac{1}{\Lambda}w_\alpha - kv = b_3, \tag{5.263}$$

where b_3 are bounded in $W_1^1(I)$.

Substituting the values of $v(\alpha)$ in (5.263), we have

$$w_\alpha + k_1\int_0^\alpha (k_1 w - b_1)d\alpha = b_3\Lambda.$$

Multiplying this equation by w and integrating, we obtain that w are bounded in $L^2(I)$. Consequently, v are bounded in $H_0^1(I)$. Thus, the following a priori estimate holds,

$$\|v\| + \|w\|_1 + \|n\| + \|m\| \le c, \tag{5.264}$$

which is uniform in $\varepsilon \le \varepsilon_0$, δ, λ. Besides, by the continuous imbedding $L^1(I) \subset M^1(I)$ and (5.262), we have one more estimate,

$$\|v_\alpha\|_{M^1(I)} + \|\left(\frac{1}{\Lambda}w_\alpha\right)_\alpha\|_{M^1(I)} \le c. \tag{5.265}$$

Now, it is possible to realize a passage to the limit as follows: $\varepsilon \to 0$, $\delta \to 0$, $\lambda \to 0$. Every time, when justifying the next passage to the limit, the solution is to be supplied with the appropriate index without mentioning the dependence on the other parameters. Denote the solution of the problem (5.255)–(5.259) by v^ε, w^ε, n^ε, m^ε. It follows from the estimates (5.264), (5.265) and the boundedness of $\sqrt{\varepsilon}v_\alpha$, $\sqrt{\varepsilon}w_{\alpha\alpha}$ in $L^2(I)$ that a subsequence can be chosen (denoted as previously) such that as $\varepsilon \to 0$

$$n^\varepsilon, m^\varepsilon \to n^\delta, m^\delta \quad \text{weakly in } L^2(I), \quad v^\varepsilon \to v^\delta \quad \text{strongly in } L^2(I),$$

$$\varepsilon v^\varepsilon \to 0 \quad \text{strongly in } H_0^1(I), \quad \varepsilon w^\varepsilon \to 0 \quad \text{strongly in } H_0^2(I),$$

$$w^\varepsilon \to w^\delta \quad \text{weakly in } H_0^1(I), \quad \text{strongly in } L^2(I),$$

$$v_\alpha^\varepsilon \to v_\alpha^\delta, \quad \left(\frac{1}{\Lambda}w_\alpha^\varepsilon\right)_\alpha \to \left(\frac{1}{\Lambda}w_\alpha^\delta\right)_\alpha \quad \star\text{- weakly in } M^1(I).$$

Here, the strong convergence of v^ε is a consequence of the compact imbedding of the space $\{u \in L^2(I),\ u_\alpha \in M^1(I)\}$ in $L^2(I)$ (see Giusti, 1984). It should be noted that the solution v^ε, w^ε, n^ε, m^ε satisfy the equations (5.255)–(5.258) in the sense of identities

$$\langle \varepsilon v_\alpha^\varepsilon + n^\varepsilon, \bar{v}_\alpha\rangle + \langle m^\varepsilon, (k\bar{v})_\alpha\rangle + \langle \frac{1}{\lambda}q_1(v^\varepsilon, w^\varepsilon) - g_1, \bar{v}\rangle = 0,$$

$$\varepsilon\langle w_{\alpha\alpha}^\varepsilon, \bar{w}_{\alpha\alpha}\rangle - \langle m^\varepsilon, \left(\frac{1}{\Lambda}\bar{w}_\alpha\right)_\alpha\rangle + \langle k_1 n + \frac{1}{\lambda}q_2(v^\varepsilon, w^\varepsilon) - f_1, \bar{w}\rangle = 0,$$

$$\langle v^\varepsilon, \bar{n}_\alpha\rangle + \langle \Lambda_1 n^\varepsilon - k_1 w^\varepsilon + \frac{1}{\delta}p_1(n^\varepsilon, m^\varepsilon), \bar{n}\rangle = 0,$$

$$\langle w^\varepsilon, \left(\frac{1}{\Lambda}\bar{m}_\alpha\right)_\alpha\rangle + \langle kv^\varepsilon, \bar{m}_\alpha\rangle + \langle \Lambda_2 m^\varepsilon + \frac{1}{\delta}p_2(n^\varepsilon, m^\varepsilon), \bar{m}\rangle = 0.$$

These identities are fulfilled for all functions

$$\bar{v} \in H_0^1(I), \quad \bar{w} \in H_0^2(I), \quad \bar{n} \in H^1(I), \quad \bar{m} \in H^2(I).$$

Besides, it can be assumed that

$$q(v^\varepsilon, w^\varepsilon) \to q(v^\delta, w^\delta) \quad \text{strongly in } L^2(I),$$

$$p(n^\varepsilon, m^\varepsilon) \to \nu^\delta \quad \text{weakly in } L^2(I).$$

It is clear that the above convergence allows us to justify a passage to the limit as $\varepsilon \to 0$ in the identities. As a result, we arrive at the relations

$$\begin{gathered}\langle n^\delta, \bar{v}_\alpha\rangle + \langle m^\delta, (k\bar{v})_\alpha\rangle + \langle \frac{1}{\lambda} q_1(v^\delta, w^\delta) - g_1, \bar{v}\rangle = 0,\\ -\langle m^\delta, \left(\frac{1}{\Lambda}\bar{w}_\alpha\right)_\alpha\rangle + \langle k_1 n^\delta + \frac{1}{\lambda} q_2(v^\delta, w^\delta) - f_1, \bar{w}\rangle = 0,\\ \langle v^\delta, \bar{n}_\alpha\rangle + \langle \Lambda_1 n^\delta - k_1 w^\delta + \frac{1}{\delta} p_1(n^\delta, m^\delta), \bar{n}\rangle = 0,\\ \langle w^\delta, \left(\frac{1}{\Lambda}\bar{m}_\alpha\right)_\alpha\rangle + \langle k v^\delta, \bar{m}_\alpha\rangle + \langle \Lambda_2 m^\delta + \frac{1}{\delta} p_2(n^\delta, m^\delta), \bar{m}\rangle = 0.\end{gathered} \tag{5.266}$$

These identities hold for the functions $\bar{v}$, $\bar{w}$, $\bar{n}$, $\bar{m}$ from the same spaces as above. The equality $\nu^\delta = p(n^\delta, m^\delta)$ is justified by the monotonicity arguments.

Now let us pass to the limit as $\delta \to 0$. From the sequence v^δ, w^δ, n^δ, m^δ, we choose a subsequence possessing the following properties:

$$n^\delta, m^\delta \to n^\lambda, m^\lambda \quad \text{weakly in } L^2(I),$$

$$w^\delta \to w^\lambda \quad \text{weakly in } H_0^1(I), \quad \text{strongly in } L^2(I),$$

$$v^\delta \to v^\lambda \text{ strongly in } L^2(I), \quad q(v^\delta, w^\delta) \to q(v^\lambda, w^\lambda) \text{ strongly in } L^2(I),$$

$$v_\alpha^\delta \to v_\alpha^\lambda, \quad \left(\frac{1}{\Lambda} w_\alpha^\delta\right)_\alpha \to \left(\frac{1}{\Lambda} w_\alpha^\lambda\right)_\alpha \quad \star\text{-weakly in } M^1(I).$$

From the first two identities (5.266), we have

$$\begin{gathered}\langle n^\lambda, \bar{v}_\alpha\rangle + \langle m^\lambda, (k\bar{v})_\alpha\rangle + \langle \frac{1}{\lambda} q_1(v^\lambda, w^\lambda) - g_1, \bar{v}\rangle = 0,\\ -\langle m^\lambda, \left(\frac{1}{\Lambda}\bar{w}_\alpha\right)_\alpha\rangle + \langle k_1 n^\lambda + \frac{1}{\lambda} q_2(v^\lambda, w^\lambda) - f_1, \bar{w}\rangle = 0.\end{gathered} \tag{5.267}$$

Substituting $\bar{n}-n^\delta$, $\bar{m}-m^\delta$ as the test functions into the third and the fourth identities of (5.266), respectively, where $(\bar{n}, \bar{m}) \in$K, $\bar{n} \in H^1(I)$, $\bar{m} \in H^2(I)$, and taking into account the monotonicity of p, we arrive at the following relation:

$$\begin{gathered}\langle v^\delta, \bar{n}_\alpha - n_\alpha^\delta\rangle + \langle \Lambda_1 n^\delta - k_1 w^\delta, \bar{n} - n^\delta\rangle + \langle \Lambda_2 m^\delta, \bar{m} - m^\delta\rangle\\ + \langle w^\delta, \left(\frac{1}{\Lambda}(\bar{m}_\alpha - m_\alpha^\delta)\right)_\alpha\rangle + \langle k v^\delta, \bar{m}_\alpha - m_\alpha^\delta\rangle \ge 0.\end{gathered} \tag{5.268}$$

The expression $\langle v^\delta, n_\alpha^\delta\rangle + \langle k v^\delta, m_\alpha^\delta\rangle$ appearing in (5.268) makes sense, and, moreover, it is possible to pass to the limit in the written terms as $\delta \to 0$. Actually, these terms can be written as $\langle v^\delta, n_\alpha^\delta + k m_\alpha^\delta\rangle$. At the same time, it follows from the first equation of (5.266) that the function $n^\delta_\alpha + km^\delta_\alpha \equiv n^\delta_\alpha + (km^\delta)_\alpha - k_\alpha m^\delta$ is bounded in the space $L^2(I)$ (generally speaking, nonuniformly in λ). Consequently, a subsequence can be chosen such that as $\delta \to 0$

$$n^\delta_\alpha + km^\delta_\alpha \;\to\; n^\delta_\alpha + km^\lambda_\alpha \quad \text{weakly in } L^2(I).$$

Since v^δ converge to v^λ strongly in $L^2(I)$, the above arguments prove the possibility of passage to the limit in the term $\langle v^\delta, n^\delta_\alpha + km^\delta_\alpha\rangle$. A similar justification can be used to prove the convergence in the term $\langle w^\delta, (\Lambda^{-1}m^\delta_\alpha)_\alpha\rangle$ of (5.268). In fact, it follows from the second equation of (5.266) that $(\Lambda^{-1}m^\delta_\alpha)_\alpha$ is bounded in $L^2(I)$ (generally speaking, nonuniformly in λ). Therefore, one can assume that as $\delta \to 0$

$$\left(\frac{1}{\Lambda}m^\delta_\alpha\right)_\alpha \;\to\; \left(\frac{1}{\Lambda}m^\lambda_\alpha\right)_\alpha \qquad \text{weakly in } L^2(I).$$

In view of the above convergence of w^δ we obtain the desired property

$$\langle w^\delta, \left(\frac{1}{\Lambda}m^\delta_\alpha\right)_\alpha\rangle \;\to\; \langle w^\lambda, \left(\frac{1}{\Lambda}m^\lambda_\alpha\right)_\alpha\rangle.$$

It is clear that

$$\liminf_{\delta\to 0}\langle \Lambda_1 n^\delta, n^\delta\rangle \ge \langle \Lambda_1 n^\lambda, n^\lambda\rangle, \qquad \liminf_{\delta\to 0}\langle \Lambda_2 m^\delta, m^\delta\rangle \ge \langle \Lambda_2 m^\lambda, m^\lambda\rangle.$$

Thus, from (5.268) we can derive that

$$\langle v^\lambda, \bar{n}_\alpha - n^\lambda_\alpha\rangle + \langle \Lambda_1 n^\lambda - k_1 w^\lambda, \bar{n} - n^\lambda\rangle + \langle \Lambda_2 m^\lambda, \bar{m} - m^\lambda\rangle \tag{5.269}$$
$$+\langle w^\lambda, \left(\frac{1}{\Lambda}(\bar{m}_\alpha - m^\lambda_\alpha)\right)_\alpha\rangle + \langle kv^\lambda, \bar{m}_\alpha - m^\lambda_\alpha\rangle \ge 0.$$

In addition, we have the inclusion $(n^\lambda, m^\lambda) \in K$ which can be proved from the two last equations of (5.266) in a standard way.

In conclusion we shall pass to the limit as $\lambda \to 0$. Let a subsequence v^λ, w^λ, n^λ, m^λ possess the property

$$n^\lambda, m^\lambda \;\to\; n, m \quad \text{weakly in } L^2(I), \qquad w^\lambda \;\to\; w \quad \text{weakly in } H^1_0(I),$$
$$v^\lambda \;\to\; v \quad \text{strongly in } L^2(I),$$
$$v^\lambda_\alpha \;\to\; v_\alpha, \quad \left(\frac{1}{\Lambda}w^\lambda_\alpha\right)_\alpha \;\to\; \left(\frac{1}{\Lambda}w_\alpha\right)_\alpha \quad \star\text{-weakly in } M^1(I).$$

We take $(\bar{v}, \bar{w}) \in B$ and substitute $\bar{v} - v^\lambda$, $\bar{w} - w^\lambda$ as test functions in (5.267). The terms containing the derivatives of n^λ, m^λ in the first equation of (5.267) are to be written down as follows, $\langle n^\lambda_\alpha + km^\lambda_\alpha, \bar{v} - v^\lambda\rangle$. As it was mentioned, the functions $n^\lambda_\alpha + km^\lambda_\alpha$ belong to the space $L^2(I)$, i.e. the scalar product makes sense. Analogously, in the second equation of (5.267) after substitution of $\bar{w}-w^\lambda$ as a test function, the term containing the derivative of m^λ can be written as $\langle(\Lambda^{-1}m^\lambda_\alpha)_\alpha, \bar{w}-w^\lambda\rangle$. Summing the obtained relations with (5.269) and using the monotonicity of q, we therefore obtain

$$\langle n^\lambda, \bar{v}_\alpha\rangle + \langle m^\lambda, (k\bar{v})_\alpha\rangle - \langle g_1, \bar{v}-v^\lambda\rangle + \langle k_1 n^\lambda, \bar{w}\rangle - \langle m^\lambda, \left(\frac{1}{\Lambda}\bar{w}_\alpha\right)_\alpha\rangle$$

$$-\langle f_1, \bar{w}-w^\lambda\rangle + \langle \Lambda_1 n^\lambda, \bar{n}-n^\lambda\rangle - \langle k_1 w^\lambda, \bar{n}\rangle + \langle v^\lambda, \bar{n}_\alpha\rangle$$

$$+\langle \Lambda_2 m^\lambda, \bar{m}-m^\lambda\rangle + \langle w^\lambda, \left(\frac{1}{\Lambda}\bar{m}_\alpha\right)_\alpha\rangle + \langle kv^\lambda, \bar{m}_\alpha\rangle \geq 0.$$

If the term $\langle\Lambda_1 n^\lambda, n^\lambda\rangle + \langle\Lambda_2 m^\lambda, m^\lambda\rangle$ is transferred into the right-hand side of this inequality and we pass to the lower limit, the inequality (5.251) follows.

Let us show that the functions v, w satisfy the restriction (5.246). Actually, $q(v^\lambda, w^\lambda) \to q(v, w)$ strongly in $L^2(I)$. Meanwhile it follows from (5.267) that $q(v^\lambda, w^\lambda) \to 0$ strongly in $H^{-1}(I) \times H^{-2}(I)$, which proves the assertion. Theorem 5.8 is completely proved.

5.5.3 The perfectly plastic problem

Let us consider the case of contact between a perfectly plastic rod and a rigid punch. This corresponds to the case when $c_1 = c_2 = 0$ in (5.247)–(5.252) or, equivalently, when $\Lambda_1 = \Lambda_2 = 0$.

Theorem 5.9. *Let all the assumptions of Theorem* 5.8 *be fulfilled and also the set* $\{(n, m) \in R^2 \mid \Phi(n, m) \leq 0\}$ *be bounded in* R^2. *Then there exists a solution of the variational inequality* (5.254) *when* $\Lambda_1 = \Lambda_2 = 0$.

PROOF. Consider the auxiliary problem (5.255)–(5.259) with $\Lambda_1 = \Lambda_2 = 0$. The inequality (5.260) can be written as follows:

$$\varepsilon\|v_\alpha\|^2 + \varepsilon\|w_{\alpha\alpha}\|^2 + \frac{1}{\delta}(\|n\|^2 + \|m\|^2)$$

$$\leq \frac{1}{\delta}\langle\pi(n,m), (n,m) - (n^0, m^0)\rangle + \frac{1}{\delta}\langle(n,m), (n^0, m^0)\rangle - \langle g_1, v^0\rangle - \langle f_1, w^0\rangle$$

$$+\langle n, v^0_\alpha\rangle - \langle m, \left(\frac{1}{\Lambda}w^0_\alpha\right)_\alpha\rangle + \varepsilon\langle v_\alpha, v^0_\alpha\rangle + \varepsilon\langle w_{\alpha\alpha}, w^0_{\alpha\alpha}\rangle$$

$$+\langle k_1 n, w^0\rangle + \langle m, (kv^0)_\alpha\rangle.$$

It is obvious that in the case under consideration the elements $\pi(n, m)$ are bounded in $L^2(I)$. Therefore, from the obtained inequality it follows that

$$\varepsilon\|v_\alpha\|^2 + \varepsilon\|w_\alpha\|^2 + \|n\|^2 + \|m\|^2 \leq c \tag{5.270}$$

uniformly in $\varepsilon \leq \varepsilon_0$, δ, λ. The estimate (5.270) provides the solvability of the problem (5.255)–(5.259) for fixed ε, δ, λ. Besides, it is possible to obtain estimates (5.264)–(5.265) uniformly in $\varepsilon \le \varepsilon_0$, $\delta \le \delta_0$, λ. We then pass to the limit as $\varepsilon \to 0$, $\delta \to 0$, $\lambda \to 0$ following the lines of those of Theorem 5.8. As a result we conclude that the limiting functions v, w, n, m satisfy the relations

$$(n,m) \in \mathrm{K}, \quad (v,w) \in L^2(I) \times H_0^1(I), \quad v_\alpha, \left(\frac{1}{\Lambda} w_\alpha\right)_\alpha \in M^1(I),$$

$$\psi(a(\alpha), b(\alpha)) + \nabla\psi \cdot \Pi \begin{pmatrix} v(\alpha) \\ w(\alpha) \end{pmatrix} \ge 0,$$

$$\langle n, \bar{v}_\alpha\rangle + \langle m, (k\bar{v})_\alpha\rangle - \langle g_1, \bar{v} - v\rangle + \langle k_1 n, \bar{w}\rangle - \langle m, \left(\frac{1}{\Lambda}\bar{w}_\alpha\right)_\alpha\rangle \tag{5.271}$$

$$- \langle f_1, \bar{w} - w\rangle - \langle k_1 w, \bar{n}\rangle + \langle v, \bar{n}_\alpha\rangle + \langle w, \left(\frac{1}{\Lambda}\bar{m}_\alpha\right)_\alpha\rangle + \langle kv, \bar{m}_\alpha\rangle \ge 0$$

$$\forall(\bar{v},\bar{w}) \in B, \quad \forall(\bar{n},\bar{m}) \in K, \quad \bar{n} \in H^1(I), \quad \bar{m} \in H^2(I).$$

This completes the proof of Theorem 5.9.

After all, we can get the answer to the question on the convergence for solutions of the elastoplastic problem to a solution of the perfectly plastic problem provided that the assumptions of Theorem 5.9 are fulfilled. To this end, consider the problem (5.255)–(5.259), where $\mu\Lambda_1$ and $\mu\Lambda_2$ are used instead of Λ_1 and Λ_2, $\mu > 0$. A priori estimates are of the same form as in the case of the perfectly plastic problem. It is important to note that the estimates are uniform not only in $\varepsilon \le \varepsilon_0$, $\delta \le \delta_0$, λ, but also in $\mu \le \mu_0$. For each fixed μ it is possible to pass to the limit as $\varepsilon \to 0$, $\delta \to 0$, $\lambda \to 0$ in (5.255)–(5.259) and to derive, therefore, that the solution v^μ, w^μ, n^μ, m^μ satisfies the inequality

$$\mu\langle\Lambda_1 n^\mu, \bar{n} - n^\mu\rangle - \langle k_1 w^\mu, \bar{n}\rangle + \langle m^\mu, (k\bar{v})_\alpha\rangle + \mu\langle\Lambda_2 m^\mu, \bar{m} - m^\mu\rangle$$

$$+ \langle n^\mu, \bar{v}_\alpha\rangle - \langle g_1, \bar{v} - v^\mu\rangle - \langle m^\mu, \left(\frac{1}{\Lambda}\bar{w}_\alpha\right)_\alpha\rangle + \langle k_1 n^\mu, \bar{w}\rangle \tag{5.272}$$

$$- \langle f_1, \bar{w} - w^\mu\rangle + \langle v^\mu, \bar{n}_\alpha\rangle + \langle w^\mu, \left(\frac{1}{\Lambda}\bar{m}_\alpha\right)_\alpha\rangle + \langle kv^\mu, \bar{m}_\alpha\rangle \ge 0$$

$$\forall(\bar{n},\bar{m}) \in K, \quad \bar{n} \in H^1(I), \quad \bar{m} \in H^2(I), \quad \forall(\bar{v},\bar{w}) \in B.$$

In the case under study the functions v^μ, w^μ are bounded in $L^2(I)$, $H_0^1(I)$, respectively, and v_α^μ, $\left(\Lambda^{-1} w_\alpha^\mu\right)_\alpha$ are bounded in $M^1(I)$ uniformly in μ. Moreover, the functions n^μ, m^μ are bounded in $L^2(I)$. Then, suppose that a subsequence v^μ, w^μ, n^μ, m^μ, denoted as before, possesses the properties

$$n^\mu, m^\mu \to n, m \quad \text{weakly in } L^2(I),$$

$$w^\mu \to w \quad \text{weakly in } H_0^1(I), \quad v^\mu \to v \quad \text{strongly in } L^2(I), \tag{5.273}$$

$$v^\mu_\alpha \to v_\alpha, \quad \left(\frac{1}{\Lambda}w^\mu_\alpha\right)_\alpha \to \left(\frac{1}{\Lambda}w_\alpha\right)_\alpha \quad \star\text{-weakly in } M^1(I).$$

We transfer the term $\mu\langle\Lambda_1 n^\mu, n^\mu\rangle + \mu\langle\Lambda_2 m^\mu, m^\mu\rangle$ to the right-hand side of (5.272). It can be ignored, due to its nonnegativeness, and then, we can pass to the limit in the inequality obtained as $\mu \to 0$. As a result, the inequality (5.271) follows. We have proved the following assertion.

Theorem 5.10. *Let all the assumption of Theorem* 5.9 *be fulfilled. Then, from the solutions v^μ, w^μ, n^μ, m^μ of the elastoplastic contact problem* (5.272) *we can choose a subsequence converging to a solution of the perfectly plastic problem in the sense* (5.273).

Note that different perfectly plastic models for three- dimensional case are considered in (Mosolov, Myasnikov, 1971).

5.6 Contact elastoplastic problem for the curvilinear Timoshenko rod

In this section we analyse the contact problem for a curvilinear Timoshenko rod. The plastic yield condition will depend just on the moments m. We shall prove that the solution of the problem satisfies all original boundary conditions, i.e., in contrast to the preceding section, we prove the existence of the solution to the original boundary value problem.

5.6.1 Existence of the solution

We shall use the notation of the preceding section. The nonpenetration condition in a linear approach is written in the form

$$\psi(a(\alpha), b(\alpha)) + (\nabla\psi(a(\alpha), b(\alpha)), \Pi h(\alpha)) \geq 0, \quad \alpha \in I, \tag{5.274}$$

where

$$\Pi = \frac{1}{\Lambda}\begin{pmatrix} a_\alpha & -b_\alpha \\ b_\alpha & a_\alpha \end{pmatrix}, \quad \Lambda = \sqrt{a_\alpha^2 + b_\alpha^2}, \quad h = \begin{pmatrix} v \\ w \end{pmatrix}.$$

The brackets $(\cdot\,,\cdot)$ mean a scalar product in R^2. On the interval I it is required to find the functions v, w, m, n, ξ such that

$$(n_\alpha + km_\alpha)(\bar{v} - v) + \left(\left(\frac{1}{\Lambda}m_\alpha\right)_\alpha - k\Lambda n\right)(\bar{w} - w) \leq 0, \tag{5.275}$$

$$v_\alpha + k\Lambda w = \Lambda n, \tag{5.276}$$

$$(kv)_\alpha - \left(\frac{1}{\Lambda}w_\alpha\right)_\alpha = -\left(\frac{1}{\Lambda}m_\alpha\right)_\alpha + \xi, \tag{5.277}$$

$$|m| \leq C^\star, \tag{5.278}$$

$$\xi(\bar{m} - m) \le 0 \quad \forall \bar{m}, \quad |\bar{m}| \le C^\star, \tag{5.279}$$

$$v = w = m = 0, \quad \alpha = 0, 1. \tag{5.280}$$

The test functions $h = \begin{pmatrix} \bar{v} \\ \bar{w} \end{pmatrix}$ in (5.275) should satisfy the restriction (5.274). The function $k \in W^1_\infty(I)$ is given (the curvature of the rod). We assume also $\psi \in H^4(G)$, $\nabla\psi(a(\alpha), b(\alpha)) \neq 0$, $\alpha \in I$, $C^\star = \text{const} > 0$.

Comparison of the models (5.247)–(5.251) and (5.275)–(5.279) show that equation (5.277) does not coincide with (5.249). Relations (5.278), (5.279) can be considered as a particular case of (5.250), (5.251).

The set of all functions $h = \begin{pmatrix} v \\ w \end{pmatrix}$ from $H^1_0(I)$ satisfying (5.274) is denoted by B_ψ. We suppose that there exists at least one function $\begin{pmatrix} v^0 \\ w^0 \end{pmatrix} \in B_\psi$. To avoid the trivial solution of the problem (5.274)–(5.280) the assumption $\begin{pmatrix} 0 \\ 0 \end{pmatrix} \notin B_\psi$ is used. Also, let

$$K = \{m \in H^1_0(I) \mid \ |m| \le C^\star \quad \text{on} \quad I\}. \tag{5.281}$$

Theorem 5.11. *Let the above assumptions be fulfilled, $a, b \in H^2(I)$, $\Lambda \ge \kappa > 0$ on I, $\kappa = \text{const}$, and the norm of k in $W^1_\infty(I)$ be small enough. Then there exist functions v, w, m, n, ξ, satisfying (5.276)–(5.277) as well as the relations*

$$\begin{pmatrix} v \\ w \end{pmatrix} \in B_\psi, \quad m \in K, \quad n \in L^2(I), \quad \xi \in H^{-1}(I), \tag{5.282}$$

$$\langle n_\alpha + k m_\alpha, \bar{v} - v\rangle + \langle \left(\frac{1}{\Lambda} m_\alpha\right)_\alpha - k\Lambda n, \bar{w} - w\rangle \le 0 \quad \forall \begin{pmatrix} \bar{v} \\ \bar{w} \end{pmatrix} \in B_\psi, \tag{5.283}$$

$$\langle \xi, \bar{m} - m\rangle \le 0 \quad \forall \bar{m} \in K. \tag{5.284}$$

Proof. Let $(q_1, q_2) : [L^2(I)]^2 \to [L^2(I)]^2$ be the penalty operator connected with the restriction (5.274), $p(m) = m - \pi m$, where π is the operator of orthogonal projection of $L^2(I)$ onto the set $\{m \in L^2(I) \mid \ |m| \le C^\star\}$. We introduce three positive parameters ε, δ, λ and consider the regularized problem

$$-\varepsilon v_{\alpha\alpha} - n_\alpha - k m_\alpha + \frac{1}{\delta} q_1(v, w) = 0, \tag{5.285}$$

$$-\varepsilon w_{\alpha\alpha} - \left(\frac{1}{\Lambda} m_\alpha\right)_\alpha + k\Lambda n + \frac{1}{\delta} q_2(v, w) = 0, \tag{5.286}$$

$$\Lambda n - v_\alpha - k\Lambda w = 0, \tag{5.287}$$

$$-(kv)_\alpha + \left(\frac{1}{\Lambda} w_\alpha\right)_\alpha - \left(\frac{1}{\Lambda} m_\alpha\right)_\alpha + \frac{1}{\lambda} p(m) = 0, \tag{5.288}$$

$$v = w = m = 0, \quad \alpha = 0, 1. \tag{5.289}$$

To obtain a priori estimates of the problem (5.285)–(5.289) we multiply (5.285)–(5.288) by $v - v^0$, $w - w^0$, n, m, respectively. Taking into account the monotonicity of the operators (q_1, q_2), p, one can derive

$$\varepsilon\|v_\alpha\|^2 + \varepsilon\|w_\alpha\|^2 + \|\sqrt{\Lambda}n\|^2 + \|\frac{1}{\sqrt{\Lambda}}m_\alpha\|^2 \leq c, \tag{5.290}$$

$$\frac{1}{\lambda}\langle p(m), m\rangle \leq c \tag{5.291}$$

with the constants c uniform in $\varepsilon \leq \varepsilon_0$, δ, λ.

In view of the properties of p, one has

$$\frac{1}{\lambda}\langle p(m), \bar{m} - m\rangle \leq 0 \quad \forall \bar{m} \in K. \tag{5.292}$$

Therefore, by (5.291), we obtain $\lambda^{-1}\langle p(m), \bar{m}\rangle \leq c \; \forall \bar{m} \in K$. Consequently,

$$\frac{1}{\lambda}\|p(m)\|_{H^{-1}(I)} \leq c \tag{5.293}$$

uniformly in $\varepsilon \leq \varepsilon_0$, δ, λ. In view of the boundary conditions for v, w, m one can conclude that all assumptions of Theorem 1.14 are fulfilled, and the solvability of the problem (5.285)–(5.289) for the fixed parameters can be stated. This means that the functions v, w, $m \in H_0^1(I)$, $n \in L^2(I)$ exist such that the equations (5.285)–(5.288) hold. Moreover, by (5.290), (5.293) and the boundary conditions (5.289), from (5.287), (5.288) we obtain one more estimate

$$\|v_\alpha\| + \|\frac{1}{\sqrt{\Lambda}}w_\alpha\| \leq c. \tag{5.294}$$

The assumption of the smallness of k in the space $W_\infty^1(I)$ is used to obtain (5.294). Indeed, with this assumption, from (5.287), (5.288) we have

$$v_\alpha + k\Lambda w \in L^2(I), \quad -(\Lambda w_\alpha)_\alpha + (kv)_\alpha \in H^{-1}(I),$$

and therefore the estimate (5.294) follows.

Let us show that the passages to the limit as $\varepsilon \to 0$, $\lambda \to 0$, $\delta \to 0$ can be justified. At every step the solution is supplied with the appropriate symbol without mentioning the dependence of the solution on the other parameters.

So, let v^ε, w^ε, m^ε, n^ε be the solution of the problem (5.285)–(5.289). By the estimates obtained we can choose a subsequence, with the previous notation, such that as $\varepsilon \to 0$

$$v^\varepsilon, w^\varepsilon, m^\varepsilon \to v^\lambda, w^\lambda, m^\lambda \quad \text{weakly in } H_0^1(I), \quad \text{strongly in } L^2(I),$$

$$n^\varepsilon \to n^\lambda \quad \text{weakly in } L^2(I).$$

Passing to the limit in (5.285)–(5.288) as $\varepsilon \to 0$ we have

$$v^\lambda,\ w^\lambda,\ m^\lambda \in H_0^1(I), \quad n^\lambda \in L^2(I),$$

$$-n_\alpha^\lambda - km_\alpha^\lambda + \frac{1}{\delta}q_1(v^\lambda, w^\lambda) = 0, \tag{5.295}$$

$$-\left(\frac{1}{\Lambda}m_\alpha^\lambda\right)_\alpha + k\Lambda n^\lambda + \frac{1}{\delta}q_2(v^\lambda, w^\lambda) = 0, \tag{5.296}$$

$$\Lambda n^\lambda - v_\alpha^\lambda - k\Lambda w^\lambda = 0, \tag{5.297}$$

$$-(kv^\lambda)_\alpha + \left(\frac{1}{\Lambda}w_\alpha^\lambda\right)_\alpha - \left(\frac{1}{\Lambda}m_\alpha^\lambda\right)_\alpha + \frac{1}{\lambda}p(m^\lambda) = 0. \tag{5.298}$$

The passage to the limit in the nonlinear terms can be justified using the strong convergence of v^ε, w^ε, m^ε. Now we can choose a subsequence v^λ, w^λ, m^λ such that as $\lambda \to 0$

$$v^\lambda,\ w^\lambda,\ m^\lambda \ \to\ v^\delta,\ w^\delta,\ m^\delta \quad \text{weakly in } H_0^1(I),$$

$$v^\lambda,\ w^\lambda \ \to\ v^\delta,\ w^\delta \quad \text{strongly in } L^2(I), \quad n^\lambda \ \to\ n^\delta \quad \text{weakly in } L^2(I),$$

$$\frac{1}{\lambda}p(m^\lambda) \ \to\ \xi^\delta \quad \text{weakly in } H^{-1}(I).$$

It is clear that after the passage to the limit in (5.295)–(5.298) we arrive at the relations

$$v^\delta,\ w^\delta \in H_0^1(I), \quad m^\delta \in K, \quad n^\delta \in L^2(I), \quad \xi \in H^{-1}(I), \tag{5.299}$$

$$-n_\alpha^\delta - km_\alpha^\delta + \frac{1}{\delta}q_1(v^\delta, w^\delta) = 0, \tag{5.300}$$

$$-\left(\frac{1}{\Lambda}m_\alpha^\delta\right)_\alpha + k\Lambda n^\delta + \frac{1}{\delta}q_2(v^\delta, w^\delta) = 0, \tag{5.301}$$

$$\Lambda n^\delta - v_\alpha^\delta - k\Lambda w^\delta = 0, \tag{5.302}$$

$$-(kv^\delta)_\alpha + \left(\frac{1}{\Lambda}w_\alpha^\delta\right)_\alpha - \left(\frac{1}{\Lambda}m_\alpha^\delta\right)_\alpha + \xi^\delta = 0, \tag{5.303}$$

$$\langle \xi^\delta, \bar{m} - m^\delta\rangle \le 0 \quad \forall \bar{m} \in K. \tag{5.304}$$

The inclusion $m^\delta \in K$ was obtained from (5.298). Now, let us pass to the limit as $\delta \to 0$. Choosing a subsequence, if necessary, we assume as $\delta \to 0$ that

$$v^\delta,\ w^\delta,\ m^\delta \ \to\ v,\ w,\ m \quad \text{weakly in } H_0^1(I),$$

$$n^\delta \ \to\ n \quad \text{weakly in } L^2(I), \quad \xi^\delta \ \to\ \xi \quad \text{weakly in } H^{-1}(I).$$

Multiplying (5.300), (5.301), (5.302) by $\bar{v}-v^\delta$, $\bar{w}-w^\delta$, n^δ, respectively, and integrating over I, the following relations are obtained:

$$\langle n_\alpha^\delta + km_\alpha^\delta, \bar{v} - v^\delta\rangle + \langle\left(\frac{1}{\Lambda}m_\alpha^\delta\right)_\alpha - k\Lambda n^\delta, \bar{w} - w^\delta\rangle \leq 0, \tag{5.305}$$

$$\langle\left(\frac{1}{\Lambda}m_\alpha^\delta\right)_\alpha + (kv^\delta)_\alpha - \left(\frac{1}{\Lambda}w_\alpha^\delta\right)_\alpha, \bar{m} - m^\delta\rangle \leq 0, \tag{5.306}$$

$$\langle\Lambda n^\delta, n^\delta\rangle - \langle v_\alpha^\delta, n^\delta\rangle - \langle k\Lambda w^\delta, n^\delta\rangle = 0. \tag{5.307}$$

In doing so, we have taken into account (5.303), (5.304). Summing relations (5.305)–(5.307), in view of the weak lower semicontinuity of the norm in $L^2(I)$ the passage to the limit can be fulfilled. The limiting relation for v, w, m, n can be written as (5.305)–(5.306). The passage to the limit in (5.302) is obvious. The inclusion $\begin{pmatrix} v \\ w \end{pmatrix} \in B_\psi$ is proved in a standard way. Therefore, we obtain the relations (5.276), (5.277), (5.282)–(5.284).

We should note that the uniqueness of m, n can be proved in (5.276), (5.277), (5.282)–(5.284). To verify this, it suffices to derive the relation for the difference of two possible solutions. The proof of Theorem 5.11 is completed.

5.6.2 Construction of a measure

In this subsection we construct a nonnegative measure characterizing the work of interacting forces. The measure is defined on the Borel subsets of I. The space of continuous functions defined on I with compact supports is denoted by $C_0(I)$.

Theorem 5.12. *A nonnegative measure ν_ψ can be defined on the σ-algebra of Borel subsets of I such that the representation*

$$\langle n_\alpha + km_\alpha, \bar{v}\rangle + \langle\left(\frac{1}{\Lambda}m_\alpha\right)_\alpha - k\Lambda n, \bar{w}\rangle = \int_0^1 (\nabla\psi, \Pi\bar{h})d\nu_\psi \tag{5.308}$$

holds for all $\bar{h} = \begin{pmatrix} \bar{v} \\ \bar{w} \end{pmatrix} \in H_0^1(I) \cap C_0(I)$.

PROOF. We introduce the notation $H = \begin{pmatrix} n_\alpha + km_\alpha \\ (\Lambda^{-1}m_\alpha)_\alpha - k\Lambda n \end{pmatrix}$. Let $\bar{h} = \begin{pmatrix} \bar{v} \\ \bar{w} \end{pmatrix} \in H_0^1(I) \cap C_0(I)$. For $\chi = (\nabla\psi, \Pi\,\bar{h})$ the functional

$$\Psi(\chi) = -\langle H, \bar{h}\rangle \tag{5.309}$$

can be considered. This functional is defined on the linear space of all the above functions χ. We first prove that the functional is well defined by the formula (5.309). To this end, we note that, if $h^0 \in H_0^1(I)$ and $(\nabla\psi, \Pi h^0) \geq 0$ a.e. on I, the inequality

$$-\langle H, h^0\rangle \geq 0 \tag{5.310}$$

holds. Really, the inequality (5.283) can be written in the form

$$\langle H, \bar{h} - h\rangle \leq 0 \quad \forall \bar{h} \in B_\psi, \tag{5.311}$$

where $h = \begin{pmatrix} v \\ w \end{pmatrix}$ is the solution satisfying (5.282)–(5.284). Obviously, $h + h^0 \in B_\psi$. Substitute $h + h^0$ in (5.311) as the test function $\bar{h}$. We obtain precisely (5.310). In particular, this fact provides a positiveness of the functional (5.309). Furthermore, if $\chi_1 - \chi_2 = 0$, then $\Psi(\chi_1 - \chi_2) = -\langle H, \bar{h}_1 - \bar{h}_2\rangle = 0$, and hence $\Psi(\chi_1) = \Psi(\chi_2)$. The correctness of the definition (5.309) is proved. The functional (5.309) can be extended to the space $C_0(I)$. Moreover, the extended functional is linear and positive. This means that a nonnegative measure ν_ψ exists such that (Landkof, 1966)

$$\Psi(\chi) = \int_0^1 \chi d\nu_\psi \quad \forall \chi \in C_0(I). \tag{5.312}$$

The representation (5.312) coincides with (5.308) for $\chi = (\nabla\psi, \Pi\bar{h})$, $\bar{h} \in H_0^1(I) \cap C_0(I)$. Theorem 5.12 is proved.

It is easy to build an explicit representation for the measure ν_ψ. Taking $\bar{h} = \begin{pmatrix} 0 \\ \bar{w} \end{pmatrix}$ it follows from (5.308) that

$$\langle \left(\frac{1}{\Lambda} m_\alpha\right)_\alpha - k\Lambda n, \bar{w}\rangle = \int_0^1 \frac{(b_\alpha\psi_x - a_\alpha\psi_y)\bar{w}}{\Lambda} d\nu_\psi.$$

Hence

$$\langle \left(\frac{1}{\Lambda} m_\alpha\right)_\alpha - k\Lambda n,\ \bar{w}(b_\alpha\psi_x - a_\alpha\psi_y)\rangle = \int_0^1 \frac{(b_\alpha\psi_x - a_\alpha\psi_y)^2}{\Lambda} \bar{w} d\nu_\psi. \tag{5.313}$$

Similarly, taking $\bar{h} = \begin{pmatrix} \bar{v} \\ 0 \end{pmatrix}$ in (5.308) we find

$$-\langle n_\alpha + km_\alpha, \bar{v}(a_\alpha\psi_x + b_\alpha\psi_y)\rangle = \int_0^1 \frac{(a_\alpha\psi_x + b_\alpha\psi_y)^2}{\Lambda} \bar{v} d\nu_\psi. \tag{5.314}$$

Let us take $\bar{v} = \bar{w}$ in (5.313), (5.314) and sum these relations. This implies

$$-\langle n_\alpha + km_\alpha, \bar{v}(a_\alpha\psi_x + b_\alpha\psi_y)\rangle + \langle \left(\frac{1}{\Lambda} m_\alpha\right)_\alpha - k\Lambda n, \bar{v}(b_\alpha\psi_x - a_\alpha\psi_y)\rangle$$

$$= \int_0^1 \Lambda |\nabla \psi|^2 \bar{v} d\nu_\psi.$$

Obviously $\Lambda(H, \Pi^* \nabla \psi) \in H^{-1}(I)$, where Π^* is the matrix conjugate to Π. Therefore, the previous equality can be written in the form

$$-\langle \Lambda(H, \Pi^* \nabla \psi), \bar{v} \rangle = \int_0^1 \Lambda |\nabla \psi|^2 \bar{v} d\nu_\psi,$$

whence we obtain the representation for the measure ν_ψ,

$$\nu_\psi = -\frac{(H, \Pi^* \nabla \psi)}{|\nabla \psi|^2}.$$

The inequality $\psi + (\nabla \psi, \Pi h) > 0$ is valid at the points $\alpha \in I$ where the contact is absent. This means that the equations

$$n_\alpha + k m_\alpha = 0, \quad \left(\frac{1}{\Lambda} m_\alpha \right)_\alpha - k \Lambda n = 0 \tag{5.315}$$

hold in the neighbourhoods of the mentioned points. This property follows immediately from the inequality (5.283). Using the notation introduced above, the equation (5.315) can be written in the form

$$H = 0.$$

There is no contact in the neighbourhoods of the points $\alpha = 0$, $\alpha = 1$ provided that $\psi(a(0), b(0)) > 0$, $\psi(a(1), b(1)) > 0$. Consequently, the equation $H = 0$ holds in the neighbourhoods of the points $\alpha = 0$, $\alpha = 1$. In particular, this implies that $\nu_\psi = 0$ near $\alpha = 0$, $\alpha = 1$. Furthermore, the measure of every compact $M \subset I$ is finite. Hence, the conditions $\psi(a(0), b(0)) > 0$, $\psi(a(1), b(1)) > 0$ provide the validity of the inequality $\nu_\psi(I) < +\infty$.

5.6.3 Optimal control problem

Let $\Phi \subset H^4(G)$ be a convex, closed and bounded set such that every element $\psi \in \Phi$ satisfies the following relations:

$$\psi(a(0), b(0)) > 0, \quad \psi(a(1), b(1)) > 0, \quad |\nabla \psi(a(\alpha), b(\alpha))| \geq 1, \quad \alpha \in I.$$

It is assumed that for each $\psi \in \Phi$, the level line $\psi(x, y) = 0$ divides G into two subdomains such that $\psi(x, y) > 0$ for the first subdomain and $\psi(x, y) < 0$ for the second one. We consider the cost functional

$$J(\psi) = \|m - m_0\| + \nu_\psi(I).$$

The optimal control problem to be analysed is formulated as follows: to find an element $\psi \in \Phi$ such that

$$J(\psi) \le J(\bar{\psi}) \quad \forall \bar{\psi} \in \Phi. \tag{5.316}$$

Theorem 5.13. *Let the above hypotheses be fulfilled and* $|a_\alpha(\alpha)| \ge c^0$ *on* I *for a given* $c^0 = const > 0$. *Then a solution to problem* (5.316) *exists.*

PROOF. We choose a minimizing sequence ψ^i, this sequence is obviously bounded in $H^4(G)$. Without loss of generality we assume that as $i \to \infty$

$$\psi^i \to \psi \quad \text{weakly in } H^4(G), \quad \text{strongly in } C^2(\bar{G}). \tag{5.317}$$

One can solve the problem (5.274)–(5.280) for every $\psi = \psi^i$ and find the functions v^i, w^i, m^i, n^i, ξ^i such that

$$h^i = \begin{pmatrix} v^i \\ w^i \end{pmatrix} \in B_\psi, \quad m^i \in K, \quad n^i \in L^2(I), \quad \xi^i \in H^{-1}(I), \tag{5.318}$$

$$\langle H^i, \bar{h}^i - h^i \rangle \le 0 \quad \forall \bar{h}^i \in B_{\psi^i}, \tag{5.319}$$

$$v^i_\alpha + k\Lambda w^i = \Lambda n^i, \tag{5.320}$$

$$(kv^i)_\alpha - \left(\frac{1}{\Lambda} w^i_\alpha\right)_\alpha = -\left(\frac{1}{\Lambda} m^i_\alpha\right)_\alpha + \xi^i, \tag{5.321}$$

$$\langle \xi^i, \bar{m} - m^i \rangle \le 0 \quad \forall \bar{m} \in \mathrm{K}. \tag{5.322}$$

It can be proved (see Lemma 5.3 below) that for every $\bar{h} \in B_\psi$ a sequence $\bar{h}^i \in B_{\psi^i}$ exists such that

$$\bar{h}^i \to \bar{h} \quad \text{strongly in } H^1_0(I).$$

Let us substitute the elements of such a strongly converging sequence in (5.319) as test functions and multiply simultaneously (5.320), (5.321) by n^i, $\bar{m} - m^i$, respectively, $\bar{m} \in \mathrm{K}$. We find that v^i, w^i, m^i are bounded in $H^1_0(I)$, n^i are bounded in $L^2(I)$, and ξ^i are bounded in $H^{-1}(I)$. Therefore, a subsequence (with the same notation) can be chosen such that as $i \to \infty$

$$v^i,\, w^i,\, m^i \to v,\, w,\, m \quad \text{weakly in } H^1_0(I),$$

$$n^i \to n \quad \text{weakly in } L^2(I), \quad \xi^i \to \xi \quad \text{weakly in } H^{-1}(I).$$

Thus, the passage to the limit can be justified in (5.318)–(5.322) using the same scheme as above. As a result, we obtain the relations like (5.276), (5.277), (5.282)–(5.284). This means that $v = v(\psi)$, $w = w(\psi)$, $m = m(\psi)$, $n = n(\psi)$, $\xi = \xi(\psi)$. In view of the above weak convergence one has

$$H^i \to H \quad \text{weakly in } H^{-1}(I).$$

Therefore

$$\frac{(H^i, \Pi^\star \nabla \psi^i)}{|\nabla \psi^i|^2} \to \frac{(H, \Pi^\star \nabla \psi)}{|\nabla \psi|^2} \quad \text{weakly in } H^{-1}(I).$$

In particular, by the formula for ν_ψ obtained in the previous subsection, this implies that $\nu_{\psi^i} \to \nu_\psi$ $\star$-weakly as measures. Hence (Landkof, 1966)

$$\liminf \nu_{\psi^i}(I) \geq \nu_\psi(I).$$

Therefore, we have proved the weak lower semicontinuity of the functional J. This provides that the function ψ is the solution of the optimal control problem (5.316). Theorem 5.13 is completely proved.

REMARK. The condition $|a_\alpha| \geq c^0$ of Theorem 5.13 actually means that the equation of the rod axis could be written in the form $y = F(x)$, where $F = ba^{-1}$.

When proving Theorem 5.13 the auxiliary statement on strong convergence of $\bar{h}^i$ was used. Here we prove the statement.

Lemma 5.3. *For every $\bar{h} \in B_\psi$ a sequence $\bar{h}^i \in B_{\psi^i}$ can be constructed such that*

$$\bar{h}^i \to \bar{h} \quad \textit{strongly in } H_0^1(I)$$

provided that all hypotheses of Theorem 5.13 *are fulfilled and ψ^i converge to ψ in the sense* (5.317).

PROOF. Let $\bar{h} \in B_\psi$, i.e.

$$\psi(a(\alpha), b(\alpha)) + (\Pi^\star \nabla \psi(a(\alpha), b(\alpha)), \bar{h}(\alpha)) \geq 0, \quad \alpha \in I.$$

By the convergence of ψ^i, we can assume that at the curve points $x = a(\alpha)$, $y = b(\alpha)$ the inequalities

$$|\psi^i - \psi| < \frac{1}{2i^2}, \quad |\Pi^\star(\nabla \psi^i - \nabla \psi)||\bar{h}| < \frac{1}{2i^2} \tag{5.323}$$

are fulfilled. There exists a constant $\beta > 0$ such that $\psi(a(0), b(0)) > \beta$, $\psi(a(1), b(1)) > \beta$. Consequently, a constant $\mu > 0$ can be found such that the inequalities

$$\psi^i > \frac{2}{3}\beta, \quad |(\Pi^\star \nabla \psi^i, \bar{h})| < \frac{\beta}{4} \tag{5.324}$$

are valid on the intervals $(0, \mu)$, $(1-\mu, 1)$. Let us take $\bar{h}^i = \bar{h} + i^{-1} \nabla \psi^i \lambda(\alpha)$, where $\lambda(\alpha)$ is a smooth finite function on I equal to sign $a_\alpha(\alpha)$ on $(\mu, 1-\mu)$, $|\lambda(\alpha)| \leq 1$. Then for sufficiently large i the inequality

$$\left| \frac{(\Pi^\star \nabla \psi^i, \nabla \psi^i)}{i} \right| \leq \frac{\beta}{4} \tag{5.325}$$

holds. Consequently, by (5.324), (5.325), the following inequality is valid throughout the intervals $(0, \mu)$, $(1 - \mu, 1)$:

$$\psi^i + (\Pi^\star \nabla \psi^i, \bar{h}^i) \geq 0. \tag{5.326}$$

The relation (5.326) is also valid beyond these intervals, because due to (5.323) for sufficiently large i and $\alpha \in (\mu, 1-\mu)$ one can obtain

$$\psi^i + (\Pi^\star \nabla \psi^i, \bar{h}^i) = (\psi^i - \psi) + \psi + (\Pi^\star \nabla \psi, \bar{h}) + (\Pi^\star (\nabla \psi^i - \nabla \psi), \bar{h})$$

$$+ \frac{\lambda(\alpha)}{i} (\Pi^\star \nabla \psi^i, \nabla \psi^i) \geq -\frac{1}{i^2} + \frac{\lambda(\alpha) a_\alpha(\alpha)}{\Lambda i} = -\frac{1}{i^2} + \frac{c^0}{\Lambda i} \geq 0.$$

This means that the inequality (5.326) takes place for all $\alpha \in I$, i.e. $\bar{h}^i \in B_{\psi^i}$. The strong convergence of $\bar{h}^i$ to $\bar{h}$ is clear, which proves Lemma 5.3.

5.7 Viscoelastoplastic curvilinear Timoshenko rod

5.7.1 Solution existence

Again, we use the notation of the two previous sections. Let the rod axes be described by the functions $x = a(\alpha)$, $y = b(\alpha)$, $\alpha \in I$. The punch shape is described by the equation $\psi(x, y) = 0$, $Q = I \times (0, T)$, $T > 0$. We consider the nonpenetration condition for the displacement vector h,

$$\psi(a(\alpha), b(\alpha)) + (\nabla \psi(a(\alpha), b(\alpha)), \Pi h(t, \alpha)) \geq 0, \quad (\alpha, t) \in Q, \tag{5.327}$$

$$\Pi = \frac{1}{\Lambda} \begin{pmatrix} a_\alpha & -b_\alpha \\ b_\alpha & a_\alpha \end{pmatrix}, \quad h = \begin{pmatrix} v \\ w \end{pmatrix}, \quad \Lambda = \sqrt{a_\alpha^2 + b_\alpha^2}.$$

As compared to the previous section the constitutive law corresponds to the viscoelastoplastic rod. The formulation of the problem is as follows. In the domain Q we want to find the functions v, w, m, n, ξ satisfying (5.327) and the relations

$$(n_\alpha + k m_\alpha)(\bar{v} - v) + \left(\left(\frac{1}{\Lambda} m_\alpha \right)_\alpha - k \Lambda n \right) (\bar{w} - w) \leq 0, \tag{5.328}$$

$$v_\alpha + k \Lambda w = \Lambda n_t, \tag{5.329}$$

$$(kv)_\alpha - \left(\frac{1}{\Lambda} w_\alpha \right)_\alpha = -\left(\frac{1}{\Lambda} m_{t\alpha} \right)_\alpha + \xi, \tag{5.330}$$

$$|m| \leq C^\star, \quad \xi(\bar{m} - m) \leq 0 \quad \forall \bar{m}, \ |\bar{m}| \leq C^\star, \tag{5.331}$$

$$m = 0, \quad n = 0, \quad t = 0, \tag{5.332}$$

$$v = w = m = 0, \quad \alpha = 0, 1. \tag{5.333}$$

The test functions $\bar{h} = \begin{pmatrix} \bar{v} \\ \bar{w} \end{pmatrix}$ in (5.328) should satisfy (5.327). Let $k \in W^1_\infty(I)$, $\psi \in H^2(G)$, where $G \subset R^2$ is a bounded domain such that the curve $x = a(\alpha)$, $y = b(\alpha)$ lies in G, $\alpha \in I$. The sets $B \equiv B_\psi$, K are introduced like those in the preceding section. We assume that there exist functions v^0, w^0 such that $v^0, w^0 \in B$, $v^0_{xx}, w^0_{xx} \in L^2(I)$. For simplicity, we assume that the exterior forces are zero. In doing so the assumption $(0,0) \notin B$ is accepted, otherwise the functions $v = 0$, $w = 0$, $m = 0$, $n = 0$, $\xi = 0$ provide the solution of the problem (5.327)–(5.333). Now we are in a position to prove the following statement.

Theorem 5.14. *Let the above hypotheses be fulfilled, $a, b \in H^2(I)$, $\Lambda \geq \kappa > 0$ on I, $\kappa = const$, and the norm of k in $W^1_\infty(I)$ be sufficiently small. Then there exist functions v, w, m, n, ξ satisfying* (5.329), (5.330), (5.332) *and the relations*

$$m,\ m_t \in L^\infty(0,T;H^1_0(I)),\ n,\ n_t \in L^\infty(0,T;L^2(I)),$$

$$v,\ w \in L^2(0,T;H^1_0(I)),$$

$$\xi \in L^2(0,T;H^{-1}(I)),\quad m(t) \in K,\quad (v(t),w(t)) \in B \quad \text{a.e. on } (0,T),$$

$$\int_0^T \langle n_\alpha + km_\alpha, \bar{v} - v\rangle dt + \int_0^T \langle \left(\frac{1}{\Lambda}m_\alpha\right)_\alpha - k\Lambda n, \bar{w} - w\rangle dt \leq 0$$

$$\forall \bar{v},\ \bar{w} \in L^2(0,T;H^1_0(I)),\quad (v(t),w(t)) \in B,$$

$$\int_0^T \langle \xi, \bar{m} - m\rangle dt \leq 0 \quad \forall \bar{m} \in L^2(0,T;H^1_0(I)),\quad \bar{m} \in K.$$

PROOF. Let the operators (q_1, q_2), p be the same as in the preceding section. In the domain Q we consider the auxiliary boundary value problem with three positive parameters ε, δ, λ,

$$\varepsilon v_t - \varepsilon v_{\alpha\alpha} - n_\alpha - km_\alpha + \frac{1}{\delta}q_1(v,w) = 0, \tag{5.334}$$

$$\varepsilon w_t - \varepsilon w_{\alpha\alpha} - \left(\frac{1}{\Lambda}m_\alpha\right)_\alpha + k\Lambda n + \frac{1}{\delta}q_2(v,w) = 0, \tag{5.335}$$

$$\Lambda n_t - v_\alpha - k\Lambda w = 0, \tag{5.336}$$

$$-\left(\frac{1}{\Lambda}m_{t\alpha}\right)_\alpha - (kv)_\alpha + \left(\frac{1}{\Lambda}w_\alpha\right)_\alpha + \frac{1}{\lambda}p(m) = 0, \tag{5.337}$$

$$v = v^0,\quad w = w^0,\quad m = 0,\quad n = 0,\quad t = 0, \tag{5.338}$$

$$v = w = m = 0,\quad x = 0, 1. \tag{5.339}$$

First of all the solvability of the problem (5.334)–(5.339) is proved for fixed ε, δ, λ. In the sequel the passages to limits as $\varepsilon \to 0$, $\lambda \to 0$, $\delta \to 0$ will be justified, which provide the proof of the theorem. Let us derive an a priori estimate of the solution. To this end, we multiply (5.334)–(5.337) by

$v - v^0$, $w - w^0$, n, m, respectively, integrate over I and sum. This implies the inequality

$$\frac{1}{2}\frac{d}{dt}\left(\varepsilon\|v(t)\|^2 + \varepsilon\|w(t)\|^2 + \|\Lambda^{-1/2}m_\alpha(t)\|^2 + \|\Lambda^{1/2}n(t)\|^2\right)$$

$$+\frac{\varepsilon}{2}\|v_\alpha(t)\|^2 + \frac{\varepsilon}{2}\|w_\alpha(t)\|^2 \le \varepsilon\langle v_t(t), v^0\rangle + \varepsilon\langle w_t(t), w^0\rangle \tag{5.340}$$

$$+\frac{1}{2}\|\Lambda^{-\frac{1}{2}}m_\alpha(t)\|^2 + \frac{1}{2}\|\Lambda^{\frac{1}{2}}n(t)\|^2 + c$$

with the constant c uniform in $\varepsilon \le \varepsilon_0$, δ, λ. By integrating (5.340) we arrive at the estimates

$$\max_{0\le t\le T}\left(\varepsilon\|v(t)\|^2 + \varepsilon\|w(t)\|^2\right) \le c, \tag{5.341}$$

$$\|\Lambda^{-1/2}m_\alpha\|_{L^2(Q)} + \|\Lambda^{1/2}n\|_{L^2(Q)} \le c, \tag{5.342}$$

$$\varepsilon\|v_\alpha\|^2_{L^2(Q)} + \varepsilon\|w_\alpha\|^2_{L^2(Q)} \le c. \tag{5.343}$$

As before, the constants in (5.341)–(5.343) are uniform in $\varepsilon \le \varepsilon_0$, δ, λ. Next, it follows from (5.334)–(5.337) that

$$v_t(0) = v^0_{\alpha\alpha}, \quad w_t(0) = w^0_{\alpha\alpha}, \quad n_t(0) \in L^2(I), \quad m_t(0) \in H^2(I). \tag{5.344}$$

Now we can differentiate the equations (5.334)–(5.337) with respect to t and multiply by v_t, w_t, n_t, m_t, respectively. The penalty terms are nonnegative and, therefore, they can be neglected. As a result, the following differential inequality is derived:

$$\frac{1}{2}\frac{d}{dt}\left(\varepsilon\|v_t(t)\|^2 + \varepsilon\|w_t(t)\|^2 + \|\Lambda^{-1/2}m_{t\alpha}\|^2 + \|\Lambda^{1/2}n_t\|^2\right)$$

$$+\varepsilon\|v_{t\alpha}(t)\|^2 + \varepsilon\|w_{t\alpha}(t)\|^2 \le 0.$$

Taking into account the initial conditions (5.344) we find

$$\max_{0\le t\le T}\Big(\varepsilon\|v_t(t)\|^2 + \varepsilon\|w_t(t)\|^2 + \|\Lambda^{-1/2}m_{t\alpha}(t)\|^2 \tag{5.345}$$

$$+\|\Lambda^{1/2}n_t(t)\|^2\Big) \le c,$$

$$\varepsilon\|v_{t\alpha}\|^2_{L^2(Q)} + \varepsilon\|w_{t\alpha}\|^2_{L^2(Q)} \le c. \tag{5.346}$$

Let us establish one more estimate. By the estimates obtained, it is seen that the multiplication of (5.334)–(5.337) by $v - v^0$, $w - w^0$, n, m provides the boundedness of $\lambda^{-1}\langle p(m), m\rangle$ in $L^2(0,T)$ uniformly in $\varepsilon \le \varepsilon_0$, δ, λ. Let us choose a function $\hat{m} \in L^\infty(Q)$ such that $|\hat{m}(t)| \le C^\star$ a.e. on $(0,T)$.

Since the derivative of the convex functional $P(m) = 1/2\,\|m - \pi m\|^2$ can be found by the formula $P'_m = p(m)$ we have for almost all $t \in (0,T)$

$$\frac{1}{\lambda}\langle p(m), \hat{m}\rangle \le \frac{1}{\lambda}\langle p(m), m\rangle + \frac{1}{\lambda}P(\hat{m}) - \frac{1}{\lambda}P(m).$$

Meantime $P(\hat{m}) = 0$ for almost all t, hence, by the above boundedness of $\lambda^{-1}\langle p(m), m\rangle$ in $L^2(0,T)$, one obtains that $\lambda^{-1}p(m)$ are bounded in the space $L^2(0,T;L^1(I))$. In view of the continuous imbedding $L^1(I) \subset H^{-1}(I)$ the ultimate estimate of the penalty term has the form

$$\frac{1}{\lambda}p(m) \quad \text{are bounded in} \quad L^2(0,T;H^{-1}(I)). \tag{5.347}$$

Hence, by (5.341)–(5.343), (5.345)–(5.347), the equations (5.336), (5.337) can be rewritten as follows,

$$v_\alpha + k\Lambda w = h_1, \qquad -\left(\frac{1}{\Lambda}w_\alpha\right)_\alpha + (kv)_\alpha = h_2, \tag{5.348}$$

where h_1 are bounded in $L^2(Q)$, and h_2 are bounded in $L^2(0,T;H^{-1}(I))$. We can express the function v from the first equation of (5.348) and substitute it in the second equation. In this case, it is seen that the estimate for w follows, when the norm of k in $W^1_\infty(I)$ is sufficiently small. Then, the first equation provides the boundedness of v. So, from (5.348) we have that

$$v,\ w \quad \text{are bounded in} \quad L^2(0,T;H^1_0(I)). \tag{5.349}$$

To prove the solvability of the problem (5.334)–(5.339) we can use the Galerkin approach. In this case the estimates (5.341)–(5.343), (5.345), (5.346) will provide an existence of the solution for fixed ε, δ, λ. To obtain the additional estimate (5.349) the above reasonings can be repeated after proving the solvability. Thus the problem (5.334)–(5.339) possesses the solution with the following regularity:

$$n,\ n_t \in L^\infty(0,T;L^2(I)), \quad v,\ v_t,\ w,\ w_t \in L^\infty(0,T;H^1_0(I)),$$

$$m,\ m_t \in L^\infty(0,T;H^1_0(I)).$$

Of course, these inclusions are not uniform in the parameters, in general, i.e. the norms of the functions are not bounded uniformly with respect to the parameters $\varepsilon, \delta, \lambda$. Now, let us justify the passages to the limit as the parameters tend to zero. At the first step we denote the solution by v^ε, w^ε, m^ε, n^ε omitting the dependence on the other parameters. Then, choosing a subsequence, if necessary, we suppose that as $\varepsilon \to 0$

$$\varepsilon v^\varepsilon_t,\ \varepsilon w^\varepsilon_t \ \to\ 0 \quad \star\text{-weakly in} \quad L^\infty(0,T;L^2(I)),$$

$$m^\varepsilon,\ m^\varepsilon_t \ \to\ m^\lambda,\ m^\lambda_t \quad \star\text{-weakly in} \quad L^\infty(0,T;H^1_0(I)),$$

$$\begin{aligned} n^\varepsilon,\, n_t^\varepsilon &\to n^\lambda,\, n_t^\lambda \quad \star\text{-weakly in } L^\infty(0,T;L^2(I)), \\ v^\varepsilon,\, w^\varepsilon &\to v^\lambda,\, w^\lambda \quad \text{weakly in } L^1(0,T;H_0^1(I)), \\ m^\varepsilon &\to n^\lambda \quad \text{strongly in } L^2(Q). \end{aligned} \tag{5.350}$$

Consequently, the passage to the limit can be justified in (5.334)–(5.339). As a result we arrive at the relations

$$-n_\alpha^\lambda - km_\alpha^\lambda + \frac{1}{\delta}q_1(v^\lambda, w^\lambda) = 0, \tag{5.351}$$

$$-\left(\frac{1}{\Lambda}m_\alpha^\lambda\right)_\alpha + k\Lambda n^\lambda + \frac{1}{\delta}q_2(v^\lambda, w^\lambda) = 0, \tag{5.352}$$

$$\Lambda n_t^\lambda - v_\alpha^\lambda - k\Lambda w^\lambda = 0, \tag{5.353}$$

$$-\left(\frac{1}{\Lambda}m_{t\alpha}^\lambda\right)_\alpha - (kv^\lambda)_\alpha + \left(\frac{1}{\Lambda}w_\alpha^\lambda\right)_\alpha + \frac{1}{\lambda}p(m^\lambda) = 0. \tag{5.354}$$

The following step is the justification of the passage to the limit as $\lambda \to 0$. Let a subsequence, denoted as before, possess the following properties as $\lambda \to 0$:

$$\begin{aligned} v^\lambda,\, w^\lambda &\to v^\delta,\, w^\delta \quad \text{weakly in } L^2(0,T;H_0^1(I)), \\ m^\lambda,\, m_t^\lambda &\to m^\delta,\, m_t^\delta \quad \star\text{-weakly in } L^\infty(0,T;H_0^1(I)), \\ n^\lambda,\, n_t^\lambda &\to n^\delta,\, n_t^\delta \quad \star\text{-weakly in } L^\infty(0,T;L^2(I)), \\ \frac{1}{\lambda}p(m^\lambda) &\to \xi^\delta \quad \text{weakly in } L^2(0,T;H^{-1}(I)). \end{aligned} \tag{5.355}$$

Meantime from (5.352) it is clear that m^λ are bounded in $L^2(0,T;H^2(I)\cap H_0^1(I))$ nonuniformly in δ, in general. Hence, by (5.345), we can assume that for every fixed δ

$$m^\lambda \to m^\delta \quad \text{strongly in } L^2(0,T;H_0^1(I)). \tag{5.356}$$

Next, the equation (5.354) implies

$$-\left(\frac{1}{\Lambda}m_{t\alpha}^\delta\right)_\alpha - (kv^\delta)_\alpha + \left(\frac{1}{\Lambda}w_\alpha^\delta\right)_\alpha + \xi^\delta = 0. \tag{5.357}$$

Besides, we obtain from the same equation the following inequality,

$$\int_0^T \langle \left(\frac{1}{\Lambda}m_{t\alpha}^\lambda\right)_\alpha + (kv^\lambda)_\alpha - \left(\frac{1}{\Lambda}w_\alpha^\lambda\right)_\alpha, \bar{m} - m^\lambda\rangle dt \le 0,$$

which holds for all $\bar{m} \in L^2(0,T;H_0^1(I))$, $\bar{m}(t) \in K$. Hence, by (5.355), (5.356), we derive

$$\int_0^T \langle \xi^\delta, \bar{m} - m^\delta\rangle dt \le 0 \quad \forall \bar{m} \in L^2(0,T;H_0^1(I)), \quad \bar{m}(t) \in K \tag{5.358}$$

and moreover, $m^\delta \in K$ a.e. on $(0,T)$. Consequently, in the limit, the equations (5.351)–(5.353) take the form

$$-n^\delta_\alpha - km^\delta_\alpha + \frac{1}{\delta}q_1(v^\delta, w^\delta) = 0, \tag{5.359}$$

$$-\left(\frac{1}{\Lambda}m^\delta_\alpha\right)_\alpha + k\Lambda n^\delta + \frac{1}{\delta}q_2(v^\delta, w^\delta) = 0, \tag{5.360}$$

$$\Lambda n^\delta_t - v^\delta_\alpha - k\Lambda w^\delta = 0. \tag{5.361}$$

In conclusion we choose a subsequence, with the same notation, possessing the following properties as $\delta \to 0$:

$$v^\delta,\, w^\delta \rightharpoonup v,\, w \quad \text{weakly in } L^2(0,T;H^1_0(I)),$$

$$\xi^\delta \rightharpoonup \xi \quad \text{weakly in } L^2(0,T;H^{-1}(I)).$$

$$m^\delta,\, m^\delta_t \rightharpoonup m,\, m_t \quad \star-\text{weakly in } L^\infty(0,T;H^1_0(I)),$$

$$n^\delta,\, n^\delta_t \rightharpoonup n,\, n_t \quad \star-\text{weakly in } L^\infty(0,T;L^2(I)).$$

In this case from (5.357), (5.361) one has

$$-\left(\frac{1}{\Lambda}m_{t\alpha}\right)_\alpha - (kv)_\alpha + \left(\frac{1}{\Lambda}w_\alpha\right)_\alpha + \xi = 0, \tag{5.362}$$

$$\Lambda n_t - v_\alpha - k\Lambda w = 0. \tag{5.363}$$

Let us multiply (5.359), (5.360) by $\bar{v} - v^\delta$, $\bar{w} - w^\delta$, respectively, integrate over Q and sum, assuming $\bar{v}, \bar{w} \in L^2(0,T;H^1_0(I))$, $(\bar{v}(t), \bar{w}(t)) \in B$. This implies

$$\int_0^T \langle n^\delta_\alpha + km^\delta_\alpha, \bar{v} - v^\delta\rangle\, dt + \int_0^T \langle \left(\frac{1}{\Lambda}m^\delta_\alpha\right)_\alpha - k\Lambda n^\delta, \bar{w} - w^\delta\rangle\, dt \le 0.$$

We can sum this inequality with (5.358), where ξ^δ is found from (5.357). By that, the terms

$$\int_0^T \langle n^\delta_\alpha, v^\delta\rangle\, dt, \quad \int_0^T \langle \left(\frac{1}{\Lambda}m^\delta_\alpha\right)_\alpha, w^\delta\rangle\, dt$$

vanish. This provides the passage to the limit in the above relation, and next we can add and subtract the terms

$$\int_0^T \langle n_\alpha, v\rangle\, dt, \quad \int_0^T \langle \left(\frac{1}{\Lambda}m_\alpha\right)_\alpha, w\rangle\, dt.$$

The limiting relation, therefore, can be written in the form

$$\int_0^T \langle n_\alpha + km_\alpha, \bar{v} - v\rangle \, dt + \int_0^T \langle (\frac{1}{\Lambda} m_\alpha)_\alpha - k\Lambda n, \bar{w} - w\rangle \, dt \tag{5.364}$$

$$+ \int_0^T \langle \xi, \bar{m} - m\rangle \, dt \le 0,$$

where $\bar{v}$, $\bar{w}$, $\bar{m} \in L^2(0, T; H_0^1(I))$, $\bar{m}(t) \in K$, $(\bar{v}(t), \bar{w}(t)) \in B$ a.e. on $(0, T)$. The function ξ is defined from (5.362). Since $(\bar{v}, \bar{w})$, $\bar{m}$ are independent the inequality (5.364) yields the two written in the formulation of Theorem 5.14. The initial conditions for m, n are fulfilled since the limiting passages $\varepsilon \to 0$, $\lambda \to 0$, $\delta \to 0$ keep the conditions (5.332). By (5.362), (5.363), we complete the proof of Theorem 5.14.

5.8 Curvilinear rod under creep conditions

5.8.1 Problem formulation and existence of solutions

The contact problem for a rod under creep conditions is considered in this section. Our goal is to prove an existence theorem. We use the notations of the preceding sections. For convenience, introduce the notations

$$Ms(t, x) = \Lambda s(t, x) + \int_0^t \Lambda s(\tau, x) \, d\tau,$$

$$M_1 s(t, x) = \left(\frac{1}{\Lambda} s_\alpha\right)_\alpha (t, x) + \int_0^t \left(\frac{1}{\Lambda} s_\alpha\right)_\alpha (\tau, x) \, d\tau.$$

The formulation of the problem is as follows. In the domain $Q = I \times (0, T)$ we have to find the functions v, w, m, n, ξ satisfying the following relations:

$$\psi(a(\alpha), b(\alpha)) + (\nabla\psi(a(\alpha), b(\alpha)), \Pi \begin{pmatrix} v(t, \alpha) \\ w(t, \alpha) \end{pmatrix}) \ge 0 \quad \text{in } Q, \tag{5.365}$$

$$(n_\alpha + km_\alpha)(\bar{v} - v) + (\left(\frac{1}{\Lambda} m_\alpha\right)_\alpha - k\Lambda n)(\bar{w} - w) \le 0, \tag{5.366}$$

$$v_\alpha + k\Lambda w = Mn, \tag{5.367}$$

$$(kv)_\alpha - \left(\frac{1}{\Lambda} w_\alpha\right)_\alpha = M_1 m + \xi, \tag{5.368}$$

$$|m| \le C^\star, \quad \xi(\bar{m} - m) \le 0 \quad \forall \bar{m}, \quad |m| \le C^\star, \tag{5.369}$$

$$v = w = m = 0, \quad \alpha = 0, 1. \tag{5.370}$$

The inequality (5.366) should be fulfilled for all functions $(\bar{v}, \bar{w})$ satisfying (5.365). We assume the existence of at least one element $(v^0, w^0) \in B$. The point $(0, 0)$ is supposed not to belong to B.

Theorem 5.15. *Assume that $a, b \in H^2(I)$, $\Lambda \geq \kappa > 0$ on I, $\kappa = const$, and the norm of k in the space $W^1_\infty(I)$ is sufficiently small. Then there exist functions v, w, m, n, ξ satisfying* (5.367), (5.368) *and*

$$v,\ w,\ m,\ m_t \in L^\infty(0, T; H^1_0(I)), \tag{5.371}$$

$$n,\ n_t \in L^2(G), \quad \xi \in L^2(0, T; H^{-1}(I)), \tag{5.372}$$

$$m(t) \in K, \quad (v(t), w(t)) \in B \quad \text{a.e. on } (0, T), \tag{5.373}$$

$$\int_0^T \langle n_\alpha + k m_\alpha, \bar{v} - v \rangle \, dt + \int_0^T \langle \left(\frac{1}{\Lambda} m_\alpha \right)_\alpha - k\Lambda n, \bar{w} - w \rangle \, dt \leq 0 \tag{5.374}$$

$$\forall\, \bar{v}, \bar{w} \in L^2(0, T; H^1_0(I)), \quad (\bar{v}(t), \bar{w}(t)) \in B,$$

$$\int_0^T \langle \xi, \bar{m} - m \rangle \, dt \leq 0, \quad \forall\, \bar{m} \in L^2(0, T; H^1_0(I)), \quad \bar{m}(t) \in K. \tag{5.375}$$

PROOF. Consider the regularized problem with three positive parameters ε, δ, λ,

$$-\varepsilon v_{\alpha\alpha} - n_\alpha - k m_\alpha + \frac{1}{\delta} q_1(v, w) = 0, \tag{5.376}$$

$$-\varepsilon w_{\alpha\alpha} - \left(\frac{1}{\Lambda} m_\alpha \right)_\alpha + k\Lambda n + \frac{1}{\delta} q_2(v, w) = 0, \tag{5.377}$$

$$-v_\alpha - k\Lambda w + M n = 0, \tag{5.378}$$

$$-M_1 m + \left(\frac{1}{\Lambda} w_\alpha \right)_\alpha - (kv)_\alpha + \frac{1}{\lambda} p(m) = 0, \tag{5.379}$$

$$v = w = m = 0, \quad \alpha = 0, 1. \tag{5.380}$$

To obtain a priori estimates of the solution of the auxiliary problem (5.376)–(5.380) we multiply (5.376)–(5.379) by $v - v^0$, $w - w^0$, n, m, respectively, integrate over Q and sum. The simple calculations lead to the following estimates:

$$\varepsilon \|v_\alpha\|^2_{L^2(Q)} + \varepsilon \|w_\alpha\|^2_{L^2(Q)} \leq c, \tag{5.381}$$

$$\|\Lambda^{-1/2} m_\alpha\|_{L^2(Q)} + \|\Lambda^{1/2} n\|_{L^2(Q)} \leq c, \tag{5.382}$$

$$\| \int_0^T \Lambda^{-1/2} m_\alpha \, d\tau \| + \| \int_0^T \Lambda^{1/2} n \, d\tau \| \leq c. \tag{5.383}$$

The equations (5.376)–(5.379) could be considered when $t = 0$. In this case we see that the obtained equations with the boundary condition (5.380) exactly coincide with the elliptic boundary value problem (5.285)–(5.289). The a priori estimate of the corresponding solution v_1, w_1, m_1, n_1 is as follows,

$$\|m_1\|_1 + \|n_1\| + \varepsilon\|v_1\|_1^2 + \varepsilon\|w_1\|_1^2 \leq c, \tag{5.384}$$

with the constant c uniform in $\varepsilon \leq \varepsilon_0$, δ, λ. This estimate can be derived by multiplying (5.376)–(5.379), taken for $t = 0$, by $v_1 - v^0$, $w_1 - w^0$, n_1, m_1, respectively.

Now, we can differentiate the equations (5.376)–(5.379) with respect to t and multiply by v_t, w_t, n_t, m_t. The penalty terms are nonnegative, hence the following estimates can be derived:

$$\varepsilon\|v_{t\alpha}\|_{L^2(Q)}^2 + \varepsilon\|w_{t\alpha}\|_{L^2(Q)}^2 \leq c, \tag{5.385}$$

$$\|\Lambda^{-1/2} m_{t\alpha}\|_{L^2(Q)} + \|\Lambda^{1/2} n_t\|_{L^2(Q)} \leq c, \tag{5.386}$$

$$\|\Lambda^{-1/2} m_\alpha(T)\| + \|\Lambda^{1/2} n(T)\| \leq c. \tag{5.387}$$

In doing so the functions m_1, n_1 satisfying (5.384) are taken as the initial values for m, n when $t = 0$. Despite the dependence of v_1, w_1, m_1, n_1 on the parameters ε, δ, λ the estimate (5.384) is uniform in $\varepsilon \leq \varepsilon_0$, δ, λ. As a result, the estimates (5.385)–(5.387) are also uniform in $\varepsilon \leq \varepsilon_0$, δ, λ. To obtain an estimate of $\lambda^{-1}p(m)$ the arguments of Section 5.7 can be used; see the estimate (5.347). So, we can assume that $\lambda^{-1}p(m)$ are bounded in $L^2(0,T;H^{-1}(I))$. By that, the equations (5.378), (5.379) can be written as follows,

$$v_\alpha + k\Lambda w = h_1, \tag{5.388}$$

$$-\left(\frac{1}{\Lambda} w_\alpha\right)_\alpha + (kv)_\alpha = h_2, \tag{5.389}$$

where h_1 are bounded in $L^2(Q)$, and h_2 are bounded in $L^2(0,T;H^{-1}(I))$. By the boundary conditions (5.380), from (5.388) one has

$$v(\alpha) = \int_0^\alpha (h_1 - k\Lambda w)\, d\alpha.$$

It is seen that after a substitution of this value in (5.389) the estimate

$$\|w_\alpha\|_{L^2(Q)} \leq c \tag{5.390}$$

follows provided that the norm of k in the space $W_\infty^1(I)$ is sufficiently small. Consequently, from (5.388) we obtain

$$\|v_\alpha\|_{L^2(Q)} \leq c. \tag{5.391}$$

To establish the solvability of the problem (5.376)–(5.380) for fixed ε, δ, λ, we can use Theorem 1.14. In this case it suffices, actually, to take into account the estimates (5.381)–(5.383). Let the solution of the problem (5.376)–(5.380) be denoted by v^ε, w^ε, m^ε, n^ε. Choosing a subsequence, if necessary, we can assume that as $\varepsilon \to 0$

$$\varepsilon v^\varepsilon,\ \varepsilon w^\varepsilon \ \to 0 \quad \text{weakly in } L^2(0,T;H_0^1(I)),$$

$$m^\varepsilon \ \to m^\lambda \quad \text{strongly in } L^2(Q),$$

$$v^\varepsilon,\ w^\varepsilon,\ m^\varepsilon,\ m_t^\varepsilon \ \to v^\lambda,\ w^\lambda,\ m^\lambda,\ m_t^\lambda \quad \text{weakly in } L^2(0,T;H_0^1(I)),$$

$$Mn^\varepsilon,\ n_t^\varepsilon,\ \int_0^t \frac{1}{\Lambda} m_\alpha^\varepsilon \, d\tau \ \to Mn^\lambda,\ n_t^\lambda,\ \int_0^t \frac{1}{\Lambda} m_\alpha^\lambda \, d\tau \quad \text{weakly in } L^2(Q).$$

Consequently, the passage to the limit as $\varepsilon \to 0$ in (5.376)–(5.379) implies

$$-n_\alpha^\lambda - km_\alpha^\lambda + \frac{1}{\delta} q_1(v^\lambda, w^\lambda) = 0, \tag{5.392}$$

$$-\left(\frac{1}{\Lambda} m_\alpha^\lambda\right)_\alpha + k\Lambda n^\lambda + \frac{1}{\delta} q_2(v^\lambda, w^\lambda) = 0, \tag{5.393}$$

$$-v^\lambda - k\Lambda w^\lambda + Mn^\lambda = 0, \tag{5.394}$$

$$-M_1 m^\lambda + \left(\frac{1}{\Lambda} w_\alpha^\lambda\right)_\alpha - (kv^\lambda)_\alpha + \frac{1}{\lambda} p(m^\lambda) = 0. \tag{5.395}$$

As usual, we use the monotonicity of the operator (q_1, q_2) to justify the passage to the limit of the terms $q_i(v^\varepsilon, w^\varepsilon)$. The next step is the passage to the limit as $\lambda \to 0$. For a subsequence, with the same notation, the following convergence can be assumed as $\lambda \to 0$:

$$v^\lambda,\ w^\lambda,\ m^\lambda,\ m_t^\lambda \ \to v^\delta,\ w^\delta,\ m^\delta,\ m_t^\delta \quad \text{weakly in } L^2(0,T;H_0^1(I)),$$

$$Mn^\lambda,\ n_t^\lambda,\ \int_0^t \frac{1}{\Lambda} m_\alpha^\lambda \, d\tau \ \to Mn^\delta,\ n_t^\delta,\ \int_0^t \frac{1}{\Lambda} m_\alpha^\delta \, d\tau \ \text{weakly in } L^2(Q), \tag{5.396}$$

$$\frac{1}{\lambda} p(m^\lambda) \ \to \xi^\delta \quad \text{weakly in } L^2(0,T;H^{-1}(I)).$$

It follows from (5.393) that m^λ are bounded in $L^2(0,T;H^2(I) \cap H_0^1(I))$ nonuniformly in δ, in general. Since m_t^λ are bounded in $L^2(0,T;H_0^1(I))$, we can additionally suppose that for every fixed δ

$$m^\lambda \ \to m^\delta \quad \text{strongly in } L^2(0,T;H_0^1(I)). \tag{5.397}$$

Consequently, it is clear that, on the one hand, the equations (5.394), (5.395) provide

$$-v_\alpha^\delta - k\Lambda w^\delta + Mn^\delta = 0, \tag{5.398}$$

$$-M_1 w^\delta + \left(\frac{1}{\Lambda} w_\alpha^\delta\right)_\alpha - (kv^\delta)_\alpha + \xi^\delta = 0. \tag{5.399}$$

On the other hand, the equation (5.395) can be multiplied by $\bar{m} - m^\lambda$, $\bar{m} \in L^2(0,T; H_0^1(I))$, $\bar{m}(t) \in K$, and the inequality can be obtained. By (5.396), (5.397), we can pass to the limit $\lambda \to 0$ in this inequality, which implies

$$\int_0^T \langle \xi^\delta, \bar{m} - m^\delta \rangle dt \leq 0 \quad \forall \bar{m} \in L^2(0,T; H_0^1(I)), \quad \bar{m}(t) \in K,$$

where ξ^δ is defined from (5.399). In this case, the inclusion $m^\delta(t) \in K$ a.e. on $(0,T)$ follows. In addition to this, the limiting functions v^δ, w^δ, m^δ, n^δ satisfy the equations

$$-n_\alpha^\delta - km_\alpha^\delta + \frac{1}{\delta} q_1(v^\delta, w^\delta) = 0, \quad -\left(\frac{1}{\Lambda} m_\alpha^\delta\right)_\alpha + k\Lambda n^\delta + \frac{1}{\delta} q_2(v^\delta, w^\delta) = 0.$$

Similar arguments can be used to justify the convergence as $\delta \to 0$ which lead to the relations (5.367), (5.368), (5.371)–(5.375). Theorem 5.15 is completely proved.

5.9 Conclusion

To complete the book we have to glance at the results presented. The principal point enveloping the analysed problems consists in studying nonpenetration conditions of inequality type at the crack faces. Different models are considered for two- and three-dimensional bodies, plates and shells with various constitutive laws: elastic, viscoelastic, thermoelastic and elastoplastic. The main focus is on qualitative properties of solutions of boundary value problems. We prove the existence of solutions and derive the solution properties resulting from the mathematical problem formulation. Meanwhile there are many open questions related to the problems considered in the monograph, for example, it is interesting to know the precise asymptotics near the crack tips of sulutions to boundary value problems describing bodies with cracks. We do hope that the book will give rise to many investigations allied to the problems discussed.

Notation

$R^m, \quad m = 1, 2, 3:$ space of m-tuples $x = (x_1, ..., x_m)$ of real numbers

$(x_1, x_2, x_3); \ (x, y, z):$ Descartes coordinates in R^3

$t \in [0, T], \ T > 0:$ time variable

$\Omega:$ domain in R^m (body, plate, bar)

$Q = \Omega \times (0, T):$ cylinder $\{x \in \Omega, \ t \in (0, T)\}$

$\Gamma:$ boundary of Ω

$\overline{\Omega} = \Omega \cup \Gamma:$ domain with its boundary

$\Gamma_c; \ \Xi_l; \ \Gamma_\psi, \ y = \psi(x):$ crack surface inside Ω

$\partial\Gamma_c:$ boundary of Γ_c

$\Sigma_c = \Gamma_c \setminus \partial\Gamma_c, \ \Gamma_c = \overline{\Sigma}_c:$ crack surface without its boundary

$\Sigma: \ \Gamma_c \subset \Sigma, \ \overline{\Omega} = \overline{\Omega}_1 \cup \overline{\Omega}_2, \ \Sigma = \overline{\Omega}_1 \cap \overline{\Omega}_2:$ extension of Γ_c

$\Gamma_c^\pm:$ positive and negative faces of Γ_c

$\Omega_c = \Omega \setminus \Gamma_c:$ domain without the crack surface

$\partial\Omega_c = \Gamma \cup \Gamma_c^+ \cup \Gamma_c^-:$ boundary of Ω_c

$(\,\cdot\,)|_{\Gamma_c^\pm}; \ (\,\cdot\,)^\pm:$ traces of a function at the crack faces

$[\,\cdot\,] = (\,\cdot\,)|_{\Gamma_c^+} - (\,\cdot\,)|_{\Gamma_c^-}:$ jump of a function at the crack

$n = (n_1, n_2, n_3); \ \nu = (\nu_1, \nu_2):$ unit outer normal vector at a boundary

$\tau = (\tau_1, \tau_2, \tau_3); \ s = (s_1, s_2):$ unit tangential vector at a boundary

$z = \phi(x):$ punch shape

$u = (u_1, u_2, u_3):$ displacements in a body

$\chi = (W, w), \ W = (w_1, w_2):$ displacements in a shell (plate)

(u_n, u_τ); (u_ν, u_s) : normal and tangential components of the displacement vector at a boundary

$\varepsilon = \{\varepsilon_{ij}\}$, $i,j = 1,2,3$: strain tensor

$\sigma = \{\sigma_{ij}\}$, $i,j = 1,2,3$: stress tensor

(σ_n, σ_τ); (σ_ν, σ_s) : normal and tangential components of the stress vector at a boundary

ε_{ij}; e_{ij}, $i,j = 1,2$: strains in a shell

N_{ij}; σ_{ij}, $i,j = 1,2$: integrated stresses in a shell

k_{ij}, $i,j = 1,2$: curvatures of a shell

$M = \{M_{ij}\}$; $m = \{m_{ij}\}$, $i,j = 1,2$: bending moments

Q_i, $i = 1,2$: transverse forces

$M(\cdot)$; $m(\cdot)$: bending moment at a boundary

$R(\cdot)$; $t(\cdot)$: transverse force at a boundary

λ, μ : Lamé parameters

E : Young's modulus

κ, $0 < \kappa < 1/2$: Poisson ratio

h; ε : thickness of a shell

$G = 2Eh/(1-\kappa^2)$; $D = 2Eh^3/3(1-\kappa^2)$: constants

$f = (f_1, f_2, f_3)$: external forces

θ : temperature function

β_{ij}, $i,j = 1,2,3$: thermal expansion coefficients

$\xi = \{\xi_{ij}\}$, $i,j = 1,2,3$: plastic deformations

$\Phi(\cdot)$: yield surface

$\Pi(\cdot)$: potential energy of a deformed body

$B(\cdot,\cdot)$; $b(\cdot,\cdot)$: bilinear forms in the plate theory

K : set of admissible displacements

a.e. : almost everywhere

$J(\cdot)$: cost functional

$p(\cdot)$: penalty operator

$\mu(A)$: measure of the set A

$X^\star$: space dual of X

$\mathcal{O}(x^0)$: neighbourhood of the point x^0

$R_\delta(x^0)$; $B_\delta(x^0)$: ball of the radius δ centred at the point x^0

$\dot{v}$: derivative of v with respect to t

D^α, $\alpha = (\alpha_1, ..., \alpha_m)$: derivatives of the order $|\alpha| = \alpha_1 + ... + \alpha_m$

Δ : Laplace operator

$C^k(\Omega)$, $k = 0, 1, 2, ...$; $C^0(\Omega) = C(\Omega)$: space of functions having k continuous derivatives in Ω

$C^{k,1}(\Omega)$, $k = 0, 1, 2, ...$: space of functions having k Lipschitz continuous derivatives in Ω

$C_0^k(\Omega)$, $k = 0, 1, 2, ...$: space of functions from $C^k(\Omega)$ with compact support in Ω

$C_0^\infty(\Omega)$: space of infinitely differentiable functions with compact supports in Ω

$L^1(\Omega)$: space of functions, absolutely integrable in Ω

$W_1^k(\Omega)$, $k = 1, 2, ...$: space of functions having k derivatives from $L^1(\Omega)$

$L^\infty(\Omega)$: space of functions, bounded almost everywhere in Ω

$L^2(\Omega)$: space of functions, integrable with square in Ω

$L^2_{loc}(\Omega)$: space of functions, integrable with square in any compact subdomain of Ω

$H^k(\Omega)$, $k = 0, 1, ...$; $H^0(\Omega) = L^2(\Omega)$: Sobolev space of functions having k derivatives from $L^2(\Omega)$

$H^k_{loc}(\Omega)$, $k = 0, 1, ...$: space of functions having k derivatives which are integrable with square in any compact subdomain of Ω

$H_0^k(\Omega)$, $k = 1, 2, ...$: completion of $C_0^\infty(\Omega)$ in $H^k(\Omega)$-norm

$H^{-k}(\Omega)$, $k = 1, 2, ...$: space dual of $H_0^k(\Omega)$

$H^{k-1/2}(\Gamma)$, $k = 1, 2, ...$: space of traces of functions from $H^k(\Omega)$ at the boundary Γ

$H^{-k+1/2}(\Gamma) = H^{k-1/2}(\Gamma)^\star$, $k = 1, 2, ...$: space dual of $H^{k-1/2}(\Gamma)$

$\langle \cdot , \cdot \rangle_{\alpha,\Gamma}$, $\alpha \geq 0$: duality pairing between $H^\alpha(\Gamma)$ and $H^\alpha(\Gamma)^\star$

$H^{k,0}(\Omega_c)$; $H^{k,\Gamma}(\Omega_c)$, $k = 1, 2, ...$: space of functions u from $H^k(\Omega_c)$ with $u = \partial u/\partial n = ... = \partial^{k-1}u/\partial n^{k-1} = 0$ on Γ

$H_{00}^{k-1/2}(\Sigma_c)$, $k = 1, 2, ...$: functions from $H^{k-1/2}(\Sigma_c)$ with compact support in Σ_c

$L^1(0, T; H)$: space of functions $(0, T) \mapsto H$, absolutely integrable on $(0, T)$

$L^\infty(0, T; H)$: space of functions bounded almost everywhere on $(0, T)$ with values in the space H

$H^k(0, T; H)$, $k = 0, 1, ...$: space of functions from $H^k(0, T)$ with values in the space H

$M^1(\Omega)$: space of measures, bounded in Ω

$R(\Omega)$: space of rigid displacements in Ω

$LD(\Omega)$: Banach space of integrable deformations in $L^1(\Omega)$

$BD(\Omega)$: Banach space of bounded deformations in $M^1(\Omega)$

References

Adams R.A. (1975): Sobolev spaces. Academic Press, New York.

Annin B.D., Cherepanov G.P. (1983): Elasto-plastic problem. Nauka, Novosibirsk (in Russian).

Antontsev S.N., Hoffmann K.-H., Khludnev A.M. (Eds.) (1992): Free boundary problems in continuum mechanics. Inter. Ser. Numer. Math., Birkhäuser Verlag.

Antontsev S.N., Kazhikhov A.V., Monakhov V.N. (1990): Boundary value problems in mechanics of homogeneous fluid. Studies in Math. Their Appl. **22**, North-Holland.

Anzellotti G. (1983): On the existence of the rates of stress and displacement for Prandtl-Reuss plasticity. Quart. Appl. Math. J. **41**, 181–208.

Anzellotti G., Giaquinta M. (1982): On the existence of the fields of stresses and displacements for an elasto-perfectly plastic body in static equilibrium. J. Math. Pure Appl. **61**, 219–244.

Arutunyan N.H., Drozdov A.D., Naumov V.E. (1987): Mechanics of growing visco-elasto-plastic bodies. Nauka, Moscow (in Russian).

Athanasopoulos I. (1981): Stability of the coincidence set for the Signorini problem. Ind. Univ. Math. J. **30** (2), 235–247.

Attouch H., Picard C. (1983): Variational inequalities with varying obstacles: The general form of the limit problem. J. Funct. Anal. **50** (3), 329–386.

Baiocchi C., Capelo A. (1984): Variational and quasivariational inequalities. Applications to free boundary problems. Wiley, Chichester.

Banichuk N.V. (1970): The small parameter method in finding a curvilinear crack shape. Izvestiya USSR Acad. Sci., Mechanics of Solid (2), 130–137 (in Russian).

Banichuk N.V. (1980): Shape optimization of elastic bodies. Nauka, Moscow (in Russian).

Barbu V. (1984): Optimal control of variational inequality. Research Notes Math. **100**. Pitman, Boston.

Bock I., Lovišek J. (1987): Optimal control problems for variational inequalities with controls in coefficients and in unilateral constraints. Apl. Mat. **32** (4), 301–314.

Brokate M., Khludnev A.M. (1998): Regularization and existence of solutions of three-dimensional elastoplastic problems. Math. Methods in the Applied Sciences **21**, 551–564.

Bui H.D., Ehrlacher A. (1997): Developments of fracture mechanics in France in the last decades. In: Fracture Research in Retrospect. H.P. Rossmanith (Ed.), A.A.Balkema, Rotterdam, 369–387.

Caffarelli L.A., Friedman A. (1979): The obstacle problem for the biharmonic operator. Ann. Scuola Norm. Sup. Pisa, serie IV **6**, 151–184.

Caffarelli L.A., Friedman A., Torelli A. (1982): The two-obstacle problem for the biharmonic operator. Pacific J. Math. **103** (3), 325–335.

Carstensen C. (1994): Interface problems in viscoplasticity and plasticity. SIAM J. Math. Anal. **25** (6), 1468–1487.

Céa J. (1971): Optimisation, théorie et algorithms. Dunod, Paris.

Cherepanov G.P. (1979): Mechanics of brittle fracture. McGraw-Hill.

Cherepanov G.P. (1983): Fracture mechanics of composite materials. Nauka, Moscow (in Russian).

Chernous'ko F.L., Banichuk N.V. (1973): Variational problems of mechanics and control. Nauka, Moscow (in Russian).

Ciarlet P.G., Sanchez-Palencia F. (1996): An existence and uniqueness theorem for the two-dimensional linear membrane shell equations. J. Math. Pures Appl. **75** (1), 51–67.

Costabel M., Dauge M. (1994): Stable asymptotics for elliptic systems on plane domains with corners. Comm. in Part. Diff. Eqs. **19** (9-10), 1677–1726.

Dauge M. (1988): Elliptic boundary value problems on the corner domains. Lect. Notes in Math. **1341**, Springer-Verlag.

Demengel F. (1983): Problèmes variationels en plasticité parfaite des plaques. Numer. Func. Anal. and Optimiz. **6**, 73–119.

Demkowicz L., Oden J.T. (1982): On some existence results in contact problems with nonlocal friction. Nonlinear Anal. Theory, Meth. and Appl. **6** (10), 1075–1093.

Destuynder P., Jaoua M. (1981): Sur une interprétation mathématique de l'intégrale de Rice en théorie de la rapture fragile. Math. Meth. in Appl. Sci. **3**, 70–87.

Donnell L.H. (1976): Beams, plates and shells. McGraw-Hill.

Duduchava R., Wendland W. (1995): The Wiener-Hopf method for system of pseudodifferential equations with applications to crack problems. Integr. Eqs. and Oper. Theory **23**, 294–335.

Duvaut G., Lions J.-L. (1972): Les inéquations en mécanique et en physique. Dunod, Paris.

Ekeland I., Temam R. (1976): Convex analysis and variational problems. North-Holland, Amsterdam, Oxford.

Elliot C.M., Ockendon J.R. (1982): Weak and variational methods for moving boundary problems. Pitman, Research Notes Math. **59**.

Erkhov M.I. (1978): Theory of perfectly plastic bodies and structures. Nauka, Moscow (in Russian).

Fichera G. (1972): Boundary value problems of elasticity with unilateral constraints. In: Handbuch der Physik, Band 6a/2, Springer-Verlag.

Frehse J. (1973): On the regularity of the solution of the biharmonic variational inequality. Manuscr. Math. **9**, 91–103.

Frehse J. (1975): Two dimensional variational problems with thin obstacles. Math. Z. **143** (3), 279–288.

Friedman A., Lin Y. (1996): Propagation of cracks in elastic media. Arch. Rat. Mech. Anal. **136**, 235–290.

Gajewski H., Gröger K., Zacharias K. (1974): Nichtlineare operatorgleichungen und operatordifferentialgleichungen. Akademie-Verlag, Berlin.

Galin L.A. (1980): Contact problems in elasticity and viscoelasticity. Nauka, Moscow (in Russian).

Gilbert R.P., Shi P., Shillor M. (1990): A quasistatic contact problem in linear thermoelasticity. Rend. Mat. Appl. **7** (10), 785–808.

Giusti E. (1984): Minimal surfaces and functions of bounded variations. Birkhäuser, Boston, Basel, Stuttgart.

Glowinski R., Lions J.L., Tremolieres R. (1976): Analyse numérique des inéquations variationnelles. Dunod, Paris.

Goldshtein R., Entov V. (1989): Qualitative methods in continuum mechanics. Nauka, Moscow (in Russian).

Grigolyuk E.I., Tolkachev V.M. (1980): Contact problems in the theory of plates and shells. Mashinostroenie, Moscow (in Russian).

Grisvard P. (1985): Elliptic problems in nonsmooth domains. Pitman, Boston, London, Melbourne.

Grisvard P. (1991): Singularities in boundary value problems. Masson, Springer-Verlag.

Haslinger J., Neittaanmäki P., Tiihonen T. (1986): Shape optimization in contact problems based on penalization of the state inequality. Apl. Mat. **31** (1), 54-77.

Haslinger J., Panagiotopoulos P.D. (1984): The reciprocal variational approach to the Signorini problem with friction. Approximation results. Proc. Roy. Soc. Edinburgh **98A**, 365–383.

Hlavaček I., Haslinger J., Nečas J., Lovišek J. (1988): Solution of variational inequalities in mechanics. Springer-Verlag, New York.

Hlavaček I., Nečas J. (1970): On inequalities of Korn's type. Arch. Rat. Mech. Anal. **36** (4), 305–334.

Hoffmann K.-H., Sprekels J. (Eds.) (1990): Free boundary value problems. Inter. Ser. Numer. Math. Birkhäuser Verlag.

Johnson C. (1976): Existence theorems for plasticity problems. J. Math. Pures Appl. **55**, 431–444.

Kachanov L.M. (1974): Foundations of fracture mechanics. Nauka, Moscow (in Russian).

Kantorovich L.V., Akilov G.P. (1984): Functional analysis. Nauka, Moscow (in Russian).

Khludnev A.M. (1983): A contact problem of a linear elastic body and a rigid punch (variational approach). Appls. Maths. Mechs. **47** (6), 999–1005 (in Russian).

Khludnev A.M. (1988): On variational inequalities in contact plastic problems. Diff. Eqs. **24** (9), 1622–1628 (in Russian).

Khludnev A. M. (1992): Contact viscoelastoplastic problem for a beam. In: Free boundary problems in continuum mechanics. S.N.Antontsev, K.-H. Hoffmann, A.M.Khludnev (Eds.). Int. Series of Numerical Mathematics **106**, Birkhäuser Verlag, Basel, 159–166.

Khludnev A.M. (1993a): Contact elastoplastic problem for a beam. Siberian Math. J. **34**, 173–179.

Khludnev A.M. (1993b): Contact elastoplastic problem for a rod under the creep condition. In: Nonlinear Math. Prob. in Industry, Math. Sci. and Appl. **1**, 311–319.

Khludnev A.M. (1994): Existence of extreme unilateral cracks in a plate. Control and Cybernetics **23** (3), 453–460.

Khludnev A.M. (1995a): On contact problem for a plate having a crack. Control and Cybernetics **24** (3), 349–361.

Khludnev A.M. (1995b): The contact problem for a shallow shell with a crack. Appls. Maths. Mechs. **59** (2), 299–306.

Khludnev A.M. (1996a): Contact problem for a plate having a crack of minimal opening. Control and Cybernetics **25** (3), 605–620.

Khludnev A.M. (1996b): On a Signorini problem for inclusions in shells. Euro. Jnl. Appl. Math. **7**, 499–510.

Khludnev A.M. (1996c): On equilibrium problem for a plate having a crack under the creep condition. Control and Cybernetics **25** (5), 1015–1029.

Khludnev A.M. (1996d): The equilibrium problem for a thermoelastic plate with a crack. Siberian Math. J. **37** (2), 394–404.

Khludnev A.M. (1997a): Extreme crack shapes in a shallow shell. Adv. Mat. Sci. Appl. **7** (1), 213–221.

Khludnev A.M. (1997b): On equilibrium problem for the inclined crack in a plate. Appl. Mech. Tech. Phys. **38** (5), 117–121 (in Russian).

Khludnev A.M. (1997c): Contact of two plates, one of which contains a crack. J. Appl. Maths Mechs. **61** (5), 851–862.

Khludnev A.M. (1998): Regularization and solution existence in the equilibrium problem for elastoplastic plate. Siberian Math. J. **39** (3), 670–682.

Khludnev A.M., Hoffmann K.-H. (1992): A variational inequality in a contact elastoplastic problem for a bar. Adv. Math. Sci. Appl. **1** (1), 127–136.

Khludnev A.M., Sokolowski J. (1997): Modelling and control in solid mechanics. Birkhäuser, Basel, Boston, Berlin.

Khludnev A.M., Sokolowski J. (1998a): On solvability of boundary value problems in elastoplasticity. Control and Cybernetics **27** (2), 311–330.

Khludnev A.M., Sokolowski J. (1998b): Griffith formula and Rice integral for elliptic equations in nonsmooth domains. Les prepublications de l'Institute Elie Cartan (7), 19p.

Khludnev A.M., Sokolowski J. (1998c): Griffith formulae for elasticity systems with unilateral conditions. Rapport de recherche. INRIA (3447), 22p.

Khludneva E. Yu. (1990a): Optimal control of external forces in contact problems for a viscoelastic plate. In: Algebra and Math. Analysis. Novosibirsk, 8–14 (in Russian).

Khludneva E.Yu. (1990b): Contact of a nonelastic plate with a rigid punch. In: Dynamics of Continuum Media. Novosibirsk (98), 67–79 (in Russian).

Kikuchi N., Oden J.T. (1988): Contact problem in elasticity. SIAM, Philadelphia.

Kinderlehrer D. (1982): Estimates for the solution and its stability in Signorini's problem. Appl. Math. and Optim. **8** (2), 159–188.

Kinderlehrer D., Nirenberg L., Spruck J. (1979): Regularity in elliptic free boundary problems. II: Equations of higher order. Ann. Scuola Norm. Sup. Pisa **6**, 637–687.

Kinderlehrer D., Stampacchia G. (1980): An introduction to variational inequalities and their applications. New York, London, Toronto, Sydney, San Francisco.

Kondrat'ev V.A., Kopaček J., Oleinik O.A. (1982): On behaviour of solutions to the second order elliptic equations and elasticity equations in a neighbourhood of boundary points. Trudy Petrovsky Sem., Moscow Univ. **8**, 135–152 (in Russian).

Kondrat'ev V.A., Oleinik O.A. (1983): Boundary value problems for partial differential equations in nonsmooth domains. Uspekhi Mat. Nauk **38** (2), 3–76 (in Russian).

Kovtunenko V.A. (1993): An iterative methods for solving variational inequalities of the contact elastoplastic problem by the penalty method. Comp. Maths. Math. Phys. **33** (9), 1245–1249.

Kovtunenko V.A. (1994a): Convergence of the variational inequality solutions for a plate contacting a nonregular obstacle. Diff. Eqs. **30** (3), 488–492 (in Russian).

Kovtunenko V.A. (1994b): An iteration penalty method for variational inequalities with strongly monotonous operators. Siberian Math. J. **35** (4), 735–738.

Kovtunenko V.A. (1994c): Iteration penalty method for the contact elastoplastic problem. Control and Cybernetics **23** (4), 803–808.

Kovtunenko V.A. (1996a): Numerical solution of a contact problem for the Timoshenko bar model. Izvestiya Rus. Acad. Sci. Mechanics of Solid **5**, 79–84 (in Russian).

Kovtunenko V.A. (1996b): An iterative penalty method for a problem with constraints on the inner boundary. Siberian Math. J. **37** (3), 508–512.

Kovtunenko V.A. (1996c): Solution of the problem of a beam with a cut. Appl. Mech. Tech. Phys. **37** (4), 595–600.

Kovtunenko V.A. (1996d): Analytical solution of a variational inequality for a cutted bar. Control and Cybernetics **25** (4), 801–808.

Kovtunenko V.A. (1997a): Iterative approximations of penalty operators. Numer. Funct. Anal. and Optimiz. **18** (3&4), 383–387.

Kovtunenko V.A. (1997b): Iterative penalty method for plate with a crack. Adv. Math. Sci. Appl. **7** (2), 667–674.

Kovtunenko V.A. (1998): Variational and boundary value problems with a friction on the internal boundary. Siberian Math. J. **39** (5), 1060–1073 (in Russian).

Kovtunenko V.A., Leontyev A.N., Khludnev A.M. (1998): On equilibrium problem for a plate with an inclined cut. Appl. Mech. Tech. Phys. **39** (2), 164–174 (in Russian).

Kravchuk A.S. (1997): Variational and quasivariational inequalities in mechanics. MGAPI, Moscow (in Russian).

Landkof N.S. (1966): Foundations of a modern theory of a potential. Nauka, Moscow (in Russian).

Lions J.-L. (1968a): Contrôle optimal de systèmes gouvernés par des equations aux dérivées partielles. Dunod Gauthier-Villars, Paris.

Lions J.-L. (1968b): Contrôle des systêmes distribués singuliers. Dunod Gauthier-Villars, Paris.

Lions J.-L. (1969): Quelques méthodes de résolution des problémes aux limites non linéaires. Dunod Gauthier-Villars, Paris.

Lions J.-L., Magenes E. (1968): Problémes aux limites non homogénes et applications, volume 1. Dunod, Paris.

Litvinov V.G. (1987): An optimization in elliptic boundary problems with applications to mechanics. Nauka, Moscow (in Russian).

Lukasiewicz S. (1979): Local loads in plates and shells. PWN-Polish, Warszawa. Noordhoff, Leyden.

Maz'ya V.G. (1985): Spaces of S.L. Sobolev. Leningrad Univ. (in Russian).

Maz'ya V.G., Nazarov S.A. (1987): Asymptotics of energy integrals for small perturbations of the boundary near corner and conical points. Trudy Moscow Math. Soc. **50**, 79–129 (in Russian).

Mignot F. (1976): Contróle dans les inéquations elliptiques. J. Func. Anal. **22**, 130–185.

Mikhailov B.K. (1980): Plates and shells with break parameters. Leningrad (in Russian).

Mikhailov V.P. (1976): Partial differential equations. Nauka, Moscow (in Russian).

Moet H.J.K. (1982): Asymptotic analysis of the boundary in singularly perturbed elliptic variational inequalities. Lect. Notes Math. **942**, 1–17.

Morel J.-M., Solimini S. (1995): Variational methods in image segmentation. Birkhäuser, Boston, Basel, Berlin.

Morozov N.F. (1984): Mathematical foundation of the crack theory. Nauka, Moscow (in Russian).

Mosco U. (1969): Convergence of convex sets and of solutions of variational inequalities. Adv. Math. **3** (4), 510–585.

Mosolov V.P., Myasnikov P.P. (1971): Variational methods in flow theory of perfect-visco-plastic media. Moscow Univ. (in Russian).

Movchan A.B., Movchan N.V. (1995): Mathematical modelling of solids with nonregular boundaries. CRC Press, Boca Raton, New York, London, Tokyo.

Mróz Z. (1963): Limit analysis of plastic structures subject to boundary variations. Arch. Mech. Stos **15**, 63–76.

Namm R.V. (1995): On uniqueness of smooth solution to the static problem with the Coulomb friction condition and contact. Appls. Maths. Mechs. **59** (2), 330–335 (in Russian).

Nazarov S.A. (1989): Derivation of variational inequality for the small increment of a crack shape. Izvestiya USSR Acad. Sci. Mechanics of Solid (2), 152–160 (in Russian).

Nazarov S.A., Plamenevskii B.A. (1991): Elliptic problems in domains with piecewise smooth boundaries. Nauka, Moscow (in Russian).

Nečas J. (1975): On regularity of solutions to nonlinear variational inequalities for second-order elliptic systems. Rend. Math. **8**, 481–495.

Nicaise S. (1992): About the Lamé system in a polygonal or polyhedral domain and coupled problem between the Lamé system and the plate equation. 1. Regularity of the solutions. Ann. Scuola Norm. Super. Pisa, Serie IV **19**, 327–361.

Nowacki W. (1962): Problems of thermoelasticity. USSR Acad. Sci., Moscow (in Russian).

Ohtsuka K. (1986): Generalized G-integral and three-dimensional fracture mechanics. Surface crack problems. Hirosima Math. J. **16** (2), 327–352.

Ohtsuka K. (1994): Mathematical aspects of fracture mechanics. Lect. Notes in Num. Appl. Anal. **13**, 39–59.

Oleinik O.A., Kondratiev V.A., Kopaček J. (1981): Asymptotic properties of solutions of a biharmonic equation. Diff. Eqs. **17** (10), 1886–1889 (in Russian).

Oleinik O.A., Shamaev A.S., Yosifian G.A. (1992): Mathematical problems in elasticity and homogenization. North-Holland, Amsterdam, London, New York, Tokyo.

Osadchuk V.A. (1985): Stress-strain state and limiting equilibrium of shells with cuts. Naukova Dumka, Kiev (in Russian).

Panagiotopoulos P.D. (1985): Inequality problems in mechanics and applications. Birkhäuser, Boston, Basel, Stuttgart.

Panasyuk V.V., Andreikiv A.V., Kovchik S.E. (1977): Methods of estimating the fracture toughness of structural materials. Kiev (in Russian).

Parton V.Z., Morozov E.M. (1985): Mechanics of elastoplastic fracture. Nauka, Moscow (in Russian).

Parton V.Z., Perlin P.I. (1981): Methods of the mathematical theory of elasticity. Nauka, Moscow (in Russian).

Pironneau O. (1984): Optimal shape design for elliptic systems. Springer, New York.

Plotnikov P.I. (1995): On a class of curves arising in a free boundary problem for Stokes flow. Siberian Math. J. **36** (3), 619–627.

Puel J.P. (1987): Some results on optimal control for unilateral problems. Cont. Partial Diff. Equat. Lect. Notes in Control and Infor. Sci. **114**, Springer-Verlag, 225–235.

Rabotnov Yu.N. (1979): Mechanics of a deformed solid body. Nauka, Moscow (in Russian).

Reshetnyak Y.G. (1982): Stability theorems in geometry and analysis. Nauka, Novosibirsk (in Russian).

Rice J.R. (1968): A path-independent integral and the approximate analysis of strain concentration by notches and cracks. J. Appl. Mech. **35**, 379–386.

Rice J.R., Drucker D. (1967): Energy changes in stressed bodies due to void and crack growth. Int. J. Fracture Mech. **3** (1), 19–27.

Roubiček T. (1997): Relaxation in optimization theory and variational calculus. W de Gruyter, Berlin.

Sadovskii V.M. (1992): To the problem of constructing weak solutions in dynamic elastoplasticity. Int. Ser. Numer. Math. **106**, 283–291.

Sadovskii V.M. (1997): Discontinuous solutions in dynamic elastic-plastic problems. Nauka, Moscow (in Russian).

Sändig A.M., Richter U., Sändig R. (1989): The regularity of boundary value problem for the Lamé equations in a polygonal domain. Rostock. Math. Kolloq. **36**, 21–50.

Schaeffer D.G. (1975): A stability theorem for the obstacle problem. Adv. Math. **16**, 34–47.

Schild B. (1984): A regularity result for polyharmonic variational inequalities with thin obstacles. Ann. Scuola Norm. Sup. Pisa **11** (1), 87–122.

Schuss Z. (1976): Singular perturbations and the transition from thin plate to membrane. Proc. Am. Math. Soc. **58**, 139–147.

Schwartz L. (1967): Analyse mathematique. Vol. 1, Hermann.

Shi P., Shillor M. (1992): Existence of a solution to the N-dimensional problem of thermoelastic contact. Comm. Partial Diff. Eqs. **17** (9&10), 1597–1618.

Sokolowski J., Zolesio J.-P. (1992): Introduction to shape optimization. Shape sensitivity analysis. Springer-Verlag.

Strang G., Temam R. (1980): Functions of bounded deformation. Arch. Rat. Mech. Anal. **75**, 7–21.

Suquet P.M. (1980): Mathematical problems in the theory of elastoplastic plates. Letters in Appl. and Eng. Sci. **18**, 527–539.

Telega J.J. (1987): Variational methods in contact problems of mechanics. Adv. Mech. **10** (2), 3–95 (in Russian).

Telega J.J., Lewinski T. (1994): Mathematical aspects of modelling the macroscopic behaviour of cross-ply laminates with intralaminar cracks. Control and Cybernetics **23** (3), 773–792.

Temam R. (1979): Navier-Stokes equations. Theory and numerical analysis. North-Holland, Amsterdam, New-York, Oxford.

Temam R. (1983): Problemes mathematiques en plasticite. Gauthier-Villars, Paris.

Temam R. (1986): A generalized Norton-Hoff model and the Prandtl-Reuss law of plasticity. Arch. Rat. Mech. Anal. **95**, 137–181.

Temam R., Strang G. (1980): Duality and relaxation in the variational problems of plasticity. J. de Mécanique **19** (3), 493–527.

Timoshenko S.P., Goodier J.N. (1951): Theory of elasticity. McGraw-Hill, New York.

Vainberg M.M. (1972): A variational method and monotonous operators method. Nauka, Moscow (in Russian).

Vladimirov V.S. (1981): Equations of mathematical physics. Nauka, Moscow (in Russian).

Vol'mir A.S. (1972): Nonlinear dynamics of plates and shells. Nauka, Moscow (in Russian).

Vorovich I.I., Lebedev L.P. (1972): Existence of solutions in nonlinear theory of shallow shells. Appls. Maths. Mechs. **36** (4), 691–704 (in Russian).

Washizu K. (1968): Variational methods in elasticity and plasticity. Pergamon Press.

Yakunina G.V. (1981): Smoothness of solutions of variational inequalities. Partial differential equations. Spectral theory. Leningrad Univ. (8), 213–220 (in Russian).

Zuazua E. (1995): Controllability of the linear system of thermoelasticity. J. Math. Pures Appl. **74** (4), 291–315.

Fracture of Rock

Editor: **M.H. ALIABADI**, Queen Mary College, University of London, UK.

Bringing together the latest research on fracture of rocks, this state-of-the-art volume covers a wide range of subjects including hydraulic fracturing, blasting and fragmentation, transport problems and creep.

All of the chapters featured have been contributed by leading scientists in the field and focus on modelling and analysis techniques. The following topics are covered: Boundary Element Analysis for Rock Fracture; Numerical Models of Shear Crack Propagation using the Displacement Discontinuity Method; The Displacement Discontinuity Method for the Analysis of Rock Structures – a Fracture Mechanics Approach; Computationally Efficient Models for the Growth of Large Fracture Systems; Fracture, Fragmentation and Rock Blasting Models in the Combined Finite-discrete Element Method; Rock Fragmentation and Optimisation of Drilling Tools; Modelling of Microcrack Induced Damage and Poroelasticity in Brittle Rocks; Modelling of Hydraulic Fracturing of Porous Materials; A General Analysis of Transports in Fracture Networks; Transport of Colloids in Saturated Fractures; Creep Deformation and Fracture in Rock Salt; Application of the Theory of Plasticity to Analysis of Bearing Capacity Problems in Fissured Materials.

Series: Advances in Fracture Mechanics, Vol 5

ISBN: 1-85312-542-3 **1998**

440pp **£132.00/US$218.00**

Computational Fracture Mechanics in Concrete Technology

Editors: **A. CARPINTERI**, Politecnico di Torino, Italy and **M.H. ALIABADI**, Queen Mary College, University of London, UK.

This book describes the most recent computational approaches, based on fracture mechanics, for the structural analysis of concrete and reinforced concrete. Both smeared and localized numerical fracture models which simulate damage zones and discrete cracks developing in loaded concrete members are considered. Particular loading conditions are also discussed where the use of fracture mechanics is appropriate.

Written by prominent researchers in the field, **Computational Fracture Mechanics in Concrete Technology** covers the following topics: Materials Engineering of Cement-Based Composites using Lattice Type Models; Computational Damage Mechanics; Continuum Damage Applied to Concrete Modelling; Integral Equation Method for Modeling Cracking Concrete; A Discrete Crack Numerical Model; Boundary Element Method for Analysis of Cracking in Plain and Reinforced Concrete; Creep Crack Growth in Concrete Structures – Cohesive Crack Model in Mode I and Mixed-Mode Loading Conditions.

Series: Advances in Fracture Mechanics, Vol 3

ISBN: 1853125075 **1998**

240pp **£88.00/$141.00**

Linear and Nonlinear Crack Growth using Boundary Elements

A. CISILINO, Universidad Nacional de Mar del Plata, Argentina.

This book brings together descriptions of three-dimensional boundary element methods for the analysis of fatigue crack problems in linear and nonlinear fracture mechanics. In order to overcome the mathematical degeneration associated with the solitary use of the displacement boundary integral equation for cracked bodies, the methods depicted rely on formulations based on two independent boundary integral equations: the dual boundary element method. The author demonstrates the effective implementation of the methods, and devotes special attention to the description of accurate algorithms for the evaluation of singular and near-singular integrals in the dual equations.
Contents: Introduction; Solid and Fracture Mechanics Fundamentals; The Dual Boundary Element Method for Three-Dimensional Cracked Bodies; Three-Dimensional DBEM Analysis for Fatigue Crack Growth; A BEM for Three-Dimensional Elastoplastic Problems; The Elastoplastic Dual Boundary Element Method in Three Dimensions; BEM Analysis of Fracture Problems using the Energy Domain Integral; Full-Penetration Welded Joint.

Series: Topics in Engineering, Vol 36

ISBN: 1-85312-700-0 1999

208pp £79.00/US$126.00

A Dual Boundary Element Formulation for Three-Dimensional Fracture Analysis

A. WILDE, Wessex Institute of Technology, UK.

This monograph is concerned with the study of the dual boundary element formulation using continuous elements in three dimensions, together with its application to the analysis of fracture problems and crack growth. Formulations for modelling geomechanical fracture are also given.
The author achieves effective implementation of the method through the use of accurate algorithms for the singular and near-singular integrals in the dual equations and demonstrates an improved technique for the evaluation of near-singular and near-hypersingular integrals.
Two new approaches are considered to enable discretisation of the crack surfaces with continuous elements. The first raises the degree of continuity of the displacement and traction fields at the source point by introducing a new interpolation scheme based on the nodal values of all elements containing the source point. The second forms an enriched element in which the continuity conditions are directly incorporated via a special interpolation of the field variables.

Series: Topics in Engineering, Vol 37

ISBN: 1-85312-679-9 1999

apx 240pp apx £84.00/US$135.00

Linear Elastic Fracture Mechanics for Engineers

Theory and Applications

L.P. POOK, University College, London.

A concise introduction to linear elastic fracture mechanics and some of its applications, this new textbook has two main objectives. Firstly to explain the main ideas underlying present-day linear elastic fracture mechanics and how these have been developed. Secondly to show how these ideas may be used to carry out calculations which answer the question *"Does this crack matter?"* from the viewpoint of an engineering designer. It is assumed that readers will be familiar with the mechanical properties of metallic materials and the linear elastic stress analysis of uncracked bodies.

The author explains many of the difficulties and inconsistencies encountered by newcomers to the subject and places emphasis on brittle fractures and fatigue crack growth in metallic materials at ambient temperature.

ISBN: 1-85312-703-5 **1999**
apx 150pp **apx £51.00/US$83.00**

Lecturers - Price reductions are available when you purchase multiple copies of this text. Please contact the Marketing Department at WIT Press for details.

Thermomechanical Crack Growth using Boundary Elements

N.N.V. PRASAD, CADS Software, India.

Thermal and mechanical fatigue problems are encountered in many engineering components such as pressure vessels, high-temperature engines and interfaces in computer technology. This book describes modelling of thermal fatigue using the Dual Boundary Element Method. In this method both thermal and elasticity equations are written as regular and hyper singular equations, then applied on the crack surfaces. For thermal stress intensity factors a new one point displacement extrapolation technique is described.

Partial Contents: Thermoelasticity and Fracture Mechanics; Boundary Integral Equations; Dual Boundary Element Method Applied to Steady State Thermoelasticity; Dual Boundary Element Method; Effect of Thermal Singularities on Stress Intensity Factors; Thermo-Mechanical Fatigue Crack Growth.

Series: Topics in Engineering, Vol 34
ISBN: 1-85312-541-5 **1998**
216pp **£74.00/$115.00**

Stress Intensity Factors and Weight Functions

T. FETT, Research Center, Karlsruhe, Germany and D. MUNZ, University of Karlsruhe, Germany.

"...if someone has a crack problem at hand and wishes to know if a solution exists for that problem, **Stress Intensity Factors and Weight Functions** will prove to be a valuable resource." APPLIED MECHANICS REVIEWS

In this book the authors describe methods for the calculation of weight functions. In the first part a detailed discussion of the accuracy and convergence behaviour of methods allowing the determination of stress intensity factors and weight functions for one- and two-dimensional cracks is given. They then move on to provide solutions for cracks subjected to mode-I and mode-II loading.

Series: Advances in Fracture Mechanics, Vol 1

ISBN: 1-85312-497-4 1997

408pp £110.00/US$169.00

Look for more information about WIT Press on the internet: http://www.witpress.com

On Wave Propagation in Elastic Solids with Cracks

C. ZHANG, Hochschule für Technik, Zittau, Germany and D. GROSS, Institut für Mechanik, Darmstadt, Germany.

Containing reports on wave propagation in elastic solids with cracks, this book looks at problems and treatments, and features numerical examples to show the accuracy and efficiency of the methods employed.

Topics covered include: Boundary Integral Methods; Elastodynamic Stress Intensity Factors; Time-Harmonic Wave Propagation; Anti-Plane Interface Cracks; Wave Attenuation and Dispersion; and Damage Mechanics.

Series: Advances in Fracture Mechanics, Vol 2

ISBN: 1-85312-535-0 1997

272pp £78.00/US$125.00

All prices correct at time of going to press. All books are available from your bookseller or in case of difficulty direct from the Publisher.

WITPress

Ashurst Lodge, Ashurst, Southampton, SO40 7AA, UK.

Tel: 44 (0) 238 029 3223

Fax: 44 (0) 238 029 2853

E-Mail: witpress@witpress

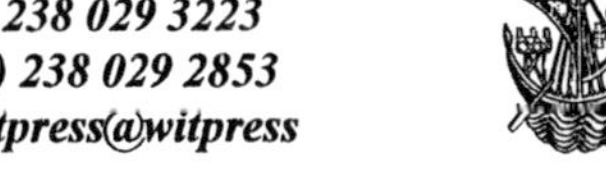